W9-BUY-061

Digital Fundamentals

3rd Edition

Thomas L. Floyd

Charles E. Merrill Publishing Company
A Bell & Howell Company
Columbus Toronto London Sydney

**To my mother, Mary Floyd, and
to my daughters, Debbie and Cindy**

Cover photo: A plasma of ionized oxygen cleans a photomask (center) during processing. The bright jets are oxygen entering the chamber. (Courtesy of Hewlett-Packard Company)

Published by
Charles E. Merrill Publishing Company
A Bell & Howell Company
Columbus, Ohio 43216

This book was set in Times Roman.
Production Coordination and Text Design: Constantina Geldis
Cover Designer: Cathy Watterson

Copyright © 1986, 1982, 1977 by Bell & Howell Company. All rights reserved. No part of this book may be reproduced in any form, electronic or mechanical, including photocopy, recording, or any information storage and retrieval system, without permission in writing from the publisher. "Bell + Howell" is a trademark of the Bell & Howell Company. "Charles E. Merrill Publishing Co." and "Merrill" are registered trademarks of the Bell & Howell Company.

Library of Congress Catalog Card Number: 85-61827
International Standard Book Number: 0-675-20517-4
Printed in the United States of America
 2 3 4 5 6 7 8 9 — 90 89 88 87 86

Merrill's International Series in Electrical and Electronics Technology

Preface

Digital Fundamentals is a text that provides not only a thorough coverage of the principles and applications necessary to understand the complex and diverse world of digital electronics, but also the foundation upon which to build and develop skills in this exciting field. All of the features in the second edition have been retained and enhanced in this third edition. Based on the comments and suggestions made by instructors in surveys and extensive reviews, additional coverage has been incorporated.

The new features in this edition are

1. Chapter objectives.
2. Reviews after each section with answers at the end of the chapter.
3. Chapter summaries.
4. Self-tests at the end of each chapter with solutions at the end of the book.
5. Increased emphasis on applications.
6. Improved and expanded troubleshooting coverage.
7. Introduction to ANSI/IEEE Std. 91-1984 logic symbols with dependency notation.
8. More emphasis on specific devices.
9. Improved coverage of flip-flops.
10. Introduction to sequential circuit design (state machines).
11. Counters and registers covered in two chapters.
12. Improved coverage of memories.
13. Improved coverage of interfacing and data transmission including a treatment of modems.

14. A concise introduction to both the 6800 and the 8085A microprocessors.
15. Availability of two laboratory manuals and transparencies.

Content and Organization

This text provides a thorough introduction to digital concepts and microprocessor-related topics for students in electronics or computer technology programs as well as those in computer science or other nonelectronic courses of study. Because little or no background in electronics is required, this book can be used, if necessary, very early in a program. For those few areas where a previous knowledge of electronics is helpful, the text has been carefully designed to permit a light treatment or complete omission without affecting the flow of material.

A chapter-like coverage of digital integrated-circuit technology containing an introduction to specific "inside-the-chip" circuitry has been placed in Appendix A. This section provides teaching flexibility and makes the detailed coverage of component-level circuitry optional. Those who wish to include it may insert it at an appropriate point in the chapter sequence.

The ANSI/IEEE Std. 91-1984 logic symbols with dependency notation are introduced gradually and conservatively at appropriate points while the use of the more traditional symbols is retained throughout. It is necessary for the student to become familiar with the new symbols because U.S. military contracts now require this notation, and much of the electronics industry has adopted this standard and incorporated it in their documentation. However, because the new symbols and notation represent a significant departure from many of the traditional logic symbols and because the more traditional symbols will be in use for some time in existing documentation, a gradual and limited introduction is appropriate at this time.

The following is a chapter-by-chapter capsulized version of the major features and changes in this edition.

Chapter 1 now includes a discussion of duty cycle and an introduction to the oscilloscope. Chapter 2, "Number Systems and Codes," has been modified only slightly in order to strengthen a few areas. The Gray code is now introduced in this chapter. Chapter 3 now includes an introduction to the ANSI/IEEE alternative logic symbols for gates, although the distinctive-shape gate symbols (also ANSI/IEEE approved) continue to be used. A strengthened coverage of IC logic families, data sheet interpretation, and an introduction to the application of oscilloscopes in troubleshooting is also presented. In Chapter 4, the coverage of DeMorgan's theorems has been strengthened and the section on Karnaugh maps has been totally revised and improved. The coverage of combinational logic in Chapter 5 has been strengthened in several areas, particularly by the use of truth tables in the analysis of combinational logic circuits. New or completely revised areas of combinational logic functions in Chapter 6 include LCDs, data selectors/multiplexers, logic function generators, parity checkers/generators, troubleshooting, and applications. Chapter 7 has been partially revised to achieve a clearer and more complete coverage of latches and flip-flops. The coverage of flip-flops now includes edge-triggered, pulse-triggered (master-slave), and data lock-out categories with more emphasis placed on D and JK flip-flops. Flip-flop symbols with dependency notation are introduced, and there is a stronger emphasis on specific devices. Sections on applications and troubleshooting are totally new. Chapter 8 is completely devoted to counters. An entire section covers the design of sequential circuits using state diagrams and Karnaugh maps. In addition, counter applications are greatly expanded. Because there is a significant departure from traditional logic symbols in the area of counters (among other areas), a section is devoted to the dependency notation and symbology recommended in the ANSI/IEEE Std. 91-1984. Chapter 9 is completely devoted to the coverage of registers. There is a complete section covering shift register applications including time delay, serial-to-parallel conversion, and a keyboard encoder. The material on troubleshooting includes an introduction to signature analysis, and more material on dependency notation is presented. Chapter 10, ''Memories,'' has been completely revised to reflect more current technology and to provide a more in-depth understanding of memories and their applications. Emphasis is placed on specific devices as well as general concepts. Memory cycle timing is treated in detail, and the coverage of MBMs, PLAs, and CCDs has been expanded. The application of a logic analyzer in troubleshooting a specific memory problem is used as an introduction to system troubleshooting. Chapter 11 on interfacing and data transmission has been significantly revised to include TTL and CMOS logic family interfacing, the use of open-collector logic, three-state interfacing with buses including the GPIB, and A/D and D/A conversion. An entire section is also devoted to modems. Chapter 12 combines signed arithmetic processes, including coverage of a specific ALU device, with a brief introduction to both the 6800 and the 8085A microprocessors.

Suggestions for Use

In a two-term course sequence, this text may be easily divided into combinational circuits (Chapters 1–6) and sequential circuits (Chapters 7–12).

For the programs which need to cover this material in one term, the following topics are suggested for possible deletion, depending on program emphasis.

1. Full chapters which may be considered for deletion:
   Chapter 1   (Introduction)
   Chapter 10   (Memories)
   Chapter 11   (Interfacing and Data Transmission)
   Chapter 12   (Arithmetic Processes with an Introduction to Microprocessors)
2. Sections within other chapters which may be considered for deletion:
   2–10   (Digital Codes)
   3–10, 3–11   (I.C. Logic Families and Data Sheet Interpretation)

4–6, 4–7	(Boolean Simplification and Karnaugh Maps)
5–2	(Designing Combinational Circuits)
6–3	(Speeding Addition)
7–4	(Data Lock-Out Flip-Flops)
8–4	(Design of Sequential Circuits)
8–8, 9–10	(Dependency Notation)

The numerous illustrations, boxed examples, and other pedagogical features, as well as the student-oriented writing style, make this text easier to complete than the number of pages might suggest. With careful planning, use in a one-term course is quite feasible.

Acknowledgments

The third edition of *Digital Fundamentals* is the result of the efforts of many people. In particular, I want to express my appreciation to Don Thompson, Chris Conty, and Connie Geldis at Charles E. Merrill Publishing Company for their enthusiasm and dedication to quality. My thanks also go to the many users of the second edition who offered constructive suggestions, especially Gary Spencer, Ted Harris, Pete Zawasky, Warren Foxwell, Barry Brey, Gary House, John W. Berry, Max Goldstein, Edward Kerly, and D. L. Kerr; to Mark Moran for checking the accuracy of problem sets and answers; and to the following instructors who reviewed the manuscript and provided many valuable comments:

John L. Morgan	DeVry Institute of Technology, Dallas
Vincent Loizzo	DeVry Institute of Technology, Chicago
Pete Holsberg	Mercer County Community College
Thomas Stultz	Sylvania Technical Institute
Dennis Hoffman	Heald College
John Chaudhry	Heald College
Peter Chesebrough	Valencia Community College
David Buchla	Yuba College

I am grateful to Hewlett-Packard, Tektronix, and Texas Instruments for their contributions of photos and other materials for use in the book. Finally, a special note of appreciation to my wife, Sheila, for her patience and support during the many months I have spent on this project.

Thomas L. Floyd

Contents

3 LOGIC GATES   75

4 BOOLEAN ALGEBRA   125

5 COMBINATIONAL LOGIC   157

6   FUNCTIONS OF COMBINATIONAL LOGIC   199

7   FLIP-FLOPS AND OTHER MULTIVIBRATORS   285

8   COUNTERS   349

12   ARITHMETIC PROCESSES WITH AN INTRODUCTION TO MICROPROCESSORS   579

A   INTEGRATED CIRCUIT TECHNOLOGIES   609

B   DATA SHEETS   629

C   ERROR DETECTION AND CORRECTION   652

D CONVERSIONS   659

Digital electronics began in 1946 with an electronic digital computer called *ENIAC*, which was implemented with vacuum-tube circuits. The concept of the digital computer can be traced to Charles Babbage, who developed a mechanical computation device in the 1830s. The first digital computer was built in 1944 at Harvard University, but it was electromechanical, not electronic.

The term *digital* is derived from the way in which computers perform operations by counting *digits*. For years, applications of digital electronics were confined to computer systems. Today, digital techniques are applied in many diverse areas, such as telephony, data processing, radar, navigation, military systems, medical instruments, process control, and consumer products. Digital technology has progressed from vacuum-tube circuits to integrated circuits and microprocessors.

Digital electronics involves circuits and systems in which there are only two possible states that are typically represented by voltage levels. Other circuit conditions, such as current levels, open or closed switches, and on or off lamps, can also represent the two states. In digital systems, the two states are used to represent numbers, symbols, alphabetic characters, and other types of information. In the two-state number system, called *binary*, the two digits are 0 and 1. These binary digits are called *bits*.

In this chapter, you will learn

- ☐ The meaning of positive and negative logic.
- ☐ The characteristics of digital waveforms.
- ☐ The principles of the basic logic elements: NOT, AND, and OR.
- ☐ The basic concepts of several logic functions, such as arithmetic, comparison, encoding, decoding, multiplexing, demultiplexing, counting, and data storage.
- ☐ What digital integrated circuits are and how they are classified in terms of their complexity.
- ☐ The definition of *microprocessor* and the basic components of a microcomputer.
- ☐ How to recognize several types of testing and troubleshooting instruments, including the oscilloscope.

1

Introduction

1-1   LOGIC LEVELS AND PULSE WAVEFORMS

In digital systems two voltage levels represent the two binary digits, 1 and 0. If the higher of the two voltages represents a 1 and the lower voltage represents a 0, the system is called a *positive logic* system. On the other hand, if the lower voltage represents a 1 and the higher voltage represents a 0, we have a *negative logic* system. To illustrate, suppose that we have +5 V and 0 V as our logic-level voltages. We will designate the +5 V as the HIGH level and the 0 V as the LOW level, so positive and negative logic can be defined as

Positive logic   *Negative logic*

HIGH = 1   HIGH = 0

LOW = 0   LOW = 1

Both positive and negative logic are used in digital systems, but positive logic is the more common. For this reason we will use only positive logic in this text.

Pulses are very important in the operation of digital circuits and systems because voltage levels are normally changing back and forth between the HIGH and LOW states. Figure 1–1(a) shows that a single *positive* pulse is generated when the voltage (or current) goes from its *normally* LOW level to its HIGH level and then back to its LOW level. The *negative* pulse in Figure 1–1(b) is generated when the voltage goes from its *normally* HIGH level to its LOW level and back to its HIGH level.

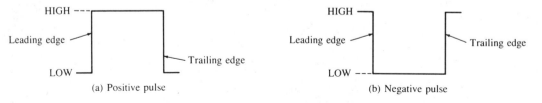

(a) Positive pulse   (b) Negative pulse

FIGURE 1–1 *Ideal pulses.*

As indicated in Figure 1–1, the pulse has two edges: a *leading* edge and a *trailing* edge. For a positive pulse, the leading edge is a *positive-going* transition (*rising* edge), and the trailing edge is a *negative-going* transition (*falling* edge). The pulses in Figure 1–1 are ideal because the rising and falling edges change in zero time (instantaneously). Actually, these transitions never occur instantaneously, although for most digital work we can assume ideal pulses. Figure 1–2 shows a nonideal pulse. The time required for the pulse to go from its LOW level to its HIGH level is called the *rise time* (t_r), and the time required for the transition from the HIGH level to the LOW level is called the *fall time* (t_f). In actual practice it is common to measure rise time from 10% of the pulse amplitude to 90% of the pulse amplitude, and to measure the fall time from 90% to 10% of the pulse amplitude. This is due to nonlinearities that commonly occur near the bottom and the top of the pulse, as indicated in Figure 1–2.

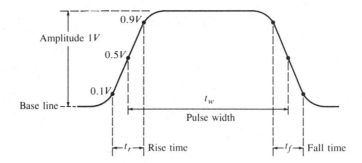

FIGURE 1–2 *Nonideal pulse characteristics.*

The pulse width (t_w) is a measure of the duration of the pulse and is often defined as the time between the 50% points on the rising and falling edges, as indicated in Figure 1–2. This definition is used throughout this text.

Most waveforms encountered in digital systems are composed of series of pulses and can be classified as either *periodic* or *nonperiodic*. A periodic pulse waveform is one that repeats itself at a fixed interval called the *period* (*T*). The frequency is the rate at which it repeats itself and is measured in pulses per second (pps) or Hertz (Hz). A nonperiodic pulse waveform, of course, does not repeat itself at fixed intervals and may be composed of pulses of differing pulse widths and/or differing time intervals between the pulses. An example of each type is shown in Figure 1–3.

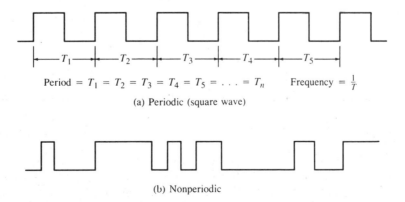

FIGURE 1–3 *Examples of pulse waveforms.*

The frequency (f) of a pulse waveform (sometimes called the *pulse repetition rate*, PRR) is the reciprocal of the period. The relationship between frequency and period is expressed as follows:

$$f = \frac{1}{T}$$

$$\text{(1–1)}$$

$$T = \frac{1}{f}$$

$$\text{(1–2)}$$

An important characteristic of a periodic pulse waveform is its *duty cycle*. The duty cycle is defined as the *ratio of the pulse width (t_W) to the period (T) expressed as a percentage.*

$$\text{Duty cycle} = \left(\frac{t_W}{T}\right) 100\%$$

$$\text{(1–3)}$$

EXAMPLE 1–1   A portion of a periodic pulse waveform is shown in Figure 1–4. Determine the duty cycle.

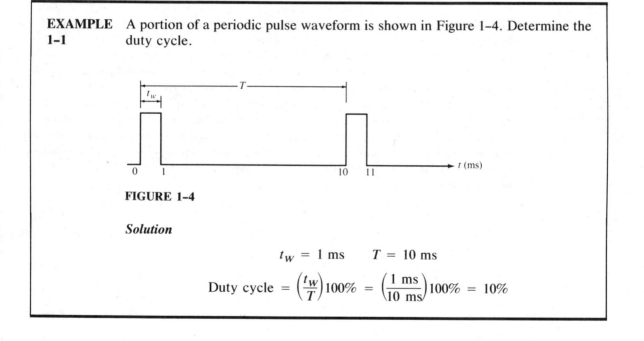

FIGURE 1–4

Solution

$$t_W = 1 \text{ ms} \qquad T = 10 \text{ ms}$$

$$\text{Duty cycle} = \left(\frac{t_W}{T}\right) 100\% = \left(\frac{1 \text{ ms}}{10 \text{ ms}}\right) 100\% = 10\%$$

SECTION REVIEW 1–1

$\mu = 10^{-6}$

1. The rise time of a certain pulse is 5 μs. If the amplitude of the pulse is 5 V and its base line is 0 V, determine the actual change in voltage during the rise-time interval using the definition of rise time.
2. A certain periodic pulse waveform has a pulse width of 15 μs and a frequency of 16.67 kHz. Find the duty cycle.

1–2   ELEMENTS OF DIGITAL LOGIC

In its basic form, *logic* is the realm of human reasoning that tells us a certain proposition (declarative statement) is true if certain conditions or premises are true. "The light is on" is an example of a proposition that can be only true or false. "The bulb is not burned out" and "The switch is on" are other examples of propositions that can be classified as true or false.

Several propositions, when combined, form *propositional* or *logic* functions. For example, the propositional statement "The light is on" will be true if "The bulb is not burned out" and if "The switch is on." Therefore, this logical statement can be made:

The light is on if and only if the bulb is not burned out and the switch is on.

In this example the first statement is true only if the last two statements are true. The first statement ("The light is on") is then the basic proposition, and the other two statements are the conditions or premises upon which the proposition depends.

Many situations, problems, and processes that we encounter in our daily lives can be expressed in the form of propositional or logic functions. Since these functions are true/false or yes/no statements, digital circuits with their two-state characteristics are extremely applicable.

A mathematical system for formulating logical statements with symbols, so that problems can be written and solved in a manner similar to ordinary algebra, was developed by the Irish logician and mathematician George Boole in the 1850s. *Boolean algebra,* as it is known today, finds application in the design and analysis of digital systems and will be covered in Chapter 4.

The term *logic* is applied to digital circuits used to implement logical functions. Several digital circuits are the basic *elements* forming the building blocks for complex digital systems such as the computer. We will now look at these elements and discuss their functions in a very general way. Later chapters will cover these circuits in full detail.

The first basic logic element is the NOT circuit. The primary function of this circuit is to produce one logic level from the opposite logic level. If a HIGH level is applied to the input of the NOT circuit, a LOW level appears on the output. If a LOW level is applied to the input, a HIGH level appears on the output. This circuit is commonly called an *inverter.*

The second basic logic element is the AND gate. The primary function of this circuit is to produce a true condition on its output if and only if *all* of its input conditions are true. For instance, let us assume that a HIGH level represents a true condition and that a LOW level represents a not-true (false) condition. Let us also use the example statement "The light is on" as our logic function. The AND gate can be used to tell us if this statement is true or false by applying the conditional functions "The switch is on" and "The bulb is not burned out" to the gate inputs. If both of these conditional statements are true, then both inputs to the AND gate are HIGH, and, as a result, the output is also HIGH. If either condition is false, as represented by a LOW, then the output is LOW. In other words, the AND gate "tells" us that *the light is on if and only if the switch is on AND the bulb is not burned out.*

The third basic logic element is the OR gate. The primary function of this circuit is to produce a true indication on its output when *one or more* of its input conditions are true. Again, let us assume that a HIGH level represents a true condition and that a LOW level represents a false condition. Let us take as an example a door that is controlled from a remote transmitter or from a wall switch. A HIGH level is required to energize the motor in order to open the door. Our propositional statement for this example is "The door opens," and the conditional statements are "The transmitter is on" and "The wall switch is on." The OR gate can be used to "tell" us when the propositional statement is true or when it is false by applying the conditional functions to its inputs. If either one or both of these conditional statements are true, then the output of the OR gate will be HIGH. If neither conditional statement is true, the output of the gate is LOW. For this example, the OR gate "tells" us that *the door opens if the transmitter is on, OR the wall switch is on, OR both are on.* The gate output can be used to open the door by properly energizing the activating motor.

The fourth basic logic element, which belongs to a class of circuits called *multivibrators*, is the bistable multivibrator, or *flip-flop.* The primary function of this circuit is to store or "memorize" a binary digit for an indefinite period of time. This device is distinguished from those previously discussed by its ability to retain either logic level after input conditions have been removed. Actually, the flip-flop can be constructed from combinations of basic gates, but it is treated as a distinct logic element because of its importance in digital systems.

The basic elements—the inverter, gates, and flip-flop—are used to construct more complex logic circuits such as counters, registers, decoders, and memories. These more complex logic functions are then combined to form complete digital systems designed to perform specified tasks.

SECTION REVIEW 1–2

1. List four basic logic elements.
2. A logic gate has HIGHs on both of its two inputs, and its output is HIGH. What type of gate is it?
   (a) AND gate   **(b)** OR gate   **(c)** Either AND or OR gate

1–3   FUNCTIONS OF DIGITAL LOGIC

The inverter, the basic gates, and the flip-flop combine to form more complex logic circuits that perform many operations. Some of the more common logic functions are comparison, arithmetic operations, decoding, encoding, code conversion, counting, memory, and multiplexing. We will see what these functions mean in terms of their "block" operations; many are already quite familiar to you because they are common functions.

Comparison

The *comparison* function is performed by a logic circuit called a *comparator*. Its function is to compare two quantities and to indicate if they are equal or not equal. For example, suppose we have two numbers and we wish to know if they are equal or not equal, and, if unequal, which is greater. Figure 1-5 is a block diagram

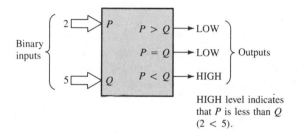

FIGURE 1-5 *The comparator.*

of a comparator. One number is applied to input P, and the other to input Q. The outputs indicate the relation of the two numbers by producing a HIGH level on the proper output line. Suppose that a binary representation of the number 2 is applied to input P, and a binary representation of the number 5 is applied to input Q. (We will discuss the representation of numbers and symbols in detail later.) A HIGH level will appear on the "$P < Q$" output, indicating the relationship between the two numbers (2 is less than 5). The wide arrows represent a group of parallel lines.

Arithmetic Operations

The *addition* function is performed by a logic circuit called an *adder*. Its function is to add two numbers with an *input carry* (CI) and generate a sum (Σ) and an *output carry* (CO). The block diagram in Figure 1-6 indicates the addition of the

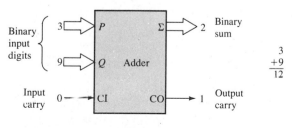

FIGURE 1-6 *The adder.*

digit 3 and the digit 9. We all know that the sum is 12, and the adder indicates this result by producing the digit 2 on the *sum* output and the digit 1 on the *carry* output. Note that we assume the input carry in this example to be 0. In later chapters you will learn how numbers can be represented by the logic levels of a digital circuit.

Subtraction is the second arithmetic function that can be performed by digital logic circuits. A *subtractor* requires three inputs: the two numbers to be subtracted and an *input borrow*. The two outputs are the *difference* and the *output borrow*. When, for instance, 5 is subtracted from 8 with no input borrow, the difference is 3 with no output borrow. We will see later how subtraction can actually be performed by an adder, because subtraction is simply a special case of addition.

The third arithmetic operation that can be performed by logic circuits is *multiplication*. Since numbers are always multiplied two at a time, two inputs are required. The output of the multiplier is the *product*. Since multiplication is simply a series of additions with shifts in the positions of the partial products, it can be performed using an adder.

Division, the fourth type of arithmetic operation, is a series of subtractions, comparisons, and shifts, and thus it can also be performed using an adder. Two inputs to the *divider* are required, and the outputs generated are called the *quotient* and the *remainder*.

Encoding

The *encoder* converts information, such as a decimal number or an alphabetic character, into some coded form. For example, a certain type of encoder converts each of the decimal digits, 0 through 9, to a binary code as shown in Figure 1–7. A HIGH level on a given input corresponding to a specific decimal digit produces the proper four-bit code on the output lines.

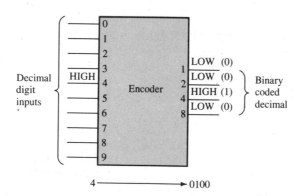

FIGURE 1–7 *Example of an encoder.*

Decoding

The *decoder* converts coded information, such as binary, into a recognizable form, such as decimal. For example, a particular type of decoder converts a four-bit binary code into the appropriate decimal digit, as illustrated in Figure 1–8.

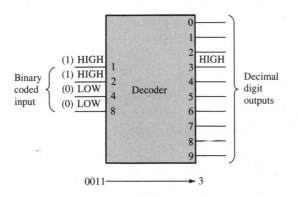

0011 ──────────→ 3

FIGURE 1–8 *Example of a decoder.*

Counting

The *counting* operation is very important in digital systems. There are many types of digital counters, but their basic function is to count events represented by changing levels or pulses or to generate a particular sequence of numbers. In order to count, the counter must "remember" the present number so that it can go to the next proper number in sequence. Therefore, storage or memory capability is an important characteristic of all counters, and flip-flops are generally used to

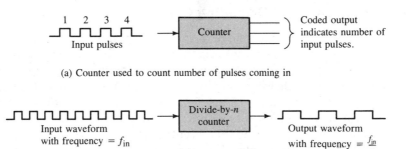

(a) Counter used to count number of pulses coming in

(b) Counter used to divide an input frequency by a factor *n*

FIGURE 1–9 *Two examples of counter applications.*

implement these devices. Another important counter application is frequency division. Figure 1–9 illustrates two simple counter applications.

Registers

Registers are digital circuits used for the temporary storage and shifting of information. For instance, a number in binary form can be stored in a register, and then its position within the register can be changed by shifting it one way or the other. Figure 1–10 illustrates a simple form of register operation. Flip-flops are common storage elements used in registers.

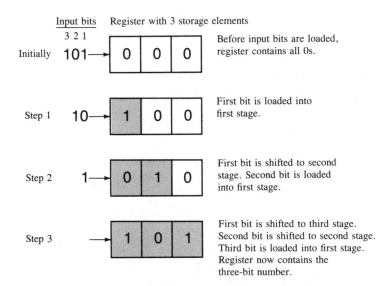

FIGURE 1–10   *Example of a simple shift register operation.*

Multiplexing and Demultiplexing

Multiplexing is an operation performed with digital logic circuits called *multiplexers*. Multiplexing allows information to be switched from several lines onto a single line in a specified sequence. A simple multiplexer can be represented by a switch operation that sequentially connects each of the input lines with the output, as illustrated in Figure 1–11. Assume that we have logic levels as indicated on the three inputs (a multiplexer can have any number of inputs). During time interval t_1, input A is connected to the output; during interval t_2, input B is connected to the output; and during interval t_3, input C is connected to the output. As a result of this *multiplexing action,* we have the three logic levels on the input lines appearing in sequence on the output line. The switching action, of course, is accomplished with logic circuits, as you will learn later.

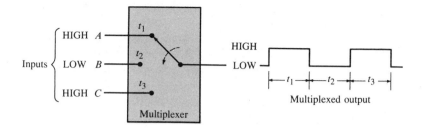

FIGURE 1-11 *Example of simple multiplexer operation.*

The inverse of the multiplexing function is called *demultiplexing*. Here, logic data from a single input line are sequentially switched onto several output lines, as shown in Figure 1-12.

Digital circuits can be used to perform a large variety of tasks limited only by the imagination. In order to give you a general picture of some of the basic aspects of digital logic, we have touched on only a few of the more basic and common functions that can be used as building blocks for increasingly complex systems.

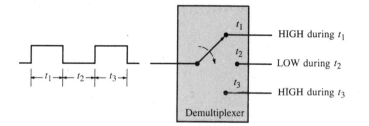

FIGURE 1-12 *Example of simple demultiplexer operation.*

SECTION REVIEW 1-3

1. List seven primary functions of digital logic.
2. Identify the function that best performs the following operations:
    (a) Determines which of two numbers is the greater.
    (b) Determines if two numbers are equal.
    (c) Produces the sum of two numbers.
    (d) Divides the frequency of a pulse waveform by 4.

1-4   DIGITAL INTEGRATED CIRCUITS

All of the logic functions that we have discussed and many more are available in integrated circuit (IC) form. Modern digital systems utilize ICs to a large extent in their designs. In most cases, ICs have size, power, reliability, and cost advantages over discrete circuitry (except in very specialized applications where a circuit must be "custom made" to meet unique requirements).

A monolithic integrated circuit is an electronic circuit that is constructed entirely on a single small chip of silicon. All of the components that make up the circuit—transistors, diodes, resistors, and capacitors—are an integral part of this single chip.

Typical chip sizes range from about 40 × 40 mils (a mil is 0.001 inch) to about 300 × 300 mils, depending on the complexity of the circuit. Anything from a few to thousands of components can be fabricated on a single chip.

Small-Scale Integration (SSI)

The least complex digital ICs are placed in the *small-scale integration* (SSI) category. These are circuits with up to 12 equivalent gate circuits on a single chip, and they include basic gate functions and flip-flops. Each SSI function is packaged in one of two main configurations: the dual-in-line package (DIP) or the flat pack. Figure 1–13 illustrates these packages in both 14- and 16-pin versions.

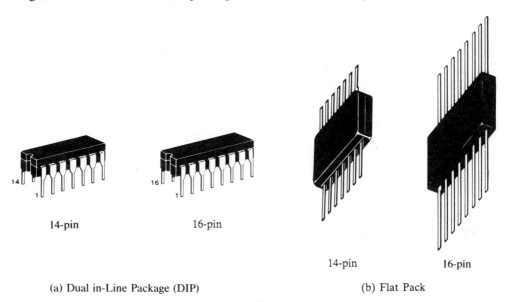

14-pin   16-pin

14-pin   16-pin

(a) Dual in-Line Package (DIP)   (b) Flat Pack

FIGURE 1–13   *Some common IC packages. (Courtesy of Motorola Semiconductor Products)*

Figure 1–14 shows a cutaway view of a DIP with the IC chip within the package. Leads from the chip are connected to the package pins to allow input and output connections to the outside world.

FIGURE 1-14 *Cutaway view of a dual-in-line 14-pin package showing the IC chip mounted inside with leads to input and output pins.*

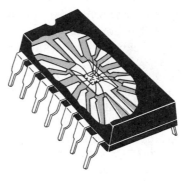

Medium-Scale Integration (MSI)

The next classification according to digital circuit complexity is called *medium-scale integration* (MSI). These are circuits with complexities ranging from 12 to 100 equivalent gates on a chip. MSI circuits include the more complex logic functions such as encoders, decoders, counters, registers, multiplexers, arithmetic circuits, small memories, and others. Figure 1-15 illustrates a 24-pin DIP and flat pack and a 28-pin chip carrier in which MSI functions are packaged.

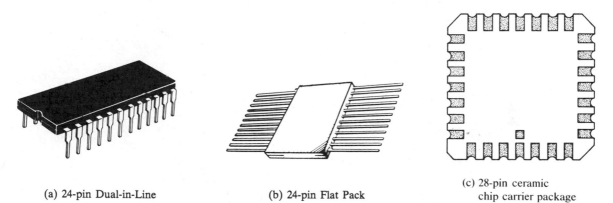

(a) 24-pin Dual-in-Line   (b) 24-pin Flat Pack   (c) 28-pin ceramic chip carrier package

FIGURE 1-15 *(a) and (b) Two types of 24-pin ICs (courtesy of Motorola Semiconductor Products). (c) A 28-pin chip carrier package (courtesy of Texas Instruments).*

Large-Scale Integration (LSI)

Circuits with complexities of 100 to 1000 equivalent gates per chip, including memories and some microprocessors, generally fall into the LSI category.

Very Large-Scale Integration (VLSI)

Integrated circuits with complexities ranging from 1000 to 100,000 equivalent gates per chip and beyond are generally considered VLSI. Very large memories, larger microprocessor systems, and single-chip computers are in this category.

The complexity figures stated here for SSI, MSI, LSI, and VLSI are generally accepted, but definitions may vary from one source to another.

---

1. What is an integrated circuit?
2. Define the terms SSI, MSI, LSI, and VLSI.
3. Generally, in what classification does an IC with the following number of equivalent gates fall?
   (a) 75   **(b)** 500   **(c)** 10   **(d)** 10,000

1-5   MICROPROCESSORS

The microprocessor is a VLSI device that can be programmed to perform arithmetic and logic operations and other functions in a prescribed sequence for the movement and processing of data. A typical microprocessor package is shown in Figure 1-16.

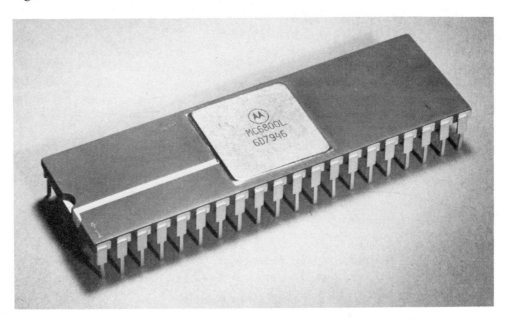

FIGURE 1-16 *Microprocessor in a 40-pin dual-in-line package. (Courtesy of Motorola)*

The microprocessor is used as the central processing unit (CPU) in microcomputer systems where it is connected with other ICs such as memories and input/output interface circuits.

The arrangement of circuits within the microprocessor (called its *architecture*) permits the system to respond correctly to each of many different *instruc-*

tions and, in addition to arithmetic and logic operations, controls the flow of signals into and out of the computer, routing each to its proper destination in the required sequence to accomplish a specified task.

The interconnections or paths along which signals flow are called *buses*. Figure 1–17 shows a simplified diagram of a microcomputer system. The memories are used to store binary information before and after processing and the sequence of instructions called the *program*. The input/output (I/O) interface block properly connects external devices—keyboards, video terminals, and printers—to the microcomputer.

FIGURE 1–17 *Microcomputer block diagram.*

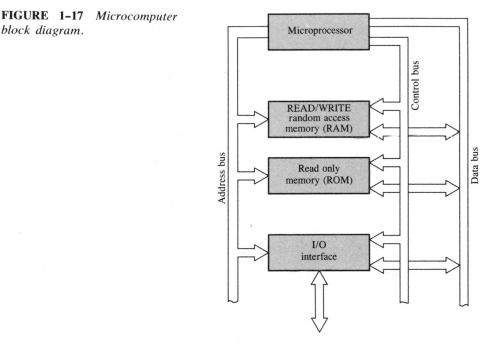

SECTION REVIEW 1–5

1. What is a microprocessor?
2. List the basic components of a microcomputer system.
3. Define the following:
   (a) bus   (b) program   (c) CPU   (d) ROM   (e) RAM

1–6   DIGITAL TESTING AND TROUBLESHOOTING INSTRUMENTS

Troubleshooting is the technique of systematically isolating and identifying a fault in a circuit or system. A variety of special instruments is available for use in digital troubleshooting and testing. Some typical equipment is presented in this section.

The Oscilloscope

Figure 1–18 shows a typical dual-channel oscilloscope. Pulse waveforms can be displayed on the screen, and parameters such as amplitude, rise and fall times, pulse width, period, and duty cycle can be measured. Also, abnormalities in the shape or characteristics of a pulse can be seen and analyzed. Two or more digital waveforms can be displayed simultaneously in order to determine and analyze their time relationships. Many specialized functions and performance levels are normally available with interchangeable plug-in modules. For these reasons, the oscilloscope is considered to be one of the most versatile instruments for testing and troubleshooting.

FIGURE 1–18   *A typical oscilloscope. (Courtesy of Tektronix, Inc.)*

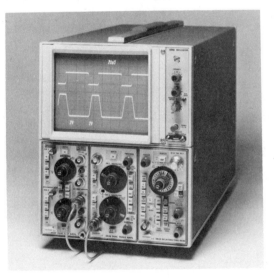

The Logic Analyzer

Figure 1–19 shows a typical *logic analyzer*. This instrument is basically a multichannel oscilloscope with the ability to detect and display logic levels in several forms.

Most logic analyzers can display data in several formats. The *timing diagram* format displays several *pulse waveforms* with the proper time relationships. The *bit* format displays a bit pattern of 1s and 0s on the screen. Bit patterns from a functioning unit can be displayed and compared to the patterns of a faulty unit to detect errors.

A third format is a display using hexadecimal code. The *hexadecimal code,* as you will learn later, is a convenient way to represent binary information. Some analyzers also provide for octal and other code displays. Octal is another code representation for binary and is discussed in the next chapter.

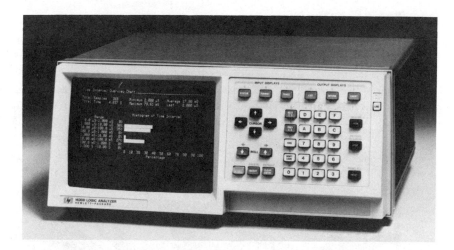

FIGURE 1–19 *The 1630A/D logic analyzer. (Courtesy of Hewlett-Packard)*

The Signature Analyzer

Another useful digital troubleshooting instrument is the *signature analyzer,* such as the one shown in Figure 1–20. Signature analysis is a troubleshooting technique that locates faults to the component level in microprocessor-based systems.

A signature analyzer converts a pattern of 1s and 0s at a given test point in the circuit under test and displays the pattern as a hexadecimal code called the *signature* for that particular point. By comparing the indicated signature with a known correct signature, a technician can identify a faulty device.

FIGURE 1–20 *The 5006 signature analyzer. (Courtesy of Hewlett-Packard)*

The Logic Probe

The *logic probe* provides a simple means for digital troubleshooting by detecting voltage levels or pulses at a given point. A typical logic probe is shown in Figure 1–21.

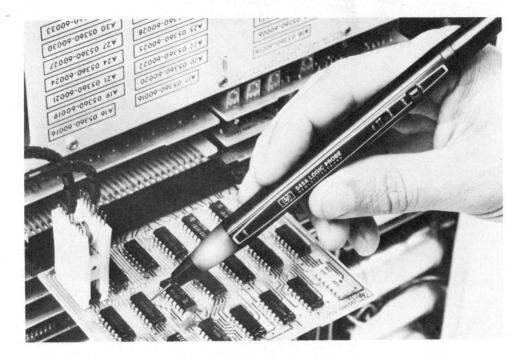

FIGURE 1–21   *The 545A logic probe. (Courtesy of Hewlett-Packard)*

The logic probe uses a single lamp or multiple lamps to indicate the various states possible on a digital signal path, such as HIGHs, LOWs, single pulses, pulse trains, and open circuits. A typical indication format is

Lamp on:	HIGH
Lamp off:	LOW
Lamp dim:	Open or bad level
One flash:	Single pulse
Repetitive flashes:	Pulse train

Although the logic probe is a very useful instrument, it alone cannot solve all troubleshooting problems. Often it must be used in conjunction with oscilloscopes, logic analyzers, signature analyzers, and other types of probes.

The Current Tracer and Logic Pulser

These instruments are shown in Figure 1–22. *Current tracing* is a very effective troubleshooting technique in many cases. It is particularly difficult to isolate a bad element when a given circuit node (point of connection) is *stuck* in one logic state, and several elements are common to the node. In this situation, current tracing is very useful.

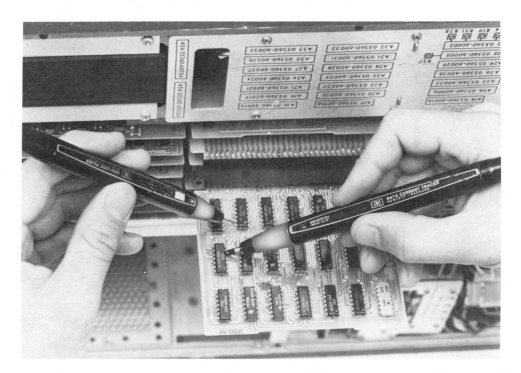

FIGURE 1–22 *The 546A logic pulser and the 547A current tracer. (Courtesy of Hewlett-Packard)*

The hand-held *current tracer* has a one-lamp indicator that glows when its tip is held over a *pulsing* current path. This instrument can detect whether current is flowing at all and, most importantly, *where* the current is flowing. For instance, if a node is stuck in the LOW state due to a shorted input in one of the devices connected to the node, a very strong current exists between the circuit driving the node and the faulty component. For purposes of detection, the current has to be *pulsing;* if it is not, a *logic pulser* can be used to force pulses on the node, and the current can then be traced to the bad component.

The Logic Clip

A typical *logic clip* is shown in Figure 1–23. It is "clipped" to an integrated circuit where it makes contact with each pin on the IC. The lamps on the clip then indicate the logic levels of each pin:

Lamp on: HIGH
Lamp off: LOW
Lamp dim: Pulses

FIGURE 1–23   *The 548A logic clip. (Courtesy of Hewlett-Packard)*

SECTION REVIEW 1-6

1. Select the proper instrument for measuring the rise time of a pulse:
    (a) logic probe   **(b)** oscilloscope
    (c) logic analyzer   **(d)** pulser
2. List the conditions at a given point in a logic system that can be detected using a logic probe.

SUMMARY

☐ In digital systems, there are two logic levels representing the two binary digits (*bits*).
☐ In positive logic, a HIGH level represents a 1, and a LOW level represents a 0.

☐ The *amplitude* is the "height" of a pulse usually expressed in volts.

☐ The *rise time* of a pulse is the time interval from 10% to 90% of the pulse amplitude on the rising edge.

☐ The *fall time* of a pulse is the time interval from 90% to 10% of the pulse amplitude on the falling edge.

☐ In this text, the *pulse width* is defined to be the time interval from the 50% point on the leading edge to the 50% point on the trailing edge. (Sometimes *pulse width* is defined as the time interval of the flat portion at the top of the pulse.)

☐ The *duty cycle* of a periodic pulse waveform is the ratio of the pulse width to the period.

☐ The output of a NOT circuit (inverter) is opposite to the input.

☐ The output of an AND gate is HIGH if and only if all the inputs are HIGH.

☐ The output of an OR gate is HIGH if and only if one or more inputs are HIGH.

☐ The flip-flop is a storage device.

☐ Several common functions of digital logic are the comparator, the arithmetic functions, the encoder, the decoder, the counter, the register, the multiplexer, and the demultiplexer.

☐ An *integrated circuit* (IC) is an electronic circuit constructed entirely on a tiny silicon chip. All of the circuit components are an integral part of the chip.

☐ Integrated circuits are classified in terms of the complexity or number of circuits on a chip as SSI, MSI, LSI, or VLSI.

☐ A *microprocessor* is an LSI/VLSI device that can be programmed to perform various digital functions in any specified sequence.

☐ Several important digital test instruments are the oscilloscope, the logic analyzer, the signature analyzer, the logic probe, the current tracer, the logic pulser, and the logic clip.

SELF-TEST

1. Name the binary digits.
2. What does a HIGH level represent in negative logic?
3. Define the term *bit*.
4. For a negative pulse, the leading edge corresponds to the _____ edge.
5. For a positive pulse, the trailing edge corresponds to the _____ edge.
6. Explain the difference between periodic and nonperiodic pulse waveforms.
7. Define *period*.
8. Define *frequency* and specify its unit.
9. For a given period, when the pulse width is increased, does the duty cycle increase or decrease?
10. What is the mathematical system for formulating logical statements?
11. A NOT circuit is also called an _____ .
12. Explain the difference between an AND gate and an OR gate.
13. Define one characteristic that distinguishes a flip-flop from a gate.
14. Briefly describe the purpose of each of the seven functions of digital logic discussed in this chapter.
15. Define the term *DIP*.
16. List three block functions found in a microcomputer system.

17. What does the term *architecture* mean in reference to a microprocessor?

18. What does *I/O* mean?

19. Define the term *troubleshooting*.

20. List at least five types of digital testing and troubleshooting instruments.

PROBLEMS

Section 1–1

1–1 Define the sequence of bits repesented by each of the following sequences of levels in a positive logic system:

(a) HIGH, HIGH, LOW, HIGH, LOW, LOW, LOW, HIGH

(b) LOW, LOW, LOW, HIGH, LOW, HIGH, LOW, HIGH, LOW

1–2 Define the sequence of bits represented by each of the following sequences of levels in a negative logic system:

(a) HIGH, LOW, HIGH, HIGH, HIGH, LOW, HIGH

(b) HIGH, HIGH, HIGH, LOW, HIGH, LOW, LOW, HIGH

1–3 For the pulse shown in Figure 1–24, determine the following:

(a) rise time   (b) fall time

(c) pulse width   (d) amplitude

FIGURE 1–24

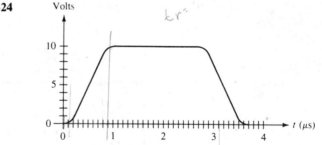

1–4 Determine the period of the waveform in Figure 1–25.

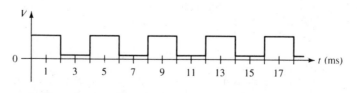

FIGURE 1–25

1–5 What is the frequency for the pulse waveform in Figure 1–25?

1–6 Is the pulse waveform in Figure 1–25 periodic or nonperiodic?

1–7 Determine the duty cycle in Figure 1–25.

Section 1–2

1–8 A basic logic element requires HIGHs on all of the inputs to make the output HIGH. What type of logic circuit is this?

1–9 A basic two-input logic element has a HIGH on one input and a LOW on the other input, and the output is LOW. Identify the element.

1–10 A basic two-input logic element has a HIGH on one input and a LOW on the other input, and the output is HIGH. What type of logic element is it?

Section 1–3

1–11 Name the function of each block in Figure 1–26.

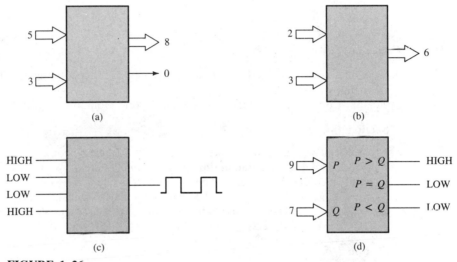

FIGURE 1–26

1–12 A pulse waveform with a frequency of 10 kHz is applied to the input of a divide-by-5 counter. What is the frequency of the ouptut waveform?

Section 1–4

1–13 An integrated circuit chip has a complexity of 100 equivalent gates. How is it classified?

1–14 Some ICs have 14, 16, or 24 pins (T or F).

1–15 Name two types of IC packages.

Section 1–5

1–16 Sketch the basic block diagram for a microcomputer.

1–17 The interface block properly connects the microcomputer to external devices such as _____.

Section 1–6

1–18 A pulse is displayed on the screen of an oscilloscope, and you measure the base line to be at 1 V and the top of the pulse to be at 8 V. What is the amplitude?

1-19   A logic probe is applied to a contact point on an IC that is operating in a system. The lamp on the probe flashes repeatedly. What does this indicate?

ANSWERS TO SECTION REVIEWS

Section 1-1
1. 4 V   2. 25%

Section 1-2
1. AND gate, OR gate, inverter, flip-flop.
2. (c)

Section 1-3
1. Comparison, arithmetic, encoding, decoding, counting, storage, multiplexing, and demultiplexing.
2. (a) Comparison   (b) Comparison   (c) Addition   (d) Counting

Section 1-4
1. A circuit with all components integrated on a single silicon chip.
2. SSI: small-scale integration (up to 12 equivalent gates).
    MSI: medium-scale integration (12 to 100 equivalent gates).
    LSI: large-scale integration (100 to 1000 equivalent gates).
    VLSI: very large-scale integration (greater than 1000 equivalent gates).
3. (a) MSI   (b) LSI   (c) SSI   (d) VLSI

Section 1-5
1. A VLSI device that can be programmed to perform arithmetic functions, logic functions, and other operations involving the movement and processing of digital data.
2. Microprocessor, memory, input/output devices.
3. (a) Bus: An interconnection specification for the communication of two or more digital devices.
    (b) Program: A sequence of computer instructions designed to tell the computer how to carry out a specified task.
    (c) CPU: central processing unit.
    (d) ROM: read-only memory.
    (e) RAM: read/write random access memory.

Section 1-6
1. (b)   2. HIGH, LOW, open, single pulse, pulse train.

The binary number system and digital codes are fundamental to digital electronics. In this chapter, the *binary number system* and its relationship to other number systems such as decimal, octal, and hexadecimal are studied. Arithmetic operations with binary numbers are covered to provide a basis for understanding how computers and other digital systems operate. The following important digital codes are introduced: BCD, Gray, Excess-3, ASCII, and EBCDIC.

In this chapter, you will learn

☐ The meaning and usefulness of 9's and 10's complements in the decimal system.
☐ How to count in binary.
☐ How to convert between binary and decimal.
☐ How to add, subtract, multiply, and divide in binary.
☐ The meaning and usefulness of 1's and 2's complements in the binary system.
☐ The octal system and how to convert between binary and octal.
☐ The hexadecimal system and how to convert between binary and hexadecimal.
☐ How to add in hexadecimal.
☐ The binary coded decimal (BCD) code.
☐ How to represent decimal numbers in BCD.
☐ How to add in BCD.
☐ The Gray code and conversion between Gray and binary.
☐ The Excess-3 code.
☐ The American Standard Code for Information Interchange (ASCII).
☐ The Extended Binary Coded Decimal Interchange Code (EBCDIC).

2

Number Systems and Codes

2–1   DECIMAL NUMBERS

We begin our study of number systems with the one with which we are all familiar, the decimal system. Because human anatomy is characterized by four fingers and a thumb on each hand, it was only natural that our method of counting involve the use of ten digits, that is, a system with a *base of ten*. Each of the ten decimal digits, 0 through 9, represents a certain quantity. The ten symbols (digits) do not limit us to expressing only ten different quantities because we use the various digits in appropriate positions within a number to indicate the magnitude of the quantity. We can express quantities up through nine before we run out of digits; if we wish to express a quantity greater than nine, we use two or more digits, and the position of each digit within the number tells us the magnitude it represents. If, for instance, we wish to express the quantity twenty-three, we use (by their respective positions in the number) the digit 2 to represent the quantity twenty and the digit 3 to represent the quantity three. Therefore, the position of each of the digits in the decimal number indicates the magnitude of the quantity represented and can be assigned a "weight."

The value of a decimal number is the sum of the digits times their respective column weights. The following examples will illustrate.

EXAMPLE 2–1

$$23 = 2 \times 10 + 3 \times 1 = 20 + 3$$

The digit 2 has a weight of 10, as indicated by its position, and the digit 3 has a weight of 1, as indicated by its position.

EXAMPLE 2–2

$$568 = 5 \times 100 + 6 \times 10 + 8 \times 1 = 500 + 60 + 8$$

The digit 5 has a weight of 100, the digit 6 has a weight of 10, and the digit 8 has a weight of 1.

SECTION REVIEW 2–1

1. What weight does the digit 7 have in each of the following numbers?
   (a) 1370   (b) 6725   (c) 7051   (d) 75,898
2. Express each of the following decimal numbers as a sum of the products of each digit and its appropriate weight:
   (a) 51   (b) 137   (c) 1492   (d) 10,658

2-2 9'S AND 10'S COMPLEMENTS

9's Complement

The 9's complement of a decimal number is found by subtracting each digit in the number from 9. The 9's complement of each of the decimal digits is as follows:

Digit	9's Complement
0	9
1	8
2	7
3	6
4	5
5	4
6	3
7	2
8	1
9	0

The following examples illustrate how to determine the 9's complement of a decimal number.

EXAMPLE 2-3 Find the 9's complement of each of the following decimal numbers:

(a) 12   (b) 28   (c) 56   (d) 115   (e) 562   (f) 3497

Solutions   To get the 9's complement of a decimal number, we subtract *each* digit in the number from 9.

(a)
$$
\begin{array}{r}
99 \\
-\ 12 \\
\hline
87
\end{array}
$$
9's complement of 12

(b)
$$
\begin{array}{r}
99 \\
-\ 28 \\
\hline
71
\end{array}
$$
9's complement of 28

(c)
$$
\begin{array}{r}
99 \\
-\ 56 \\
\hline
43
\end{array}
$$
9's complement of 56

(d)
$$
\begin{array}{r}
999 \\
-\ 115 \\
\hline
884
\end{array}
$$
9's complement of 115

(e)
$$
\begin{array}{r}
999 \\
-562 \\
\hline
437
\end{array}
$$
9's complement of 562

(f)
$$
\begin{array}{r}
9999 \\
-3497 \\
\hline
6502
\end{array}
$$
9's complement of 3497

9's Complement Subtraction

The usefulness of the 9's complement stems from the fact that subtraction of a smaller decimal number from a larger one can be accomplished by *adding* the 9's complement of the subtrahend (in this case the smaller number) to the minuend and then adding the carry to the result (this is called an *end-around carry*). When subtracting a larger number from a smaller one, there is no carry, and the result is in 9's complement form and negative. This procedure has a distinct advantage in certain types of arithmetic logic. A few examples will demonstrate decimal subtraction using the 9's complement method.

EXAMPLE 2–4   Perform the following subtractions by using the 9's complement method:

Regular Subtraction   *9's Complement Subtraction*

(a)

```
        8                          8
      − 3                        + 6        9's complement of 3
      ───                       ⓵  4
        5                        ↳+ 1       Add carry to result.
                                 ───
                                   5
```

(b)

```
       13                         13
      − 07                       + 92       9's complement of 07
      ────                      ⓵  05
        06                       ↳+ 1       Add carry to result.
                                 ────
                                    6
```

(c)

```
       54                         54
      − 21                       + 78       9's complement of 21
      ────                      ⓵  32
        33                       ↳+ 1       Add carry to result.
                                 ────
                                   33
```

(d)

```
       15                         15
      − 28                       + 71       9's complement of 28
      ────                        86        9's complement of result
      − 13                         ↓         (No carry indicates that
                                 − 13        the answer is negative
                                             and in complement form.)
```

10's Complement

The 10's complement of a decimal number is equal to the *9's complement plus 1*. The following example illustrates how to determine the 10's complement of a decimal number.

EXAMPLE 2–5 Convert the following decimal numbers to their 10's complement form:

(a) 8   (b) 17   (c) 52   (d) 428

Solutions First find the 9's complement, and then add 1.

(a)
```
    9
  − 8
 ─────
    1    9's complement
  + 1    of 8; add 1.
 ─────
    2    10's complement
         of 8
```

(b)
```
   99
  −17
 ─────
   82    9's complement
  + 1    of 17; add 1.
 ─────
   83    10's complement
         of 17
```

(c)
```
   99
  −52
 ─────
   47    9's complement
  + 1    of 52; add 1.
 ─────
   48    10's complement
         of 52
```

(d)
```
  999
 −428
 ─────
  571    9's complement
  + 1    of 428; add 1.
 ─────
  572    10's complement
         of 428
```

10's Complement Subtraction

The 10's complement can be used to perform subtraction by adding the minuend to the 10's complement of the subtrahend and *dropping* the carry. The following example illustrates.

EXAMPLE 2–6 Perform the following subtractions by using the 10's complement method:

| | *Regular Subtraction* | *10's Complement Subtraction* | |

(a)
```
      8              8
    −3             +7      10's complement of 3
   ────           ────
      5             1̸5      Drop carry.
```

(b)
```
     13             13
   −07            +93      10's complement of 07
   ────           ────
      6            1̸06      Drop carry.
```

(c)
```
     54             54
   −21            +79      10's complement of 21
   ────           ────
     33            1̸33      Drop carry.
```

(d)
```
    196            196
  −155           +845      10's complement of 155
  ─────          ─────
     41           1̸041      Drop carry.
```

Note that in the 10's complement, we do not have to add the carry to the result.

1. Find the 9's complements of the following decimal numbers:
   (a) 34   (b) 72   (c) 4580   (d) 6109
2. Find the 10's complements of the following decimal numbers:
   (a) 93   (b) 102   (c) 2738   (d) 9300

2-3   BINARY NUMBERS

The binary number system is simply another way to count. It is less complicated than the decimal system because it is composed of only *two* digits. It may seem more difficult at first because it is unfamiliar to you.

Just as the decimal system with its ten digits is a base-ten system, the binary system with its two digits is a *base-two* system. The two binary digits (bits) are 1 and 0. The position of the 1 or 0 in a binary number indicates its "weight" or value within the number, just as the position of a decimal digit determines the magnitude of that digit. The weight of each successively higher position (to the left) in a binary number is an increasing power of two.

Counting in Binary

To learn to count in binary, let us first look at how we count in decimal. We start at 0 and count up to 9 before we run out of digits. We then start another digit position (to the left) and continue counting 10 through 99. At this point we have exhausted all two-digit combinations, so a third digit is needed in order to count from 100 through 999.

A comparable situation occurs when counting in binary, except that we have only two digits. We begin counting, 0, 1. At this point we have used both digits, so we include another digit position and continue, 10, 11. We have now exhausted all combinations of two digits, so a third is required. With three digits we can continue to count, 100, 101, 110, and 111. Now we need a fourth digit to continue, and so on. A binary count of 0 through 31 is shown in Table 2–1.

An easy way to remember how to write a binary sequence such as in Table 2–1 for a five-bit example is as follows:

1. The right-most column in the binary number begins with a 0 and alternates each bit.
2. The next column begins with two 0s and alternates every two bits.
3. The next column begins with four 0s and alternates every four bits.
4. The next column begins with eight 0s and alternates every eight bits.
5. The next column begins with sixteen 0s and alternates every sixteen bits.

As you have seen, it takes at least five bits to count from 0 to 31. The following formula tells us how high we can count in decimal with n bits, beginning with zero:

$$\text{Highest decimal number} = 2^n - 1 \qquad \textbf{(2–1)}$$

For instance, with two bits we can count from 0 through 3.

$$2^2 - 1 = 4 - 1 = 3$$

With four bits, we can count from 0 to 15.

$$2^4 - 1 = 16 - 1 = 15$$

A table of powers of two (2^n) is given on page 661.

TABLE 2–1

Count	Decimal Number	Binary Number
zero	0	00000
one	1	00001
two	2	00010
three	3	00011
four	4	00100
five	5	00101
six	6	00110
seven	7	00111
eight	8	01000
nine	9	01001
ten	10	01010
eleven	11	01011
twelve	12	01100
thirteen	13	01101
fourteen	14	01110
fifteen	15	01111
sixteen	16	10000
seventeen	17	10001
eighteen	18	10010
nineteen	19	10011
twenty	20	10100
twenty-one	21	10101
twenty-two	22	10110
twenty-three	23	10111
twenty-four	24	11000
twenty-five	25	11001
twenty-six	26	11010
twenty-seven	27	11011
twenty-eight	28	11100
twenty-nine	29	11101
thirty	30	11110
thirty-one	31	11111

Binary-to-Decimal Conversion

A binary number is a weighted number, as mentioned previously. The value of a given binary number in terms of its decimal equivalent can be determined by adding the products of each bit and its weight. The right-most bit is the *least significant bit (LSB)* in a binary number and has a weight of $2^0 = 1$. The weights increase by a power of two for each bit from right to left. The method of converting a binary number to decimal is illustrated by the following example.

EXAMPLE 2–7

Convert the binary number 1101101 to decimal.

Solution

Binary weight:	2^6	2^5	2^4	2^3	2^2	2^1	2^0
Weight value:	64	32	16	8	4	2	1
Binary number:	1	1	0	1	1	0	1

$$1 \times 64 + 1 \times 32 + 0 \times 16 + 1 \times 8 + 1 \times 4 + 0 \times 2 + 1 \times 1$$
$$= \quad 64 \quad + \quad 32 \quad + \quad 0 \quad + \quad 8 \quad + \quad 4 \quad + \quad 0 \quad + \quad 1$$
$$= \quad 109_{10}$$

This is the equivalent decimal value of the binary number. The subscript 10 identifies this as a decimal number to avoid confusion with other number systems.

The binary numbers we have seen so far have been whole numbers. Fractional numbers can also be represented in binary by placing bits to the right of the *binary point*, just as fractional decimal digits are placed to the right of the decimal point.

The column weights of a binary number are

$$2^n \ldots 2^3 2^2 2^1 2^0 . 2^{-1} 2^{-2} \ldots 2^{-n}$$
$$\underset{\text{binary point}}{\uparrow}$$

This indicates that all the bits to the left of the binary point have weights that are positive powers of two, as previously discussed. All bits to the right of the binary point have weights that are negative powers of two, or fractional weights ($2^{-1} = \frac{1}{2^1}$, etc.), as illustrated in the following example.

EXAMPLE 2–8

Determine the decimal value of the fractional binary number 0.1011.

Solution First determine the weight of each bit, and then sum the weights times the bit.

Binary weight:	2^{-1}	2^{-2}	2^{-3}	2^{-4}
Weight value:	0.5	0.25	0.125	0.0625

Binary number: 0.1   0   1   1

$$1 \times 0.5 + 0 \times 0.25 + 1 \times 0.125 + 1 \times 0.0625$$
$$= \quad 0.5 \quad + \quad 0 \quad + \quad 0.125 \quad + \quad 0.0625$$
$$= 0.6875_{10}$$

Note that to determine the decimal value of a binary number, fractional or whole, we simply *add the weights of each 1 and ignore each 0* because the product of a 0 and its weight is 0.

Another method of evaluating a binary fraction is to determine the *whole-number value* of the bits and divide by the *total possible combinations* of the number of bits appearing in the fraction. For instance, for the binary fraction in Example 2–8,

1. If we neglect the binary point, the value of 1011_2 is 11_{10} in decimal.
2. With four bits, there are $2^4 = 16$ possible combinations.
3. Dividing, we obtain $11/16 = 0.6875_{10}$.

The following examples illustrate the evaluation of binary numbers in terms of their equivalent decimal values. The subscript 2 identifies binary numbers.

EXAMPLE 2–9

Determine the decimal value of 11101.011_2.

Solution

$$(16 + 8 + 4 + 1).(0.25 + 0.125) = 29.375_{10}$$

Using the alternate method to evaluate the fraction, we have $3/8 = 0.375_{10}$ (same result).

EXAMPLE 2–10

Determine the decimal value of 110101.11_2.

Solution

$$(32 + 16 + 4 + 1).(0.5 + 0.25) = 53.75_{10}$$

Using the alternate method to evaluate the fraction gives us $3/4 = 0.75_{10}$ (same result).

SECTION REVIEW 2–3

1. What is the weight of the 1 in 10000_2?
2. Convert each of the following binary numbers to decimal:
   (a) 1010_2 (b) 11010_2 (c) 10111101_2 (d) 110.101_2

2-4   BINARY ARITHMETIC

Addition

There are four basic rules for adding binary digits:

$$0 + 0 = 0$$
$$0 + 1 = 1$$
$$1 + 0 = 1$$
$$1 + 1 = 10_2$$

Notice that three of the addition rules result in a single bit, and that the addition of two 1s yields a binary two (10_2). When binary numbers are added, the latter condition creates a sum of 0 in a given column and a carry of 1 over to the next higher column, as illustrated below:

$$
\begin{array}{r}
^{1\ 1} \\
011 \\
+\,001 \\
\hline
100
\end{array}
$$

In the right column, $1 + 1 = 0$ with a carry of 1 to the next column to the left. In the next column, $1 + 1 + 0 = 0$ with a carry of 1 to the next column to the left. In the left column, $1 + 0 + 0 = 1$.

When there is a carry, we have a situation where three bits are being added (a bit in each of the two numbers and a carry bit). The rules for this are as follows:

$$
\text{Carry bits}
\begin{cases}
1 + 0 + 0 = 01_2 & \text{1 with a carry of 0} \\
1 + 1 + 0 = 10_2 & \text{0 with a carry of 1} \\
1 + 0 + 1 = 10_2 & \text{0 with a carry of 1} \\
1 + 1 + 1 = 11_2 & \text{1 with a carry of 1}
\end{cases}
$$

The following example will illustrate binary addition with the equivalent decimal addition also shown.

EXAMPLE 2-11	(a)	$\begin{array}{r} 11_2 \\ +11_2 \\ \hline 110_2 \end{array}$	$\begin{array}{r} 3_{10} \\ +3_{10} \\ \hline 6_{10} \end{array}$	(b)	$\begin{array}{r} 100_2 \\ +10_2 \\ \hline 110_2 \end{array}$	$\begin{array}{r} 4_{10} \\ +2_{10} \\ \hline 6_{10} \end{array}$
	(c)	$\begin{array}{r} 111_2 \\ +11_2 \\ \hline 1010_2 \end{array}$	$\begin{array}{r} 7_{10} \\ +3_{10} \\ \hline 10_{10} \end{array}$	(d)	$\begin{array}{r} 110_2 \\ +100_2 \\ \hline 1010_2 \end{array}$	$\begin{array}{r} 6_{10} \\ +4_{10} \\ \hline 10_{10} \end{array}$
	(e)	$\begin{array}{r} 1111_2 \\ +1100_2 \\ \hline 11011_2 \end{array}$	$\begin{array}{r} 15_{10} \\ +12_{10} \\ \hline 27_{10} \end{array}$	(f)	$\begin{array}{r} 11100_2 \\ +10011_2 \\ \hline 10111_2 \end{array}$	$\begin{array}{r} 28_{10} \\ +19_{10} \\ \hline 47_{10} \end{array}$

Subtraction

There are four basic rules for subtracting binary digits:

$$0 - 0 = 0$$
$$1 - 1 = 0$$
$$1 - 0 = 1$$
$$10_2 - 1 = 1$$

When subtracting numbers, we sometimes have to borrow from the next higher column. A borrow is required in binary only when we try to subtract a 1 from a 0. In this case, when a 1 is borrowed from the next higher column, a 10_2 is created in the column being subtracted, and the last of the four basic rules listed above must be applied. The following examples illustrate binary subtraction with the equivalent decimal subtraction shown.

EXAMPLE 2–12

(a)
$$\begin{array}{cc} 11_2 & 3_{10} \\ -01_2 & -1_{10} \\ \hline 10_2 & 2_{10} \end{array}$$

(b)
$$\begin{array}{cc} 11_2 & 3_{10} \\ -10_2 & -2_{10} \\ \hline 01_2 & 1_{10} \end{array}$$

No borrows were required in this example. 01_2 is the same as 1_2.

EXAMPLE 2–13

$$\begin{array}{cc} 101_2 & 5_{10} \\ -011_2 & -3_{10} \\ \hline 010_2 & 2_{10} \end{array}$$

Let us examine exactly what was done to subtract the two binary numbers.

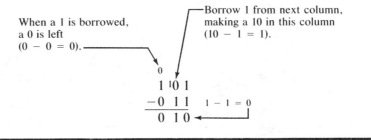

When a 1 is borrowed, a 0 is left ($0 - 0 = 0$).

Borrow 1 from next column, making a 10 in this column ($10 - 1 = 1$).

$$\begin{array}{l} 0 \\ 1\ {}^1 0\ 1 \\ -0\ 1\ 1 \quad 1 - 1 = 0 \\ \hline 0\ 1\ 0 \end{array}$$

Multiplication

The following are four basic rules for multiplying binary digits:

$$0 \times 0 = 0$$
$$0 \times 1 = 0$$

$$1 \times 0 = 0$$
$$1 \times 1 = 1$$

Multiplication is performed in binary in the same manner as with decimal numbers. It involves forming the partial products, shifting each successive partial product left one place, and then adding all the partial products. A few examples will illustrate the procedure and the equivalent decimal multiplication.

EXAMPLE 2–14

(a)
$$\begin{array}{r} 11_2 \\ \times\, 1_2 \\ \hline 11_2 \end{array} \qquad \begin{array}{r} 3_{10} \\ \times\, 1_{10} \\ \hline 3_{10} \end{array}$$

(b)
$$\begin{array}{r} 11_2 \\ \times\, 11_2 \\ \hline 11_2 \\ 11_2 \\ \hline 1001_2 \end{array} \qquad \begin{array}{r} 3_{10} \\ \times\, 3_{10} \\ \hline 9_{10} \end{array}$$

(c)
$$\begin{array}{r} 111_2 \\ \times\, 101_2 \\ \hline 111_2 \\ 000_2 \\ 111_2 \\ \hline 100011_2 \end{array} \qquad \begin{array}{r} 7_{10} \\ \times\, 5_{10} \\ \hline 35_{10} \end{array}$$

(d)
$$\begin{array}{r} 1011_2 \\ \times\, 1001_2 \\ \hline 1011_2 \\ 0000_2 \\ 0000_2 \\ 1011_2 \\ \hline 1100011_2 \end{array} \qquad \begin{array}{r} 11_{10} \\ \times\, 9_{10} \\ \hline 99_{10} \end{array}$$

Division

Division in binary follows the same procedure as division in decimal.

EXAMPLE 2–15

(a)
$$\begin{array}{r} 10_2 \\ 11\overline{)110} \\ \underline{11} \\ 000 \end{array} \qquad \begin{array}{r} 2_{10} \\ 3\overline{)6} \\ \underline{6} \\ 0 \end{array}$$

(b)
$$\begin{array}{r} 11_2 \\ 10\overline{)110} \\ \underline{10} \\ 10 \\ \underline{10} \\ 00 \end{array} \qquad \begin{array}{r} 3_{10} \\ 2\overline{)6} \\ \underline{6} \\ 0 \end{array}$$

(c)
$$\begin{array}{r} 11_2 \\ 100\overline{)1100} \\ \underline{100} \\ 100 \\ \underline{100} \\ 000 \end{array} \qquad \begin{array}{r} 3_{10} \\ 4\overline{)12} \\ \underline{12} \\ 0 \end{array}$$

(d)
$$\begin{array}{r} 10.1_2 \\ 110\overline{)1111.0} \\ \underline{110} \\ 11\,0 \\ \underline{11\,0} \\ 00\,0 \end{array} \qquad \begin{array}{r} 2.5_{10} \\ 6\overline{)15.0} \\ \underline{12} \\ 3\,0 \\ \underline{3\,0} \\ 0 \end{array}$$

SECTION REVIEW 2–4

1. Perform the following additions:
    (a) $1101_2 + 1010_2$ (b) $10111_2 + 01101_2$

2. Perform the following subtractions:
   (a) $1101_2 - 0100_2$ (b) $1001_2 - 0111_2$
3. Perform the indicated operation:
   (a) $110_2 \times 111_2$ (b) $1100_2 \div 011_2$

2-5   1'S AND 2'S COMPLEMENTS

Obtaining the 1's Complement of a Binary Number

The 1's complement of a binary number is found by simply changing all 1s to 0s and all 0s to 1s, as illustrated by a few examples.

Binary Number	1's Complement
10101	01010
10111	01000
111100	000011
11011011	00100100

1's Complement Subtraction

Subtraction of binary numbers can be accomplished by the direct method described in Section 2–4 or by using the 1's complement method, which *allows us to subtract using only addition.*

To subtract a smaller number from a larger number, the 1's complement method is as follows:

1. Determine the 1's complement of the smaller number.
2. Add the 1's complement to the larger number.
3. Remove the carry and add it to the result. This is called *end-around carry.*

EXAMPLE 2–16   Subtract 10011_2 from 11001_2 using the 1's complement method. Show direct subtraction for comparison.

Solution

Direct Subtraction	1's Complement Method	
11001	11001	
− 10011	+ 01100	1's complement of 10011
00110	①00101	
	→ + 1	Add end-around carry.
	00110	Final answer

To subract a larger number from a smaller one, the 1's complement method is as follows:

1. Determine the 1's complement of the larger number.
2. Add the 1's complement to the smaller number.
3. The answer has an opposite sign and is the 1's complement of the result. There is no carry.

EXAMPLE 2–17

Subtract 1101_2 from 1001_2 using the 1's complement method. Show direct subtraction for comparison.

Solution

Direct Subtraction	*1's Complement Method*	
1001	1001	
-1101	$+0010$	1's complement of 1101
-0100	1011	Answer in 1's complement form and opposite in sign
	$\rightarrow$ -0100	Final answer

The 1's complement method is particularly useful in arithmetic logic circuits because subtraction can be accomplished with an adder. Also, the 1's complement of a number is easily obtained by inverting each bit in the number.

Obtaining the 2's Complement of a Binary Number

The 2's complement of a binary number is found by adding 1 to the 1's complement. The following example shows how this is done.

EXAMPLE 2–18

Binary Number		*2's Complement*
110	001	1's complement of 110
	$+1$	Add 1.
	010	2's complement of 110
10101	01010	1's complement of 10101
	$+1$	Add 1.
	01011	2's complement of 10101

An alternative method of obtaining the 2's complement is demonstrated for the binary number 1101011101000.

1. Start at the right and write the bits *as they are* up to and including the first 1.
2. Take the 1's complements of the remaining bits.

$$\underbrace{001010001}_{\substack{\text{1's complements of}\\\text{original bits}}} \underbrace{\overset{\Large\downarrow\ \text{first 1 going right to left}}{1000}}_{\text{These bits remain as they were.}}$$

2's Complement Subtraction

To subtract a smaller number from a larger one, the 2's complement method is applied as follows:

1. Determine the 2's complement of the smaller number.
2. Add the 2's complement to the larger number.
3. Discard the carry (there is always a carry in this case).

EXAMPLE 2–19   Subtract 1011_2 from 1100_2 using the 2's complement method. Show direct subtraction for comparison.

Solution

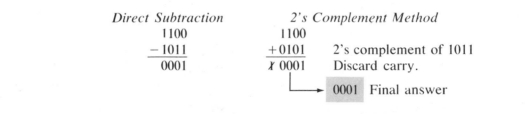

Direct Subtraction		*2's Complement Method*	
1100		1100	
− 1011		+0101	2's complement of 1011
0001		̸1 0001	Discard carry.
		0001	Final answer

To subtract a larger number from a smaller one, the 2's complement method is as follows:

1. Determine the 2's complement of the larger number.
2. Add the 2's complement to the smaller number.
3. There is no carry. The result is in 2's complement form and is negative.
4. To get an answer in true form, take the 2's complement and change the sign.

EXAMPLE 2–20   Subtract 11100_2 from 10011_2 using the 2's complement method.

Solution

Direct Subtraction		*2's Complement Method*	
10011		10011	
− 11100		+00100	2's complement of 11100
− 01001		10111	No carry; 2's complement of answer
		− 01001	Final answer

At this point, both the 1's and 2's complement methods of subtraction may seem excessively complex compared with direct subtraction. However, as mentioned before, they both have distinct advantages when implemented with logic circuits because they allow subtraction to be done by using only addition. Both the 1's and the 2's complements of a binary number are relatively easy to accomplish with logic circuits; and the 2's complement has an advantage over the 1's complement in that an end-around carry operation does not have to be performed. You will see later how arithmetic operations are implemented with logic circuits.

SECTION REVIEW 2–5

1. Determine the 1's complement of each binary number.
   (a) 11010_2 (b) 10111_2 (c) 001101_2
2. Perform the following subtractions using the 1's complement method.
   (a) $1101_2 - 0111_2$ (b) $10110_2 - 11011_2$
3. Determine the 2's complement of each number.
   (a) 10111_2 (b) 11111_2 (c) 010001_2
4. Perform the following subtractions using the 2's complement method.
   (a) $1110_2 - 1001_2$ (b) $1001_2 - 1110_2$

2–6   DECIMAL-TO-BINARY CONVERSION

In Section 2–3 we discussed how to determine the equivalent decimal value of a binary number. Now, you will learn two ways of converting from a decimal number to a binary number.

Sum-of-Weights Method

One way to find the binary number equivalent to a given decimal number is to determine the set of binary weight values whose sum is equal to the decimal number. For instance, the decimal number 9 can be expressed as the sum of binary weights as follows:

$$9 = 8 + 1 = 2^3 + 2^0$$

By placing a 1 in the appropriate weight positions, 2^3 and 2^0, and a 0 in the other positions, we have the binary number for decimal 9.

$$
\begin{array}{cccc}
2^3 & 2^2 & 2^1 & 2^0 \\
1 & 0 & 0 & 1_2
\end{array} \quad \text{binary nine}
$$

EXAMPLE 2–21   Convert the following decimal numbers to binary:

(a) 12   (b) 25   (c) 58   (d) 82

Solutions

(a) $12_{10} = 8 + 4 = 2^3 + 2^2$ ⟶ 1100_2
(b) $25_{10} = 16 + 8 + 1 = 2^4 + 2^3 + 2^0$ ⟶ 11001_2
(c) $58_{10} = 32 + 16 + 8 + 2 = 2^5 + 2^4 + 2^3 + 2^1$ ⟶ 111010_2
(d) $82_{10} = 64 + 16 + 2 = 2^6 + 2^4 + 2^1$ ⟶ 1010010_2

Repeated Division-by-2 Method

A more systematic method of converting from decimal to binary is the *repeated division-by-2* process. For example, to convert the decimal number 12 to binary, we begin by dividing 12 by 2 and then dividing each resulting quotient by 2 until there is a 0 quotient. The remainders generated by each division form the binary number. The first remainder to be produced is the least significant bit (LSB) in the binary number. This procedure is shown in the following steps.

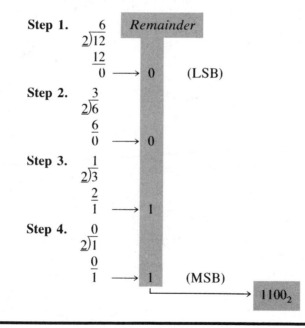

EXAMPLE 2–22   Convert the following decimal numbers to binary:
(a) 19   (b) 45

Solutions

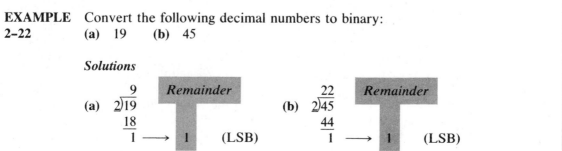

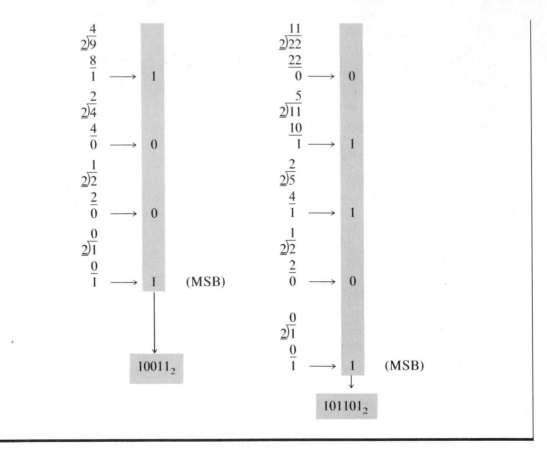

Converting Decimal Fractions to Binary

The previous examples demonstrated whole number conversions. Now, fractional conversions are examined.

The *sum-of-weights* method can be applied to fractional decimal numbers as shown in the following example:

$$0.625_{10} = 0.5 + 0.125 = 2^{-1} + 2^{-3} = 0.101_2$$

There is a 1 in the 2^{-1} position, a 0 in the 2^{-2} position, and a 1 in the 2^{-3} position.

As you have seen, decimal whole numbers can also be converted to binary by repeated division-by-2. Decimal fractions can be converted to binary by *repeated multiplication-by-2*. For example, to convert the decimal fraction 0.3125 to binary, we begin by multiplying 0.3125 by 2 and then multiplying each resulting fractional part of the product by 2 until the fractional product is zero. The carries generated by each multiplication form the binary number. The first carry produced is the MSB. This procedure is shown in the following steps:

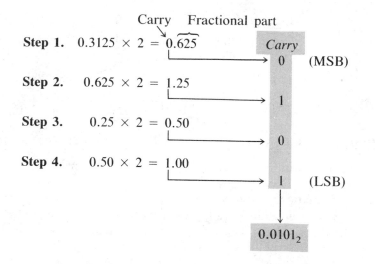

Carry   Fractional part

Step 1. $0.3125 \times 2 = 0.\overbrace{625}$ *Carry*

→ 0   (MSB)

Step 2. $0.625 \times 2 = 1.25$

→ 1

Step 3. $0.25 \times 2 = 0.50$

→ 0

Step 4. $0.50 \times 2 = 1.00$

→ 1   (LSB)

0.0101_2

SECTION REVIEW 2–6

1. Convert each decimal number to binary by using the sum-of-weights method.
   (a) 23   (b) 57   (c) 45.5
2. Convert each decimal number to binary by using the repeated division-by-2 method (repeated multiplication-by-2 for fractions).
   (a) 14   (b) 21   (c) 0.375

2–7   OCTAL NUMBERS

The octal number system is composed of eight digits, which are

$$0, 1, 2, 3, 4, 5, 6, 7$$

To count above 7, we begin another column and start over.

$$10, 11, 12, 13, 14, 15, 16, 17, 20, 21, \text{etc.}$$

Counting in octal is the same as counting in decimal, except any number with an 8 or a 9 is omitted. To distinguish octal numbers from decimal numbers, we will use the subscript 8 to indicate an octal number. For instance, 15_8 is equivalent to 13_{10}.

Octal-to-Decimal Conversion

Since the octal number system has a base of eight, each successive digit position is an increasing power of eight, beginning in the right-most column with 8^0. The evaluation of an octal number in terms of its *decimal equivalent* is accomplished by multiplying each digit by its weight and summing the products, as illustrated below for octal 2374_8.

Weight:		8^3	8^2	8^1	8^0
Decimal value:		512	64	8	1
Octal number:		2	3	7	4

$$2374_8 = 2 \times 8^3 + 3 \times 8^2 + 7 \times 8^1 + 4 \times 8^0$$
$$= 2 \times 512 + 3 \times 64 + 7 \times 8 + 4 \times 1$$
$$= 1024 + 192 + 56 + 4 = 1276_{10}$$

Decimal-to-Octal Conversion

A method of converting a decimal number into an octal number is the *repeated divison-by-8* method, which is similar to the method used in conversion of decimal to binary. To show how it works, we convert the decimal number 359_{10} to octal. Each successive division by 8 yields a remainder that becomes a digit in the equivalent octal number. The first remainder generated is the least significant digit (LSD).

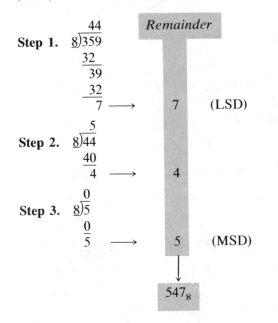

Step 1.

$$8\overline{)359} \quad \frac{44}{}$$

with remainder 7 ⟶ 7 (LSD)

Step 2.

$$8\overline{)44} \quad \frac{5}{}$$

with remainder 4 ⟶ 4

Step 3.

$$8\overline{)5} \quad \frac{0}{}$$

with remainder 5 ⟶ 5 (MSD)

$$547_8$$

The octal numbers to this point have been whole numbers. Fractional octal numbers are represented by digits to the right of the octal point.

The column weights of an octal number are

$$8^n \ldots 8^3 8^2 8^1 8^0 . \: 8^{-1} 8^{-2} 8^{-3} \ldots 8^{-n}$$

This shows that all digits to the left of the octal point have weights that are positive powers of eight, as seen previously. All digits to the right of the octal point have fractional weights, or negative powers of eight, as illustrated in the following example.

**EXAMPLE
2-23**   Determine the decimal value of the octal fraction 0.325_8.

Solution   First, determine the weights of each digit and then sum the weight times the digit.

Octal weight:	8^{-1}	8^{-2}	8^{-3}
Decimal value:	0.125	0.015625	0.001953
Octal number:	0.3	2	5

$$0.325_8 = 3(0.125) + 2(0.015625) + 5(0.001953)$$
$$= 0.375 + 0.03125 + 0.009765 = 0.416015_{10}$$

Octal-to-Binary Conversion

The primary application of octal numbers is in the representation of binary numbers. Since it takes only one octal digit to represent three bits, octal numbers are much easier to "read" than binary numbers.

Because all three-bit binary numbers are required to represent the eight octal digits, it is very easy to convert from octal to binary and from binary to octal. The octal number system is often used in digital systems, especially for input/output applications.

Each octal digit is represented by three bits as indicated:

Octal Digit	*Binary*
0	000
1	001
2	010
3	011
4	100
5	101
6	110
7	111

To convert an octal number to a binary number, simply replace each octal digit by the appropriate three bits. This is illustrated in the following example.

**EXAMPLE
2-24**   Convert each of the following octal numbers to binary:
(a) 13_8 (b) 25_8 (c) 47_8 (d) 170_8 (e) 752_8 (f) 5276_8
(g) 37.12_8

Solutions

(a) 13_8 (b) 25_8 (c) 47_8 (d) 170_8

001011_2 010101_2 100111_2 001111000_2

(e) 752_8 **(f)** 5276_8 **(g)** 37.12_8

111101010_2 101010111110_2 011111.001010_2

Binary-to-Octal Conversion

Conversion of a binary number to an octal number is also a straightforward process. Beginning at the binary point, simply break the binary number into groups of three bits and convert each group into the appropriate octal digit.

EXAMPLE 2–25   Represent each of the following binary numbers by its octal equivalent:

(a) 110101_2 **(b)** 101111001_2 **(c)** 1011100110_2 **(d)** 1001101.1011_2

Solutions

(a) 110101_2 **(b)** 101111001_2 **(c)** 001011100110_2 **(d)** 001001101.101100

    $6 \quad 5_8$ $5 \quad 7 \quad 1_8$ $1 \quad 3 \quad 4 \quad 6_8$ $1 \quad 1 \quad 5. \quad 5 \quad 4_8$

SECTION REVIEW 2–7

1. Convert the octal numbers to decimal.
      (a) 73_8 **(b)** 125_8
2. Convert the decimal numbers to octal.
      (a) 98_{10} **(b)** 163_{10}
3. Convert the octal numbers to binary.
      (a) 46_8 **(b)** 723_8 **(c)** 5624.37_8
4. Convert the binary numbers to octal.
      (a) 110101111_2 **(b)** 100110001.101111_2 **(c)** 10111111.0011_2

2–8   HEXADECIMAL NUMBERS

The hexadecimal system has a base of sixteen; that is, it is composed of 16 digits and characters. Many digital systems process binary data in groups that are multiples of four bits, making the hexadecimal number very convenient because each hexadecimal digit represents a four-bit binary number (as listed in Table 2–2).

    Ten digits and six alphabetic characters make up this number system. A subscript 16 indicates a hexadecimal number.

TABLE 2-2

Decimal	Binary	Hexadecimal
0	0000	0
1	0001	1
2	0010	2
3	0011	3
4	0100	4
5	0101	5
6	0110	6
7	0111	7
8	1000	8
9	1001	9
10	1010	A
11	1011	B
12	1100	C
13	1101	D
14	1110	E
15	1111	F

Binary-to-Hexadecimal Conversion

Converting a binary number to hexadecimal is a straightforward procedure. Simply break the binary number into four-bit groups starting at the binary point, and replace each group with the equivalent hexadecimal symbol.

EXAMPLE 2-26

Convert the following binary numbers to hexadecimal:

(a) 1100101001010111_2 (b) 111111000101101001_2

(c) 1110011000.111_2

Solutions

(a) 1100101001010111_2

$\quad$ C $\quad$ A $\quad$ 5 $\quad$ 7_{16}

(b) 0011111000101101001_2

$\quad$ 3 $\quad$ F $\quad$ 1 $\quad$ 6 $\quad$ 9_{16}

(c) 001110011000.1110

$\quad$ 3 $\quad$ 9 $\quad$ 8 $\quad$ E_{16}

Hexadecimal-to-Binary Conversion

To convert from a hexadecimal number to a binary number, reverse the process and replace each hexadecimal symbol with the appropriate four bits.

EXAMPLE 2–27   Determine the binary numbers for the following hexadecimal numbers:
(a) $10A4_{16}$ (b) $CF83_{16}$ (c) 9742_{16} (d) $D2E.8_{16}$

Solutions

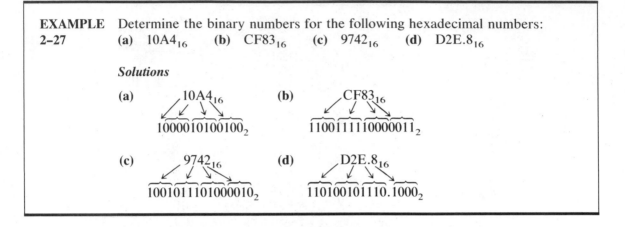

(a)
$$10A4_{16}$$
$$1000\,0101\,0010\,0100_2$$

(b)
$$CF83_{16}$$
$$1100\,1111\,1000\,0011_2$$

(c)
$$9742_{16}$$
$$1001\,0111\,0100\,0010_2$$

(d)
$$D2E.8_{16}$$
$$1101\,0010\,1110.1000_2$$

The use of letters to represent quantities may seem strange at first, but keep in mind that *any* number system is only a set of sequential symbols. If we understand what these symbols mean in terms of quantities represented, then the form of the symbols themselves is unimportant once we get accustomed to using them.

It should be clear that it is much easier to write the hexadecimal number than the equivalent binary number, and since conversion is so easy, the hexadecimal system is a ''natural'' for representing binary numbers and digital codes.

Counting in Hexadecimal

How do we count in hexadecimal once we get to F? Simply start over with another column and continue as follows:

10, 11, 12, 13, 14, 15, 16, 17, 18, 19, 1A, 1B, 1C, 1D, 1E, 1F,
20, 21, 22, 23, 24, 25, 26, 27, 28, 29, 2A, 2B, 2C, 2D, 2E, 2F,
30, 31, and so forth

With two hexadecimal digits, we can count up to FF_{16}, which is 255_{10}. To count beyond this, three hexadecimal digits are needed. For instance, 100_{16} is decimal 256_{10}, 101_{16} is decimal 257_{10}, and so forth. The maximum three-digit hexadecimal number is FFF_{16}, or 4095_{10}. The maximum four-digit hexadecimal number is $FFFF_{16}$, which is $65,535_{10}$.

Hexadecimal-to-Decimal Conversion

One way to evaluate a hexadecimal number in terms of its decimal equivalent is to first convert the hexadecimal number to binary and then convert from binary to decimal. The following example illustrates this procedure.

EXAMPLE 2-28

Convert the following hexadecimal numbers to decimal:

(a) $1C_{16}$ (b) $A85_{16}$

Solutions

(a) $1C_{16}$

$\overline{0001}\ \overline{1100}_2$

$00011100_2 = 2^4 + 2^3 + 2^2$
$\qquad\qquad = 16 + 8 + 4$
$\qquad\qquad = 28_{10}$

(b) $A85_{16}$

$\overline{1010}\ \overline{1000}\ \overline{0101}_2$

$101010000101_2 = 2^{11} + 2^9 + 2^7 + 2^2 + 2^0$

$\qquad\qquad\qquad = 2048 + 512 + 128 + 4 + 1$
$\qquad\qquad\qquad = 2693_{10}$

Another way to convert a hexadecimal number to its decimal equivalent is by multiplying each hexadecimal digit by its weight and then taking the sum of these products. The weights of a hexadecimal number are increasing powers of 16 (from right to left). For a four-digit hexadecimal number the weights are

$$16^3 \qquad 16^2 \qquad 16^1 \qquad 16^0$$
$$4096 \qquad 256 \qquad 16 \qquad 1$$

The following example shows this conversion method.

EXAMPLE 2-29

Convert (a) $E5_{16}$ and (b) $B2F8_{16}$ to decimal.

Solutions

(a) $E5_{16} = E \times 16 + 5 \times 1 = 14 \times 16 + 5 \times 1 = 224 + 5 = 229_{10}$

(b) $B2F8_{16} = B \times 4096 \ + 2 \times 256 + F \times 16 \ + 8 \times 1$
$\qquad\qquad = 11 \times 4096 \ + 2 \times 256 + 15 \times 16 + 8 \times 1$
$\qquad\qquad = 45,056 \qquad + 512 \qquad + 240 \qquad + 8$
$\qquad\qquad = 45,816_{10}$

Decimal-to-Hexadecimal Conversion

Repeated division of a decimal number by 16 will produce the equivalent hexadecimal number formed by the remainders of each division. This is similar to the repeated division-by-2 for decimal-to-binary conversion and repeated division-by-8 for decimal-to-octal conversion.

The following example illustrates this procedure.

**EXAMPLE
2–30**

Convert 650_{10} to hexadecimal by repeated division-by-16_{10}.

Solution

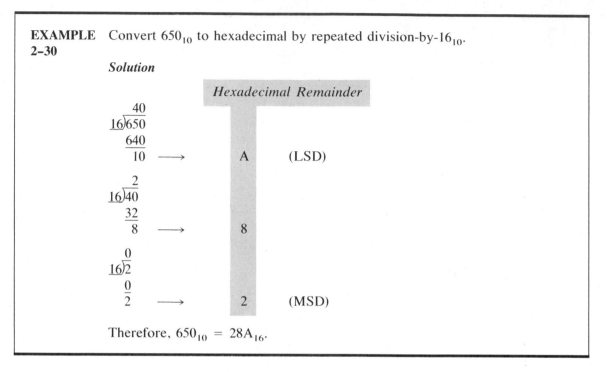

$$\begin{array}{r} 40 \\ 16\overline{)650} \\ 640 \\ \hline 10 \end{array} \longrightarrow \quad A \quad \text{(LSD)}$$

$$\begin{array}{r} 2 \\ 16\overline{)40} \\ 32 \\ \hline 8 \end{array} \longrightarrow \quad 8$$

$$\begin{array}{r} 0 \\ 16\overline{)2} \\ 0 \\ \hline 2 \end{array} \longrightarrow \quad 2 \quad \text{(MSD)}$$

Hexadecimal Remainder

Therefore, $650_{10} = 28A_{16}$.

Hexadecimal Addition

Addition can be done directly with hexadecimal numbers by remembering that the hexadecimal digits 0 through 9 are equivalent to decimal digits 0 through 9 and that hexadecimal digits A through F are equivalent to decimal numbers 10 through 15.

When adding two hexadecimal numbers, use the following rules:

1. In any given column of an addition problem, think of the two hexadecimal digits in terms of their decimal value. For instance, $5_{16} = 5_{10}$ and $C_{16} = 12_{10}$.
2. If the sum of these two digits is 15_{10} or less, bring down the corresponding hexadecimal digit.
3. If the sum of these two digits is greater than 15_{10}, bring down the amount of the sum that exceeds 16_{10} and carry a 1 to the next column.

**EXAMPLE
2–31**

Add the following hexadecimal numbers:

(a) $23_{16} + 16_{16}$ (b) $58_{16} + 22_{16}$ (c) $2B_{16} + 84_{16}$

(d) $DF_{16} + AC_{16}$

Solution

(a) $\quad 23_{16}$ right column: $\quad 3_{16} + 6_{16} = 3_{10} + 6_{10} = 9_{10} = 9_{16}$

$\quad \underline{+16_{16}}$ left column: $\quad 2_{16} + 1_{16} = 2_{10} + 1_{10} = 3_{10} = 3_{16}$

$\quad 39_{16}$

(b) $\quad 58_{16}$ right column: $\quad 8_{16} + 2_{16} = 8_{10} + 2_{10} = 10_{10} = A_{16}$

$\quad \underline{+22_{16}}$ left column: $\quad 5_{16} + 2_{16} = 5_{10} + 2_{10} = 7_{10} = 7_{16}$

$\quad 7A_{16}$

(c) $\quad 2B_{16}$ right column: $\quad B_{16} + 4_{16} = 11_{10} + 4_{10} = 15_{10} = F_{16}$

$\quad \underline{+84_{16}}$ left column: $\quad 2_{16} + 8_{16} = 2_{10} + 8_{10} = 10_{10} = A_{16}$

$\quad AF_{16}$

(d) $\quad DF_{16}$ right column: $\quad F_{16} + C_{16} = 15_{10} + 12_{10} = 27_{10}$

$\quad \underline{+AC_{16}}$ $\qquad\qquad\qquad 27_{10} - 16_{10} = 11_{10} = B_{16}$ with a 1 carry

$\quad 18B_{16}$ left column: $\quad D_{16} + A_{16} + 1_{16} = 13_{10} + 10_{10} + 1_{10} = 24_{10}$

$\qquad\qquad\qquad\qquad\qquad 24_{10} - 16_{10} = 8_{10} = 8_{16}$ with a 1 carry

Hexadecimal Subtraction Using 2's Complement Method

Since a hexadecimal number can be used to represent a binary number, it can also be used to represent the 2's complement of a binary number. For instance, the hexadecimal representation of 11001001_2 is $C9_{16}$. The 2's complement of this binary number is 00110111, which is written in hexadecimal as 37_{16}.

As you have learned, the 2's complement allows us to subtract by *adding* binary numbers. We can also use this method for hexadecimal subtraction, as the following example shows.

EXAMPLE 2–32

Subtract the following hexadecimal numbers:

(a) $84_{16} - 2A_{16}$ $\qquad$ (b) $C3_{16} - 0B_{16}$

Solution

(a) $\quad 2A_{16} = 00101010_2$

$\quad$ 2's complement of $2A_{16} = 11010110 = D6_{16}$

$\qquad 84_{16}$

$\qquad \underline{+D6_{16}}$ $\quad$ Add.

$\qquad \cancel{1}5A_{16}$

$\qquad 5A_{16}$ $\quad$ Drop carry as in 2's complement addition.

(b) $0B_{16} = 00001011_2$

2's complement of $0B_{16} = 11110101 = F5_{16}$

$$
\begin{array}{r}
C3_{16} \\
+ F5_{16} \quad \text{Add.} \\
\hline
\cancel{1}B8_{16} \\
\end{array}
$$

$B8_{16}$ Drop carry.

---

1. Convert the binary numbers to hexadecimal.
   (a) 10110011_2 **(b)** 110011101000_2
2. Convert the hexadecimal numbers to binary.
   (a) 57_{16} **(b)** $3A5_{16}$ **(c)** $F80B_{16}$
3. Add the hexadecimal numbers directly.
   (a) $18_{16} + 34_{16}$ **(b)** $3F_{16} + 2A_{16}$
4. Subtract the hexadecimal numbers.
   (a) $75_{16} - 21_{16}$ **(b)** $94_{16} - 5C_{16}$

2–9   BINARY CODED DECIMAL (BCD)

As you have learned, decimal, octal, and hexadecimal numbers can be represented by binary digits. Not only numbers, but letters and other symbols, can be represented by 1s and 0s. In fact, *any* entity expressible as numbers, letters, or other symbols can be represented by binary digits, and therefore can be processed by digital logic circuits.

Combinations of binary digits that represent numbers, letters, or symbols are *digital codes*. In many applications, special codes are used for such auxiliary functions as error detection.

The 8421 Code

The 8421 code is a type of *binary coded decimal* (BCD) code and is composed of four bits representing the decimal digits 0 through 9. The designation 8421 indicates the binary weights of the four bits (2^3, 2^2, 2^1, 2^0). The ease of conversion between 8421 code numbers and the familiar decimal numbers is the main advantage of this code. The 8421 code is the predominant BCD code, and when we refer to BCD, we always mean the 8421 code unless otherwise stated.

Binary coded decimal means that *each* decimal digit is represented by a binary code of four bits. All you have to remember are the ten binary combinations that represent the ten decimal digits as shown in Table 2–3.

You should realize that with four bits, sixteen numbers (2^4) can be represented, and that in the 8421 code only ten of these are used. The six code

TABLE 2-3 *The 8421 BCD code.*

8421 (BCD)	Decimal
0000	0
0001	1
0010	2
0011	3
0100	4
0101	5
0110	6
0111	7
1000	8
1001	9

combinations that are not used—1010, 1011, 1100, 1101, 1110, and 1111—are *invalid* in the 8421 BCD code.

To express any decimal number in BCD, simply replace each decimal digit by the appropriate four-bit code, as shown by the following example.

EXAMPLE 2-33

Convert each of the following decimal numbers into BCD:

(a) 3   (b) 9.2   (c) 18   (d) 34.8   (e) 65   (f) 92
(g) 150   (h) 321   (i) 1472

Solutions

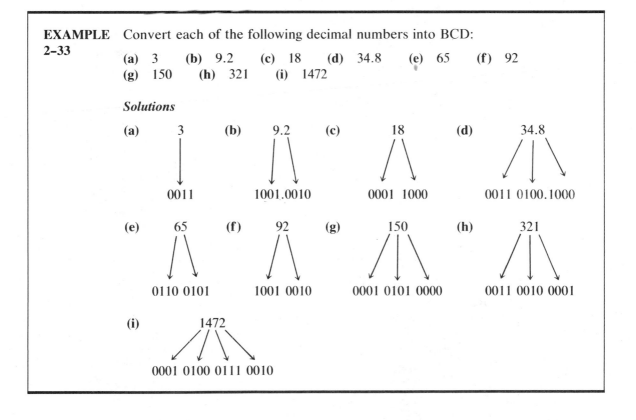

It is equally easy to determine a decimal number from a BCD number. Start at the decimal point and break the code into groups of four bits. Then write the decimal digit represented by each four-bit group. An example illustrates.

EXAMPLE 2–34

Find the decimal numbers represented by the following BCD codes:

(a) 10000110   (b) 00110001   (c) 01010011   (d) 100101110100

(e) 0001100001100000.0111

Solutions

(a)   1000   0110   (b)   0011   0001   (c)   0101   0011
       ↓   ↓   ↓   ↓   ↓   ↓
       8   6   3   1   5   3

(d)   1001   0111   0100   (e)   0001   1000   0110   0000.0111
       ↓   ↓   ↓   ↓   ↓   ↓   ↓   ↓
       9   7   4   1   8   6   0   7

BCD Addition

BCD is a numerical code, and many applications require that arithmetic operations be performed. Addition is the most important operation because the other three operations (subtraction, multiplication, and division) can be accomplished using addition. Here is how to add two BCD numbers:

1. Add the two numbers, using the rules for binary addition in Section 2–4.
2. If a four-bit sum is equal to or less than 9, it is a *valid* BCD number.
3. If a four-bit sum is greater than 9, or if a carry-out of the group is generated, it is an *invalid* result. Add 6 (0110_2) to the four-bit sum in order to skip the six invalid states and return the code to 8421. If a carry results when 6 is added, simply add the carry to the next four-bit group.

Several examples illustrate BCD addition for the case in which the sum of any four-bit column does not exceed 9.

EXAMPLE 2–35

Add the following BCD numbers (the decimal addition is shown for comparison):

(a)   0011   3   (b)   0110   6
       +0100   +4   +0010   +2
        0111 7_{10} 1000 8_{10}

(c)   0010   0011   23   (d)   1000   0110   86
       +0001   0101   +15   +0001   0011   +13
        0011   1000 38_{10} 1001   1001 99_{10}

(e)	0100	0101	0000	450	(f)	1000	0111	0011	873
	+0100	0001	0111	+417		+0001	0001	0010	+112
	1000	0110	0111	867_{10}		1001	1000	0101	985_{10}

Note that in each case the sum in any four-bit column does not exceed 9, and the results are valid BCD numbers.

Next, we deal with the case of an invalid sum (greater than 9 or a carry) by illustrating the procedure with several examples.

EXAMPLE 2–36   Add the following BCD numbers:

(a)
```
              1001                                        9
            + 0100                                      + 4
              1101      Invalid BCD number (> 9)         13₁₀
            + 0110      Add 6.
      0001    0011      Valid BCD number
        ↓       ↓
        1      3₁₀
```

(b)
```
              1001                                        9
            + 1001                                      + 9
        1     0010      Invalid because of carry         18₁₀
            + 0110      Add 6.
      0001    1000      Valid BCD number
        ↓       ↓
        1      8₁₀
```

(c)
```
      0001    0110                                       16
    + 0001    0101                                      +15
      0010    1011      Right group is invalid (> 9),    31₁₀
                          left group valid. Add 6
            + 0110      to invalid code (add
                          carry, 0001, to next group).
      0011    0001      Valid BCD number
        ↓       ↓
        3      1₁₀
```

(d)
```
      0110    0111                                       67
    + 0101    0011                                      +53
      1011    1010      Both groups are invalid (> 9).   120₁₀
    + 0110  + 0110      Add 6 to both groups.
 0001 0010   0000      Valid BCD number
   ↓    ↓      ↓
   1    2      0₁₀
```

1. What are the binary weights of each 1 in the following BCD numbers?
   (a) 0010   (b) 1000   (c) 0001   (d) 0100
2. Convert the following decimal numbers to BCD:
   (a) 6   (b) 15   (c) 273   (d) 84.9
3. What decimal numbers are represented by each BCD code?
   (a) 10001001   (b) 001001111000   (c) 00010101.0111

2–10   DIGITAL CODES

The Gray Code

The Gray code is an *unweighted* code, which means that there are no specific weights assigned to the bit positions. The Gray code exhibits only a *single-bit change* from one code number to the next. This property is important in many applications, such as shaft position encoders, where error susceptibility increases with the number of bit changes between adjacent numbers in a sequence. The Gray code is not an arithmetic code.

Table 2–4 is a listing of four-bit Gray code numbers for decimal numbers 0 through 9. Like binary, the Gray code can have any number of bits. Notice the single-bit change between successive code numbers. Binary numbers are shown for reference. For instance, in going from decimal 3 to 4, the Gray code changes from 0010 to 0110, while the binary code changes from 0011 to 0100, a change of three bits. The only bit change is in the third bit from the right in the Gray code; the others remain the same.

Binary-to-Gray conversion   By representing the ten decimal digits with a four-bit Gray code, we have another form of BCD code. The Gray code, however, can be extended to any number of bits, and conversion between binary code and Gray

TABLE 2–4 *Four-bit Gray code.*

Decimal	Binary	Gray
0	0000	0000
1	0001	0001
2	0010	0011
3	0011	0010
4	0100	0110
5	0101	0111
6	0110	0101
7	0111	0100
8	1000	1100
9	1001	1101

code is sometimes useful. First, we will discuss how to convert from a binary number to a Gray code number. The following rules apply:

1. The most significant digit (left-most) in the Gray code is the same as the corresponding digit in the binary number.
2. Going from left to right, add each adjacent pair of binary digits to get the next Gray code digit. Disregard carries.

For example, let us convert the binary number 10110_2 to Gray code.

Step 1. The left-most Gray digit is the same as the left-most binary digit.

$$
\begin{array}{ccccc}
1 & 0 & 1 & 1 & 0 \quad \text{Binary} \\
\downarrow & & & & \\
1 & & & & \quad \text{Gray}
\end{array}
$$

Step 2. Add the left-most binary digit to the adjacent one.

$$
\begin{array}{ccccc}
\boxed{1 + 0} & 1 & 1 & 0 & \quad \text{Binary} \\
\downarrow & & & & \\
1 \quad 1 & & & & \quad \text{Gray}
\end{array}
$$

Step 3. Add the next adjacent pair.

$$
\begin{array}{ccccc}
1 & \boxed{0 + 1} & 1 & 0 & \quad \text{Binary} \\
& \downarrow & & & \\
1 \quad 1 \quad 1 & & & & \quad \text{Gray}
\end{array}
$$

Step 4. Add the next adjacent pair and discard carry.

$$
\begin{array}{ccccc}
1 & 0 & \boxed{1 + 1} & 0 & \quad \text{Binary} \\
& & \downarrow & & \\
1 \quad 1 \quad 1 \quad 0 & & & & \quad \text{Gray}
\end{array}
$$

Step 5. Add the last adjacent pair.

$$
\begin{array}{ccccc}
1 & 0 & 1 & \boxed{1 + 0} & \quad \text{Binary} \\
& & & \downarrow & \\
1 \quad 1 \quad 1 \quad 0 \quad 1 & & & & \quad \text{Gray}
\end{array}
$$

The conversion is now complete and the Gray code is 11101.

Gray-to-binary conversion To convert from Gray code to binary, a similar method is used, but there are some differences. The following rules apply:

1. The most significant digit (left-most) in the binary code is the same as the corresponding digit in the Gray code.

2. Add each binary digit generated to the Gray digit in the next adjacent position. Disregard carries.

For example, the conversion of the Gray code number 11011 to binary is as follows:

Step 1.   The left-most digits are the same.

Step 2.   Add the last binary digit just generated to the Gray digit in the next position. Discard carry.

Step 3.   Add the last binary digit generated to the next Gray digit.

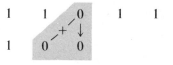

Step 4.   Add the last binary digit generated to the next Gray digit.

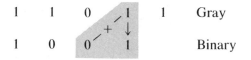

Step 5.   Add the last binary digit generated to the next Gray digit. Discard carry.

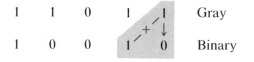

This completes the conversion. The final binary number is 10010_2.

The Excess-3 Code

The Excess-3 is a digital code that is derived by adding 3 to *each* decimal digit and then converting the result to four-bit binary. Since no definite weights can be assigned to the four digit positions, Excess-3 is an unweighted code. For instance, the Excess-3 code for the decimal 2 is

$$\begin{array}{r} 2 \\ +3 \\ \hline 5 \end{array} \rightarrow 0101$$

The Excess-3 code for the decimal 9 is

$$\begin{array}{r} 9 \\ +3 \\ \hline 12 \end{array} \rightarrow 1100$$

The Excess-3 code for each decimal digit is found by the same procedure, and the entire code is shown in Table 2-5.

TABLE 2-5 *Excess-3 code.*

Decimal	BCD	Excess-3
0	0000	0011
1	0001	0100
2	0010	0101
3	0011	0110
4	0100	0111
5	0101	1000
6	0110	1001
7	0111	1010
8	1000	1011
9	1001	1100

Notice that ten of a possible 16 code combinations are used in the Excess-3 code. The six *invalid* combinations are 0000, 0001, 0010, 1101, 1110, and 1111.

Self-Complementing Property

The key feature of the Excess-3 code is that it is *self-complementing*. That is, the *1's complement of an Excess-3 number is the Excess-3 code for the 9's complement of the corresponding decimal number*. For example, the Excess-3 code for decimal 4 is 0111. The 1's complement of this is 1000, which is the Excess-3 code for the decimal 5 (and 5 is the 9's complement of 4).

It should be noted that the 1's complement is easily produced with digital logic circuits by simply inverting each bit. The self-complementing property makes the Excess-3 code useful in some arithmetic operations, because subtraction can be performed using the 9's complement method.

Table 2-6 shows that the 1's complement of each Excess-3 number is the Excess-3 code for the 9's complement of the corresponding decimal digit.

EXAMPLE 2-37 Convert each of the following decimal numbers to Excess-3 code:
(a) 13   (b) 159   (c) 35   (d) 430   (e) 87

Solutions

First, add 3 to *each* digit in the decimal number, and then convert each resulting sum to its equivalent binary code.

(a)

$$
\begin{array}{cc}
1 & 3 \\
+3 & +3 \\
\hline
4 & 6 \\
\downarrow & \downarrow \\
0100 & 0110
\end{array}
$$

Excess-3 for 13_{10}

(b)

$$
\begin{array}{ccc}
1 & 5 & 9 \\
+3 & +3 & +3 \\
\hline
4 & 8 & 12 \\
\downarrow & \downarrow & \downarrow \\
0100 & 1000 & 1100
\end{array}
$$

Excess-3 for 159_{10}

(c)

$$
\begin{array}{cc}
3 & 5 \\
+3 & +3 \\
\hline
6 & 8 \\
\downarrow & \downarrow \\
0110 & 1000
\end{array}
$$

Excess-3 for 35_{10}

(d)

$$
\begin{array}{ccc}
4 & 3 & 0 \\
+3 & +3 & +3 \\
\hline
7 & 6 & 3 \\
\downarrow & \downarrow & \downarrow \\
0111 & 0110 & 0011
\end{array}
$$

Excess-3 for 430_{10}

(e)

$$
\begin{array}{cc}
8 & 7 \\
+3 & +3 \\
\hline
11 & 10 \\
\downarrow & \downarrow \\
1011 & 1010
\end{array}
$$

Excess-3 for 87_{10}

Alphanumeric Codes

In order to communicate, we need not only numbers, but also letters and other symbols. In the strictest sense, codes that represent numbers and alphabetic characters (letters) are called *alphanumeric* codes. Most of these codes, however, also represent symbols and various instructions necessary for conveying intelligible information.

TABLE 2–6

Excess-3	1's Complement	Decimal Digit	9's Complement
0011	1100	0	9
0100	1011	1	8
0101	1010	2	7
0110	1001	3	6
0111	1000	4	5
1000	0111	5	4
1001	0110	6	3
1010	0101	7	2
1011	0100	8	1
1100	0011	9	0

TABLE 2-7 *American Standard Code for Information Interchange.*

LSBs	MSBs							
	000 (0)	001 (1)	010 (2)	011 (3)	100 (4)	101 (5)	110 (6)	111 (7)
0000 (0)	NUL	DLE	SP	0	@	P	`	p
0001 (1)	SOH	DC$_1$	!	1	A	Q	a	q
0010 (2)	STX	DC$_2$	"	2	B	R	b	r
0011 (3)	ETX	DC$_3$	#	3	C	S	c	s
0100 (4)	EOT	DC$_4$	$	4	D	T	d	t
0101 (5)	ENQ	NAK	%	5	E	U	e	u
0110 (6)	ACK	SYN	&	6	F	V	f	v
0111 (7)	BEL	ETB	'	7	G	W	g	w
1000 (8)	BS	CAN	(	8	H	X	h	x
1001 (9)	HT	EM	)	9	I	Y	i	y
1010 (A)	LF	SUB	*	:	J	Z	j	z
1011 (B)	VT	ESC	+	;	K	[	k	{
1100 (C)	FF	FS	,	<	L	\	l	|
1101 (D)	CR	GS	–	=	M	]	m	}
1110 (E)	SO	RS	.	>	N	↑	n	~
1111 (F)	SI	US	/	?	O	←	o	DEL

Definitions of control abbreviations:

ACK	Acknowledge	FS	Form separator
BEL	Bell	GS	Group separator
BS	Backspace	HT	Horizontal tab
CAN	Cancel	LF	Line feed
CR	Carriage return	NAK	Negative acknowledge
DC$_1$–DC$_4$	Direct control	NUL	Null
DEL	Delete idle	RS	Record separator
DLE	Data link escape	SI	Shift in
EM	End of Medium	SO	Shift out
ENQ	Enquiry	SOH	Start of heading
EOT	End of transmission	STX	Start text
ESC	Escape	SUB	Substitute
ETB	End of transmission block	SYN	Synchronous idle
ETX	End text	US	Unit separator
FF	Form feed	VT	Vertical tab

NOTE: The hexadecimal digit representing each bit pattern is shown in parentheses.

At a minimum, an alphanumeric code must represent ten decimal digits and 26 letters of the alphabet, for a total of 36 items. This requires six bits in each code combination because five bits are insufficient ($2^5 = 32$). There are 64 total combinations of six bits, so we have 28 unused code combinations. Obviously, in many applications, symbols other than just numbers and letters are necessary to communicate completely. We need spaces to separate words, periods to mark the end of sentences or for decimal points, instructions to tell the receiving system what to do with the information, and more. So, with codes that are six bits long, we can handle decimal numbers, the alphabet, and 28 other symbols. This should give you an idea of the requirements for a basic alphanumeric code.

ASCII   One standardized alphanumeric code, called the *American Standard Code for Information Interchange* (ASCII), is perhaps the most widely used type. It is a seven-bit code in which the decimal digits are represented by the 8421 BCD code preceded by 011. The letters of the alphabet and other symbols and instructions are represented by other code combinations as shown in Table 2–7. For instance, the letter *A* is represented by 1000001 (41_{16}), the comma by 0101100 ($2C_{16}$), and ETX (end of text) by 0000011 (03_{16}).

EBCDIC   Another alphanumeric code also frequently encountered is called the *Extended Binary Coded Decimal Interchange Code* (EBCDIC). This is an eight-bit code in which the decimal digits are represented by the 8421 BCD code preceded by 1111. Both lowercase and uppercase letters are represented in addition to numerous other symbols and commands, as shown in Table 2–8. For example, uppercase *A* is 11000001, and lowercase *a* is 10000001.

The coverage of codes in this section is not intended to be comprehensive, but it is intended to serve as an introduction to how information can be represented by combinations of bits. Code applications, including codes for error detection, are introduced at various points throughout the book, and you will gain additional exposure at that time. Also, Appendix C provides an introduction to error-detection and error-correction codes as optional material.

SECTION REVIEW 2–10

1. Convert the following binary numbers to the Gray code:
   (a) 1100_2 (b) 1010_2 (c) 11010_2 (d) 101110_2
2. Convert the following Gray codes to binary:
   (a) 1000   (b) 1010   (c) 11101   (d) 00110
3. Convert the following decimal numbers to Excess-3 code:
   (a) 3_{10} (b) 15_{10} (c) 87_{10} (d) 349_{10}
4. What is the ASCII representation for each of the following characters? Express each as a bit pattern and in hexadecimal notation.
   (a) K   (b) r   (c) \$ (d) +
5. What does 11000010 represent in EBCDIC? The left-most bit is position 0 in this case.

TABLE 2-8 *Partial EBCDIC table.*

Bit positions 4, 5, 6, 7	00				01				10				11			
Bit positions 2, 3 →	00	01	10	11	00	01	10	11	00	01	10	11	00	01	10	11
0000	NUL		DS		SP	&	-									0
0001			SOS				/		a	j			A	J		1
0010			FS						b	k	s		B	K	S	2
0011		TM							c	l	t		C	L	T	3
0100	PF	RES	BYP	PN					d	m	u		D	M	U	4
0101	HT	NL	LF	RS					e	n	v		E	N	V	5
0110	LC	BS	EOB	UC					f	o	w		F	O	W	6
0111	DL	IL	PRE	EOT					g	p	x		G	P	X	7
1000		·							h	q	y		H	Q	Y	8
1001									i	r	z		I	R	Z	9
1010		CC	SM		¢	!		:								
1011						$	,	#								
1100					<	*	%	@								
1101					(	)	_	'								
1110					+	;	>	=								
1111	CU1	CU2	CU3	¬	|		?	"								

Header: Bit positions 0, 1 → 00, 01, 10, 11 (each spanning four columns).

Definitions of control abbreviations:

BS	Backspace	LC	Lowercase
BYP	Bypass	LF	Line feed
CC	Cursor control	NL	New line
CU1	Customer use	PF	Punch off
CU2	Customer use	PN	Punch on
CU3	Customer use	PRE	Prefix
DL	Delete	RES	Restore
DS	Digit select	RS	Reader stop
EOB	End of block	SM	Set mode
EOT	End of transmission	SP	Space
FS	Field separator	TM	Tape mark
HT	Horizontal tab	UC	Uppercase
IL	Idle		

Meanings of unfamiliar symbols:

\| Vertical bar: logical OR
¬   Logical NOT
−   Hyphen or minus sign
—   Underscore (01101101)

Example of code format:
01234567   Bit positions
11000110   is F

SUMMARY

☐ The 9's complement of a decimal number is derived by subtracting each digit in the decimal number from 9.

☐ The 10's complement of a decimal number is the 9's complement plus 1.

☐ Subtraction can be accomplished by addition using the 9's complement or 10's complement methods.

☐ The binary number system has a base of two and consists of two digits (called *bits*), 1 and 0.

☐ A binary number is a weighted number with the weight of each whole number digit from least significant (2^0) to most significant being an increasing positive power of two. The weight of each fractional digit beginning at 2^{-1} is an increasing negative power of two.

☐ The basic rules for binary addition are

$$0 + 0 = 0$$
$$0 + 1 = 1$$
$$1 + 0 = 1$$
$$1 + 1 = 10_2$$

☐ The basic rules for binary subtraction are

$$0 - 0 = 0$$
$$1 - 1 = 0$$
$$1 - 0 = 1$$
$$10_2 - 1 = 1$$

☐ The 1's complement of a binary number is derived by changing 1s to 0s and 0s to 1s.

☐ The 2's complement of a binary number is derived by adding 1 to the 1's complement.

☐ Binary subtraction can be accomplished by addition using the 1's or 2's complement methods.

☐ A decimal whole number can be converted to binary by using the *sum-of-weights* or the *repeated division-by-2* methods.

☐ A decimal fraction can be converted to binary by using the *sum-of-weights* or the *repeated mutiplication-by-2* methods.

☐ The octal number system has a base of eight and consists of eight digits (0 through 7).

☐ A decimal whole number can be converted to octal by using the *repeated division-by-8* method.

☐ Octal-to-binary conversion is accomplished by simply replacing each octal digit with its three-bit binary equivalent. The process is reversed for binary-to-octal conversion.

☐ The hexadecimal number system has a base of sixteen and consists of 16 digits and characters (0 through 9 and A through F).

☐ One hexadecimal digit represents a four-bit binary number, and its primary usefulness is simplifying bit patterns by making them easier to "read."

☐ BCD (Binary Coded Decimal) represents each decimal digit by a four-bit binary number. Arithmetic operations can be performed in BCD.

☐ The main feature of the Gray code is the *single-bit change* going from one number in sequence to the next.

☐ The ASCII (American Standard Code for Information Interchange) is a seven-bit alphanumeric code. The EBCDIC (Extended Binary Coded Decimal Interchange Code) is an eight-bit code. These codes are commonly used in data transfer and computer interface applications.

SELF-TEST

1. What is the base of the decimal number system? This means that there are _____ digits in the system.
2. Break each decimal number down into a sum of products of the digits and their appropriate weights.
   (a) 28_{10} (b) 389_{10} (c) 1473_{10} (d) $10,956_{10}$
3. Find the 9's complement of each decimal number.
   (a) 57_{10} (b) 892_{10} (c) 3347_{10}
4. Find the 10's complement of each decimal number.
   (a) 8_{10} (b) 57_{10} (c) 1964_{10} (d) 8840_{10}
5. Write the sequence of binary numbers for the decimal numbers 32 through 42.
6. Find the decimal equivalent of each binary number.
   (a) 1101_2 (b) 100101_2 (c) 11011101_2
   (d) 1011.011_2 (e) 1110.1011_2
7. Perform the indicated operations.
   (a) $110_2 + 011_2$ (b) $11010_2 + 01111_2$
   (c) $110_2 - 010_2$ (d) $11011_2 - 10110_2$
   (e) $101_2 \times 111_2$ (f) $1110_2 \div 0111_2$
8. Convert each binary number to its 1's complement and then to its 2's complement.
   (a) 110101_2 (b) 0001101_2
9. Express each of the following decimal numbers in binary:
   (a) 17_{10} (b) 102_{10} (c) 555_{10} (d) 72.66_{10}
10. What is the base of the octal number system? This means that there are _____ digits in the system.
11. Convert each decimal number to octal.
    (a) 16_{10} (b) 45_{10} (c) 380_{10}
12. Express each octal number in binary.
    (a) 7041_8 (b) 65315_8
13. Express the binary number 10010111000010110010_2 in octal.
14. What is the hexadecimal representation for $10001100010111110011110100010000101 1_2$?
15. What binary number does $5A34F_{16}$ represent?
16. Write the hexadecimal sequence for decimal 0_{10} through 32_{10}.
17. Convert each decimal number to BCD.
    (a) 23_{10} (b) 79_{10} (c) 718_{10} (d) 954.61_{10}
18. What decimal number does the BCD sequence 0100100110000110 represent?
19. Convert 101011011_2 to Gray code.
20. What binary number does the Gray code 1100100 represent?
21. What is the key feature of the Excess-3 code?
22. Determine the ASCII representations for H, X, :, =, and %.

PROBLEMS

Section 2-2

2-1 Determine the 9's complement of each decimal number.
(a) 3_{10} (b) 5_{10} (c) 8_{10} (d) 12_{10}
(e) 17_{10} (f) 25_{10} (g) 49_{10} (h) 86_{10}
(i) 127_{10} (j) 381_{10} (k) 690_{10} (l) 1354_{10}

2-2 Perform the following subtractions using the 9's complement method:
(a) $6_{10} - 2_{10}$ (b) $15_{10} - 7_{10}$ (c) $23_{10} - 14_{10}$ (d) $48_{10} - 33_{10}$
(e) $69_{10} - 68_{10}$ (f) $91_{10} - 70_{10}$ (g) $98_{10} - 59_{10}$ (h) $100_{10} - 82_{10}$

2-3 Determine the 10's complement of each decimal number.
(a) 7_{10} (b) 19_{10} (c) 36_{10}
(d) 52_{10} (e) 84_{10} (f) 90_{10}

2-4 Perform the following subtractions using the 10's complement method:
(a) $115_{10} - 92_{10}$ (b) $159_{10} - 125_{10}$ (c) $298_{10} - 200_{10}$
(d) $561_{10} - 443_{10}$ (e) $846_{10} - 709_{10}$ (f) $1024_{10} - 837_{10}$

Section 2-3

2-5 Convert the following binary numbers to decimal:
(a) 11_2 (b) 100_2 (c) 111_2 (d) 1000_2
(e) 1001_2 (f) 1100_2 (g) 1011_2 (h) 1111_2

2-6 Convert the following binary numbers to decimal:
(a) 1110_2 (b) 1010_2 (c) 11100_2 (d) 10000_2
(e) 10101_2 (f) 11101_2 (g) 10111_2 (h) 11111_2

2-7 Convert each binary number to decimal.
(a) 110011.11_2 (b) 101010.01_2 (c) 1000001.111_2
(d) 1111000.101_2 (e) 1011100.10101_2 (f) 1110001.0001_2
(g) 1011010.1010_2 (h) 1111111.11111_2

2-8 What is the highest decimal number that can be represented by the following number of binary digits (bits)?
(a) 2_{10} (b) 3_{10} (c) 4_{10} (d) 5_{10}
(e) 6_{10} (f) 7_{10} (g) 8_{10} (h) 9_{10}
(i) 10_{10} (j) 11_{10}

2-9 How many bits are required to represent the following decimal numbers?
(a) 17_{10} (b) 35_{10} (c) 49_{10} (d) 68_{10}
(e) 81_{10} (f) 114_{10} (g) 132_{10} (h) 205_{10}
(i) 271_{10}

2-10 Generate the binary count sequences for each decimal sequence.
(a) 0_{10} through 7_{10} (b) 8_{10} through 15_{10} (c) 16_{10} through 31_{10}
(d) 32_{10} through 63_{10} (e) 64_{10} through 75_{10}

Section 2-4

2-11 Add the binary numbers.
(a) $11_2 + 1_2$ (b) $10_2 + 10_2$ (c) $101_2 + 11_2$
(d) $111_2 + 110_2$ (e) $1001_2 + 101_2$ (f) $1101_2 + 1011_2$

2–12 Use direct subtraction on the following binary numbers:
(a) $11_2 - 1_2$ (b) $101_2 - 100_2$ (c) $110_2 - 101_2$
(d) $1110_2 - 11_2$ (e) $1100_2 - 1001_2$ (f) $11010_2 - 10111_2$

2–13 Perform the following binary multiplications:
(a) $11_2 \times 11_2$ (b) $100_2 \times 10_2$ (c) $111_2 \times 101_2$
(d) $1001_2 \times 110_2$ (e) $1101_2 \times 1101_2$ (f) $1110_2 \times 1101_2$

2–14 Divide the binary numbers as indicated.
(a) $100_2 \div 10_2$ (b) $1001_2 \div 11_2$ (c) $1100_2 \div 100_2$

Section 2–5

2–15 Determine the 1's complement of each binary number.
(a) 101_2 (b) 110_2 (c) 1010_2
(d) 11010111_2 (e) 1110101_2 (f) 00001_2

2–16 Perform the following subtractions using the 1's complement method:
(a) $11_2 - 10_2$ (b) $100_2 - 11_2$ (c) $1010_2 - 111_2$
(d) $1101_2 - 1010_2$ (e) $11100_2 - 1101_2$ (f) $100001_2 - 1010_2$
(g) $1001_2 - 1110_2$ (h) $10111_2 - 11111_2$

2–17 Determine the 2's complement of each binary number.
(a) 10_2 (b) 111_2 (c) 1001_2
(d) 1101_2 (e) 11100_2 (f) 10011_2

2–18 Perform the following subtractions using the 2's complement method:
(a) $10_2 - 01_2$ (b) $111_2 - 110_2$ (c) $1101_2 - 1001_2$
(d) $1111_2 - 1101_2$ (e) $10111_2 - 10011_2$ (f) $10001_2 - 11100_2$
(g) $10101_2 - 10111_2$ (h) $1111000_2 - 1111111_2$

Section 2–6

2–19 Convert each decimal number to binary using the sum-of-weights method.
(a) 10_{10} (b) 17_{10} (c) 24_{10} (d) 48_{10}
(e) 61_{10} (f) 93_{10} (g) 125_{10} (h) 186_{10}
(i) 298_{10}

2–20 Convert each decimal number to binary by repeated division-by-2.
(a) 15_{10} (b) 21_{10} (c) 28_{10} (d) 34_{10}
(e) 40_{10} (f) 59_{10} (g) 65_{10} (h) 73_{10}
(i) 99_{10}

Section 2–7

2–21 Convert each octal number to decimal.
(a) 12_8 (b) 27_8 (c) 56_8
(d) 64_8 (e) 103_8 (f) 557_8
(g) 163_8 (h) 1024_8 (i) 7765_8

2–22 Convert each decimal number to octal by repeated division-by-8.
(a) 15_{10} (b) 27_{10} (c) 46_{10} (d) 70_{10}
(e) 100_{10} (f) 142_{10} (g) 219_{10} (h) 435_{10}
(i) 791_{10}

2–23 Convert each octal number to binary.
(a) 13_8 (b) 57_8 (c) 101_8

    (d) 321_8 **(e)** 540_8 **(f)** 4653_8

    (g) 13271_8 **(h)** 45600_8 **(i)** 100213_8

    (j) 103.45_8

2-24 Convert each binary number to octal.

    (a) 111_2 **(b)** 10_2 **(c)** 110111_2

    (d) 101010_2 **(e)** 1100_2 **(f)** 1011110_2

    (g) 101100011001_2 **(h)** 10110000011_2 **(i)** 111111101111000_2

    (j) 10011.011_2

Section 2-8

2-25 Convert each hexadecimal number to binary.

    (a) 38_{16} **(b)** 59_{16} **(c)** $A14_{16}$ **(d)** $5C8_{16}$

    (e) 4100_{16} **(f)** $FB17_{16}$ **(g)** $8A.9_{16}$

2-26 Convert each binary number to hexadecimal.

    (a) 1110_2 **(b)** 10_2 **(c)** 10111_2

    (d) 10100110_2 **(e)** 1111110000_2 **(f)** 100110000010_2

2-27 Convert each hexadecimal number to decimal.

    (a) 23_{16} **(b)** 92_{16} **(c)** $1A_{16}$ **(d)** $8D_{16}$

    (e) $F3_{16}$ **(f)** EB_{16} **(g)** $5C2_{16}$ **(h)** 700_{16}

2-28 Convert each decimal number to hexadecimal.

    (a) 8_{10} **(b)** 14_{10} **(c)** 33_{10} **(d)** 52_{10}

    (e) 284_{10} **(f)** 2890_{10} **(g)** 4019_{10} **(h)** 6500_{10}

2-29 Perform the following additions.

    (a) $37_{16} + 29_{16}$ **(b)** $A0_{16} + 6B_{16}$ **(c)** $FF_{16} + BB_{16}$

2-30 Perform the following subtractions.

    (a) $51_{16} - 40_{16}$ **(b)** $C8_{16} - 3A_{16}$ **(c)** $FD_{16} - 88_{16}$

Section 2-9

2-31 Convert each of the following decimal numbers to 8421 BCD:

    (a) 10_{10} **(b)** 13_{10} **(c)** 18_{10}

    (d) 21_{10} **(e)** 25_{10} **(f)** 36_{10}

    (g) 44_{10} **(h)** 57_{10} **(i)** 69_{10}

    (j) 98_{10} **(k)** 125_{10} **(l)** 156_{10}

2-32 Convert each of the decimal numbers in Problem 2-31 to straight binary, and compare the number of bits required with that required in BCD.

2-33 Convert the following decimal numbers to BCD:

    (a) 104_{10} **(b)** 128_{10} **(c)** 132_{10}

    (d) 150_{10} **(e)** 186_{10} **(f)** 210_{10}

    (g) 359_{10} **(h)** 547_{10} **(i)** 1051_{10}

    (j) 2563_{10}

2-34 Convert each of the BCD code numbers to decimal.

    (a) 0001   **(b)** 0110   **(c)** 1001

    (d) 00011000   **(e)** 11001   **(f)** 00110010

    (g) 1000101   **(h)** 10011000   **(i)** 100001110000

    (j) 011000011001

2-35 Convert each of the BCD code numbers to decimal.

    (a) 10000000   **(b)** 1000110111

    (c) 1101000110   **(d)** 10000100001

        (e) 11101010100   (f) 100000000000
        (g) 100101111000   (h) 1011010000011
        (i) 1001000000011000   (j) 0110011001100111

2–36 Add the following BCD numbers:

      (a) 0010 + 0001   (b) 0101 + 0011
      (c) 0111 + 0010   (d) 1000 + 0001
      (e) 00011000 + 00010001 (f) 01100100 + 00110011
      (g) 01000000 + 01000111 (h) 10000101 + 00010011

2–37 Add the following BCD numbers:

      (a) 1000 + 0110   (b) 0111 + 0101
      (c) 1001 + 1000   (d) 1001 + 0111
      (e) 00100101 + 00100111 (f) 01010001 + 01011000
      (g) 10011000 + 10010111 (h) 010101100001 + 011100001000

2–38 Convert each pair of decimal numbers to BCD and add as indicated.

      (a) $4_{10} + 3_{10}$ (b) $5_{10} + 2_{10}$ (c) $6_{10} + 4_{10}$
      (d) $17_{10} + 12_{10}$ (e) $28_{10} + 23_{10}$ (f) $65_{10} + 58_{10}$
      (g) $113_{10} + 101_{10}$ (h) $295_{10} + 157_{10}$

Section 2–10

2–39 In a certain application, a four-bit binary sequence recycles from 1111 to 0000 periodically. There are four bit changes, and these changes may not occur at the same instant due to circuit delays. For example, if the LSB changes first, the number will appear as 1110 during the transition from 1111 to 0000 and may be misinterpreted by the system. Illustrate how the Gray code avoids this problem.

2–40 Convert each binary number to Gray code.

      (a) 11011_2 (b) 1001010_2 (c) 1111011101110_2

2–41 Convert each Gray code to binary.

      (a) 1010   (b) 00010   (c) 11000010001

2–42 Convert each of the following decimal numbers to Excess-3 code.

      (a) 1_{10} (b) 3_{10} (c) 6_{10}
      (d) 10_{10} (e) 18_{10} (f) 29_{10}
      (g) 56_{10} (h) 75_{10} (i) 107_{10}

2–43 Convert each Excess-3 code number to decimal.

      (a) 0011   (b) 1001
      (c) 0111   (d) 01000110
      (e) 01111100 (f) 10000101

2–44 Decode the following ASCII coded message to English. Bit 7 of the first character is the left-most bit in the first row.

    10010011010101000011000100100010000011010011001100 0
    10000101000101100010110011100011000101001110000 01
    10010011000100001100010101001001000100000110101 00
    00110001000111100010110011101001001101010110100 11
    00110001001001101001100110000111001011100100110 00
    01001010011000101000010000101101001010100111010 000
    10010011010010100000110101001001001100111110011 10
    00110001000001100111011000100001100001100010011 000
    01001010011000100100110011101010011101000010010 01
    10100101000001101010010010011001111110011100000 011

2-45 Convert your name and address to ASCII code.

2-46 Convert your name and address to EBCDIC.

ANSWERS TO SECTION REVIEWS

Section 2-1
1. (a) 10   (b) 100   (c) 1000   (d) 10,000
2. (a) $5 \times 10 + 1 \times 1$ (b) $1 \times 100 + 3 \times 10 + 7 \times 1$
    (c) $1 \times 1000 + 4 \times 100 + 9 \times 10 + 2 \times 1$
    (d) $1 \times 10,000 + 6 \times 100 + 5 \times 10 + 8 \times 1$

Section 2-2
1. (a) 65   (b) 27   (c) 5419   (d) 3890
2. (a) 07   (b) 898   (c) 7262   (d) 0700

Section 2-3
1. 16   2. (a) 10_{10} (b) 26_{10} (c) 189_{10} (d) 6.625_{10}

Section 2-4
1. (a) 10111   (b) 100100
2. (a) 1001   (b) 0010
3. (a) 101010   (b) 100

Section 2-5
1. (a) 00101   (b) 01000   (c) 110010
2. (a) 0110   (b) -00101
3. (a) 01001   (b) 00001   (c) 101111
4. (a) 0101   (b) -0101

Section 2-6
1. (a) 10111_2 (b) 111001_2 (c) 101101.1_2
2. (a) 1110_2 (b) 10101_2 (c) 0.011_2

Section 2-7
1. (a) 59_{10} (b) 85_{10}
2. (a) 142_8 (b) 243_8
3. (a) 100110_2 (b) 111010011_2 (c) 101110010100.011111_2
4. (a) 657_8 (b) 461.57_8 (c) 277.14_8

Section 2-8
1. (a) $B3_{16}$ (b) $CE8_{16}$
2. (a) 01010111_2 (b) 001110100101_2
    (c) 1111100000001011_2
3. (a) $4C_{16}$ (b) 69_{16}
4. (a) 54_{16} (b) 38_{16}

Section 2–9

1. (a) 2   (b) 8   (c) 1   (d) 4
2. (a) 0110   (b) 00010101   (c) 001001110011   (d) 10000100.1001
3. (a) 89_{10} (b) 278_{10} (c) 15.7_{10}

Section 2–10

1. (a) 1010   (b) 1111   (c) 10111   (d) 111001
2. (a) 1111_2 (b) 1100_2 (c) 10110_2 (d) 00100_2
3. (a) 0110   (b) 01001000   (c) 10111010   (d) 011001111100
4. (a) $1001011 \longrightarrow 4B_{16}$ (b) $1110010 \longrightarrow 72_{16}$
   (c) $0100100 \longrightarrow 24_{16}$ (d) $0101011 \longrightarrow 2B_{16}$
5. $11000010 \longrightarrow B$

In this chapter *logic gates*, the basic elements that make up a digital system, are introduced. The emphasis is on the logical operation, characteristics, and operating parameters. It is important to know what the output of a gate should be for all possible input combinations and to understand how the electrical characteristics of a gate affect the operation.

The logic symbols used to represent the various types of gates are in accordance with ANSI/IEEE Std. 91–1984, which was briefly discussed in the Preface. Primarily, distinctive shape symbols are used to represent the logic gates, although the newer rectangular outline symbols are also provided. This new ANSI/IEEE standard has been adopted by major corporations such as Texas Instruments, Honeywell, General Electric, Signetics, Hewlett-Packard, and Control Data, to name a few, for use in their internal documentation as well as in their published literature. All documentation under military contract will also require the standard symbols. Additionally, CAD/CAM (computer-aided design/computer-aided manufacture) libraries containing these symbols are being made available.

Because of the universal use of digital integrated circuits, from the user's standpoint the logic function is more important than the actual component-level circuit design. A detailed coverage of the circuits at the component level can be treated as an optional topic. For those who need to cover this material, a chapterlike coverage of digital circuit technologies is presented in Appendix A. This material can be inserted following this chapter or used in conjunction with it if desired. Omission of IC technologies will not affect the flow of material throughout the rest of the book.

Specific devices introduced in this chapter are 7404 hex inverter, 7408 quad 2-input AND gates, 7411 triple 3-input AND gates, 7421 dual 4-input AND gates, 7400 quad 2-input NAND gates, 7410 triple 3-input NAND gates, 7420 quad 4-input NAND gates, 7430 single 8-input NAND gates, 74133 single 13-input NAND gates, 7432 quad 3-input OR gates, 7402 quad 2-input NOR gates, and 7427 triple 3-input NOR gates.

In this chapter, you will learn

- ☐ The logical operation of the inverter, AND gate, OR gate, NAND gate, and NOR gate.
- ☐ The standard symbols for each element as specified in ANSI/IEEE Std. 91–1984.
- ☐ Why the NAND gate and the negative-OR gate are equivalent.
- ☐ Why the NOR gate and the negative-AND gate are equivalent.
- ☐ The definition of *propagation delay time* in a logic gate and how it is measured.
- ☐ How the power dissipation of a logic gate is determined.
- ☐ The definition of *noise margin* of a logic gate.
- ☐ The definition of *fan-out*.
- ☐ The types of integrated circuit logic families and how they compare in terms of their performance characteristics.
- ☐ The various gates available in the 54/74 TTL series.
- ☐ What precautions to take in handling CMOS devices.
- ☐ How to use a data sheet to determine the performance of a logic gate.
- ☐ How to use logic probes and logic pulsers in troubleshooting logic gates.
- ☐ How to use the oscilloscope to measure propagation delay time.

3

Logic Gates

3-1   THE INVERTER

The inverter (NOT circuit) performs a basic logic function called *inversion* or *complementation*. The purpose of the inverter is to change one logic level to the opposite level. In terms of bits, it changes a 1 to a 0 and a 0 to a 1.

    Standard logic symbols for the inverter are shown in Figure 3–1. Parts (a) and (b) of the figure show the *distinctive shape* symbols, and parts (c) and (d) show the *rectangular outline* symbols. In this text, distinctive shape symbols are used; however, the rectangular outline symbols are commonly found in many industry publications, and you should become familiar with them as well.

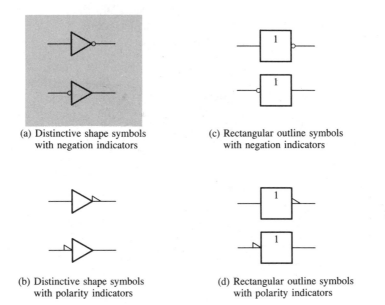

(a) Distinctive shape symbols
    with negation indicators

(c) Rectangular outline symbols
    with negation indicators

(b) Distinctive shape symbols
    with polarity indicators

(d) Rectangular outline symbols
    with polarity indicators

FIGURE 3–1   *Standard logic symbols for the inverter (ANSI/IEEE Std. 91–1984).*

The Negation and Polarity Indicators

The *negation indicator* is a "bubble" (○) appearing on the input or output of a logic element, as shown in Figure 3–1(a) and (c). When appearing on the input, the bubble means that an external 0 produces an internal 1. When appearing on the output, the bubble means that an internal 1 produces an external 0. Typically, inputs are on the left of a logic symbol and outputs are on the right.

    The *polarity indicator* is a "triangle" (◿) appearing on the input or output of a logic element, as shown in Figure 3–1(b) and (d). When appearing on the input, it means that an external LOW level produces an internal HIGH. When appearing on the output, it means that an internal HIGH produces an external

LOW level. Either indicator (bubble or triangle) can be used on both distinctive shape symbols and rectangular outlines as indicated. The shaded portion of the figure [part (a)] indicates the principal inverter symbol used in this text. The placement of the negation or polarity indicator does not imply a change in the way an inverter operates.

Because positive logic (HIGH = 1, LOW = 0) is used in this text, both indicators are equivalent and can be interchanged. However, we will use primarily the negation indicator (bubble) throughout this book to represent an inversion.

The placement of the bubble on the input or output of a logic element is determined by the *active state* of the input signal (pulse or level). The active state is the state (1 or 0) when the signal is considered to be present on the input. When the active state of the input is a 0, the bubble is placed on the input. When the active output state is a 0, the bubble is placed on the output.

Inverter Operation

When a HIGH level is applied to an inverter input, a LOW level will appear on its output. When a LOW level is applied to its input, a HIGH will appear on its output. This operation is summarized in Table 3–1, which shows the output for each possible input in terms of levels and bits. These tables are called *truth tables*.

TABLE 3–1 *Inverter truth tables.*

Input	Output		Input	Output
LOW	HIGH		0	1
HIGH	LOW		1	0

Pulsed Operation

Figure 3–2 shows the output of an inverter for a pulse input where t_1 and t_2 indicate the corresponding points on the input and output pulse waveforms. Note that when the input is LOW, the output is HIGH, and when the input is HIGH, the output is LOW, thereby producing an *inverted* output pulse.

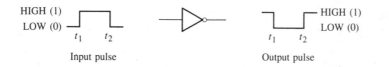

FIGURE 3–2 *Inverter with pulse input.*

EXAMPLE 3–1   A pulse waveform is applied to an inverter in Figure 3–3(a). Determine the output waveform corresponding to the input.

FIGURE 3-3

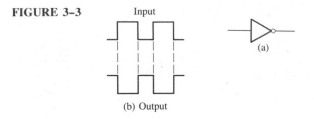

Input

(a)

(b) Output

Solution The output waveform is exactly opposite to the input (inverted) at each point, as shown in Figure 3-3(b).

SECTION REVIEW 3-1

1. When a 1 is on the input of an inverter, what is the output?
2. An active HIGH pulse (HIGH level when present, LOW level when absent) is required on an inverter input.
   (a) Draw the appropriate logic symbol using the distinctive shape and negation indicator for the inverter in this application.
   (b) Describe the output signal when the pulse is present.

3-2   THE AND GATE

The AND gate performs logical multiplication, more commonly known as the *AND function*. The mathematical aspects of this function are discussed in Chapter 4.

The AND gate is composed of two or more inputs and a single output, as indicated by the standard logic symbols shown in Figure 3-4. Inputs are on the left and the output is on the right in each symbol. Gates with two and four inputs are shown; however, an AND gate can have any number of inputs greater than one. Although examples of both distinctive shape symbols and rectangular outline

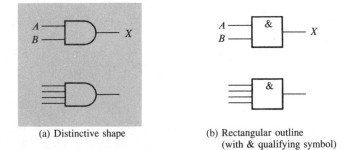

(a) Distinctive shape

(b) Rectangular outline
(with & qualifying symbol)

FIGURE 3-4 *Standard logic symbols for the AND gate showing two and four inputs (ANSI/IEEE Std. 91-1984).*

symbols are shown, the distinctive shape [shaded, part (a)] is predominantly used in this book.

Logical Operation of the AND Gate

The operation of the AND gate is such that the output is HIGH only when *all* of the inputs are HIGH. When *any* of the inputs are LOW, the output is LOW. Therefore, the basic purpose of an AND gate is to determine when certain conditions are simultaneously true, as indicated by HIGH levels on all of its inputs, and to produce a HIGH on its output indicating this condition. The inputs of the two-input AND gate in Figure 3–4 are labeled *A* and *B,* and the output is labeled *X.* We can express the gate operation with the following description:

> If *A* AND *B* are HIGH, then *X* is HIGH. If *A* is LOW, or if *B* is LOW, or if both *A* and *B* are LOW, then *X* is LOW.

The HIGH level is the *active* output level for the AND gate. Figure 3–5 illustrates a two-input AND gate with all four possibilities of input combinations, and the resulting output for each.

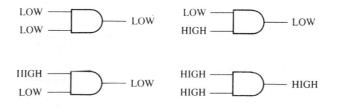

FIGURE 3–5 *All possible logic levels for a two-input AND gate.*

We generally express the logical operation of a gate with a table that lists all input combinations and the corresponding outputs. This table of combinations is also called a *truth table* and is illustrated in Table 3–2 for a two-input AND gate. The truth table can be expanded for any number of inputs.

TABLE 3–2 *Truth table for a two-input AND gate.*

Inputs		Output
A	B	X
0	0	0
0	1	0
1	0	0
1	1	1

NOTE: 1 ≡ HIGH, 0 ≡ LOW.

Although the terms HIGH and LOW tend to give a "physical" sense to input and output states, the truth table is shown with 1s and 0s since a HIGH is equivalent to a 1 and a LOW is equivalent to a 0 in positive logic.

For any AND gate, regardless of the number of inputs, the output is HIGH *only* when *all* inputs are HIGH.

The total number of possible combinations of binary inputs is determined by the following formula:

$$N = 2^n \qquad\qquad (3\text{--}1)$$

where N is the total possible combinations and n is the number of input variables. To illustrate, the following calculations are made using Equation (3–1):

$$\begin{aligned}
\text{For two input variables:} &\quad 2^2 = 4 \\
\text{For three input variables:} &\quad 2^3 = 8 \\
\text{For four input variables:} &\quad 2^4 = 16
\end{aligned}$$

This is how we determine the number of combinations for gates with any number of inputs.

Pulsed Operation

In a majority of applications, the inputs to a gate are not stationary levels but are voltages that change frequently between two logic levels and that can be classified as pulse waveforms. We will now look at the operation of AND gates with pulsed input waveforms. Keep in mind that an AND gate obeys the truth table operation regardless of whether its inputs are constant levels or pulsed levels.

In examining the pulsed operation of the AND gate, we will look at the inputs with respect to each other in order to determine the output level at any given time. For example, in Figure 3–6, the inputs are both HIGH (1) during the interval t_1, making the output HIGH (1) during this interval. During interval t_2, input A is LOW (0) and input B is HIGH (1), so the output is LOW (0). During interval t_3, both inputs are HIGH (1) again, and therefore the output is HIGH (1). During interval t_4, input A is HIGH (1) and input B is LOW (0), resulting in a LOW (0) output. Finally, during interval t_5, input A is LOW (0), input B is LOW (0), and the output is therefore LOW (0). A diagram of input and output waveforms showing time relationships is called a *timing diagram*.

FIGURE 3–6 *Example of pulsed AND gate operation.*

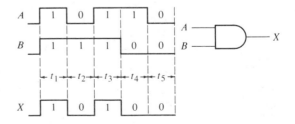

EXAMPLE 3-2   If the two waveforms are applied to the AND gate as in Figure 3–7(a), what is the resulting output waveform?

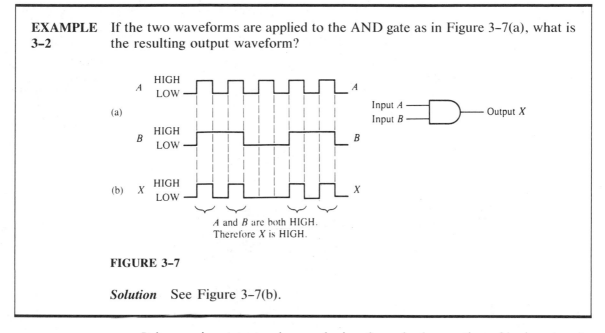

FIGURE 3-7

Solution   See Figure 3–7(b).

It is very important, when analyzing the pulsed operation of logic gates, to pay very careful attention to the time relationships of all the inputs with respect to each other and with respect to the output.

EXAMPLE 3-3   For the two input waveforms graphed in Figure 3–8(a), sketch the output waveform showing its proper relation to the inputs for a two-input AND gate.

FIGURE 3-8

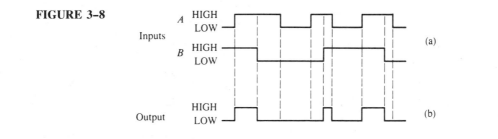

Solution   The output is HIGH only when both of the inputs are HIGH. See Figure 3–8(b).

EXAMPLE 3-4   For the three-input AND gate in Figure 3–9(a), determine the output waveform in proper relation to the inputs.

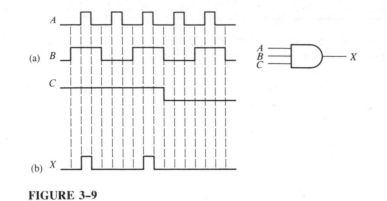

FIGURE 3-9

Solution   See Figure 3-9(b). The output of a three-input AND gate is HIGH only when all three inputs are HIGH.

**SECTION
REVIEW
3-2**
1.   When is the output of an AND gate HIGH?
2.   When is the output of an AND gate LOW?
3.   Develop the truth table for a three-input AND gate.

3-3   THE OR GATE

The OR gate performs logical addition, more commonly known as the *OR function*. The mathematical aspects of this operation will be covered in Chapter 4 on Boolean algebra.

An OR gate has two or more inputs and one output, as indicated by the standard logic symbols in Figure 3-10 where OR gates with two and four inputs

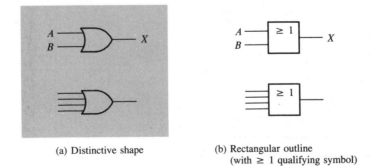

(a) Distinctive shape

(b) Rectangular outline
(with ≥ 1 qualifying symbol)

FIGURE 3-10   *Standard logic symbols for the OR gate showing two and four inputs (ANSI/IEEE Std. 91-1984).*

are illustrated. An OR gate can have any number of inputs greater than one. Although both distinctive shape and rectangular outline symbols are shown for familiarization, the distinctive shape OR gate symbol predominantly will be used.

Logical Operation of the OR Gate

The operation of the OR gate is such that a HIGH on the output is produced when *any* of the inputs are HIGH. The output is LOW only when *all* of the inputs are LOW. Therefore, the purpose of an OR gate is to determine when one or more of its inputs are HIGH and to produce a HIGH on its output to indicate this condition. The inputs of the two-input OR gate in Figure 3–10(a) are labeled A and B, and the output is labeled X. We can express the operation of the gate as follows:

If either A OR B OR both are HIGH, then X is HIGH. If both A and B are LOW, then X is LOW.

The HIGH level is the *active* output level for the OR gate. Figure 3–11 illustrates the logic operation for a two-input OR gate for all four possible input combinations.

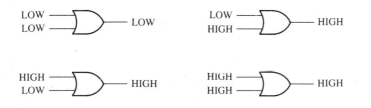

FIGURE 3–11 *All possible logic levels for a two-input OR gate.*

The logical operation of the two-input OR gate is described in the truth table shown in Table 3–3. The truth table can be expanded for any number of inputs; but regardless of the number of inputs, the output is HIGH when *any* of the inputs are HIGH.

TABLE 3–3 *Truth table for a two-input OR gate.*

Inputs		Output
A	B	X
0	0	0
0	1	1
1	0	1
1	1	1

NOTE: 1 ≡ HIGH, 0 ≡ LOW.

Pulsed Operation

Let us now turn our attention to the operation of an OR gate with pulsed inputs, keeping in mind what we have learned about its logical operation.

Again, the important thing in analysis of gate operation with pulsed waveforms is the relationship of all the waveforms involved. For example, in Figure 3–12, the inputs A and B are both HIGH (1) during interval t_1, making the output HIGH (1). During interval t_2, input A is LOW (0), but because input B is HIGH (1), the output is HIGH (1). Both inputs are LOW (0) during interval t_3, and we have a LOW (0) output during this time. During t_4, the output is HIGH (1) because input A is HIGH (1).

In this illustration, we have simply applied the truth table operation of the OR gate to each of the intervals during which the levels are nonchanging. A few examples further illustrate OR gate operation with pulse waveforms on the inputs.

FIGURE 3–12   *Example of pulsed OR gate operation.*

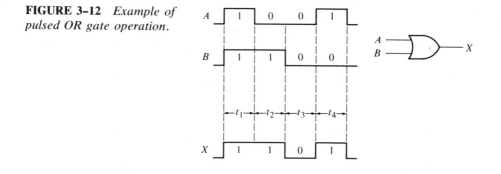

EXAMPLE 3–5   If the two waveforms are applied to the OR gate as in Figure 3–13(a), what is the resulting output waveform?

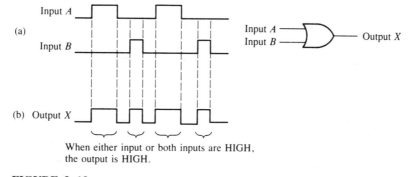

When either input or both inputs are HIGH, the output is HIGH.

FIGURE 3–13

Solution   See Figure 3–13(b). The output of a two-input OR gate is HIGH when either or both inputs are HIGH.

EXAMPLE 3-6

For the two input waveforms in Figure 3-14(a), sketch the output waveform showing its proper relation to the inputs for a two-input OR gate.

FIGURE 3-14

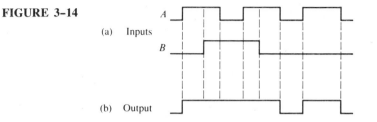

(a) Inputs

(b) Output

Solution When either input or both inputs are HIGH, the output is HIGH. See Figure 3-14(b).

EXAMPLE 3-7

For the three-input OR gate in Figure 3-15(a), determine the output waveform in proper relation to the inputs.

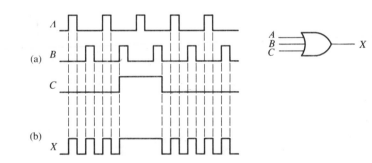

(a)

(b)

FIGURE 3-15

Solution See Figure 3-15(b). The output is HIGH when any of the inputs are HIGH.

SECTION REVIEW 3-3

1. When is the output of an OR gate HIGH?
2. When is the output of an OR gate LOW?
3. Develop the truth table for a three-input OR gate.

3-4   THE NAND GATE

The term NAND is a contraction of NOT-AND and implies an AND function with a complemented (inverted) output. A standard logic symbol for a two-input NAND gate and its equivalency to an AND gate followed by an inverter are shown in Figure 3–16(a). A rectangular outline representation is shown in part (b).

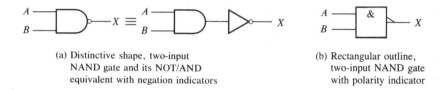

(a) Distinctive shape, two-input
    NAND gate and its NOT/AND
    equivalent with negation indicators

(b) Rectangular outline,
    two-input NAND gate
    with polarity indicator

FIGURE 3–16   *Standard NAND gate logic symbols (ANSI/IEEE Std. 91–1984).*

The NAND gate is a very popular logic function because it is a "universal" function; that is, it can be used to construct an AND gate, an OR gate, an inverter, or any combination of these functions. In Chapter 5 we will examine this "universal" property of the NAND gate. In this chapter we are going to look at the logical operation of the NAND gate.

Logical Operation of the NAND Gate

The logical operation of the NAND gate is such that a LOW output occurs only when *all* inputs are HIGH. When *any* of the inputs are LOW, the output will be HIGH. For the specific case of a two-input NAND gate, as shown in Figure 3–16 with the inputs labeled *A* and *B* and the output labeled *X,* we can state the operation as follows:

> If *A* AND *B* are HIGH, then *X* is LOW. If *A* is LOW, or *B* is LOW, or if both *A* and *B* are LOW, then *X* is HIGH.

Note that this operation is opposite that of the AND as far as output is concerned. In a NAND gate, the LOW level is the *active* output level. The bubble on the output indicates that the output is *active-0.* Figure 3–17 illustrates the logical operation of a two-input NAND gate for all four input combinations, and the truth

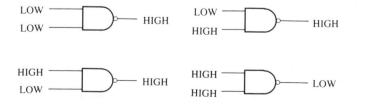

FIGURE 3–17   *Logical operation of a two-input NAND gate.*

table summarizing the logical operation of the two-input NAND gate is shown in Table 3–4.

TABLE 3–4 *Truth table for a two-input NAND gate.*

Inputs		Output
A	*B*	*X*
0	0	1
0	1	1
1	0	1
1	1	0

NOTE: 1 ≡ HIGH, 0 ≡ LOW.

Pulsed Operation

We will now look at the pulsed operation of the NAND gate. Remember from the truth table that any time *all* of the inputs are HIGH, the output will be LOW, and this is the *only* time a LOW output occurs. The following examples illustrate pulsed operation.

EXAMPLE 3–8   If the two waveforms shown in Figure 3–18(a) are applied to the NAND gate, determine the resulting output waveform.

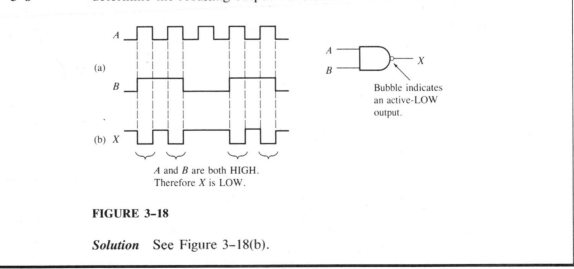

(a)

A and *B* are both HIGH. Therefore *X* is LOW.

Bubble indicates an active-LOW output.

FIGURE 3–18

Solution   See Figure 3–18(b).

EXAMPLE 3–9   Sketch the output waveform for the three-input NAND gate in Figure 3–19(a), showing its proper relationship to the inputs.

FIGURE 3–19

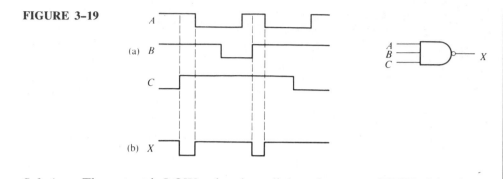

(a)

(b)

Solution The output is LOW only when all three inputs are HIGH. See Figure 3–19(b).

The NAND Gate as an Active-LOW Input OR Gate

Inherent in the NAND gate's operation are the conditions in which one or more LOW inputs produce a HIGH output. If you look at Table 3–4, you will see that the output is HIGH when any of the inputs are LOW. These conditions can be stated as follows:

If *A* is LOW OR *B* is LOW, OR if both *A* and *B* are LOW, then *X* is HIGH.

Here we have an OR operation that requires one or more LOW inputs to produce a HIGH output and that is referred to as *negative-OR*. When the NAND gate is looking for one or more LOWs on its inputs rather than for all HIGHs, it is acting as a negative-OR gate and is represented by the standard logic symbol in Figure 3–20(b). The two symbols in Figure 3–20 represent the same gate, but they also serve to define its role in a particular application, as illustrated by the following examples.

FIGURE 3–20 *Standard symbols representing the two equivalent functions of the NAND gate.*

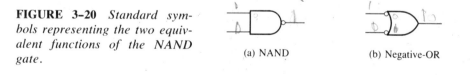

(a) NAND

(b) Negative-OR

EXAMPLE 3–10

The simultaneous occurrence of two HIGH level voltages must be detected and indicated by a LOW level output that is used to illuminate a LED. Sketch the operation.

Solution This application requires a NAND function, since the output must be active-LOW in order to produce current through the LED when two HIGHs occur on its inputs. The NAND symbol is therefore used to show the operation. See Figure 3–21.

FIGURE 3-21 *Example of a simple NAND gate application.*

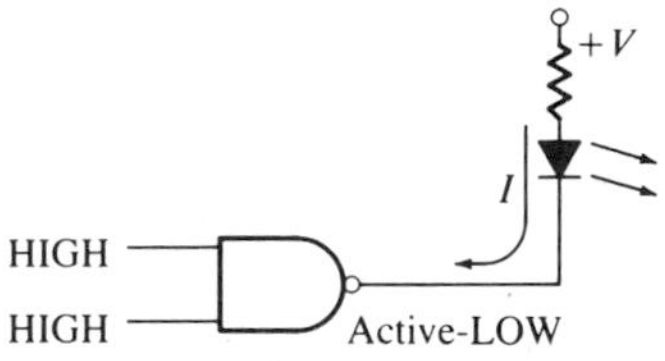

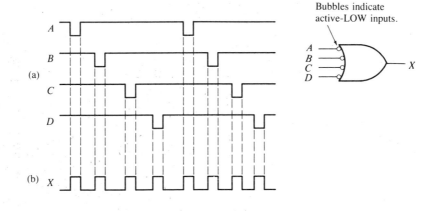

EXAMPLE 3-11

A gate is required to monitor two lines and to generate a HIGH level output used to activate an electric motor whenever either or both lines are LOW. Sketch the operation.

Solution   This application requires an active-LOW input OR function because the output has to be active-HIGH in order to produce an indication of the occurrence of one or more LOW levels on its inputs. In this case, the gate functions as a negative-OR and is represented by the appropriate symbol shown in Figure 3-22. A LOW on either input or both inputs causes an *active-HIGH* output to activate the motor through an appropriate interface circuit.

FIGURE 3-22

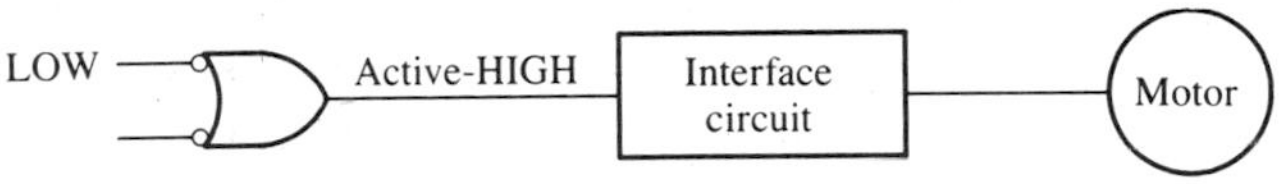

EXAMPLE 3-12

For the four-input NAND gate in Figure 3-23(a), operating as a negative-OR, determine the output with respect to the inputs.

FIGURE 3-23

Solution   The output is HIGH any time an input is LOW. See Figure 3–23(b).

**SECTION
REVIEW
3–4**

1. When is the output of a NAND gate LOW?
2. When is the output of a NAND gate HIGH?
3. Describe the functional differences between a NAND gate and a negative-OR gate. Do they both have the same truth table?

3–5   THE NOR GATE

The term NOR is a contraction of NOT-OR and implies an OR function with an inverted output. A standard logic symbol for a two-input NOR gate and its equivalent OR gate followed by an inverter are shown in Figure 3–24(a). A rectangular outline symbol is shown in part (b).

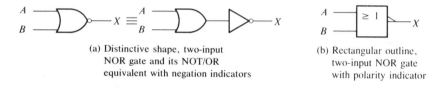

(a) Distinctive shape, two-input
  NOR gate and its NOT/OR
  equivalent with negation indicators

(b) Rectangular outline,
  two-input NOR gate
  with polarity indicator

FIGURE 3–24   *Standard NOR gate logic symbols (ANSI/IEEE Std. 91–1984).*

The NOR gate, like the NAND, is a very useful logic gate because of its universal property. We will examine the universal property of this gate in detail in Chapter 5.

Logical Operation of the NOR Gate

The logical operation of the NOR gate is such that a LOW output occurs when *any* of its inputs are HIGH. Only when *all* of its inputs are LOW is the output HIGH. For the specific case of a two-input NOR gate, as shown in Figure 3–24 with the inputs labeled A and B and the output labeled X, we can state the operation as follows:

If A OR B OR both are HIGH, then X is LOW. If both A and B are LOW, then X is HIGH.

Note that this operation results in an output opposite that of the OR gate. In a NOR gate, the LOW output is the *active* output level. As was pointed out for the NAND gate, the bubble on the output indicates that the function is *active-0*. Figure 3–25 illustrates the logical operation of a two-input NOR gate for all four possible input combinations, and the truth table for the two-input NOR gate is given in Table 3–5.

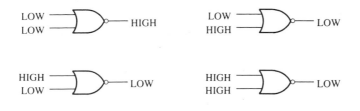

FIGURE 3-25 *Logical operation of a two-input NOR gate.*

TABLE 3-5 *Truth table for a two-input NOR gate.*

Inputs		Output
A	B	X ·
0	0	1
0	1	0
1	0	0
1	1	0

NOTE: 1 ≡ HIGH, 0 ≡ LOW.

Pulsed Operation

The next three examples illustrate the logical operation of the NOR gate with pulsed inputs. Again, as with the other types of gates, we will simply follow the truth table operation in order to determine the output waveforms.

EXAMPLE 3-13

If the two waveforms shown in Figure 3-26(a) are applied to the NOR gate, what is the resulting output waveform?

FIGURE 3-26

(a)

A

B

A
B
X

Whenever any input is HIGH, the output is LOW.

(b) X

Solution Whenever any input of a NOR gate is HIGH, the output is LOW. See Figure 3-26(b).

EXAMPLE 3-14

Sketch the output waveform for the three-input NOR gate in Figure 3-27(a), showing the proper relation to the inputs.

FIGURE 3–27

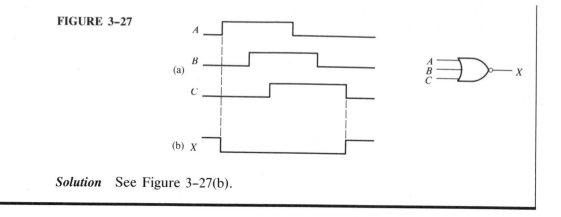

(a)

(b) X

Solution See Figure 3–27(b).

The NOR Gate as an Active-LOW Input AND Gate

The NOR gate, like the NAND, displays another aspect of its operation that is inherent in the way it logically functions. Table 3–5 shows that a HIGH is produced on the gate output only if *all* of the inputs are LOW. In reference to Figure 3–28(a), this aspect of NOR operation is stated as follows:

> If both A AND B are LOW, then X is HIGH.

Here we have essentially an AND operation that requires all LOW inputs to produce a HIGH output. This is called *negative-AND* and is represented by the standard symbol in Figure 3–28(b). It is important to remember that the two symbols in this figure represent the same gate and serve only to distinguish between the two facets of the logical operation. The following two examples illustrate this.

FIGURE 3–28 *Standard symbols representing the two equivalent functions of the NOR gate.*

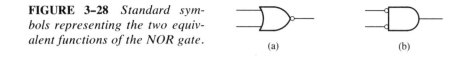

(a)   (b)

EXAMPLE 3–15   A certain application requires that two lines be monitored for the occurrence of a HIGH level voltage on either or both lines. Upon detection of a HIGH level, the circuit must provide a LOW voltage to energize a particular indicating device. Sketch the operation.

Solution   This application requires a NOR function, since the output must be active-LOW in order to give an indication of at least one HIGH on its inputs. The NOR symbol is therefore used to show the operation. See Figure 3–29.

FIGURE 3-29 *Example of a NOR gate application.*

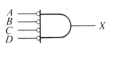

Active-LOW energizes the device.

EXAMPLE 3-16   A device is needed to indicate when two LOW levels occur simultaneously on its inputs and to produce a HIGH output as an indication. Sketch the operation.

Solution   Here, an active-LOW input AND function is required, as shown in Figure 3-30.

FIGURE 3-30

EXAMPLE 3-17   For the four-input NOR gate operating as an active-LOW input AND in Figure 3-31(a), determine the output relative to the inputs.

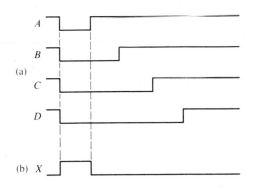

FIGURE 3-31

Solution   Any time *all* of the inputs are LOW, the output is HIGH. See Figure 3-31(b).

SECTION REVIEW 3-5

1. When is the output of a NOR gate HIGH?
2. When is the output of a NOR gate LOW?
3. Describe the functional difference between a NOR gate and a negative-AND gate. Do they both have the same truth table?

3-6   GATE PROPAGATION DELAY TIME

Propagation delay is a very important characteristic of logic circuits because it limits the speed (frequency) at which they can operate. The terms *low speed* and *high speed,* when applied to logic circuits, refer to the propagation delays; the shorter the propagation delay, the higher the speed of the circuit.

The propagation delay of a gate is basically the *time interval between the application of an input pulse and the occurrence of the resulting output pulse.* There are two propagation delays associated with a logic gate:

1. t_{PHL}: The time between a specified reference point on the input pulse and a corresponding reference point on the output pulse, *with the output changing from the HIGH level to the LOW level.*
2. t_{PLH}: The time between a specified reference point on the input pulse and a corresponding reference point on the output pulse, *with the output changing from the LOW level to the HIGH level.*

Figure 3–32 illustrates these propagation delay times for both inverted and non-inverted outputs.

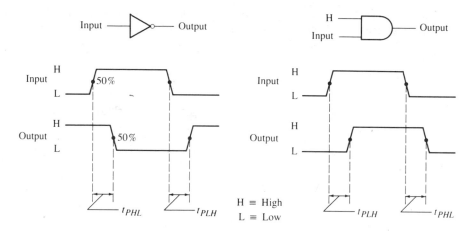

FIGURE 3–32   *Propagation delay times.*

Notice that (in this case) the two delay times are measured between the 50% points on the leading edges of the input and output pulses and between the 50% points on the trailing edges of the input and output pulses. When t_{PHL} and t_{PLH} are not equal, the larger value is the "worst-case" delay.

SECTION REVIEW 3-6

1. Define t_{PLH} and t_{PHL}.
2. A positive pulse is applied to an inverter input. The time from the leading edge

of the input to the leading edge of the output is 10 ns. The time from the trailing edge of the input to the trailing edge of the output is 8 ns. What are the values of t_{PLH} and t_{PHL}? What is the worst-case propagation delay time?

3-7   POWER DISSIPATION

The power dissipation of a logic gate equals the dc supply voltage V_{CC} times the *average* supply current I_{CC}. Normally, the value of I_{CC} for a LOW gate output is higher than for a HIGH output. The manufacturer's data sheet usually specifies both of these values as I_{CCL} and I_{CCH}. The average I_{CC} is then determined based on a 50% duty cycle operation of the gate (LOW half of the time and HIGH half of the time).

EXAMPLE 3-18

A certain gate draws 2 mA when its output is HIGH and 3.6 mA when its output is LOW. What is its average power dissipation if V_{CC} is 5 V and it is operated on a 50% duty cycle?

Solution   The average I_{CC} is

$$I_{CC} = \frac{I_{CCH} + I_{CCL}}{2} - \frac{2.0 \text{ mA} + 3.6 \text{ mA}}{2} = 2.8 \text{ mA}$$

The average power is

$$P_{AVG} = V_{CC}I_{CC} = (5 \text{ V})(2.8 \text{ mA}) = 14 \text{ mW}$$

In a logic system, the total average supply current is determined by the sum of the average supply currents (I_{CC}) of each gate in the system. This establishes the requirements for the power supply or battery used to power the system. For example, a certain logic system consists of 1000 equivalent gates, each having an average I_{CC} of 2.75 mA. The total current is 1000 × 2.75 mA = 2.75 A. A battery must be selected that is capable of providing at least 2.75 A for a specified number of hours. Multiplying the amperes and the number of hours establishes the ampere-hour (Ah) rating of the battery.

SECTION REVIEW 3-7

1. Define I_{CCL} and I_{CCH}.
2. Determine the minimum ampere-hour rating of a battery required to power a certain digital system that draws an average supply current of 2.75 A if the system must operate for at least 40 hours.

3-8   NOISE IMMUNITY

dc Noise Margins

The *dc noise margin* of a logic gate is a measure of its *noise immunity,* a gate's ability to withstand fluctuations of the voltage levels (noise) at its inputs. Common sources of noise are variations of the dc supply voltage, ground noise, magnetically coupled voltages from adjacent lines, and radiated signals. The term *dc noise margin* applies to noise voltages of relatively long duration compared to a gate's response time.

To learn about dc noise margins, we will define several voltages that are typically specified on a manufacturer's data sheet for an IC logic gate:

$$V_{IL} = \text{LOW level input voltage}$$
$$V_{IH} = \text{HIGH level input voltage}$$
$$V_{OL} = \text{LOW level output voltage}$$
$$V_{OH} = \text{HIGH level output voltage}$$

Every logic circuit has certain limits on the values of these voltages within which it will operate properly. To illustrate, let us discuss a logic gate that operates with 0 V as its ideal LOW and +5 V as its ideal HIGH. These voltages will vary because of circuit parameters, and the gate must be able to tolerate variations within certain specified limits. If, for example, a LOW level input voltage range of 0 V to 0.5 V is acceptable to a gate, then the *maximum* LOW level input voltage $[V_{IL(\text{max})}]$ for that gate is +0.5 V. Any voltage above this value can appear as a HIGH to the gate. The effect of a fluctuation due to noise of the LOW level input voltage on the output of the gate is illustrated in Figure 3–33.

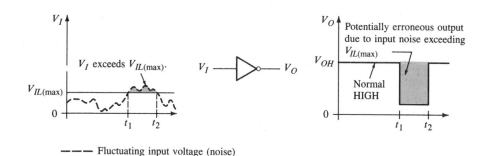

————— Fluctuating input voltage (noise)

FIGURE 3–33 *Potential effect on an inverter (or gate) output of LOW level noise on the input.*

Now, if the type of logic gate we are using has a maximum LOW level *output* voltage $[V_{OL(\text{max})}]$ of +0.2 V, then there is a "safety" margin of 0.3 V between the maximum LOW level that a gate puts out and the maximum LOW level input that a gate being driven can tolerate. This is called the LOW level *noise margin* and is expressed as

$$V_{NL} = V_{IL(max)} - V_{OL(max)} \qquad (3\text{--}2)$$

Figure 3–34 illustrates the LOW level dc noise margin. In essence, there can be fluctuations due to noise on the line between the two gates without affecting the output of the second gate, as long as the peak value of these two fluctuations does not exceed the LOW level noise margin (V_{NL}).

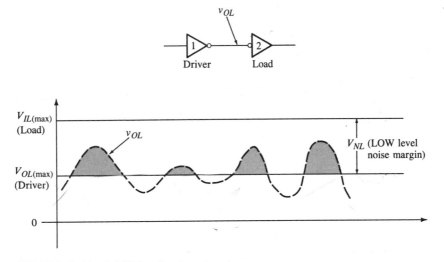

FIGURE 3–34 *LOW level noise margin.*

Now let us consider the HIGH level dc noise margin of a gate. If, for example, a HIGH level input voltage between $+5$ V and $+4$ V is acceptable as a HIGH level to the gate, then the *minimum* HIGH level input voltage [$V_{IH(min)}$] is $+4.0$ V. Any voltage above this value is acceptable as a HIGH to the gate, and any voltage below this value can appear as a LOW, as illustrated in Figure 3–35.

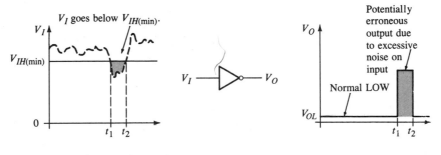

FIGURE 3–35 *Potential effect on an inverter (or gate) output of HIGH level noise on the input.*

Now, if the driving gate has a minimum HIGH level output voltage [$V_{OH(min)}$] of $+4.5$ V, then there is a HIGH level dc noise margin of 0.5 V, expressed as

$$V_{NH} = V_{OH(min)} - V_{IH(min)} \qquad (3\text{-}3)$$

Figure 3–36 illustrates the HIGH level noise margin and shows that there can be fluctuations due to noise on the line between the two gates without affecting the output of the second gate, as long as the peak value of the noise does not exceed V_{NH}, the HIGH level noise margin.

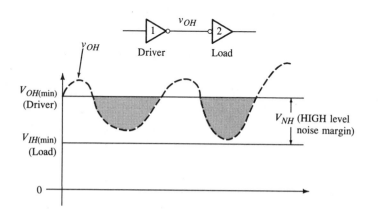

FIGURE 3–36 *HIGH level noise margin.*

ac Noise Margin

The term *ac noise margin* applies to the noise immunity of a gate to noise of very short duration. A typical gate is much more immune to this type of noise because it often cannot respond fast enough to be affected. Therefore, ac noise margins are considerably higher than dc noise margins.

SECTION REVIEW 3–8

1. Define *noise margin*.
2. Determine V_{NL} (LOW level noise margin) when $V_{IL(max)} = 0.4$ V and $V_{OL(max)} = 0.1$ V.

3–9   LOADING CONSIDERATIONS

In a digital system, you will typically find many types of digital ICs interconnected to perform various functions. In these situations, the output of a logic gate may be connected to the inputs of several other similar gates so the *load* on the driving gate becomes an important factor.

Fan-Out

The *fan-out* of a gate is the maximum number of inputs of the same IC family that the gate can drive while maintaining its output levels within specified limits. That is, the fan-out specifies the maximum loading that a given gate is capable of handling.

We will use TTL (transistor-transistor logic) to illustrate the concept of fan-out. A TTL gate is classified as *current-sinking* because when its output state is LOW, it *accepts* current from the input of the gate that it is driving.

TTL loading is illustrated in Figure 3–37, where one TTL gate is driving another TTL gate. For simplicity, only the output and input portions of TTL gate circuits are shown. A further discussion of TTL is presented in the next section.

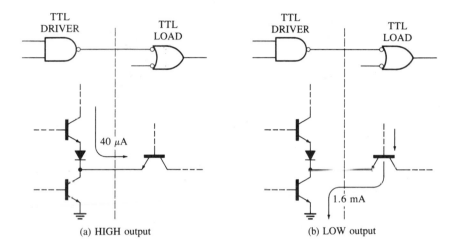

FIGURE 3–37 *TTL loading with maximum current values shown.*

When the driving gate is in its HIGH output state, the input to the load gate is like a reverse-biased diode. That is, there is practically no current required from the driving gate. Actually, for a TTL gate, a maximum of 40 μA of reverse leakage current flows from the driving gate into the load gate input, as shown in Figure 3–37(a).

When the driving gate is in its LOW output state, the input to the load gate is like a forward-biased diode. That is, there is a certain amount of current flowing *out* (conventional direction) of the load gate input into the driving gate output, as shown in Figure 3–37(b). In this case, the driving gate is *sinking* current from the load. For a TTL gate, this current is a maximum of 1.6 mA.

There is a limitation on the total amount of current that a TTL gate can sink, and this sets the limit on the number of other TTL gate inputs that it can drive. One TTL input is termed a *unit load* (UL) and represents 1.6 mA that the driving gate must sink in the LOW state. The maximum number of unit loads that a TTL gate can handle is called its *fan-out*.

For example, a standard TTL gate has a fan-out of 10 UL. This means that it can drive no more than 10 inputs of other standard TTL gates and still operate reliably. If the fan-out is exceeded, specified operation is not guaranteed. Figure 3–38 shows a gate driving 10 other gates.

FIGURE 3–38 *A NAND gate driving 10 unit loads.*

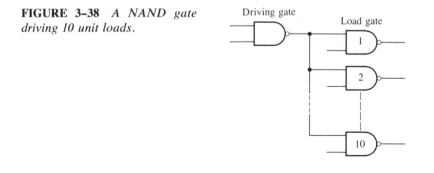

SECTION REVIEW 3–9

1. Define *fan-out*.
2. A standard TTL gate is driving eight standard TTL gate inputs. In the HIGH output state, how much output current must the driving gate be able to provide to the load gates? In the LOW output state, how much output current must the driving gate be able to sink?

3–10 INTEGRATED CIRCUIT LOGIC FAMILIES

In the last section, TTL (transistor-transistor logic) was introduced to illustrate the concept of loading. TTL and another family of logic circuits called CMOS (complementary metal-oxide semiconductor) are two of the most widely used digital integrated circuit technologies. In this section, a general comparison of TTL and CMOS is given, and some specific logic gate devices are introduced.

The term TTL actually applies to a series of logic circuits: *standard TTL, high-speed TTL, low-power TTL, Schottky TTL, low-power Schottky TTL, advanced low-power Schottky TTL,* and *advanced Schottky TTL.* The differences in these various types of TTL are in their performance characteristics, such as propagation delay times, power dissipation, and fan-out. The TTL family has a number prefix of 54 or 74 followed by a letter or letters that specify the series, as shown in Table 3–6. The prefix 54 indicates an operating temperature range of −55°C to 125°C (generally for military usage). The prefix 74 indicates a temperature range of 0°C to 70°C (for commercial use).

TTL and CMOS differ in that TTL uses bipolar transistors in its circuit technology and CMOS uses field-effect transistors. Logic functions are the same, however, whether the device is implemented with TTL or CMOS technologies. The circuit technologies make a difference not in logic function, but only in

TABLE 3–6 *TTL series designations.*

TTL Series	Prefix Designation	Example of Device
Standard TTL	54 or 74 (no letter)	7400 (quad NAND gates)
High-speed TTL	54H or 74H	74H00 (quad NAND gates)
Low-power TTL	54L or 74L	74L00 (quad NAND gates)
Schottky TTL	54S or 74S	74S00 (quad NAND gates)
Low-power Schottky TTL	54LS or 74LS	74LS00 (quad NAND gates)
Advanced low-power		
Schottky TTL	54ALS or 74ALS	74ALS00 (quad NAND gates)
Advanced Schottky TTL	54AS or 74AS	74AS00 (quad NAND gates)

performance characteristics. Generally, CMOS uses less power, is slower (longer propagation delays), and has less fan-out capability than TTL. There are exceptions to this general comparison, as you can see in Table 3–7.

Several series of CMOS logic are available, and they fall basically into two process technology categories: *metal-gate CMOS* and *silicon-gate CMOS*. The most recent high-speed CMOS series is designated with the prefixes 54HC or 74HC. The 54/74 indicates that the series is *pin compatible* with the TTL family. That is, each pin on the CMOS IC package is the same function (input, output, supply voltage, or ground) as the corresponding pin on a TTL IC package. A detailed coverage of TTL and CMOS interfacing is given in Chapter 11, and IC technologies are presented in Appendix A. Table 3–7 provides a comparison of some performance characteristics of CMOS and TTL.

TABLE 3–7 *Comparison of performance characteristics of CMOS and TTL.*

Technology	CMOS* (silicon-gate)	CMOS* (metal-gate)	TTL Std.	TTL LS	TTL S	TTL ALS	TTL AS
Device series	74HC	4000	74	74LS	74S	74ALS	74AS
Power dissipation (mW per gate): Static @ 100 kHz	2.5 nW 0.17 mW	1 μW 0.1 mW	10 mW 10 mW	2 mW 2 mW	19 mW 19 mW	1 mW 1 mW	8.5 mW 8.5 mW
Propagation delay time	8 ns	105 ns	10 ns	10 ns	3 ns	4 ns	1.5 ns
Fan-out (LS loads)	10	4	40	20	50	20	50

* Propagation delay is dependent on V_{CC}. Power dissipation is a function of frequency.

Speed/Power Product

The *speed/power product* is sometimes specified by the manufacturer as a measure of the performance of a logic circuit based on the product of the propagation delay time and the power dissipation at a specified frequency. The speed/power product is expressed in units of joules, symbolized J. For example, the speed/power product (SPP) of a 74HC CMOS gate at 100 kHz is

$$SPP = (8 \text{ ns})(0.17 \text{ mW}) = 1.36 \text{ pJ}$$

Specific Devices

A wide variety of SSI logic gate configurations are available in both the TTL and the CMOS families. When specific devices are referenced in the SSI and MSI categories, standard TTL generally will be used for illustrative purposes in this book, although other types will occasionally be used. Keep in mind that most of the specific devices are also available in the other TTL series as well as in CMOS. Later, when LSI devices are covered, MOS devices will be referenced to a greater extent.

A selection of typical logic gate devices are now presented. These devices are commonly housed in the dual-in-line package (DIP), as shown in Figure 3–39. The pin arrangements for both the 14-pin and the 16-pin DIPs are shown. Notice that pin 1 is to the left of the notched end of the DIP looking from the top. The dc supply voltage V_{CC} and ground are connected to pins 14 and 7, respectively (16 and 8 for the 16-pin package). There are other pin arrangements, but this is the most common.

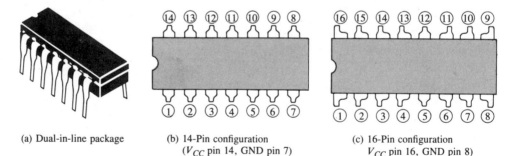

(a) Dual-in-line package

(b) 14-Pin configuration
(V_{CC} pin 14, GND pin 7)

(c) 16-Pin configuration
V_{CC} pin 16, GND pin 8)

FIGURE 3–39 *Dual-in-line package (DIP) and two typical pin configurations.*

For simplicity, V_{CC} and ground connections to each gate are normally not shown in a logic diagram. In Figures 3–40 through 3–44, each device is represented by a distinctive shape logic diagram with the pin numbers indicated in parentheses. Additionally, each device is shown as a rectangular outline logic symbol. These two representations are equivalent.

Hex inverter The 7404 hex inverter is a standard TTL device consisting of six inverters in a 14-pin package, as shown in Figure 3–40.

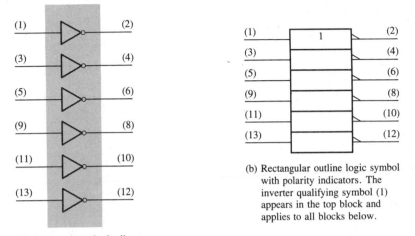

(a) Distinctive shape logic diagram

(b) Rectangular outline logic symbol with polarity indicators. The inverter qualifying symbol (1) appears in the top block and applies to all blocks below.

FIGURE 3–40 *7404 hex inverter (pin numbers in parentheses).*

AND gates Several configurations of AND gates are available in IC form. The 7408 has four 2-input AND gates (quad 2-input AND); the 7411 has three 3-input AND gates (triple 3-input AND); and the 7421 has two 4-input AND gates (dual 4-input AND). These gates are shown in Figure 3–41 (pp. 103–104).

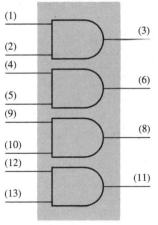

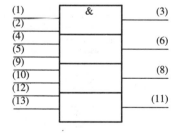

(a) 7408 Quad 2-input AND

FIGURE 3–41 *Standard TTL AND gates.*

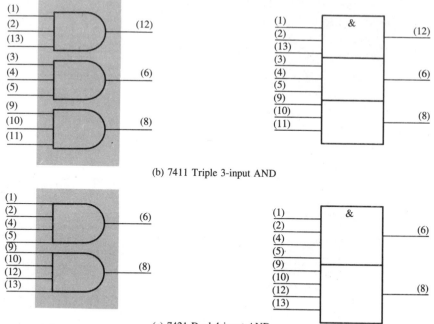

(b) 7411 Triple 3-input AND

(c) 7421 Dual 4-input AND

FIGURE 3–41   *(Continued)*

NAND gates   A variety of NAND gates are available, including the 7400 with four 2-input gates, the 7410 with three 3-input gates, the 7420 with two 4-input gates, the 7430 with one 8-input gate, and the 74133 with one 13-input gate. These gates are shown in Figure 3–42. Notice that the 74133 is in a 16-pin DIP.

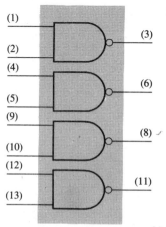

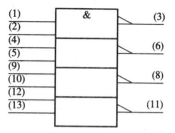

(a) 7400 Quad 2-input NAND

FIGURE 3–42   *Standard TTL NAND gates.*

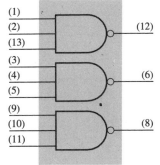
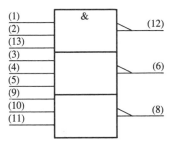

(b) 7410 Triple 3-input NAND

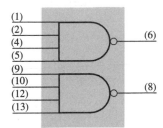
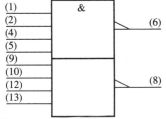

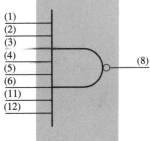
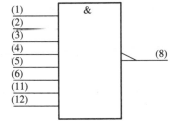

(c) 7420 Dual 4-input NAND

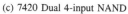

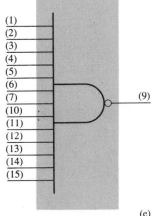

(d) 7430 Single 8-input NAND

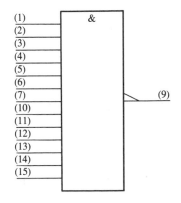

(e) 74133 Single 13-input NAND

FIGURE 3–42 *(Continued)*

OR gates   The 7432 has four 2-input OR gates, as shown in Figure 3–43.

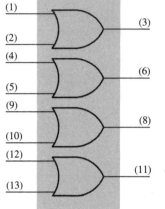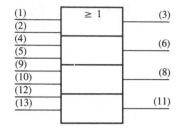

FIGURE 3–43   *7432 Quad 2-input OR gates.*

NOR gates   Examples of NOR gate configurations are shown in Figure 3–44. The 7402 has four 2-input gates, and the 7427 has three 3-input gates.

Unused Inputs

An unconnected input on a TTL gate acts as a HIGH because an open input results in a reverse-biased emitter junction on the input transistor just as a HIGH level does. However, due to noise sensitivity, it is best not to leave unused TTL inputs unconnected (or open). There are several possible ways of handling unused inputs:

1. Connect unused inputs to a used input if the maximum fan-out of the driving gate will not be exceeded. Each additional input represents a unit load to the driving gate.
2. Connect unused inputs of AND and NAND gates to the dc supply voltage (V_{CC}) through a 1-kΩ resistor. Connect unused inputs of OR and NOR gates to ground.
3. Connect unused inputs to the output of a gate that is not being used. The unused gate output must be a constant HIGH for unused inputs of AND and NAND gates.

Unused inputs of CMOS gates must be connected either to the dc supply voltage or to ground, depending on the type of gate.

Handling of CMOS Devices

CMOS devices are *electrostatic-discharge sensitive* and must be handled with caution because they can be easily damaged by electrostatic charges that are

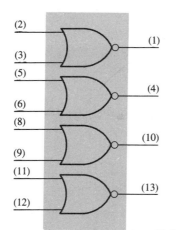
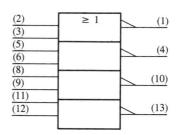

(a) 7402 Quad 2-input NOR

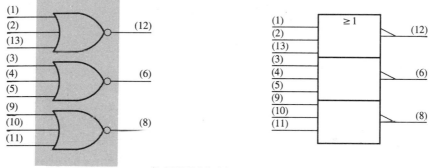

(b) 7427 Triple 3-input NOR

FIGURE 3-44 *Standard TTL NOR gates.*

generated and stored on surfaces of ordinary plastics, most clothing items, ungrounded human bodies, and other sources. You should consult the manufacturer's recommended procedures before working with these devices. Several general precautions are as follows:

1. CMOS devices should be shipped and stored in conductive foam.
2. All instruments and metal benches used in assembly or test should be connected to earth ground.
3. The handler's wrist should be connected to earth ground by a length of wire and a high-value resistor.
4. Do not remove a CMOS device (or any other type) from the circuit while the dc power is on.
5. Do not apply signals to the device while the dc power is off.

1. Name the type and series of IC technology that exhibits each of the following performance characteristics:
    (a) Fastest switching time (lowest propagation delay)
    (b) Lowest power dissipation
    (c) Highest fan-out (into TTL LS load)
2. Calculate the speed/power product for an advanced Schottky (74AS) gate.

3–11   DATA SHEET INTERPRETATION

Specific information about operating characteristics of a particular IC family can be determined from *data sheets* published by the manufacturer.

A typical data sheet is divided into three main sections: *recommended operating conditions, electrical characteristics,* and *switching characteristics.* As an example, Figure 3–45 shows the arrangement of a typical data sheet for the 5400/7400 quad 2-input NAND gate. For additional data sheets, see Appendix B.

Explanation of Data Sheet Parameters

The following list describes the parameters of the data sheet in Figure 3–45:

1. V_{CC}: The dc voltage that supplies power to the device. Below the specified minimum, reliable operation cannot be guaranteed. Above the specified maximum, damage to the device may occur.
2. I_{OH}: The maximum output current that the gate can provide (source) to a load and operate reliably when the output is at the HIGH level. By convention, the current out of a terminal is assigned a negative value. Figure 3–46(a) illustrates this parameter.
3. I_{OL}: The maximum output current that the gate can sink and operate reliably when the output is at the LOW level. By convention, the current into a terminal is assigned a positive value. Figure 3–46(b) illustrates this parameter.
4. V_{IH}: The value of input voltage that can be accepted as a HIGH level by the gate. This parameter, as well as the next three, was introduced in the discussion of noise margin.
5. V_{IL}: The value of input voltage that can be accepted as a LOW level by the gate.
6. V_{OH}: The value of HIGH level output voltage that the gate produces.
7. V_{OL}: The value of LOW level output voltage that the gate produces.
8. I_{IH}: The value of input current for a HIGH level input voltage. Figure 3–46(c) illustrates this parameter.
9. I_{IL}: The value of input current for a LOW level input voltage. Figure 3–46(d) illustrates this parameter.

Parameter	5400			7400			Units
	Minimum	Typical	Maximum	Minimum	Typical	Maximum	
Supply voltage (V_{CC})	4.5	5.0	5.5	4.75	5.0	5.25	V
Operating free-air temperature range	-55	25	125	0	25	70	°C
HIGH level output current (I_{OH})			-400			-400	μA
LOW level output current (I_{OL})			16			16	mA

(a) Recommended operating conditions

Parameter	Limits			Units	Test Conditions[1]
	Minimum	Typical[2]	Maximum		
HIGH level input voltage (V_{IH})	2.0			V	
LOW level input voltage (V_{IL})			0.8	V	
HIGH level output voltage (V_{OH})	2.4	3.4		V	V_{CC} = min., I_{OH} = 0.4 mA, V_{IN} = 0.8 V
LOW level output voltage (V_{OL})		0.2	0.4	V	V_{CC} = min., I_{OL} = 16 mA, V_{IN} = 2.0 V
HIGH level input current (I_{IH})			40	μA	V_{CC} = max., V_{IN} = 2.4 V
LOW level input current (I_{IL})			-1.6	mA	V_{CC} = max , V_{IN} — 0.4 V
Short-circuit output current[3] (I_{OS}) 5400	-20		-55	mA	V_{CC} = max.
Short-circuit output current[3] (I_{OS}) 7400	-18		-55	mA	
Total supply current with outputs HIGH (I_{CCH})		4.0	8.0	mA	V_{CC} = max.
Total supply current with outputs LOW (I_{CCL})		12	22	mA	V_{CC} = max.

(b) Electrical characteristics over operating temperature range (unless otherwise noted)

Parameter	Limits			Units	Test Conditions
	Minimum	Typical	Maximum		
Propagation delay time, LOW-to-HIGH output (t_{PLH})		11	22	ns	V_{CC} = 5.0 V C_{LOAD} = 15 pF R_{LOAD} = 400 Ω
Propagation delay time, HIGH-to-LOW output (t_{PHL})		7.0	15	ns	

(c) Switching characteristics (T_A = 25°C)

NOTES:
[1] For conditions shown as min. or max., use the appropriate value specified under recommended operating conditions for the applicable device type.
[2] Typical limits are at V_{CC} = 5.0 V, 25°C.
[3] Not more than one output should be shorted at a time. Duration of short not to exceed 1 s.

FIGURE 3–45

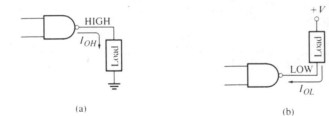

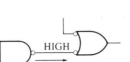

(a)

(b)

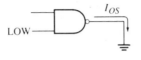

(c)

(d)

(e)

FIGURE 3–46   *Illustration of some data sheet parameters.*

10. I_{OS}: The output current when the gate output is shorted to ground and with input conditions that establish a HIGH level output. Figure 3–46(e) illustrates this parameter.

11. I_{CCH}: The total current from the V_{CC} supply when all gate outputs are at the HIGH level.

12. I_{CCL}: The total current from the V_{CC} supply when all gate outputs are at the LOW level.

SECTION REVIEW 3–11

1. Using the data sheet in Figure 3–45, calculate the worst-case noise margin for the 7400 gate.

2. What is the maximum current that a 7400 gate can supply to a load when the output is HIGH?

3–12   LOGIC GATE APPLICATIONS

In this section we will look at two simple examples to illustrate how logic gates might be applied to practical situations.

AND Gate Application

In a simple application, an AND gate can be used to detect the existence of a specified number of conditions and, in response, to initiate an appropriate action.

For example, an automobile's safety system may require that an audible signal be produced to warn the driver that the seat belt is not engaged. The conditions for this are that the ignition switch be on, the seat belt be unbuckled, and the warning signal last for a specified time and then turn off automatically. The first two conditions can be represented by switch positions and the third by a timing circuit.

Figure 3–47 shows an AND gate whose HIGH output activates a buzzer when these three conditions are met on its inputs. When the ignition switch, represented by S_1 in the figure, is on, a HIGH is connected to the gate input A. When the belt is not properly buckled, switch S_2 is off and a HIGH is connected to the gate input B. At the instant the ignition switch is turned on, the timer is activated and produces a HIGH on gate input C. The resulting HIGH gate output activates the alarm. After a specified time the timer's output goes LOW, disabling the AND gate and turning off the alarm. If the seat belt is buckled when the ignition is turned on, a LOW is applied to input B, keeping the gate output LOW, thus preventing the alarm from sounding.

FIGURE 3–47 *Example of an AND gate application.*

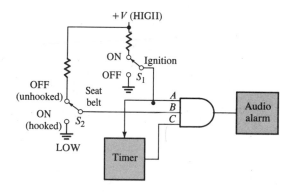

OR Gate Application

As an example application of an OR gate, let us assume that in a room with three doors, an indicator lamp must be turned on when any of the doors is not completely closed.

The sensors are switches that are open when a door is ajar or open. This open switch creates the HIGH level for the OR gate input, as shown in Figure 3–48. If any or all of the doors are open, the gate output is HIGH. This HIGH level is then used to illuminate the indicator lamp. The gate is assumed to be capable of supplying sufficient current to the lamp.

FIGURE 3–48 *Example of an OR gate application.*

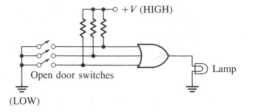

1. When is the alarm in Figure 3–47 activated?
2. What change is required in Figure 3–48 if a NOR gate replaces the OR gate?

3–13   TROUBLESHOOTING LOGIC GATES

Open gate inputs and outputs represent a large percentage of all digital IC failures. Various types of *shorts* account for the remaining failure modes. These include shorts to ground or supply voltage, shorts between traces on printed circuit boards, and shorted inputs and outputs due to internal gate failures.

Figure 3–49 illustrates an *open* failure of a gate input and how to check for it. Part (a) assumes that one of the inputs of a two-input NAND gate is open. Troubleshooting this type of failure can be accomplished with a *logic pulser* and *logic probe*. Start by pulsing one of the inputs and observing the output activity with the logic probe as in part (b). If activity is observed, then that particular input and output are good. Next, pulse the other input. No activity on the output indicates that the input is open as shown in part (c). (See Chapter 1, section 6, for a discussion of logic probe operation.)

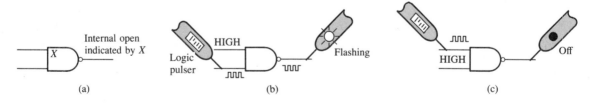

FIGURE 3–49 *Troubleshooting an open input.*

Figure 3–50 illustrates troubleshooting an open output. In part (a), one of the inputs is pulsed, and no activity is observed on the output. In part (b), the other input is pulsed, and again no activity is observed on the output. This test indicates that the gate output is open.

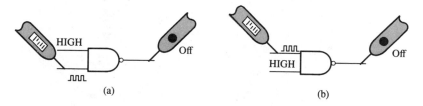

FIGURE 3-50 *Troubleshooting an open output.*

EXAMPLE 3-19

By using a logic probe and pulser, you find that the 7402 gate in Figure 3–51 is defective. Find out the nature of the fault (internal to gate or on pc board) before removing what appears to be an internally defective IC. Pin 11 is HIGH.

FIGURE 3-51

Solution

(a) Pulse pin 12 and observe with the probe that pin 13 has pulse activity. It should be LOW due to the HIGH on pin 11.

(b) Pulse pin 12 and read the current at pin 13 with a current tracer. Now pulse pin 13 and read the current at pin 12 with the current tracer. The same current is observed in both readings.

(c) Pins 12 and 13 are shorted together by a solder bridge on the back of the pc board. Although the problem was originally located with a pulser and probe, the tracer added important information that kept the IC from being removed.

Oscilloscope Measurements

Although the probe and pulser provide a quick and inexpensive method for many troubleshooting problems, more accurate measurements are sometimes needed. The oscilloscope is a very useful measurement and troubleshooting tool, especially when accurate time measurements must be made and when pulse waveforms must be observed.

An example of the application of an oscilloscope is the measurement of the propagation delay time of a gate, as illustrated in Figure 3–52 in which two waveforms are displayed on an oscilloscope screen. The 2-ns display in the upper right corner indicates that each major horizontal division represents a time interval of 2 ns. The 1-V display in the upper left corner indicates that each major vertical division represents 1 V. The upper waveform is the leading portion of the input pulse, and the lower waveform is the leading portion of the output pulse.

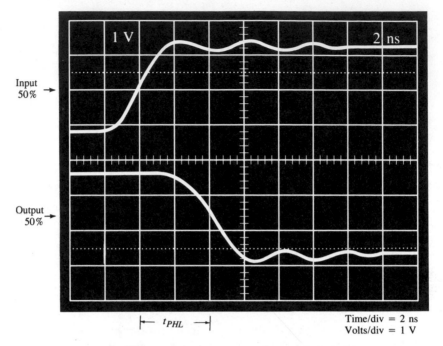

Time/div = 2 ns
Volts/div = 1 V

FIGURE 3–52   *Oscilloscope measurement of propagation delay time.*

Notice that there are two horizontal divisions between the 50% point on the input pulse and the 50% point on the output pulse. The propagation delay time, t_{PHL} in this case, is (2 ns/div)(2 div) = 4 ns.

SECTION REVIEW 3–13

1. What does a dim light on a logic probe indicate?
2. In Figure 3–52, determine the pulse amplitude of the input pulse at 14 ns past the 50% point on the leading edge.

SUMMARY

☐ An inverter changes a LOW level to a HIGH level (0 to 1), and vice versa.
☐ An AND gate produces a HIGH (1) output if and only if all of its inputs are HIGH (1).
☐ An OR gate produces a HIGH (1) output if any of its inputs are HIGH (1).
☐ A NAND gate is effectively an AND gate followed by an inverter. It produces a LOW (0) output if and only if all of its inputs are HIGH (1).
☐ A NAND gate also acts as a negative-OR gate. It produces a HIGH (1) output if any of its inputs are LOW (0).
☐ A NOR gate is effectively an OR gate followed by an inverter. It produces a LOW (0) output if any of its inputs are HIGH (1).

☐ A NOR gate also acts as a negative-AND gate. It produces a HIGH (1) output if and only if all of its inputs are LOW (0).

☐ The propagation delay time of a gate is the time interval from the leading edge of the input pulse to the leading edge of the output pulse or from the trailing edge of the input pulse to the trailing edge of the output pulse.

☐ The power dissipation of a logic gate is the dc supply voltage (V_{CC}) times the average current drawn from the supply (I_{CC}).

☐ The greater the noise margin of a gate, the better its immunity to noise.

☐ The fan-out of a gate is the number of other gate inputs of the same family that the gate can drive from its output.

☐ TTL and CMOS are two types of digital integrated circuit technologies. CMOS generally requires less power than TTL but is slower (has a longer propagation delay time).

SELF-TEST

1. What is the output of an inverter for each of the following inputs? Which input states are equivalent?
   (a) HIGH   (b) LOW   (c) 1   (d) 0
2. Develop the truth table for a four-input AND gate.
3. Develop the truth table for a four-input OR gate.
4. Two positive pulses are applied to the inputs of a two-input NAND gate. One pulse begins at $t = 0$ and ends at $t = 1$ ms The second pulse begins at $t = 0.8$ ms and ends at $t = 3$ ms. Describe the output.
5. The same two pulses described in Problem 4 are applied to a two-input NOR gate. Describe the output.
6. A positive pulse is applied to an inverter. The time interval from the leading edge of the input to the leading edge of the output is 10 ns. Is this parameter t_{PHL} or t_{PLH}?
7. If $V_{CC} = 5$ V and $I_{CC} = 2$ mA, what is the power dissipation of the device?
8. A certain series of logic gates has a HIGH level noise margin of 0.2 V. What does this mean?
9. A certain type of logic gate has a fan-out of 20. What does this mean?
10. Which TTL series has the faster switching speed, 74L or 74S? What do the prefixes L and S represent?
11. In a certain application, propagation delay is not a major consideration, but power dissipation is critical. Which logic family would you use?
12. Define the type of logic gate for each of the following:
    (a) 7400   (b) 7404   (c) 7411   (d) 7420   (e) 7432   (f) 7427

PROBLEMS

Section 3–1

3–1   The input waveform shown in Figure 3–53 is applied to an inverter. Sketch the output waveform in proper relationship to the input.

FIGURE 3–53 V_{in}

3–2 A network of cascaded inverters is shown in Figure 3–54. If a HIGH is applied to point A, determine the logic levels at points B through F.

FIGURE 3–54

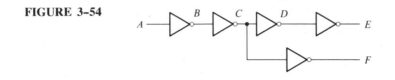

Section 3–2

3–3 Determine the output, X, for a two-input AND gate with the input waveforms shown in Figure 3–55.

FIGURE 3–55

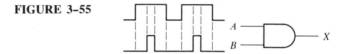

3–4 Repeat Problem 3–3 for the waveforms in Figure 3–56.

FIGURE 3–56

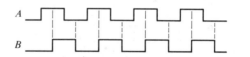

3–5 The input waveforms are applied to a three-input AND gate as indicated in Figure 3–57. Determine the output waveform in proper relationship to the inputs.

FIGURE 3–57

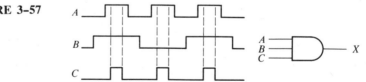

3–6 The input waveforms are applied to a four-input gate as indicated in Figure 3–58. Determine the output waveform in proper relation to the inputs.

FIGURE 3–58

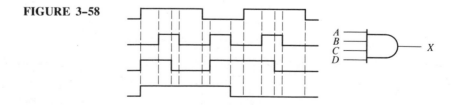

Section 3–3

3–7   Determine the output for a two-input OR gate using the input waveforms in Figure 3–56.

3–8   Repeat Problem 3–5 for a three-input OR gate.

3–9   Repeat Problem 3–6 for a four-input OR gate.

3–10 For the five input waveforms in Figure 3–59, determine the output for an AND gate and the output for an OR gate.

FIGURE 3–59

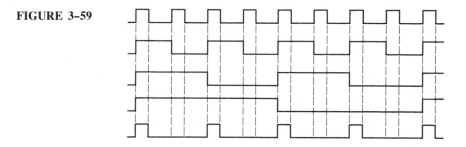

Section 3–4

3–11 For the set of input waveforms in Figure 3–60, determine the output for the gate shown.

FIGURE 3–60

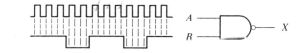

3–12 Determine the gate output for the input waveforms in Figure 3–61.

FIGURE 3–61

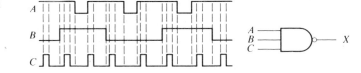

3–13 Determine the output waveform in Figure 3–62.

FIGURE 3–62

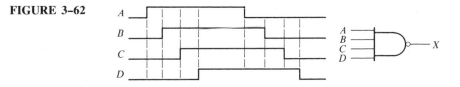

3–14 As you learned, the two logic symbols shown in Figure 3–63 are equivalent. The difference between the two is strictly a matter of how we look at them from a functional viewpoint. For the NAND symbol, we are looking for two HIGHs on the inputs to

give us a LOW output. For the negative-OR, we are looking for at least one LOW on
the inputs to give us a HIGH on the output. Using these two functional viewpoints,
show that each gate will produce the same output for the given inputs.

FIGURE 3-63

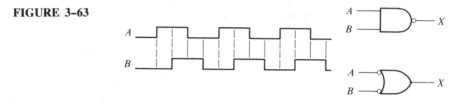

Section 3-5

3-15   Repeat Problem 3-11 for a two-input NOR gate.

3-16   Determine the output waveform in Figure 3-64.

FIGURE 3-64

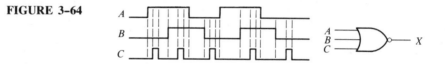

3-17   Repeat Problem 3-13 for a four-input NOR gate.

3-18   For the NOR symbol, we are looking for at least one HIGH on the inputs to give us a
LOW on the output. For the negative-AND, we are looking for two LOWs on the
inputs to give us a HIGH output. Using these two functional points of view, show
that both gates in Figure 3-65 will produce the same output for the given inputs.

FIGURE 3-65

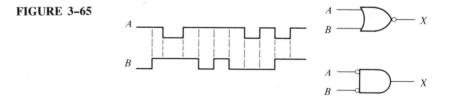

Section 3-6

3-19   Determine t_{PLH} and t_{PHL} in Figure 3-66.

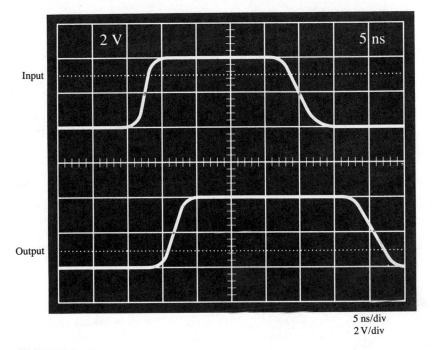

5 ns/div
2 V/div

FIGURE 3–66

3–20 Gate *A* has $t_{PLH} = 6$ ns and $t_{PHL} = 5.5$ ns. Gate *B* has $t_{PLH} = 10$ ns and $t_{PHL} = 9.2$ ns. Which gate can be operated at a higher frequency than the other?

Section 3–7

3–21 If a logic gate operates on a dc supply voltage of $+5$ V and draws an average current of 4 mA, what is its power dissipation?

3–22 I_{CCH} is the dc supply current from V_{CC} when all outputs of an IC are HIGH. I_{CCL} is the dc supply current when all outputs are LOW. For a 7400 IC determine the *typical* power dissipation when all four gate outputs are HIGH.

3–23 Determine the typical 7400 power dissipation when two outputs are HIGH and two are LOW.

Section 3–8

3–24 The minimum HIGH output of a gate is 2.4 V. The gate is driving a second gate that can tolerate a minimum HIGH input of 2.1 V. What is the HIGH level dc noise margin?

3–25 The maximum LOW output of a gate is specified as 0.3 V, and the maximum LOW input is specified as 0.6 V. What is the LOW level dc noise margin when these gates work together?

Section 3–9

3–26 If a 7400 NAND gate is driving six unit loads, determine the amount of current that it must be able to sink in the LOW output state.

3-27   A 7400 NAND gate is driving ten unit loads. How much current must it be able to source in the HIGH state?

Section 3-10

3-28   Using Table 3–7, determine which logic series offers the best performance at 100 kHz in terms of both switching speed and power dissipation. Note: Find the speed/power product of each and compare the results.

3-29   In the comparison of certain logic devices, it is noted that the power dissipation for one particular type increases as the frequency increases. Is the device TTL or CMOS?

Section 3-11

3-30   From the data sheet in Figure 3–45, what is the maximum supply voltage for a 5400 gate?

3-31   How long does it take a typical 54/74 gate output to make a transition from its LOW state to its HIGH state in response to an input? Refer to Figure 3–45.

3-32   Determine the maximum LOW level output voltage from the data sheet in Figure 3–45.

Section 3-12

3-33   Sensors are used to monitor the pressure and the temperature of a chemical solution stored in a vat. The circuitry for each sensor produces a HIGH voltage when a specified maximum value is exceeded. An alarm requiring a LOW voltage input must be activated when either the pressure or the temperature is excessive. What type of logic gate is required in this application?

3-34   In a certain automated manufacturing process, electrical components are automatically inserted in a printed circuit (pc) board. Before the insertion tool is activated, the pc board must be properly positioned and the component must be in the chamber. Each of these prerequisite conditions is indicated by a HIGH voltage. The insertion tool requires a LOW voltage to activate it. Draw a diagram showing the logic gate required to implement this process and its input and output connections to the system.

Section 3-13

3-35   Using a logic probe and pulser, a technician makes the observations indicated in Figure 3–67. For each observation determine the most likely gate failure.

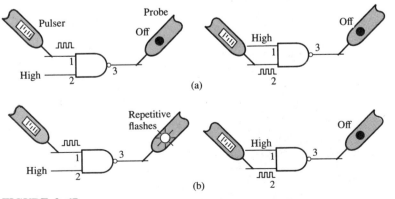

FIGURE 3-67

ANSWERS TO SECTION REVIEWS

Section 3–1
1. 0   **2.** **(a)** The negation symbol ($\bigcirc$) on the inverter output.
                (b) Output is LOW.

Section 3–2
1. When all inputs are HIGH.   **2.** When one or more inputs are LOW.
3. $X = 1$ when $ABC = 111$, and $X = 0$ for all other combinations of ABC.

Section 3–3
1. When one or more inputs are HIGH.   **2.** When all inputs are LOW.
3. $X = 0$ when $ABC = 000$, and $X = 1$ for all other combinations of ABC.

Section 3–4
1. When all inputs are HIGH.   **2.** When one or more inputs are LOW.
3. NAND: Active-LOW output for all HIGH inputs. Negative-OR: Active-HIGH output for one or more LOW inputs. Same truth tables.

Section 3–5
1. When all inputs are LOW.   **2.** When one or more inputs are HIGH.
3. NOR: Active-LOW output for one or more HIGH inputs. Negative-AND: Active-HIGH output for all LOW inputs. Same truth tables.

Section 3–6
1. t_{PLH}: Propagation delay time from input to output with output changing from LOW to HIGH. t_{PHL}: Propagation delay time from input to output with output changing from HIGH to LOW.
2. $t_{PLH} = 8$ ns; $t_{PHL} = 10$ ns; 10 ns

Section 3–7
1. I_{CCL}: Supply current with gate output LOW. I_{CCH}: Supply current with gate output HIGH.
2. 110 Ah

Section 3–8
1. A measure of noise immunity, which is the gate's ability to withstand voltage fluctuations on its inputs.
2. 0.3 V

Section 3–9
1. The maximum number of like loads that a logic circuit can drive.
2. 320 µA; 12.8 mA

Section 3–10
1. **(a)** TTL, AS   **(b)** CMOS, 74HC (static), 4000 (100 kHz)   **(c)** TTL, S, AS
2. 12.75 pJ

Section 3–11
1. $V_{NL} = 0.4$ V; $V_{NH} = 0.4$ V   **2.** 400 µA

Section 3–12
1. When the seat belt is unhooked, the ignition is on, and the timer is running.
2. The lamp must be connected to $+V$ rather than to ground.

Section 3–13
1. Open   **2.** 2.4 V

In 1854, George Boole published a book entitled *An Investigation of the Laws of Thought on Which Are Founded the Mathematical Theories of Logic and Probabilities*. It was in this publication that a "logical algebra," known today as *Boolean algebra,* was developed.

Boolean algebra is a set of rules, laws, and theorems by which logical operations can be expressed mathematically. It is a convenient and systematic way of expressing and analyzing the operation of digital circuits and systems.

The application of Boolean algebra to the analysis and design of digital logic circuits was first explored by Claude Shannon at MIT in a 1938 thesis entitled *A Symbolic Analysis of Relay and Switching Circuits*. This paper described a method by which any circuit consisting of combinations of switches and relays could be represented by mathematical expressions. Today, semiconductor circuits have largely replaced mechanical switches and relays. However, the same logical analysis is still valid, and a basic knowledge in this area is essential to the study of digital electronics.

In this chapter, you will learn

☐ The symbols used in Boolean algebra and the rules for Boolean addition and multiplication.
☐ How to express the logical functions of NOT, AND, OR, NAND, and NOR mathematically.
☐ The basic rules of Boolean algebra.
☐ DeMorgan's theorems and how to apply them.
☐ How to express logic networks mathematically in sum-of-products or product-of-sums form.
☐ How to simplify logic expressions using Boolean algebra.
☐ How to simplify logic expressions using the Karnaugh map method.
☐ How to use the Karnaugh map to implement logic functions from truth tables.

4

Boolean
Algebra

4-1   BOOLEAN OPERATIONS

Symbology

In the applications of Boolean algebra in this book, we will use *capital letters* to represent variables and functions of variables. Any single variable or a function of several variables can have either a 1 or a 0 value. In Boolean algebra, the binary digits are utilized to represent the two levels that occur within digital logic circuits. A binary 1 will represent a HIGH level, and a binary 0 will represent a LOW level in Boolean equations. This is in keeping with our use in this text of positive logic as explained in Chapter 1.

The complement of a variable is represented by a ''bar'' over the letter. For instance, for a variable represented by A, the complement of A is $\overline{A}$. So if $A = 1$, then $\overline{A} = 0$; or if $A = 0$, then $\overline{A} = 1$. The complement of a variable A is usually read ''A bar'' or ''Not A.'' Sometimes a prime symbol rather than the bar symbol is used to denote the complement. For example, the complement of A can be written as A'.

The logical AND function of two variables is represented either by writing a ''dot'' between the two variables, such as $A \cdot B$, or by simply writing the adjacent letters without the dot, such as AB. We will normally use the latter notation because it is easier to write. The logical OR function of two variables is represented by a ''$+$'' between the two variables, such as $A + B$.

Boolean Addition and Multiplication

Addition in Boolean algebra involves variables having values of either a binary 1 or a binary 0. The basic rules for Boolean addition are as follows:

$$0 + 0 = 0$$
$$0 + 1 = 1$$
$$1 + 0 = 1$$
$$1 + 1 = 1$$

In the application of Boolean algebra to logic circuits, *Boolean addition is the same as the OR*. Notice that it differs from *binary* addition in the case where two 1s are added.

Multiplication in Boolean algebra follows the same basic rules governing binary multiplication, which were discussed in Chapter 2 and are as follows:

$$0 \cdot 0 = 0$$
$$0 \cdot 1 = 0$$
$$1 \cdot 0 = 0$$
$$1 \cdot 1 = 1$$

Boolean multiplication is the same as the AND.

1. Addition in Boolean algebra is the same as the _____ function.
2. Multiplication in Boolean algebra is the same as the _____ function.
3. Represent each as a Boolean expression:
   (a) A AND B (b) A OR B

4-2   LOGIC EXPRESSIONS

NOT

The operation of an inverter (NOT circuit) can be expressed with symbols as follows: If the input variable is called A and the output variable is called X, then $X = \overline{A}$. This expression states that the output is the complement of the input, so that if $A = 0$, then $X = 1$, and if $A = 1$, then $X = 0$. Figure 4-1 illustrates this.

FIGURE 4-1 *The inverter complements an input variable.*

A ———▷o—— $X = \overline{A}$

AND

The operation of a two-input AND gate can be expressed in equation form as follows: If one input variable is A, the other input variable is B, and the output variable is X, then the Boolean expression for this basic gate function is $X = AB$. Figure 4-2(a) shows the gate with the input and output variables indicated.

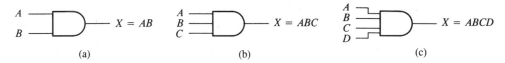

A
B $X = AB$
(a)

A
B $X = ABC$
C
(b)

A
B
C $X = ABCD$
D
(c)

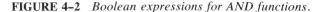

FIGURE 4-2 *Boolean expressions for AND functions.*

To extend the AND expression to more than two input variables, we simply use a new letter for each input variable. The function of a three-input AND gate, for example, can be expressed as $X = ABC$, where A, B, and C are the input variables. The expression for a four-input AND gate can be $X = ABCD$, and so on. Figures 4-2(b) and (c) show AND gates with three and four input variables, respectively.

An evaluation of AND gate operation can be made using the Boolean expressions for the output. For example, each variable on the inputs can be either a 1 or a 0, so for the two-input AND gate we can make the following substitutions in the equation for the output X:

$$A = 0, B = 0: \quad X = AB = 0 \cdot 0 = 0$$
$$A = 0, B = 1: \quad X = AB = 0 \cdot 1 = 0$$
$$A = 1, B = 0: \quad X = AB = 1 \cdot 0 = 0$$
$$A = 1, B = 1: \quad X = AB = 1 \cdot 1 = 1$$

The evaluation of this equation simply tells us that the output X of an AND gate is a 1 (HIGH) only when both inputs are 1s (HIGHs). A similar analysis can be made for any number of input variables.

OR

The operation of a two-input OR gate can be expressed in equation form as follows: If one input is A, the other input is B, and the output is X, then the Boolean expression is $X = A + B$. Figure 4–3(a) shows the gate logic symbol, with input and output variables labeled.

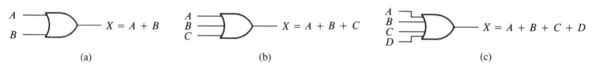

(a)   (b)   (c)

FIGURE 4–3 *Boolean expressions for OR functions.*

To extend the OR expression to more than two input variables, a new letter is used for each additional variable. For instance, the function of a three-input OR gate can be expressed as $X = A + B + C$. The expression for a four-input OR gate can be written as $X = A + B + C + D$, and so on. Figures 4–3(b) and (c) show OR gate logic symbols with three and four input variables, respectively.

OR gate operation can be evaluated using the Boolean expressions for the output X by substituting all possible combinations of 1 and 0 values for the input variables, as shown here for a two-input OR gate:

$$A = 0, B = 0: \quad X = A + B = 0 + 0 = 0$$
$$A = 0, B = 1: \quad X = A + B = 0 + 1 = 1$$
$$A = 1, B = 0: \quad X = A + B = 1 + 0 = 1$$
$$A = 1, B = 1: \quad X = A + B = 1 + 1 = 1$$

This evaluation shows that the output of an OR gate is a 1 (HIGH) when any one or more of the inputs are 1 (HIGH). A similar analysis can be extended to OR gates with any number of input variables.

NAND

The Boolean expression for a two-input NAND gate is $X = \overline{AB}$. This expression says that the two input variables, A and B, are first ANDed and then complemented, as indicated by the bar over the AND expression. This is a logical description in equation form of the operation of a NAND gate with two inputs. If we evaluate this expression for all possible values of the two input variables, the results are as follows:

$$A = 0, B = 0: \quad X = \overline{A \cdot B} = \overline{0 \cdot 0} = \overline{0} = 1$$
$$A = 0, B = 1: \quad X = \overline{A \cdot B} = \overline{0 \cdot 1} = \overline{0} = 1$$
$$A = 1, B = 0: \quad X = \overline{A \cdot B} = \overline{1 \cdot 0} = \overline{0} = 1$$
$$A = 1, B = 1: \quad X = \overline{A \cdot B} = \overline{1 \cdot 1} = \overline{1} = 0$$

Thus, once a Boolean expression is determined for a given logic function, that function can be evaluated for all possible values of the variables. The evaluation tells us exactly what the output of the logic circuit is for each of the input conditions, and therefore, it gives us a complete description of the circuit's logical operation. The NAND expression can be extended to more than two input variables by including additional letters to represent all of the variables.

NOR

Finally, the expression for a two-input NOR gate can be written as $X = \overline{A + B}$. This equation says that the two input variables are first ORed and then complemented, as indicated by the bar over the OR expression. Evaluating this expression, we get the following results:

$$A = 0, B = 0: \quad X = \overline{A + B} = \overline{0 + 0} = \overline{0} - 1$$
$$A = 0, B = 1: \quad X = \overline{A + B} = \overline{0 + 1} = \overline{1} = 0$$
$$A = 1, B = 0: \quad X = \overline{A + B} = \overline{1 + 0} = \overline{1} = 0$$
$$A = 1, B = 1: \quad X = \overline{A + B} = \overline{1 + 1} = \overline{1} = 0$$

SECTION REVIEW 4-2

1. Write the output expression for a five-input AND gate with input variables A, B, C, D, and E.
2. Write the output expression for a five-input OR gate with input variables F, G, H, I, and J.
3. (a) Repeat Problem 1 for a NAND gate.
   (b) Repeat Problem 2 for a NOR gate.

4-3 RULES AND LAWS OF BOOLEAN ALGEBRA

As in other areas of mathematics, there are certain well-developed rules and laws that must be followed in order to properly apply Boolean algebra. The most important of these are presented in this section.

Three of the basic laws of Boolean algebra are the same as in ordinary algebra: the *commutative laws,* the *associative laws,* and the *distributive law.*

Commutative Laws

The cummutative law of addition for two variables is written algebraically as

$$A + B = B + A \tag{4-1}$$

This states that the order in which the variables are ORed makes no difference. Remember, in Boolean algebra terminology as applied to logic circuits, addition and the OR function are the same. Figure 4-4 illustrates the commutative law as applied to the OR gate.

FIGURE 4-4 *Application of commutative law of addition.*

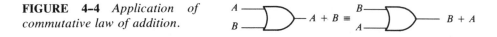

The commutative law of multiplication of two variables is

$$AB = BA \tag{4-2}$$

This states that the order in which the variables are ANDed makes no difference. Figure 4-5 illustrates this law as applied to the AND gate.

FIGURE 4-5 *Application of commutative law of multiplication.*

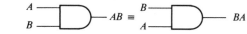

Associative Laws

The associative law of addition is stated as follows for three variables:

$$A + (B + C) = (A + B) + C \tag{4-3}$$

This law states that in the ORing of several variables, the result is the same regardless of the grouping of the variables. Figure 4-6 illustrates this law as applied to OR gates.

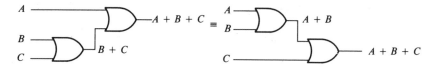

FIGURE 4-6 *Application of associative law of addition.*

The associative law of multiplication is stated as follows for three variables:

$$A(BC) = (AB)C \tag{4-4}$$

This law tells us that it makes no difference in what order the variables are grouped when ANDing several variables. Figure 4-7 illustrates this law as applied to AND gates.

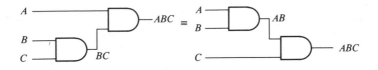

FIGURE 4-7 *Application of associative law of multiplication.*

Distributive Law

The distributive law is written for three variables as follows:

$$A(B + C) = AB + AC \qquad (4\text{-}5)$$

This law states that ORing several variables and ANDing the result with a single variable is equivalent to ANDing the single variable with each of the several variables and then ORing the products. This law and the ones previously discussed should be familiar because they are the same as in ordinary algebra. Keep in mind that each of these laws can be extended to include any number of variables. Figure 4-8 illustrates this law in terms of gate implementation.

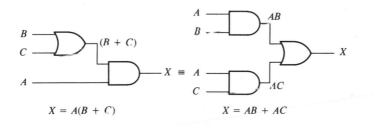

FIGURE 4-8 *Application of distributive law.*

Rules for Boolean Algebra

Table 4-1 lists several basic rules that are useful in manipulating and simplifying Boolean algebra expressions.

We will now look at rules 1 through 9 of Table 4-1 in terms of their application to logic gates. Rules 10 through 12 will be derived in terms of the simpler rules and the laws previously discussed.

You can understand *rule 1* by observing what happens when one input to an OR gate is always 0 and the other input, A, can take on either a 1 or a 0 value. If A is a 1, the output obviously is a 1, which is equal to A. If A is a 0, the output is a 0, which is also equal to A. Therefore, it follows that a variable ORed with a 0 is equal to the value of the variable $(A + 0 = A)$. This rule is further demonstrated in Figure 4-9 where the lower input is fixed at 0.

Rule 2 is demonstrated when one input to an OR gate is always 1 and the other input, A, takes on either a 1 or a 0 value. A 1 on an input to an OR gate

TABLE 4-1 *Basic rules of Boolean algebra.*

1. $A + 0 = A$
2. $A + 1 = 1$
3. $A \cdot 0 = 0$
4. $A \cdot 1 = A$
5. $A + A = A$
6. $A + \overline{A} = 1$
7. $A \cdot A = A$
8. $A \cdot \overline{A} = 0$
9. $\overline{\overline{A}} = A$
10. $A + AB = A$
11. $A + \overline{A}B = A + B$
12. $(A + B)(A + C) = A + BC$

NOTE: *A, B,* or *C* can represent a single variable or a combination of variables.

FIGURE 4-9 *Illustration of rule 1.*

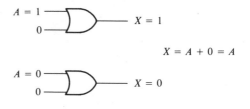

$$X = A + 0 = A$$

produces a 1 on the output, regardless of the value of the variable on the other input. Therefore, a variable ORed with a 1 is always equal to 1 ($A + 1 = 1$). This rule is illustrated in Figure 4–10 where the lower input is fixed at 1.

FIGURE 4-10 *Illustration of rule 2.*

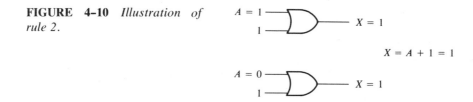

$$X = A + 1 = 1$$

Rule 3 is demonstrated when a 0 is ANDed with a variable. Of course, any time one input to an AND gate is 0, the output is 0, regardless of the value of the variable on the other input. A variable ANDed with a 0 always produces a 0 ($A \cdot 0 = 0$). This rule is illustrated in Figure 4–11 where the lower input is fixed at 0.

Rule 4 can be verified by ANDing a variable with a 1. If the variable *A* is a 0, the output of the AND gate is a 0. If the variable *A* is a 1, the output of the AND gate is a 1 because both inputs are now 1s. Therefore, the AND function of a variable and a 1 is equal to the value of the variable ($A \cdot 1 = A$). This is shown in Figure 4–12 where the lower input is fixed at 1.

FIGURE 4-11 *Illustration of rule 3.*

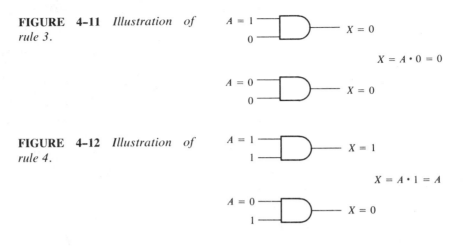

$$X = A \cdot 0 = 0$$

FIGURE 4-12 *Illustration of rule 4.*

$$X = A \cdot 1 = A$$

Rule 5 states that if a variable is ORed with itself, the output is equal to the variable. For instance, if A is a 0, then $0 + 0 = 0$, and if A is a 1, then $1 + 1 = 1$. This is shown in Figure 4-13 where both inputs are the same variable.

FIGURE 4-13 *Illustration of rule 5.*

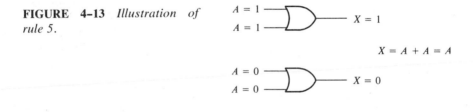

$$X = A + A = A$$

Rule 6 can be explained as follows: If a variable and its complement are ORed, the result is always a 1. If A is a 0, then $0 + \overline{0} = 0 + 1 = 1$. If A is a 1, then $1 + \overline{1} = 1 + 0 = 1$. For further illustration of this rule, see Figure 4-14 where one input is the complement of the other.

FIGURE 4-14 *Illustration of rule 6.*

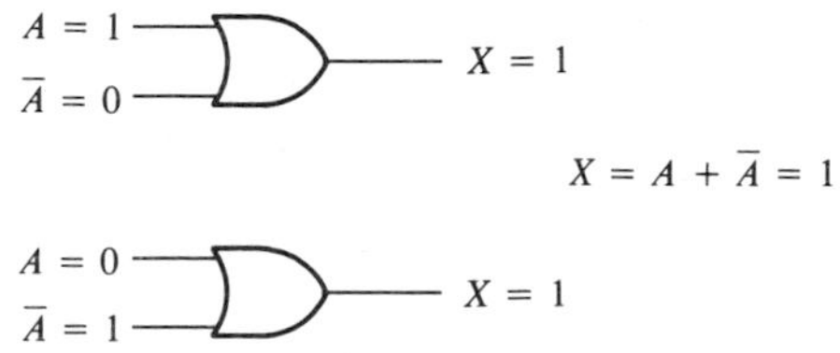

$$X = A + \overline{A} = 1$$

Rule 7 states that if a variable is ANDed with itself, the result is equal to the variable. For example, if $A = 0$, then $0 \cdot 0 = 0$, and if $A = 1$, then $1 \cdot 1 = 1$. For either case, the output of an AND gate is equal to the value of the input variable A. Figure 4-15 illustrates this rule.

FIGURE 4-15 *Illustration of rule 7.*

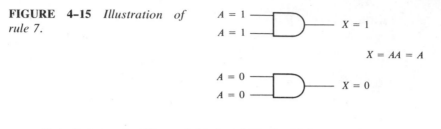

$$X = AA = A$$

Rule 8 states that if a variable is ANDed with its complement, the result is 0. This is readily seen because either A or $\overline{A}$ will always be 0, and when a 0 is applied to the input of an AND gate, it ensures that the output will be 0 also. Figure 4-16 helps illustrate this rule.

FIGURE 4-16 *Illustration of rule 8.*

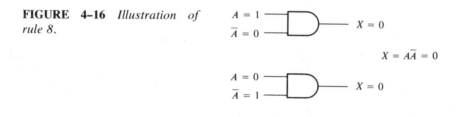

$$X = A\overline{A} = 0$$

Rule 9 simply says that if a variable is complemented twice, the result is the variable itself. If we start with the variable A and complement (invert) it once, we get $\overline{A}$. If we then take $\overline{A}$ and complement (invert) it, we get A, which is the original variable. This is shown in Figure 4-17 using inverters.

FIGURE 4-17 *Illustration of rule 9.*

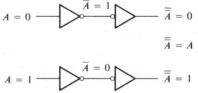

$$\overline{\overline{A}} = A$$

Rule 10 is proved by using the distributive law, rule 2, and rule 4 as follows:

$$
\begin{aligned}
A + AB &= A(1 + B) & &\text{distributive law} \\
&= A \cdot 1 & &\text{rule 2} \\
&= A & &\text{rule 4}
\end{aligned}
$$

Rule 11 is proved as follows:

$$
\begin{aligned}
A + \overline{A}B &= (A + AB) + \overline{A}B & &\text{rule 10} \\
&= (AA + AB) + \overline{A}B & &\text{rule 7} \\
&= AA + AB + A\overline{A} + \overline{A}B & &\text{rule 8 (adding } A\overline{A} = 0\text{)}
\end{aligned}
$$

$$= (A + \overline{A})(A + B) \qquad \text{by factoring}$$
$$= 1 \cdot (A + B) \qquad \text{rule 6}$$
$$= A + B \qquad \text{rule 4}$$

Rule 12 is proved as follows:

$$(A + B)(A + C) = AA + AC + AB + BC \qquad \text{distributive law}$$
$$= A + AC + AB + BC \qquad \text{rule 7}$$
$$= A(1 + C) + AB + BC \qquad \text{distributive law}$$
$$= A \cdot 1 + AB + BC \qquad \text{rule 2}$$
$$= A(1 + B) + BC \qquad \text{distributive law}$$
$$= A \cdot 1 + BC \qquad \text{rule 2}$$
$$= A + BC \qquad \text{rule 4}$$

SECTION REVIEW 4–3

1. Apply the associative law of addition to the expression $A + (B + C + D)$.
2. Apply the distributive law to the expression $A(B + C + D)$.

4-4 DEMORGAN'S THEOREMS

DeMorgan, a logician and mathematician who was acquainted with Boole, proposed two theorems that are an important part of Boolean algebra. They are stated as follows in equation form:

$$\overline{AB} = \overline{A} + \overline{B} \qquad \qquad \textbf{(4–6)}$$
$$\overline{A + B} = \overline{A}\,\overline{B} \qquad \qquad \textbf{(4–7)}$$

The theorem expressed in Equation (4–6) can be stated as follows:

The complement of a product is equal to the sum of the complements.

It really says that the complement of two or more variables ANDed is the same as the OR of the complements of each individual variable.

The theorem expressed in Equation (4–7) can be stated as follows:

The complement of a sum is equal to the product of the complements.

It says that the complement of two or more variables ORed is the same as the AND of the complements of each individual variable.

These theorems are illustrated by the gate equivalencies and truth tables in Figure 4–18.

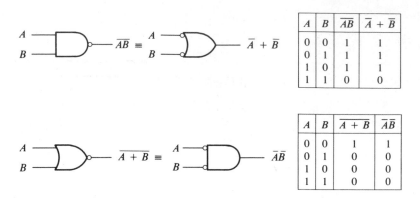

A	B	$\overline{AB}$	$\overline{A} + \overline{B}$
0	0	1	1
0	1	1	1
1	0	1	1
1	1	0	0

A	B	$\overline{A + B}$	$\overline{A}\overline{B}$
0	0	1	1
0	1	0	0
1	0	0	0
1	1	0	0

FIGURE 4–18 *Gate equivalencies and corresponding truth tables illustrating DeMorgan's theorems.*

Applying DeMorgan's Theorems

The following procedure illustrates the application of DeMorgan's theorems using a specific expression:

$$X = \overline{\overline{A + B\overline{C}} + D(\overline{E + \overline{F}})}$$

Step 1. Break the bar over the entire expression and change the sign (+ to ·) between the terms, $\overline{A + B\overline{C}}$ and $D(\overline{E + \overline{F}})$:

$$X = \overline{(A + B\overline{C})}[\overline{D(\overline{E + \overline{F}})}]$$

Step 2. Cancel the double bars over the left term:

$$X = (A + B\overline{C})[\overline{D(\overline{E + \overline{F}})}]$$

Step 3. Break the bar over the term $D(\overline{E + \overline{F}})$ and change the sign (· to +) between D and $\overline{E + \overline{F}}$:

$$X = (A + B\overline{C})[\overline{D} + (\overline{\overline{E + \overline{F}}})]$$

Step 4. Cancel the double bars over the term $E + \overline{F}$:

$$X = (A + B\overline{C})(\overline{D} + E + \overline{F})$$

EXAMPLE 4–1   Express the following complement-of-product terms as sum-of-complements using DeMorgan's theorem:

(a) $\overline{ABC}$ (b) $\overline{ABCD}$ (c) $\overline{ABCDEF}$

Solutions   Use DeMorgan's theorem as expressed in Equation (4–6).

(a) $\overline{ABC} = \overline{A} + \overline{B} + \overline{C}$

(b) $\overline{ABCD} = \overline{A} + \overline{B} + \overline{C} + \overline{D}$

(c) $\overline{ABCDEF} = \overline{A} + \overline{B} + \overline{C} + \overline{D} + \overline{E} + \overline{F}$

EXAMPLE 4–2   Apply DeMorgan's theorems to each expression.

(a) $\overline{\overline{(A + B)} + \overline{C}}$

(b) $\overline{\overline{\overline{(A + B)} + \overline{CD}}}$

(c) $\overline{(A + B)\overline{CD} + E + \overline{F}}$

Solutions

(a) $\overline{\overline{(A + B)} + \overline{C}} = \overline{\overline{(A + B)}} \cdot \overline{\overline{C}} = (A + B)C$

(b) $\overline{\overline{\overline{(A + B)}} + \overline{CD}} = \overline{\overline{\overline{(A + B)}}} \cdot \overline{\overline{CD}} = (\overline{A} + B)CD$

(c) $\overline{(A + B)\overline{CD} + E + \overline{F}} = \overline{[(A + B)\overline{CD}]}\overline{(E + \overline{F})}$

$\qquad\qquad\qquad\qquad\qquad = (\overline{A}\,\overline{B} + C + D)\overline{E}F$

SECTION REVIEW 4–4

1. Apply DeMorgan's theorems to the following expressions:

   (a) $X = \overline{ABC} + \overline{(\overline{D} + E)}$

   (b) $X = \overline{\overline{(A + B)}C}$

   (c) $X = \overline{A + B + C} + \overline{\overline{DE}}$

4–5   BOOLEAN EXPRESSIONS FOR GATE NETWORKS

The form of a given Boolean expression indicates the type of gate network it describes. For example, let us take the expression $A(B + CD)$ and determine what kind of logic circuit it represents. First, there are four variables: A, B, C, and D. C is ANDed with D, giving CD; then CD is ORed with B, giving $(B + CD)$. Then this is ANDed with A to produce the final function. Figure 4–19 illustrates the gate network represented by this particular Boolean expression, $A(B + CD)$.

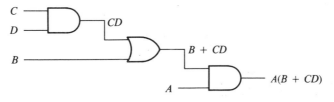

FIGURE 4–19   *Logic gate implementation of the expression $A(B + CD)$.*

As you have seen, the form of the Boolean expression does determine how many logic gates are used, what type of gates are needed, and how they are connected together; this will be explored further in Chapter 5. The more complex an expression, the more complex the gate network will be. It is therefore an advantage to simplify an expression as much as possible in order to have the simplest gate network. We will cover simplification methods in a later section of this chapter. There are also certain forms of Boolean expressions that are more commonly used than others; the two most important of these are the *sum-of-products* and the *product-of-sums* forms.

Sum-of-Products Form

What does the *sum-of-products* form mean? First, let us review *products* in Boolean algebra. A product of two or more variables or their complements is simply the AND function of those variables. The product of two variables can be expressed as *AB,* the product of three variables as *ABC,* the product of four variables as *ABCD,* and so on. Recall that a sum in Boolean algebra is the same as the OR function, so *a sum-of-products expression is two or more AND functions ORed together.* For instance, *AB + CD* is a sum-of-products expression. Several other examples of expressions in sum-of-products form are as follows:

$$AB + BCD$$
$$ABC + DEF$$
$$A\overline{B}C + D\overline{E}FG + AEG$$
$$ABC + \overline{DEF} + FGH + A\overline{F}G$$

A sum-of-products form can also contain a term with a single variable, such as $A + BCD + EFG$.

One reason the sum-of-products is a useful form of Boolean expression is the straightforward manner in which it can be implemented with logic gates. We have AND functions that are ORed, as Example 4–3 illustrates.

EXAMPLE 4–3   Implement the expression $AB + BCD + EFGH$ with logic gates.

Solution

FIGURE 4–20

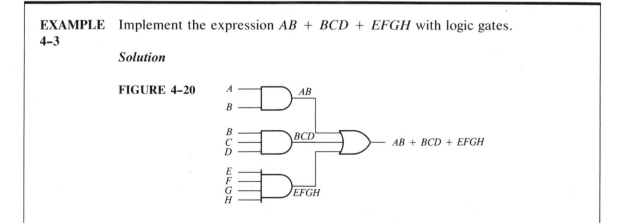

An important characteristic of the sum-of-products form is that the corresponding implementation is always a *two-level* gate network; that is, the maximum number of gates through which a signal must pass in going from an input to the output is *two,* excluding inversions.

Product-of-Sums Form

The product-of-sums form can be thought of as the dual of the sum-of-poducts. It is, in terms of logic functions, the AND of two or more OR functions. For instance, $(A + B)(B + C)$ is a product-of-sums expression. Several other examples are

$$(A + B)(B + C + D)$$
$$(A + B + C)(D + E + F)$$
$$(A + B + C)(D + E + F + G)(A + E + G)$$
$$(A + B + \overline{C})(D + \overline{E} + F)(F + G + H)(A + \overline{F} + G)$$

A product-of-sums expression can also contain a single variable term, such as $A(B + C + D)(E + F + G)$.

This form also lends itself to straightforward implementation with logic gates because it involves simply ANDing two or more OR terms. A two-level gate network will always result, as the following example will show.

EXAMPLE 4-4   Construct the following function with logic gates:

$$(A + B)(C + D + E)(F + G + H + I)$$

Solution

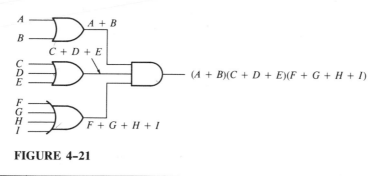

FIGURE 4-21

SECTION
REVIEW
4-5
1.   Identify each expression as either a sum-of-products or a product-of-sums.
     (a) $X = AB + CD + EF$
     (b) $X = (A + B)(C + D)(E + F)$
     (c) $X = \overline{A}BC + A\overline{B}C + ABC + \overline{A}B\overline{C}$
2.   Draw the logic gate diagram for each Boolean expression in Problem 1.

4-6   SIMPLIFICATION OF BOOLEAN EXPRESSIONS

Many times in the application of Boolean algebra, we have to reduce a particular expression to its simplest form or change its form to a more convenient one in order to implement the expression most efficiently. The approach taken in this section is to use the basic laws, rules, and theorems of Boolean algebra to manipulate and simplify an expression. This method depends on a thorough knowledge of Boolean algebra and considerable practice in its application. Examples illustrate the technique.

EXAMPLE 4-5   Simplify the expression $AB + A(B + C) + B(B + C)$ using Boolean algebra techniques.

Solution   The following is not necessarily the only approach.

Step 1.   Apply the distributive law to the *second* and *third* terms in the expression, as follows:

$$AB + AB + AC + BB + BC$$

Step 2.   Apply rule 7 ($BB = B$):

$$AB + AB + AC + B + BC$$

Step 3.   Apply rule 5 ($AB + AB = AB$):

$$AB + AC + B + BC$$

Step 4.   Factor B out of the *last two* terms:

$$AB + AC + B(1 + C)$$

Step 5.   Apply rule 2 ($1 + C = 1$):

$$AB + AC + B \cdot 1$$

Step 6.   Apply rule 4 ($B \cdot 1 = B$):

$$AB + AC + B$$

Step 7.   Factor B out of the *first* and *third* terms, as follows:

$$B(A + 1) + AC$$

Step 8.   Apply rule 2 ($A + 1 = 1$):

$$B \cdot 1 + AC$$

Step 9.   Apply rule 4 ($B \cdot 1 = B$):

$$B + AC$$

At this point we have simplified the expression as much as possible. It should be noted that once you gain experience in applying Boolean algebra, you can combine many individual steps. See if you can find an alternate approach.

EXAMPLE 4-6

Simplify the expression $[A\overline{B}(C + BD) + \overline{A}\,\overline{B}]C$ as much as possible. Note that brackets and parentheses mean the same thing: simply that the terms are multiplied (ANDed) with each other.

Solution

Step 1. Apply the distributive law (multiply out) to the terms within the brackets:

$$(A\overline{B}C + A\overline{B}BD + \overline{A}\,\overline{B})C$$

Step 2. Apply rule 8 to the second term in the parentheses:

$$(A\overline{B}C + A \cdot 0 \cdot D + \overline{A}\,\overline{B})C$$

Step 3. Apply rule 3 to the second term:

$$(A\overline{B}C + 0 + \overline{A}\,\overline{B})C$$

Step 4. Apply rule 1 within the parentheses:

$$(A\overline{B}C + \overline{A}\,\overline{B})C$$

Step 5. Apply the distributive law:

$$A\overline{B}CC + \overline{A}\,\overline{B}C$$

Step 6. Apply rule 7 to the first term:

$$A\overline{B}C + \overline{A}\,\overline{B}C$$

Step 7. Factor out $\overline{B}C$:

$$\overline{B}C(A + \overline{A})$$

Step 8. Apply rule 6:

$$\overline{B}C \cdot 1$$

Step 9. Apply rule 4:

$$\overline{B}C$$

SECTION REVIEW 4-6

1. Simplify the following expressions if possible:
   (a) $A + AB + A\overline{B}C$
   (b) $(\overline{A} + B)C + ABC$

(c) $A\overline{B}C(BD + CDE) + A\overline{C}$

2. Implement each expression in Problem 1 as originally stated with the appropriate logic gates. Then implement the simplified expression and compare the number of gates.

4-7   THE KARNAUGH MAP

Simplification of Boolean Expressions

The Karnaugh map provides a *systematic* method for simplifying a Boolean expression and, if properly used, will produce the simplest sum-of-products expression possible.

As you have seen, the effectiveness of algebraic simplification depends on your familiarity with all the laws, rules, and theorems of Boolean algebra and on your ability to apply them. Cleverness is often an important factor in algebraic simplification. The Karnaugh map, on the other hand, basically provides a "cookbook" approach.

The map format   The Karnaugh map is composed of an arrangement of adjacent "cells," each representing one particular combination of variables in product form. Since the total number of combinations of n variables and their complements is 2^n, the Karnaugh map consists of 2^n cells. For example, there are four combinations of the products of two variables (A and B) and their complements: $\overline{A}\overline{B}$, $\overline{A}B$, $A\overline{B}$, and AB. Therefore, the Karnaugh map must have four cells, with each cell representing one of the variable combinations, as illustrated in Figure 4–22(a). The variable combinations are labeled in the cells only for purposes of illustration. In practice, the map is actually arranged with the variables labeled outside the cells, as shown in Figure 4–22(b). The variable to the left of a row of cells applies to each cell in that row. The variable above a column of cells applies to each cell in that column.

FIGURE 4-22 *Format of a two-variable Karnaugh map.*

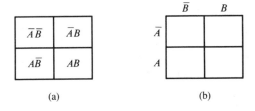

(a)   (b)

Extensions of the Karnaugh map to three and four variables are shown in Figure 4–23. Notice that the cells are arranged such that there is only a *single variable change* between any *adjacent* cells (this is the characteristic that determines adjacency). In Figure 4–23(a), the upper left cell is for $\overline{A}\,\overline{B}\,\overline{C}$, the lower left

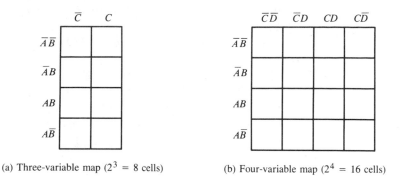

(a) Three-variable map (2^3 = 8 cells) (b) Four-variable map (2^4 = 16 cells)

FIGURE 4-23 *Formats for three- and four-variable Karnaugh maps.*

cell is for $A\overline{B}\,\overline{C}$, and so on. In part (b) of the figure, the upper left cell is $\overline{A}\,\overline{B}\,\overline{C}D$, the lower left cell is $A\overline{B}\,\overline{C}D$, and so on.

Karnaugh maps can be used for five, six, or more variables, but these are not covered in this book. Beyond six variables, the Karnaugh map is quite impractical except when implemented on a computer.

Plotting a Boolean expression Once a Boolean expression is in the *sum-of-products* form, you can "plot" it on the Karnaugh map by placing a 1 in each cell corresponding to a term in the sum-of-products expression. For example, the three-variable expression $\overline{A}BC + AB\overline{C} + ABC$ is plotted on the Karnaugh map in Figure 4-24(a), and the four-variable expression $\overline{A}\,\overline{B}\,\overline{C}D + \overline{A}BCD + A\overline{B}\,\overline{C}D + A\overline{B}CD$ is plotted on the map in part (b).

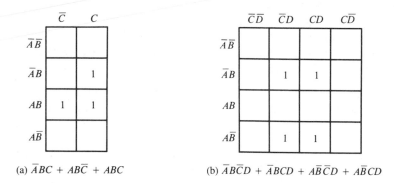

(a) $\overline{A}BC + AB\overline{C} + ABC$ (b) $\overline{A}\,\overline{B}\,\overline{C}D + \overline{A}BCD + A\overline{B}\,\overline{C}D + A\overline{B}CD$

FIGURE 4-24 *Examples of plotting a Boolean expression on a Karnaugh map.*

Grouping cells for simplification You can *group* 1s that are in adjacent cells according to the following rules by drawing a loop around those cells:

1. Adjacent cells are cells that differ by only a single variable (for example, $AB\overline{C}D$ and $ABC\overline{D}$ are adjacent).
2. The 1s in adjacent cells must be combined in groups of 1, 2, 4, 8, 16, and so on.
3. Each group of 1s should be maximized to include the largest number of adjacent cells as possible in accordance with rule 2.
4. Every 1 on the map must be included in at least one group. There can be overlapping groups if they include noncommon 1s.

Grouping is illustrated by an example in Figure 4–25. Notice how the groups are overlapped to include all the 1s in the largest possible group.

FIGURE 4–25 *Example of grouping 1s on a four-variable Karnaugh map.*

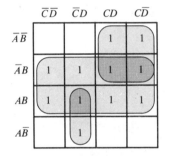

Simplifying the expression When all the 1s representing each term in the original Boolean expression are grouped, the mapped expression is ready for simplification. The following rules apply:

1. Each group of 1s creates a *product term* composed of all variables that appear in only one form (uncomplemented or complemented) within the group. Variables that appear both uncomplemented *and* complemented are eliminated.
2. The final simplified expression is formed by summing the product terms of all the groups.

For example, in Figure 4–26, the product term for the eight-cell group is B because the cells within that group contain both A and $\overline{A}$, C and $\overline{C}$, and D and $\overline{D}$,

FIGURE 4–26 *This plotted Boolean expression simplifies to $B + \overline{A}C + A\overline{C}D$.*

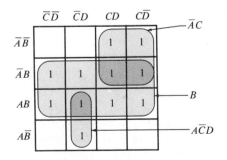

so these variables are eliminated. The four-cell group contains B, $\overline{B}$, D, and $\overline{D}$, leaving the product term $\overline{A}C$. The two-cell group contains B and $\overline{B}$, leaving $A\overline{C}D$ as the product term. The resulting Boolean expression is the sum of these product terms: $B + \overline{A}C + A\overline{C}D$.

EXAMPLE 4-7

Minimize the expression $X = A\overline{B}\overline{C} + \overline{A}BC + \overline{A}\,\overline{B}C + \overline{A}\,\overline{B}\,\overline{C} + A\overline{B}C$.

Solution Notice that this expression is already in a sum-of-products form from which the 1s can be plotted very easily, as shown in Figure 4–27.

FIGURE 4-27

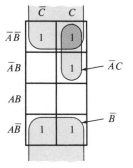

Four of the 1s appearing in adjacent cells can be grouped. The remaining 1 is absorbed in an overlapping group. The group of four 1s produces a single variable term, $\overline{B}$. This is determined by observing that within the group, $\overline{B}$ is the only variable that does not change from cell to cell. The group of two 1s produces a two-variable term, $\overline{A}C$. This is determined by observing that within the group, the variables $\overline{A}$ and C do not change from one cell to the next. To get the minimized function, the two terms that are produced are summed (ORed) as $X = \overline{B} + \overline{A}C$.

EXAMPLE 4-8

Reduce the following four-variable function to its minimum sum-of-products form:

$$X = \overline{A}\,\overline{B}\,\overline{C}\,\overline{D} + \overline{A}B\overline{C}\,\overline{D} + AB\overline{C}\,\overline{D} + A\overline{B}\,\overline{C}\,\overline{D} + \overline{A}\,\overline{B}CD$$
$$+ A\overline{B}CD + \overline{A}\,\overline{B}C\overline{D} + \overline{A}BC\overline{D} + ABC\overline{D} + A\overline{B}C\overline{D}$$

If all variables and their complements were available, this function would take ten 4-input AND gates and one 10-input OR gate to implement.

Solution A group of eight 1s can be factored as shown in Figure 4–28 because the 1s in the outer columns are adjacent. A group of four 1s is formed by the

FIGURE 4–28

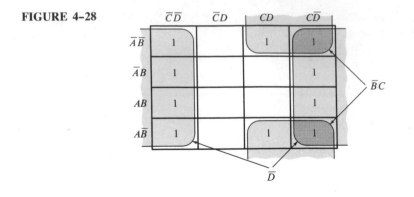

"wrap-around" adjacency of the cells to pick up the remaining two 1s. The minimum form of the original equation is $X = \overline{D} + \overline{B}C$.

$X = \overline{D} + \overline{B}C$ takes one 2-input AND gate and one 2-input OR gate. Compare this to the implementation of the original function.

EXAMPLE 4–9

Reduce the following function to its minimum sum-of-products form:

$$X = \overline{A}\,\overline{B}\,\overline{C}\,\overline{D} + \overline{A}\,\overline{B}CD + A\overline{B}\,\overline{C}\,\overline{D} + \overline{A}CD + A\overline{B}C\overline{D}$$

Solution The function is plotted on the four-variable map and factored as indicated in Figure 4–29. Notice that the four corner cells are adjacent.

FIGURE 4–29

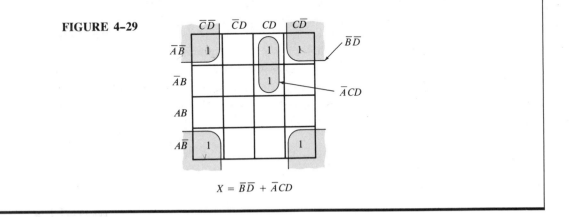

$$X = \overline{B}\,\overline{D} + \overline{A}CD$$

Implementing Truth Table Functions

Another use of the Karnaugh map is in implementing logic functions that are specified in truth table format. Since truth tables normally use 1s and 0s (HIGHs and LOWs) to represent logic states, the format of the Karnaugh map is modified slightly for this application, as shown in Figure 4–30 for two, three, and four variables; 1s and 0s are used to label the cells. The variables represented by the 1s and 0s are specified in the upper left corner of the map.

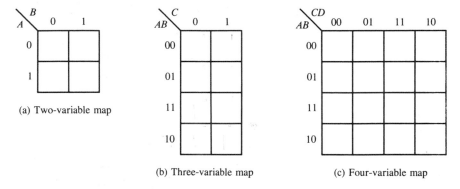

(a) Two-variable map

(b) Three-variable map

(c) Four-variable map

FIGURE 4–30 *Karnaugh maps with a 1/0 labeling format.*

Plotting the truth table   When the output variable on the truth table is a 1, a 1 is placed on the Karnaugh map in the cell corresponding to the states of the input variables. For example, if the output variable is a 1 when $A = 1$, $B = 0$, and $C = 1$, then a 1 is placed in the lower right cell of the three-variable map.

Grouping and simplification   The previously stated rules apply in grouping the 1s. Variables that are both 1 and 0 within a group are eliminated. If a variable is a 1 in all cells of a group, it appears uncomplemented in the product term. If a variable is a 0 in all cells of a group, it appears complemented in the product term.

EXAMPLE 4-10   Implement the logic function specified by the truth table in Figure 4–31 using the Karnaugh map method. X is the output variable, and A, B, and C are the input variables.

FIGURE 4–31

Inputs			Output
A	B	C	X
0	0	0	1
0	0	1	0
0	1	0	0
0	1	1	0
1	0	0	1
1	0	1	0
1	1	0	1
1	1	1	1

Solution The truth table function is plotted and simplified on the map in Figure 4–32(a). The logic gate implementation for the resulting sum-of-products expression is shown in part (b) of the figure.

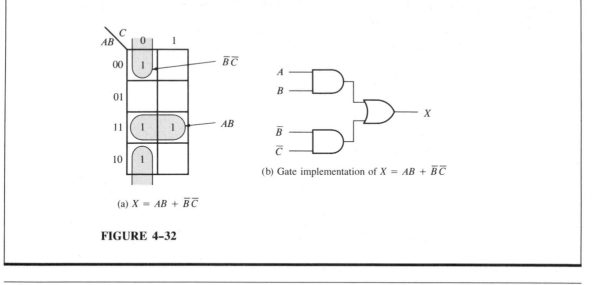

(a) $X = AB + \overline{B}\,\overline{C}$

(b) Gate implementation of $X = AB + \overline{B}\,\overline{C}$

FIGURE 4–32

---

1. Lay out a Karnaugh map for each of the following numbers of variables:
   (a) Two variables   **(b)** Three variables   **(c)** Four variables
2. Group the 1s and write the simplified expression for each Karnaugh map in Figure 4–24.
3. Write the original sum-of-products expression that is plotted on the Karnaugh map in Figure 4–26.

SUMMARY

☐ Boolean multiplication is the same as ANDing.
☐ Boolean addition is the same as ORing.
☐ A bar over a variable means complementation.
☐ Complementation is the same as inversion.
☐ The three basic laws of Boolean algebra are the commutative law, the associative law, and the distributive law.
☐ DeMorgan's theorems state that (1) *the complement of a product equals the sum of the complements* and (2) *the complement of a sum equals the product of the complements.*
☐ There are two basic forms of Boolean expressions: the *sum-of-products* and the *product-of-sums.*
☐ Boolean expressions can be simplified using the algebraic method or the Karnaugh map method.

SELF-TEST

1. (a) AB means that the two variables are _____.
    (b) $A + B$ means that the two variables are _____.
    (c) $\overline{A}$ means that the variable is _____.
2. Determine the logic gate required to implement each of the following terms:
    (a) ABC (b) $A + B + C$ (c) $\overline{ABC}$ (d) $\overline{A + B + C}$
3. Apply the commutative law, the associative law, or the distributive law to each expression as appropriate.
    (a) $A + B$ (b) AB (c) CD (d) $(A + B) + C$
    (e) $A(B + C + D)$ (f) $A(BCD)$
4. Use the basic rules of Boolean algebra to evaluate each of the following expressions:
    (a) $B + 1$ (b) $0 + B$ (c) $B \cdot 1$ (d) $B \cdot B$
    (e) $C + C$ (f) $D \cdot D$ (g) $\overline{\overline{C}}$ (h) $B + BC$
5. According to _____, $\overline{CD} = \overline{C} + \overline{D}$.
6. How many and what type of logic gates are required to implement the following expressions as they appear:
    (a) $AB + CD$ (b) $A + B + CD$
    (c) $A(B + C + DE)$
7. (a) The expression $ABC + \overline{A}BC + AB\overline{C}$ is an example of the _____ form.
    (b) The expression $(A + \overline{B} + C)(\overline{A} + B + C)(A + B + C)$ is an example of the _____ form.
8. Determine whether or not the following equalities are correct:
    (a) $ABC + A\overline{BC} = A$
    (b) $A + BC + \overline{A}C = BC$
    (c) $A(\overline{A}BC + ABCD) = ABCD$

PROBLEMS

Section 4–2

4–1 Find the value of X for all possible values of the variables.
    (a) $X = AB$ (b) $X = ABC$ (c) $X = A + B$
    (d) $X = A + B + C$ (e) $X = AB + C$ (f) $X = \overline{A} + B$
    (g) $X = A\overline{B}\overline{C}$ (h) $X = AB + \overline{A}C$ (i) $X = A(B + C)$
    (j) $X = \overline{A}(\overline{B} + \overline{C})$

4–2 Find the value of X for all possible values of the variables.
    (a) $X = (A + B)C + B$ (b) $X = (\overline{A} + B)C$
    (c) $X = A\overline{B}C + AB$ (d) $X = (A + B)(\overline{A} + B)$
    (e) $X = (A + BC)(\overline{B} + \overline{C})$

4–3   Write the Boolean expression for each of the logic gates in Figure 4–33.

FIGURE 4–33

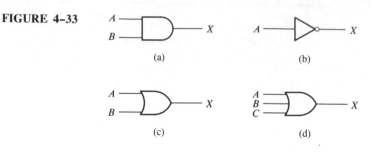

(a)

(b)

(c)

(d)

4–4   Repeat Problem 4–3 for the circuits in Figure 4–34.

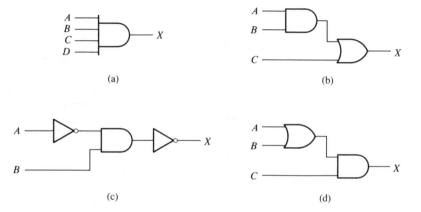

(a)

(b)

(c)

(d)

FIGURE 4–34

4–5   Construct a truth table for each of the following Boolean expressions:
(a) $A + B$ (b) AB (c) $AB + BC$
(d) $(A + B)C$ (e) $(A + B)(\overline{B} + C)$

Section 4–4

4–6   Apply DeMorgan's theorems to each expression.

(a) $\overline{A + \overline{B}}$ (b) $\overline{\overline{A}B}$ (c) $\overline{A + B + C}$
(d) $\overline{ABC}$ (e) $\overline{A(B + C)}$ (f) $\overline{AB + CD}$
(g) $\overline{AB + CD}$ (h) $\overline{(A + B)(\overline{C} + D)}$

4–7   Apply DeMorgan's theorems to each expression.

(a) $\overline{A\overline{B}(C + \overline{D})}$ (b) $\overline{AB(CD + EF)}$
(c) $\overline{(A + \overline{B} + C + \overline{D}) + ABC\overline{D}}$ (d) $\overline{(\overline{A} + B + C + D)(A\overline{B}\overline{C}D)}$
(e) $\overline{A\overline{B}(CD + \overline{E}F)(\overline{AB} + \overline{CD})}$

4-8 Apply DeMorgan's theorems to the following:

(a) $\overline{\overline{(ABC)}\overline{(EFG)} + \overline{(HIJ)}\overline{(KLM)}}$ (b) $\overline{(A + B\overline{C} + CD) + \overline{\overline{BC}}}$

(c) $\overline{\overline{(A + B)}\overline{(C + D)}\overline{(E + F)}\overline{(G + H)}}$

Section 4–5

4-9 Convert the following expressions to sum-of-product forms:

(a) $(A + B)(C + D)$ (b) $(A + \overline{B}C)D$

(c) $(A + C)(ABC + ACD)$

4-10 Convert the following expressions to sum-of-product forms:

(a) $AB + CD(A\overline{B} + CD)$ (b) $AB(\overline{B}\,\overline{C} + BC)$

(c) $A + B[AC + (B + \overline{C})D]$

4-11 Write an expression for each of the gate networks in Figure 4–35 and identify the form.

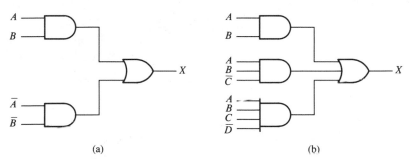

(a)

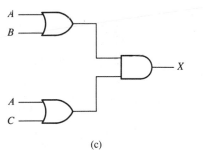

(c)

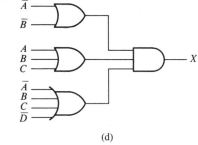

(b)

(d)

FIGURE 4–35

4-12 Write a Boolean expression for the following statement:
X is a 1 only if A is a 1 and B is a 1
or if A is a 0 and B is a 0.

4-13 Write a Boolean expression for the following statement:
X is a 1 only if A, B, and C are all 1s
or if only one of the variables is a 0.

4-14 Write a Boolean expression for the following conditions:
X is a 0 if any two of the three variables A, B, and C are 1s. X is a 1 for all other conditions.

4-15 Draw the logic circuit represented by each of the following expressions:
    (a) $A + B + C$ **(b)** ABC **(c)** $AB + C$
    (d) $AB + CD$ **(e)** $\overline{AB}(C + \overline{D})$

4-16 Draw the logic circuit represented by each expression.
    (a) $A\overline{B} + \overline{A}B$ **(b)** $AB + \overline{A}\overline{B} + \overline{A}BC$
    (c) $A + B[C + D(B + \overline{C})]$

Section 4-6

4-17 Using Boolean algebra techniques, simplify the following expressions as much as possible:
    (a) $A(A + B)$ **(b)** $A(\overline{A} + AB)$ **(c)** $BC + \overline{B}\overline{C}$
    (d) $A(A + \overline{A}B)$ **(e)** $A\overline{B}C + \overline{A}BC + \overline{A}\,\overline{B}C$

4-18 Using Boolean algebra, simplify the following expressions:
    (a) $(A + \overline{B})(A + C)$ **(b)** $\overline{A}B + \overline{A}B\overline{C} + \overline{A}BCD + \overline{A}B\overline{C}DE$
    (c) $AB + \overline{A}BC + A$ **(d)** $(A + \overline{A})(AB + AB\overline{C})$
    (e) $AB + (\overline{A} + \overline{B})C + AB$

4-19 Using Boolean algebra, simplify each expression.
    (a) $BD + B(D + E) + \overline{D}(D + F)$
    (b) $\overline{A}\,\overline{B}C + \overline{(A + B + \overline{C})} + \overline{A}\,\overline{B}\,\overline{C}D$
    (c) $(B + BC)(B + \overline{B}C)(B + D)$
    (d) $ABCD + AB(\overline{CD}) + (\overline{AB})CD$
    (e) $ABC[AB + \overline{C}(BC + AC)]$

Section 4-7

4-20 Using the Karnaugh map method, simplify the following expressions to their minimum sum-of-products form:
    (a) $X = \overline{A}\,\overline{B} + A\overline{B}$ **(b)** $X = \overline{A}\,\overline{B} + \overline{A}B$
    (c) $X = A\overline{B} + AB$ **(d)** $X = \overline{A}\,\overline{B} + A\overline{B} + AB$
    (e) $X = A(\overline{B} + AB)$ **(f)** $X = \overline{A}\,\overline{B} + AB$

4-21 Use a Karnaugh map to find the minimum sum-of-products form for each expression.
    (a) $X = \overline{A}\,\overline{B}\,\overline{C} + \overline{A}\,B\overline{C} + A\overline{B}C$
    (b) $X = AC(\overline{B} + C)$
    (c) $X = \overline{A}(BC + B\overline{C}) + A(BC + B\overline{C})$
    (d) $X = \overline{A}\,\overline{B}\,\overline{C} + A\overline{B}\,\overline{C} + \overline{A}B\overline{C} + AB\overline{C}$
    (e) $X = A + \overline{B}C$

4-22 Use a Karnaugh map to simplify each function to a minimum sum-of-products form.
    (a) $X = \overline{A}\,\overline{B}\,\overline{C} + A\overline{B}C + \overline{A}BC + AB\overline{C}$
    (b) $X = AC[\overline{B} + A(B + \overline{C})]$

(c) $X = DE\overline{F} + \overline{DE}\overline{F} + \overline{D}\overline{E}\overline{F}$

4–23 Use a Karnaugh map to reduce each expression to a minimum sum-of-products form.

(a) $X = A + B\overline{C} + CD$

(b) $X = \overline{A}\overline{B}C\overline{D} + \overline{A}BC\overline{D} + ABCD + ABC\overline{D}$

(c) $X = \overline{A}B(\overline{C}D + \overline{C}D) + AB(\overline{C}D + \overline{C}D) + A\overline{B}CD$

(d) $X = (\overline{A}\overline{B} + A\overline{B})(CD + C\overline{D})$

(e) $X = \overline{A}\overline{B} + A\overline{B} + \overline{C}D + CD$

4–24 Use the Karnaugh map method to implement the logic function specified in the following truth table:

A	B	C	D	X
0	0	0	0	0
0	0	0	1	1
0	0	1	0	1
0	0	1	1	0
0	1	0	0	0
0	1	0	1	0
0	1	1	0	1
0	1	1	1	1
1	0	0	0	1
1	0	0	1	0
1	0	1	0	1
1	0	1	1	0
1	1	0	0	1
1	1	0	1	1
1	1	1	0	0
1	1	1	1	1

ANSWERS TO SECTION REVIEWS

Section 4–1

1. OR   **2.** AND   **3.** $X = AB; X = A + B$

Section 4–2

1. $X = ABCDE$ **2.** $Y = F + G + H + I + J$

3. (a) $X = \overline{ABCDE}$ **(b)** $Y = \overline{F + G + H + I + J}$

Section 4–3

1. $(A + B + C) + D$ **2.** $AB + AC + AD$

Section 4–4

1. (a) $\overline{A} + \overline{B} + \overline{C} + D\overline{E}$ **(b)** $\overline{A}\overline{B} + \overline{C}$ **(c)** $\overline{A}\overline{B}\overline{C} + D + \overline{E}$

Section 4–5

1. **(a)** Sum-of-products   **(b)** Product-of-sums
   (c) Sum-of-products
2. **(a)** Three AND gates, one OR gate   **(b)** Three OR gates, one AND gate
   (c) Four AND gates, one OR gate

Section 4–6

1. **(a)** A **(b)** $C(\overline{A} + B)$ **(c)** $A(\overline{C} + \overline{B}DE)$
2. **(a)** Original: Three gates. Simplified: No gates (straight connection).
   (b) Original: Four gates. Simplified: Two gates.
   (c) Original: Seven gates. Simplified: Three gates.

Section 4–7

1. **(a)** Four cells   **(b)** Eight cells   **(c)** Sixteen cells
2. **(a)** $X = AB + \overline{B}\overline{C}$ **(b)** $X = \overline{A}BD + A\overline{B}D$
3. $X = \overline{A}\,\overline{B}CD + \overline{A}\,\overline{B}\overline{C}D + \overline{A}BC\overline{D} + \overline{A}BCD + \overline{A}B\overline{C}D + \overline{A}B\overline{C}\overline{D} + AB\overline{C}\,\overline{D} + AB\overline{C}D$
   $+\ ABCD + AB\overline{C}D + A\overline{B}CD$

In Chapter 3, logic gates were studied on an individual basis and in simple combinations. When logic gates are connected together to produce a specified output for certain specified combinations of input variables, with no storage involved, the resulting network is called *combinational logic*. In combinational logic, the output level is at all times dependent on the combination of input levels. In this chapter, methods of analysis and design of combinational logic circuits are examined.

Specific devices introduced in this chapter are 74H52 AND-OR logic, 7451 AND-OR-INVERT logic, and 7486 quad exclusive-OR gates.

In this chapter, you will learn

☐ How to analyze basic combinational logic circuits such as AND-OR, AND-OR-INVERT, exclusive-OR, and generalized logic circuits.

☐ Why the exclusive-OR gate is important in certain arithmetic operations.

☐ How to derive the truth table from a logic equation of the function.

☐ How to design combinational logic circuits starting with a truth table or an equation.

☐ How to simplify combinational logic to reduce the number of gates.

☐ How to use NAND gates to construct any logic function.

☐ How to use NOR gates to construct any logic function.

☐ How to analyze multilevel NAND and NOR logic.

☐ How to analyze combinational logic with pulse waveform inputs using timing diagrams.

☐ How to troubleshoot combinational logic using the logic probe, pulser, and current tracer.

5

Combinational Logic

5-1   ANALYSIS OF COMBINATIONAL LOGIC CIRCUITS

AND-OR Logic

Figure 5–1 shows a combinational logic circuit consisting of two AND gates and one OR gate. Each of the three gates has two input variables as indicated.

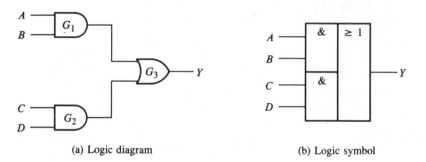

(a) Logic diagram   (b) Logic symbol

FIGURE 5-1   *AND-OR logic (ANSI/IEEE Std. 91–1984).*

Each of the input variables can be either a HIGH (1) or a LOW (0). Because there are four input variables, there are sixteen possible combinations of the input variables ($2^4 = 16$). To illustrate an analysis procedure, we will assign one of the sixteen possible input combinations and see what the corresponding output value is.

First, make each input variable a LOW, and examine the output of each gate in the network in order to arrive at the final output, Y. If the inputs to gate G_1 are both LOW, the output of gate G_1 is LOW. Also, the output of gate G_2 is LOW because its inputs are LOW. As a result of the LOWs on the outputs of gates G_1 and G_2, both inputs to gate G_3 are LOW, and therefore its output is LOW. We have determined that the output function of the logic circuit of Figure 5–1 is LOW when all of its inputs are LOW. This condition is illustrated in the first row of Table 5–1 along with the remaining fifteen input combinations. You should verify each of these conditions on the logic diagram.

Now, as a second method of analyzing the logical operation of the circuit of Figure 5–1, we can develop a logic equation for the output function and, using Boolean algebra, evaluate the equation for each of the sixteen combinations of input variables.

Since gate G_1 is an AND gate and its two inputs are A and B, its output can be expressed as AB. Gate G_2 is an AND gate and its two inputs are C and D, so its output can be expressed as CD. Gate G_3 is an OR gate and its two inputs are the outputs of gates G_1 and G_2, so its output can be expressed as $AB + CD$. The output of gate G_3 is the output function of the logic network, so $Y = AB + CD$. Figure 5–2 shows the logic functions at each point in the circuit.

TABLE 5-1 *Truth table for Figure 5-1 (1 = HIGH, 0 = LOW).*

	Inputs			G_1 Output	G_2 Output	G_3 Output
A	B	C	D	AB	CD	Y
0	0	0	0	0	0	0
0	0	0	1	0	0	0
0	0	1	0	0	0	0
0	0	1	1	0	1	1
0	1	0	0	0	0	0
0	1	0	1	0	0	0
0	1	1	0	0	0	0
0	1	1	1	0	1	1
1	0	0	0	0	0	0
1	0	0	1	0	0	0
1	0	1	0	0	0	0
1	0	1	1	0	1	1
1	1	0	0	1	0	1
1	1	0	1	1	0	1
1	1	1	0	1	0	1
1	1	1	1	1	1	1

FIGURE 5-2 *AND-OR logic with Boolean expressions.*

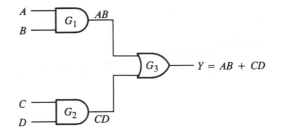

We can now evaluate the output equation by substituting into it the various combinations of input variable values. For example, when $A = 1, B = 1, C = 1$, and $D = 0$, the output expression is evaluated as follows:

$$Y = AB + CD = 1 \cdot 1 + 1 \cdot 0 = 1 + 0 = 1$$

The same procedure can be used for any of the other input variable combinations.

EXAMPLE 5-1

The 74H52 TTL AND-OR IC with pin numbers in parentheses is shown in Figure 5-3. Write the output expression and determine the logic level at pin 8 for the following input logic levels: pins 3, 4, 13, and 11 are HIGH, and pins 1, 2, 5, 9, 10, and 12 are LOW.

FIGURE 5-3   *74H52 AND-OR with expander.*

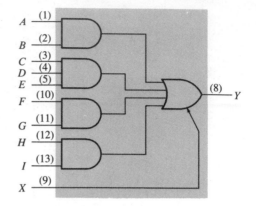

Solution   The output expression is $Y = AB + CDE + FG + HI + X$. The X input is for the connection of an *expander* AND gate when extra inputs must be handled; X is from the output of the expander gate.

The output expression is evaluated for the given input levels as follows:

$$Y = AB + CDE + FG + HI + X = 0 \cdot 0 + 1 \cdot 1 \cdot 0 + 0 \cdot 1 + 0 \cdot 1 + 0$$
$$= 0 + 0 + 0 + 0 + 0 = 0$$

AND-OR-INVERT Logic

Figure 5-4(a) shows a combinational logic circuit consisting of two AND gates, one OR gate, and an inverter. As you can see, the operation is the same as that of the AND-OR circuit in Figure 5-1 except that the output is inverted. The output expression is $Y = \overline{AB + CD}$. An evaluation of this for the inputs $A = 1$, $B = 1$, $C = 1$, $D = 0$ is as follows:

$$Y = \overline{AB + CD} = \overline{1 \cdot 1 + 1 \cdot 0} = \overline{1 + 0} = \overline{1} = 0$$

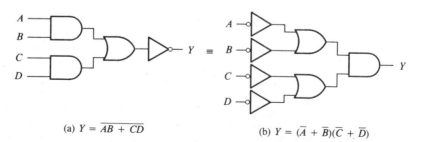

(a) $Y = \overline{AB + CD}$ (b) $Y = (\overline{A} + \overline{B})(\overline{C} + \overline{D})$

FIGURE 5-4   *AND-OR-INVERT circuit and one of its equivalent circuits.*

Notice that DeMorgan's theorem can be applied to this output expression as follows:

$$Y = \overline{AB + CD} = (\overline{A} + \overline{B})(\overline{C} + \overline{D})$$

The resulting equivalent circuit is shown in Figure 5–4(b).

EXAMPLE 5-2

The 7451 dual AND-OR-INVERT circuit is represented by a logic diagram and symbol in Figure 5–5. Notice that the inversion is indicated by a bubble on the output of the OR gate, effectively making it a NOR gate.

The following levels are applied to the inputs: pins 1, 2, 4, and 13 are HIGH, and pins 3, 5, 9, and 10 are LOW. Determine the output levels and indicate the pin designations.

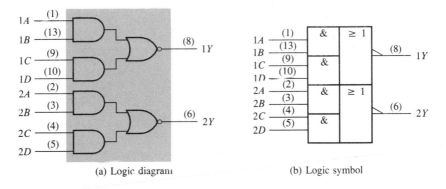

(a) Logic diagram   (b) Logic symbol

FIGURE 5–5 *The 7451 dual AND-OR-INVERT.*

Solution   The 1Y output at pin 8 is LOW because both pins 1 and 13 are HIGH. The output 2Y at pin 6 is HIGH because neither AND gate in this circuit has both of its inputs HIGH.

Exclusive-OR Logic

Figure 5–6(a) shows a combinational logic circuit known as an *exclusive-OR* (often abbreviated XOR). It is a widely used function because of special arithmetic properties which will be discussed in a later chapter. Because of its wide applications, it has a special symbol, shown in Figures 5–6(b) and (c).

The output expression for this circuit is $Y = A\overline{B} + \overline{A}B$. Evaluation of this expression results in the truth table in Table 5–2. *Notice that the output is HIGH only when the two inputs are at opposite levels.* A special exclusive-OR symbol,

$\oplus$, is often used so that the expression $Y = A\overline{B} + \overline{A}B$ can also be written as $Y = A \oplus B$, stated as "$Y = A$ exclusive-OR B."

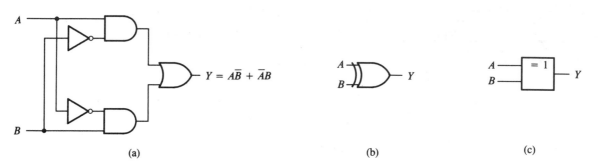

FIGURE 5–6 *Exclusive-OR logic diagram and symbols.*

TABLE 5–2 *Truth table for an exclusive-OR.*

A	B	Y
0	0	0
0	1	1
1	0	1
1	1	0

EXAMPLE 5–3 The 7486 quad exclusive-OR gate IC is shown in Figure 5–7. Certain pins are connected *externally* to positive logic levels indicated by 1s and 0s on the inputs. Determine the output at pin 6 for these conditions.

FIGURE 5–7 *7486 quad exclusive-OR.*

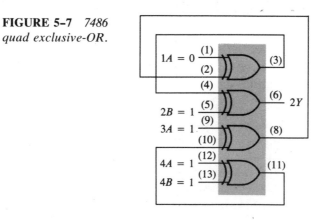

Solution The final output on pin 6 is 0 (LOW). We determined this by tracing through the logic with the use of the exclusive-OR truth table. The intermediate levels are as follows: pin 11 = 0, pin 8 = 1, and pin 3 = 1.

The complement of the exclusive-OR function is the *exclusive-NOR* which is derived as follows:

$$X = \overline{A\overline{B} + \overline{A}B} = (\overline{A\overline{B}})(\overline{\overline{A}B}) = (\overline{A} + B)(A + \overline{B}) = \overline{A}\,\overline{B} + AB$$

Notice that *the output X is HIGH only when the two inputs, A and B, are at the same level.*

A General Combinational Logic Circuit

Figure 5-8 is the fourth combinational logic circuit that we will analyze. Unlike the previous circuits, it is not available as a specific IC device but is a general combination of gates. As before, the approach to analysis is to develop the logic expression for the circuit and evaluate that expression for all of the possible input conditions. In this case, there are four input variables with sixteen combinations of input states.

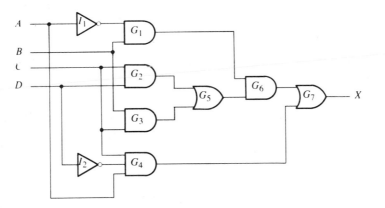

FIGURE 5-8 *A combinational logic circuit.*

The logic expression for Figure 5-8 is developed step-by-step as follows:

$$\text{Output of inverter } I_1 = \overline{A}$$
$$\text{Output of gate } G_1 = \overline{A}B$$
$$\text{Output of gate } G_2 = CD$$
$$\text{Output of gate } G_3 = BC$$
$$\text{Output of inverter } I_2 = \overline{D}$$

Output of gate $G_4 = AC\overline{D}$

Output of gate G_5 = (output of G_2) + (output of G_3)

$\qquad = CD + BC$

Output of gate G_6 = (output of G_1) · (output of G_5)

$\qquad = \overline{A}B(CD + BC)$

Final output X = output of gate G_7

$\qquad$ = (output of G_4) + (output of G_6)

$\qquad = AC\overline{D} + \overline{A}B(CD + BC)$

$\qquad = AC\overline{D} + \overline{A}BCD + \overline{A}BC$

Figure 5–9 illustrates the logic expressions at each point in the circuit. Notice that the expression for the output X can be converted to a sum-of-products form as indicated.

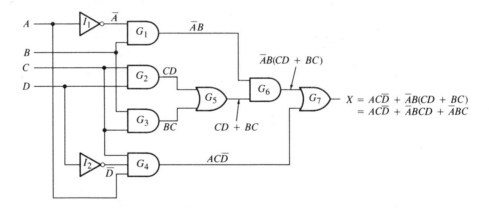

FIGURE 5–9

We will now discuss how to develop a truth table fully describing the circuit's operation from the output expression.

Expansion of the Output Expression

The output sum-of-products expression of a logic circuit can be expanded to include *all* variables in *each* term. The truth table can then be written directly from the expanded expression.

Notice that the expression $X = AC\overline{D} + \overline{A}BCD + \overline{A}BC$ in Figure 5–9 has two terms that do not include all four variables. Using Boolean algebra, expand each of the three-variable terms to include all four variables as follows:

$$AC\overline{D} = ABC\overline{D} + A\overline{B}C\overline{D}$$

$$\overline{A}BC = \overline{A}BCD + \overline{A}BC\overline{D}$$

The expanded output expression is

$$X = ABC\overline{D} + A\overline{B}C\overline{D} + \overline{A}BCD + \overline{A}BCD + \overline{A}BC\overline{D}$$

There are two $\overline{A}BCD$ terms, so one is dropped. The result is

$$X = ABC\overline{D} + A\overline{B}C\overline{D} + \overline{A}BCD + \overline{A}BC\overline{D}$$

The Karnaugh map can also be used to systematically expand a sum-of-products expression by the following steps:

Step 1. For each term in the expression, place a 1 in each cell where the variables in that term appear.

Step 2. Write the product term for each *individual* cell containing a 1 (do not group the cells).

Step 3. Write the sum of all the product terms.

To illustrate this procedure, the previous expression $X = AC\overline{D} + \overline{A}BCD + \overline{A}BC$ is expanded using the sixteen-cell (four-variable) map in Figure 5–10. In part (a) of the figure, the three terms in the expression are expanded on the map. In part (b), the four-variable terms are written for each 1, and the expanded expression is developed. Notice that it is the same as was obtained by Boolean expansion (the order is different, but that does not matter).

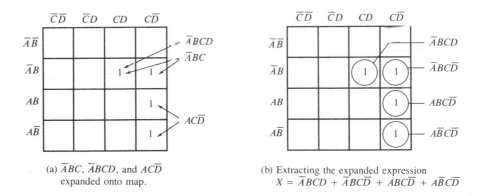

(a) $\overline{A}BC$, $\overline{A}BCD$, and $AC\overline{D}$ expanded onto map.

(b) Extracting the expanded expression
$X = \overline{A}BCD + \overline{A}BC\overline{D} + ABC\overline{D} + A\overline{B}C\overline{D}$

FIGURE 5–10 *Sum-of-products expansion by Karnaugh map.*

Deriving the Truth Table from the Expanded Expression

The expanded output expression for a logic circuit shows *all* the combinations of the input variables that make the output a 1. An uncomplemented input variable stands for a 1, and a complemented input variable stands for a 0. For the expression $X = ABC\overline{D} + A\overline{B}C\overline{D} + \overline{A}BCD + \overline{A}BC\overline{D}$, $X = 1$ for the four input variable combinations and $X = 0$ for all other input combinations as shown in Table 5–3.

TABLE 5-3 *Truth table for the circuit of Figure 5–9.*

Inputs				Output
A	*B*	*C*	*D*	*X*
0	0	0	0	0
0	0	0	1	0
0	0	1	0	0
0	0	1	1	0
0	1	0	0	0
0	1	0	1	0
0	1	1	0	1
0	1	1	1	1
1	0	0	0	0
1	0	0	1	0
1	0	1	0	1
1	0	1	1	0
1	1	0	0	0
1	1	0	1	0
1	1	1	0	1
1	1	1	1	0

SECTION REVIEW 5-1

1. Determine the output (1 or 0) of a four-variable AND-OR-INVERT circuit for each of the following input conditions:
   (a) $A = 1, B = 0, C = 1, D = 0$
   (b) $A = 1, B = 1, C = 0, D = 1$
   (c) $A = 0, B = 1, C = 1, D = 1$
2. Determine the output (1 or 0) of an exclusive-OR gate for each of the following input conditions:
   (a) $A = 1, B = 0$ (b) $A = 1, B = 1$
   (c) $A = 0, B = 1$ (d) $A = 0, B = 0$
3. Develop the truth table for a certain three-input logic circuit with the output expression $X = A\overline{B}C + \overline{A}BC + \overline{A}\,\overline{B}\,\overline{C} + AB\overline{C} + ABC$
4. Draw the logic diagram for an exclusive-NOR circuit.

5-2   DESIGNING COMBINATIONAL LOGIC CIRCUITS

In this section we start with an equation or a truth table that describes a logic function and from it determine the circuit required to implement the function. As before, several examples are used to illustrate a general procedure.

Equation (5–1) is the first in the series of examples that will be considered.

$$X = AB + CDE \qquad (5\text{–}1)$$

A brief inspection reveals that this function is composed of two terms, *AB* and *CDE*, with a total of five variables. The first term is formed by ANDing *A* with *B*, and the second term is formed by ANDing *C*, *D*, and *E*. These two terms are then

ORed to form the output X. These operations are indicated in the structure of the equation as follows:

$$X = AB + CDE$$

AND (second-level operation)

OR (first-level operation)

It should be noted that in this particular equation, the AND operations forming the two individual terms, AB and CDE, must be performed *before* the terms can be ORed. The OR operation is the last to be performed before the final output function is produced; therefore, it is called a *first-level* operation, meaning that it is performed by the first-level gate, starting at the output and working back toward the inputs. The AND operations are performed by the second-level gates from the output, and they are therefore *second-level* operations.

To implement the logic function, a two-input AND gate is required to form the term AB, and a three-input AND gate is needed to form the term CDE. A two-input OR gate is then required to combine the two AND terms. The resulting logic circuit is shown in Figure 5–11.

FIGURE 5–11 *Logic circuit for Equation (5–1).*

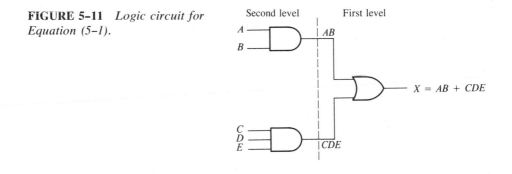

As another example, we will implement Equation (5–2).

$$X = AB(\overline{CD} + EF) \tag{5–2}$$

A breakdown of this equation shows that the term AB and the term $\overline{CD} + EF$ are ANDed. The term AB is formed by ANDing the variables A and B. The term $\overline{CD} + EF$ is formed by first ANDing C and $\overline{D}$, ANDing E and F, and then ORing these two terms. This structure is indicated in relation to the equation as follows:

$$X = AB(\overline{CD} + EF)$$

AND (second-level)

AND (first-level)

NOT (fourth-level)

OR (second-level)

AND (third-level)

Before the output function X can be formed, we must have the term AB. Also, we must have the term $C\overline{D} + EF$; but before we can get this term, we must have the terms $C\overline{D}$ and EF; but before we can get the term $C\overline{D}$, we must have $\overline{D}$. So, as you can see, there is a "chain" of logic operations that must be done in the proper order before the output function itself is realized.

The logic gates required to implement Equation (5–2) are as follows:

1. One inverter to form $\overline{D}$.
2. Two 2-input AND gates to form $C\overline{D}$ and EF.
3. One 2-input OR gate to form $C\overline{D} + EF$.
4. One 2-input AND gate to form AB.
5. One 2-input AND gate to form X.

The logic circuit that produces this function is shown in Figure 5–12(a). Notice that there is a maximum of four gates between an input and output in this circuit (from input D to output). Often, the total propagation delay time through a logic circuit is a major consideration. Propagation delays are additive, so the more gate levels between input and output, the greater the propagation delay time.

The number of gate levels (and sometimes the number of gates) can be reduced by implementing a logic function in the sum-of-products form. Equation (5–2) can be written as

$$X = ABC\overline{D} + ABEF$$

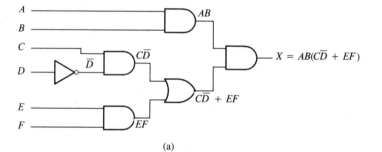

(a)

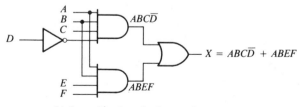

(b) Sum-of-products implementation of the circuit in part (a)

FIGURE 5–12 *Logic circuits for Equation (5–2).*

The sum-of-products implementation of this circuit is shown in Figure 5–12(b).

Now, as a third design example, we will begin with the truth table for a logic function, write the Boolean expression from the truth table, and implement the logic circuit.

TABLE 5–4

Inputs			Output	
A	B	C	X	
0	0	0	0	
0	0	1	0	
0	1	0	0	
0	1	1	1	$\longrightarrow \overline{A}BC$
1	0	0	1	$\longrightarrow A\overline{B}\,\overline{C}$
1	0	1	0	
1	1	0	0	
1	1	1	0	

Table 5–4 specifies a logic function. The expression is obtained from the truth table by ORing the product terms for each occurrence of $X = 1$ (the unshaded rows) where a 1 represents an uncomplemented variable and a 0 represents a complemented variable. In this case, the expression is

$$X = \overline{A}BC + A\overline{B}\,\overline{C} \tag{5–3}$$

This equation requires three levels of logic gates and inverters. How do we know this? First, observe that the equation is composed of two terms in sum-of-products form (the two AND terms, $\overline{A}BC$ and $A\overline{B}\,\overline{C}$, are ORed). The OR operation requires one level (one OR gate). The first term in the equation is formed by ANDing the three variables A, B, and C. The second term is formed by ANDing the three variables, A, $\overline{B}$, and $\overline{C}$. These two operations are performed at the second level (the two AND gates whose outputs are connected to the first-level OR gate inputs). The formation of the complement of each variable requires a third level consisting of inverters; these inverters are at the third level because the complements must be formed before each term is formed. The logic gates required to implement this function are as follows: three inverters to form the $\overline{A}$, $\overline{B}$, and $\overline{C}$ variables; two 3-input AND gates to form the terms $\overline{A}BC$ and $A\overline{B}\,\overline{C}$; and one 2-input OR gate to form the final output function, $\overline{A}BC + A\overline{B}\,\overline{C}$. Figure 5–13 illustrates the implementation of this logic function. Do not confuse levels discussed here with input and output levels of a gate (HIGH or LOW).

In this case where there is a three-level logic function, notice that between input A and the output X there are three logic gates; the same is true for input B and input C. In general, *the number of levels in a given network is the greatest number of gates and inverters through which a signal has to pass in going from an input to the output.*

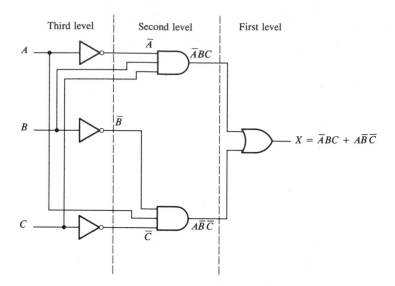

FIGURE 5–13 *Logic circuit for Equation (5–3).*

EXAMPLE 5–4 Design a logic circuit to implement the operation specified in the following truth table:

Inputs			Output
A	B	C	X
0	0	0	0
0	0	1	0
0	1	0	0
0	1	1	1
1	0	0	0
1	0	1	1
1	1	0	1
1	1	1	0

Solution Notice that $X = 1$ for three and only three of the input conditions. Therefore, the logic expression is

$$X = \overline{A}BC + A\overline{B}C + AB\overline{C}$$

The logic circuit is shown in Figure 5–14.

FIGURE 5-14

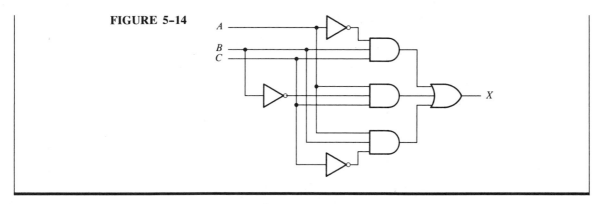

EXAMPLE 5-5   Develop a logic circuit with four input variables that will produce a 1 output when any three and only three input variables are 1s.

Solution   Out of sixteen possible combinations of four variables, the combinations in which there are three 1s are listed below, along with the corresponding product term for each.

$$
\begin{array}{cccc}
A & B & C & D \\
0 & 1 & 1 & 1 \longrightarrow \overline{A}BCD \\
1 & 0 & 1 & 1 \longrightarrow A\overline{B}CD \\
1 & 1 & 0 & 1 \longrightarrow AB\overline{C}D \\
1 & 1 & 1 & 0 \longrightarrow ABC\overline{D}
\end{array}
$$

FIGURE 5-15

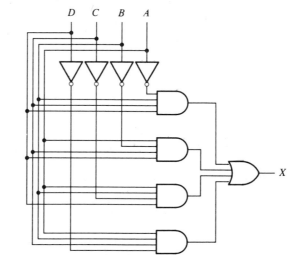

These product terms are ORed to get the logic expression.

$$X = \overline{A}BCD + A\overline{B}CD + AB\overline{C}D + ABC\overline{D}$$

This expression is implemented in Figure 5–15 (p. 171) with AND-OR logic.

Gate Minimization

In many applications it is desirable to use the minimum number of gates in the simplest configuration possible to implement a given logic function. This simplification may be desirable for several reasons, such as economy, limitations of available power, minimization of delay times by reduction of logic levels, or maximum utilization of chip area in the case of IC designs.

Here we will examine two basic methods of gate minimization. First, the rules and laws of Boolean algebra will be applied in order to simplify the logic equation for the circuit in question, and thereby reduce the number of gates required to implement the function. Second, the Karnaugh map method will be used to minimize the logic function. Both of these methods were studied in the previous chapter.

To begin, Boolean algebra will be applied to the equation for the circuit of Figure 5–8, which is redrawn for convenience in Figure 5–16. The logic equation for this circuit was developed in Section 5–1 and is restated in Equation (5–4).

$$X = AC\overline{D} + \overline{A}B(CD + BC) \tag{5-4}$$

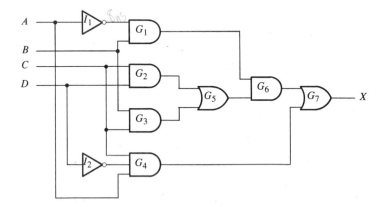

FIGURE 5–16

We know that five AND gates, two OR gates, and two inverters are needed to implement this function *as it appears* in Equation (5–4). We do not yet know whether this is the simplest form of the equation. When we apply the rules of Boolean algebra in an attempt to simplify the equation, our cleverness and imagination are major factors in determining the outcome because an equation can normally be written in many ways and still express the same logic function. In general, we must use all the rules, laws, and theorems at our command in such a way as to arrive at what appears to be the simplest form of the equation. Inspection of Equation (5–4) shows that a possible first step in simplification is to apply the distributive law to the second term by multiplying the term $CD + BC$ by $\overline{AB}$. The result is

$$X = AC\overline{D} + \overline{A}BCD + \overline{A}BBC$$

The rule that $BB = B$ can be applied to the third term, yielding the sum-of-products form

$$X = AC\overline{D} + \overline{A}BCD + \overline{A}BC$$

Notice that C is common to each term, so it can be factored out using the distributive law.

$$X = C(A\overline{D} + \overline{A}BD + \overline{A}B)$$

Now notice that $\overline{A}B$ appears in the last two terms within the parentheses and can be factored out of those two terms.

$$X = C[A\overline{D} + \overline{A}B(D + 1)]$$

Since $D + 1 = 1$,

$$X = C(A\overline{D} + \overline{A}B)$$

It appears that this equation cannot be reduced any further, although it can be written in a slightly different way (sum-of-products form) by application of the distributive law.

$$X = AC\overline{D} + \overline{A}BC \tag{5-5}$$

This equation can be implemented with two 3-input AND gates, two inverters, and one 2-input OR gate, as shown in Figure 5–17.

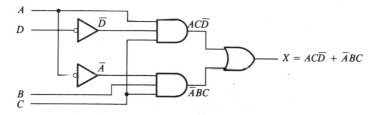

FIGURE 5-17 *Logic circuit for Equation (5–5).*

Compare this circuit with the circuit in Figure 5–16 which has nine logic gates including inverters. The minimized circuit is equivalent but has only five gates including two inverters. At this point you should verify that the logic circuit of Figure 5–17 is indeed equivalent to the logic circuit of Figure 5–16 in terms of the logical operation. Also, the number of gate *levels* has been reduced from a maximum of five in Figure 5–16 to three in Figure 5–17. Fewer levels mean a shorter propagation delay through the circuit as previously mentioned.

Karnaugh Map Approach

The second approach to minimization of Equation (5–4) is the Karnaugh map method. The function of Equation (5–4) is rewritten here for convenience.

$$X = AC\overline{D} + \overline{A}B(CD + BC)$$

To minimize this expression, convert it to the sum-of-products form.

$$X = AC\overline{D} + \overline{A}BCD + \overline{A}BC$$

This expression is expanded on the Karnaugh map in Figure 5–18 to include all variables in each term. The fully expanded sum-of-products expression is

$$X = \overline{A}BCD + \overline{A}BC\overline{D} + ABC\overline{D} + A\overline{B}C\overline{D}$$

FIGURE 5–18 *Karnaugh map for the function of Equation (5–4).*

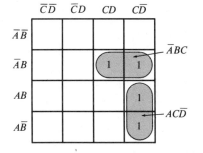

To simplify, the 1s are grouped as shown in the figure and the resulting product terms are ORed to obtain the following minimum expression:

$$X = AC\overline{D} + \overline{A}BC$$

Notice that this is the same result that was achieved using algebraic simplification.

SECTION REVIEW 5–2

1. Implement the following logic expressions with logic gates:
   (a) $X = ABC + AB + AC$
   (b) $X = AB(C + DE)$
2. Develop a logic circuit to produce a 1 on its output only when all three inputs are 1s or when all three inputs are 0s.

5-3   THE UNIVERSAL PROPERTY OF THE NAND GATE

Up to this point, combinational circuits using only AND and OR gates and inverters have been considered. In this section we will see that NAND gates can be used to produce *any* logic function. For this reason, they are referred to as *universal gates*.

The NAND gate can be used to generate the NOT function, the AND function, the OR function, and the NOR function. An inverter can be made from a NAND gate by connecting all of the inputs together and creating, in effect, a single common input, as shown in Figure 5–19(a) for a two-input gate. An AND function can be generated using only NAND gates, as shown in part (b). Also, an OR function can be produced with NAND gates, as illustrated in part (c). Finally, a NOR function is produced, as shown in part (d).

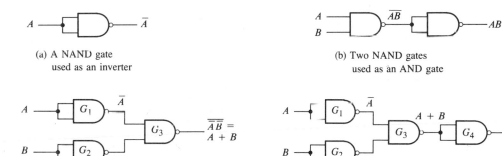

(a) A NAND gate used as an inverter

(b) Two NAND gates used as an AND gate

(c) Three NAND gates used as an OR gate

(d) Four NAND gates used as a NOR gate

FIGURE 5–19 *Universal application of NAND gates.*

In Figure 5–19(b), a NAND gate is used to invert (complement) a NAND output to form the AND function, as indicated in the following equation:

$$X = \overline{\overline{AB}} = AB$$

In Figure 5–19(c), NAND gates G_1 and G_2 are used to invert the two input variables before they are applied to NAND gate G_3. The final OR output is derived as follows by application of DeMorgan's theorem:

$$X = \overline{\overline{A}\,\overline{B}} = A + B$$

In Figure 5–19(d), NAND gate G_4 is used as an inverter to produce the NOR function $\overline{A + B}$.

Multilevel NAND Logic

As you already know, the function of the NAND gate can be expressed in two ways by applying DeMorgan's theorem. If inputs A and B are applied to a NAND gate, the output can be expressed as

$$X = \overline{AB} = \overline{A} + \overline{B}$$

Therefore, the function of the NAND gate can be stated in two ways:

1. The output of a NAND gate is equal to the complement of the AND of the input variables.
2. The output of a NAND gate is equal to the OR of the complements of the input variables (corresponding to a negative-OR).

Rules for multilevel NAND logic can be developed using these two statements of NAND operation as the basis. To develop a set of rules, we will begin with a two-level NAND circuit, as shown in Figure 5–20.

For this circuit, the logic equation is developed as follows by repeated application of DeMorgan's theorem:

$$X = \overline{(\overline{AB})C} = \overline{(\overline{A} + \overline{B})C} = AB + \overline{C}$$

Notice that input variables A and B appear as an AND term in the final output equation, and input variable C appears complemented and ORed with the term AB. Gate G_2 is a second-level gate (second gate back from the output), and gate G_1 is a first-level gate (first gate from output). Notice also that individual variables A and B are inputs to the second-level gate G_2, and that the term $\overline{AB}$ and individual variable C are the inputs to the first-level gate G_1. From this observation, two general rules for multilevel NAND logic can be stated:

1. All odd-numbered gate levels (1, 3, 5, . . .) "act" as OR gates with single input variables complemented.
2. All even-numbered gate levels (2, 4, 6, . . .) "act" as AND gates.

To illustrate how this method can be applied to multilevel NAND logic, let us take the network shown in Figure 5–20 and proceed as follows. Replace each *odd*-level NAND gate with its equivalent *negative*-OR symbol, as shown in Figure 5–21. Remember, nothing has been changed operationally in the circuit; we are

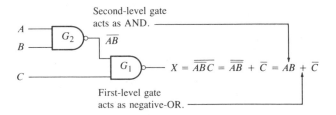

FIGURE 5–20 *Two-level NAND logic.*

FIGURE 5–21 *NAND/negative-OR equivalent of two-level NAND logic.*

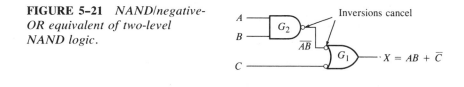

simply replacing the NAND gate symbol with an *equivalent* one. Now, rather than having to remember the two rules, we can simply write the output expression based on the logical function at each level, as indicated by the gate symbol at that level.

Notice that the bubble output of G_2 is connected to the bubble input of G_1. This, of course, indicates a double inversion that effectively cancels the inversion at each of the levels. In this approach, all connections between levels are "bubble-to-bubble," going from even levels to odd levels, and straight connections going from odd levels to even levels. Thus, all inversions can be discarded when the output expression is written, except for single input variables at odd levels.

Next, let us take the three-level network of Figure 5–22(a) to further illustrate this method. By replacing gates G_1 and G_3 (odd-level gates) with their negative-OR equivalents, we have the equivalent network of Figure 5–22(b). From this we can easily write the output expression as follows:

$$X = (\overline{A} + \overline{B})C + \overline{D}$$

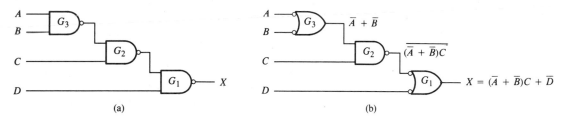

FIGURE 5–22 *Three-level NAND logic and its NAND/negative-OR equivalent.*

EXAMPLE 5–6 Redraw the logic diagram and develop the output expression for the circuit in Figure 5–23 using the procedure just discussed.

FIGURE 5–23

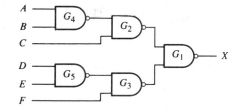

178   COMBINATIONAL LOGIC

> *Solution*   Redrawing the network in Figure 5–23 with equivalent negative-OR symbols at the odd levels, we get Figure 5–24. Writing the expression for X directly from the indicated logic operation at each level gives us
>
> $$X = (\overline{A} + \overline{B})C + (\overline{D} + \overline{E})F$$
>
>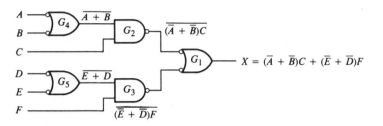
>
> **FIGURE 5–24**

SECTION REVIEW 5–3

1. Use NAND gates to implement each expression.
    (a) $X = \overline{A} + B$ **(b)** $X = A\overline{B}$
2. Implement the expression $X = \overline{(\overline{A} + \overline{B} + \overline{C})DE}$ using multilevel NAND logic.

5–4   THE UNIVERSAL PROPERTY OF THE NOR GATE

As with the NAND gate, the NOR gate can be used to generate the NOT, AND, OR, and NAND functions. A NOT circuit or inverter can be made from a NOR gate by connecting all of its inputs together to effectively create a single input, as shown in Figure 5–25(a) with a two-input example. Also, an OR gate can be produced from NOR gates, as illustrated in Figure 5–25(b).

(a) A NOR gate used as an inverter

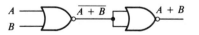

(b) Two NOR gates used as an OR gate

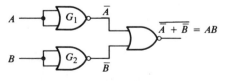

(c) Three NOR gates used as an AND gate

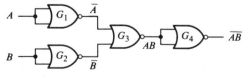

(d) Four NOR gates used as a NAND gate

FIGURE 5–25   *Universal application of NOR gates.*

An AND gate can be constructed using NOR gates, as shown in Figure 5–25(c). In this case the NOR gates G_1 and G_2 are used as inverters, and the final output is derived using DeMorgan's theorem as follows:

$$X = \overline{\overline{A} + \overline{B}} = AB$$

Figure 5–25(d) shows NOR gates used to form a NAND function.

Multilevel NOR Logic

If the inputs A and B are applied to the NOR gate in Figure 5–26, the output will be as shown. By applying DeMorgan's theorem to the output function, we get

$$X = \overline{A + B} = \overline{A}\,\overline{B}$$

FIGURE 5–26 *Dual expression of NOR output.*

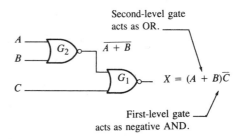

Notice that the output can be expressed in the following two ways:

1. The output of a NOR gate is equal to the complement of the OR of the input variables.
2. The output of the NOR gate is equal to the AND of the complements of input variables (corresponding to a negative-AND).

Rules for multilevel NOR logic can be developed using these two statements of NOR operation as the basis. To develop a set of rules, we will begin with a two-level NOR circuit, as shown in Figure 5–27.

FIGURE 5–27 *Two-level NOR logic.*

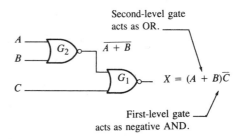

For this circuit, the logic equation is developed as follows by application of DeMorgan's theorem:

$$X = \overline{\overline{(A + B)} + C} = \overline{\overline{(A + B)}}\,\overline{C} = (A + B)\overline{C}$$

Notice that the input variables A and B appear as an OR term in the final output equation, and the input variable C appears complemented and ANDed with the term $A + B$. Gate G_2 is a second-level gate (second back from output), and gate

G_1 is a first-level gate (first gate from output). Notice also that the individual variables A and B are inputs to the second-level gate, and that the term $\overline{A + B}$ and the individual variable C are the inputs to the first-level gate. From this observation, two rules for multilevel NOR logic can be stated:

1. All odd-numbered gate levels (1, 3, 5, . . .) act as AND gates with single input variables complemented.
2. All even-numbered levels (2, 4, 6, . . .) act as OR gates.

To illustrate how this method can be applied to multilevel NOR logic, let us take the network shown in Figure 5–27 and proceed as follows. Replace each *odd*-level NOR gate with its equivalent *negative*-AND symbol, as shown in Figure 5–28. Keep in mind that nothing has been changed operationally in the circuit; we are simply replacing the NOR gate symbol with an *equivalent* one. Now we can write the expression for the output based on the logical function at each level, as indicated by the gate symbol at that level.

FIGURE 5–28   *NOR/negative-AND equivalent of two-level NOR logic.*

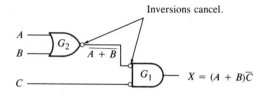

Notice that the bubble output of gate G_2 is connected to the bubble input of gate G_1. The bubble-to-bubble connection going from an even to an odd level effectively cancels the inversions. So, in this case, gate G_2 effectively looks like an OR gate. The only inversions occur for single input variables at odd levels.

Next, let us take the three-level network of Figure 5–29(a) in order to further illustrate this method of analysis. By replacing gates G_1 and G_3 (odd-level gates) with their equivalent negative-AND symbols, we have the equivalent network of Figure 5–29(b). From this we can write the output expression as follows:

$$X = (\overline{A}\,\overline{B} + C)\overline{D}$$

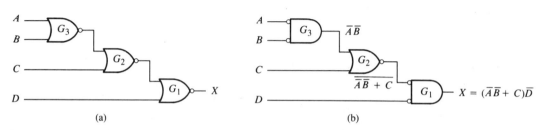

FIGURE 5–29   *Three-level NOR logic and its NOR/negative-AND equivalent.*

EXAMPLE 5-7 Using the method just discussed, redraw the logic diagram and develop the output expression for the circuit in Figure 5–30.

FIGURE 5–30

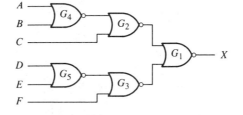

Solution Redraw the network with the equivalent negative-AND symbols at the odd levels, as shown in Figure 5–31. Writing the expression for X directly from the indicated logic operation at each level, we get

$$X = (\overline{A}\,\overline{B} + C)(\overline{D}\,\overline{E} + F)$$

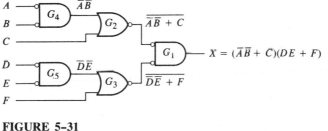

FIGURE 5–31

SECTION REVIEW 5-4

1. Use NOR gates to implement each expression.
   (a) $X = \overline{A} + B$ (b) $X = A\overline{B}$
2. Implement the expression $X = \overline{\overline{A}\,\overline{B}\,\overline{C} + (D + E)}$ with multilevel NOR logic.

5-5   PULSED OPERATION

The operation of combinational logic circuits with pulsed inputs follows the same logical operation we have discussed in the previous sections. The output of the logic network is dependent on the states of the inputs at any given time; if the inputs are changing levels according to a time-varying pattern (waveform), the output depends on what states the inputs happen to be at any instant. The following is a summary of the operation of individual gates for use in analyzing combinational circuits with time-varying inputs:

1. The output of an AND gate is HIGH only when *all* inputs are HIGH *at the same time*.
2. The output of an OR gate is HIGH any time at least *one* of its inputs is HIGH. The output is LOW only when *all* inputs are LOW *at the same time*.
3. The output of a NAND gate is LOW only when *all* inputs are HIGH *at the same time*.
4. The output of a NOR gate is LOW any time at least *one* of its inputs is HIGH. The output is HIGH only when *all* inputs are LOW *at the same time*.

The logical operation of any gate is the same regardless of whether its inputs are pulsed or constant levels. The nature of the inputs (pulsed or constant levels) does not alter the truth table operation of a gate. The following examples illustrate the analysis of combinational logic circuits with pulsed inputs.

EXAMPLE 5-8   Determine the output waveform for the circuit in Figure 5–32(a), with the inputs as shown.

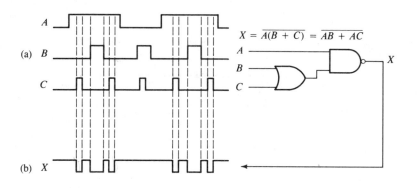

FIGURE 5-32

Solution X is shown in the proper time relationship to the inputs in Figure 5–32(b).

EXAMPLE 5-9   Determine the output waveform for the circuit in Figure 5–33(a) if the input waveforms are as indicated.

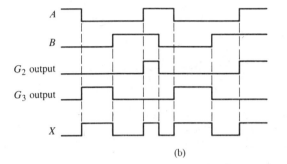

(a)

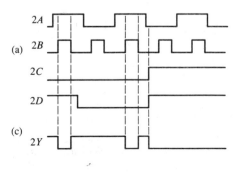

(b)

FIGURE 5-33

Solution When both inputs are HIGH or when both inputs are LOW, the output is HIGH. This is called a *coincidence circuit* or *exclusive-NOR*. The intermediate outputs of gates G_2 and G_3 are used to develop the final output. See Figure 5-33(b).

EXAMPLE 5-10

The pulse waveforms shown in Figure 5-34(a) are applied to the 7451 AND-OR-INVERT circuit in part (b). Sketch the output waveform at pin 6.

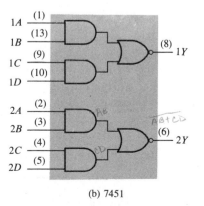

(b) 7451

FIGURE 5-34

Solution   The output waveform is shown in part (c) in the proper time relationship to the inputs.

SECTION REVIEW 5–5

1. One pulse with $t_W = 50$ μs is applied to one of the inputs of an exclusive-OR circuit. A second positive pulse with $t_W = 10$ μs is applied to the other input beginning 15 μs after the leading edge of the first pulse. Sketch the output in relation to the inputs.

2. The pulse waveforms A and B in Figure 5–32(a) are applied to the exclusive-NOR circuit in Figure 5–33(a). Sketch a complete timing diagram showing the output in relation to the inputs.

5–6   TROUBLESHOOTING GATE NETWORKS

In a typical combinational logic network, the output of one gate is connected to one or more gate inputs as shown in Figure 5–35. The interconnecting paths represent a common electrical point known as a *node*.

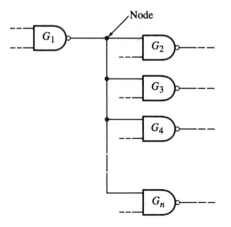

FIGURE 5–35   *Illustration of a node in a gate network.*

Gate G_1 in the figure is *driving* the node, and the other gates represent *loads* connected to the node. Several types of failures are possible in this situation. Some of these failure modes are difficult to isolate down to a bad gate because all the gates connected to the node are affected. The types of failures we will consider are as follows:

1. *Open output in driving gate.* This failure will cause a loss of signal to all load gates.
2. *Open input in a load gate.* This failure will not affect the operation of any of the other gates connected to the node, but it will result in loss of signal output from the faulty gate.
3. *Shorted output in driving gate.* This failure can cause the node to be "stuck" in the LOW state.

4. *Shorted input in a load gate.* This failure can also cause the node to be "stuck" in the LOW state.

We will now explore some approaches to troubleshooting each of the faults listed above.

Open Output in Driving Gate

In this situation, there is no pulse activity on the node. With circuit power on, an open node will normally result in a "floating" level and will be indicated by a *dim lamp* on the *logic probe*, as illustrated in Figure 5-36.

FIGURE 5-36 *Open output in driving gate.*

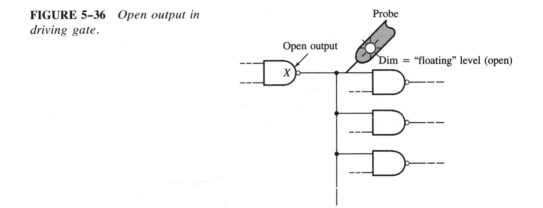

Open Input in a Load Gate

If the check for an open driver output is negative, then a check for an open input in a load gate should be performed. Apply the *logic pulser* tip to the node. Then

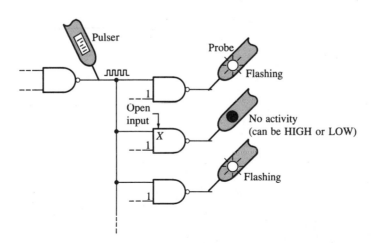

FIGURE 5-37 *Open input in a load gate.*

check the output of each gate for pulse activity with the *logic probe*, as illustrated in Figure 5–37. If one of the inputs that is normally connected to the node is open, no pulses will be detected on that gate's output.

Shorted Output in Driving Gate

This fault can cause the node to be stuck LOW as previously mentioned. A quick check with a *pulser* and a *logic probe* will indicate this, as shown in Figure 5–38(a). A short to ground in the driving gate's output or in any gate input will cause this symptom, and therefore further checks must be made to isolate the short to a particular gate.

If the driving gate's output is internally shorted to ground, then, essentially, no current activity will be present on any of the connections to the node. Thus, a *current tracer* will indicate no activity with circuit power on, as illustrated in Figure 5–38(b).

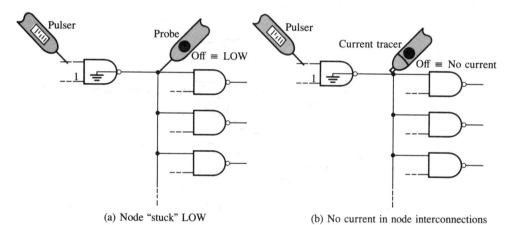

(a) Node "stuck" LOW   (b) No current in node interconnections

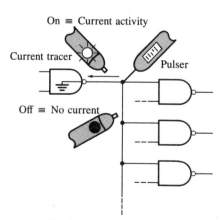

(c) Tracing current along path into short

FIGURE 5–38   *Shorted output in driving gate.*

To further verify a shorted output, a *pulser* and *current tracer* can be used with the circuit power off, as shown in Figure 5–38(c). When current pulses are applied to the node with the *pulser*, all of the current will flow into the shorted output and none will flow through the circuit paths into the load gate inputs.

Shorted Input in a Load Gate

If one of the load gate inputs is internally shorted to ground, the node will be stuck in the LOW state. Again, as in the case of a shorted output, the *logic pulser* and *current tracer* can be used to isolate the faulty gate.

When the node is pulsed with circuit power off, essentially all the current will flow into the shorted input, and tracing its path with the *current tracer* will lead to the bad input, as illustrated in Figure 5–39.

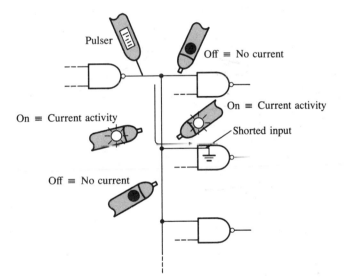

FIGURE 5–39 *Shorted input in a load gate.*

<table>
<tr><td>

**SECTION
REVIEW
5–6**

</td><td>

1. List four common internal failures in logic gates.
2. One input of a NOR gate is externally shorted to $+V_{CC}$. How does this condition affect the gate operation?

</td></tr>
</table>

SUMMARY

- [ ] A sum-of-products expression can be implemented with AND-OR logic.
- [ ] The output of an exclusive-OR gate is HIGH when the inputs have opposite states. The output of an exclusive-NOR is HIGH when the inputs have the same states.

188   COMBINATIONAL LOGIC

☐ Any logic expression can be implemented using only NAND gates.
☐ Any logic expression can be implemented using only NOR gates.
☐ In multilevel NAND logic, all odd-numbered gate levels act as OR gates with single input variables complemented, and all even-numbered levels act as AND gates.
☐ In multilevel NOR logic, all odd-numbered gate levels act as AND gates with single input variables complemented, and all even-numbered levels act as OR gates.
☐ Common internal gate failures include open input or output and shorted input or output.

SELF-TEST

1. Write the output expression for an AND-OR logic circuit having one 4-input AND gate with inputs A, B, C, and D and one 2-input AND gate with inputs E and F.
2. Sketch the logic circuit described by each of the following output expressions:
   (a) $Y = \overline{A\overline{B}C} + AB\overline{C}$ (b) $Y = \overline{A}B + A\overline{B}$
   (c) $Y = A\overline{B}\,\overline{C}D + ABC\overline{D} + \overline{A}B\overline{C}D$
3. Write the truth table for a logic circuit that produces a 1 output only when its three input variables represent a 2, 5, or 7 in binary and a 0 output for all other input states.
4. Draw the logic circuit for the truth table specified in Problem 3.
5. What is the maximum propagation delay through the circuit of Figure 5–16 if each gate and inverter has $t_{PHL} = t_{PLH} = 8$ ns?
6. Draw the logic diagram for a multilevel NOR circuit having the output expression $X = (A + B + C)(D + E)\overline{F}$.
7. Draw the logic diagram for a multilevel NAND circuit having the output expression $X = ABC + DE + \overline{F}$.
8. A certain application requires that the absence of input pulses (constant LOWs) on two input lines be indicated by a HIGH output. The condition in which both inputs have simultaneous positive pulses must also be indicated by a HIGH output. Draw the logic diagram for this circuit.

PROBLEMS

Section 5–1
5–1   Determine the truth table for each of the circuits in Figure 5–40.

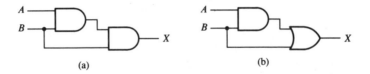

(a)   (b)

FIGURE 5–40

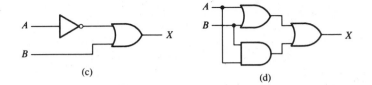

(c)   (d)

FIGURE 5-40 *(Continued)*

5-2 Determine the truth table for each circuit in Figure 5-41.

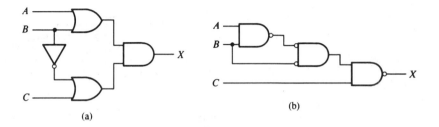

(a)   (b)

FIGURE 5-41

5-3 Determine the truth table for the circuit shown in Figure 5-42.

FIGURE 5-42

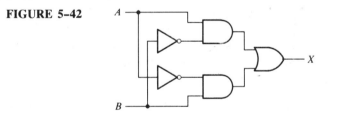

5-4 Write the logic expression for the output of each of the circuits in Figure 5-43.

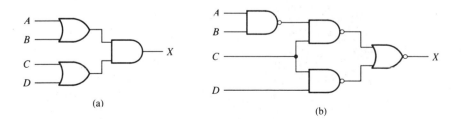

(a)   (b)

FIGURE 5-43

5-5 Write the logic expression without simplification for the output of each of the circuits in Figure 5–44.

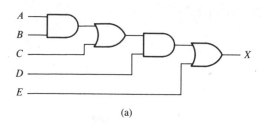

(a)

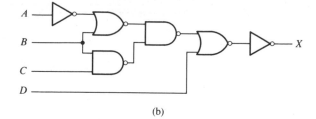

(b)

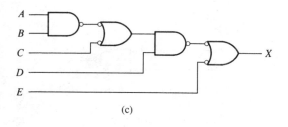

(c)

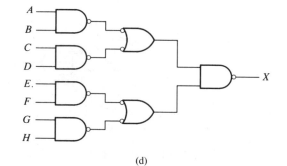

(d)

FIGURE 5-44

5-6 Develop the truth table for the circuit in Figure 5–45.

FIGURE 5-45

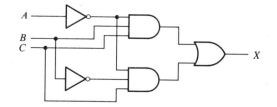

5-7 Repeat Problem 5-6 with each AND gate replaced by a NAND gate and each OR gate replaced by a NOR gate.

Section 5-2

5-8 Use AND gates, OR gates, or combinations of both to implement the following logic expressions:

(a) $X = AB$
(b) $X = A + B$
(c) $X = AB + C$
(d) $X = ABC + D$
(e) $X = A + B + C$
(f) $X = ABCD$
(g) $X = A(CD + B)$
(h) $X = AB(C + DEF) + CE(A + B + F)$

5-9 Use AND gates, OR gates, and inverters as needed to implement the following logic expressions:

(a) $X = A\overline{B} + \overline{B}C$ (b) $X = A(\overline{B} + \overline{C})$
(c) $X = \overline{A}\overline{B} + AB$ (d) $X = \overline{ABC} + B(EF + \overline{G})$
(e) $X = A[BC(A + B + C + D)]$ (f) $X = B(CDE + \overline{EFG})(\overline{AB} + C)$

5-10 Use NAND gates, NOR gates, or combinations of both to implement the following logic expressions:

(a) $X = \overline{AB} + CD + (\overline{A} + \overline{B})(ACD + \overline{BE})$
(b) $X = \overline{ABCD} + \overline{DEF} + \overline{AF}$
(c) $X = \overline{A}[B + \overline{C}(D + E)]$

5-11 Show how the following expressions can be implemented using only NAND gates:

(a) $X = ABC$ (b) $X = \overline{ABC}$
(c) $X = A + B$ (d) $X = A + B + \overline{C}$
(e) $X = \overline{AB} + \overline{CD}$ (f) $X = (A + B)(C + D)$
(g) $X = AB[C(\overline{DE} + \overline{AB}) + \overline{BCE}]$

5-12 Repeat Problem 5-11 using only NOR gates.

5-13 Develop the logic circuit necessary to meet the following requirements:

> A lamp in a room is to be operated from two switches, one at the back door and one at the front door. The lamp is to be on if the front switch is on and the back switch is off, or if the front switch is off and the back switch is on. The lamp is to be off if both switches are off or if both switches are on. Let a HIGH output represent the on condition and a LOW output represent the off condition.

5-14 Design a logic circuit to produce a HIGH output if and only if the input, represented by a 4-bit binary number, is greater than 12 or less than 3. First develop the truth table and then draw the logic diagram.

5-15 Simplify the circuit in Figure 5-46 as much as possible, and verify that the simplified circuit is equivalent to the original by showing that their truth tables are identical.

FIGURE 5-46

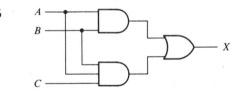

5-16 Repeat Problem 5-15 for the circuit in Figure 5-47.

FIGURE 5-47

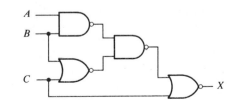

5–17   Minimize the gates required to implement the functions in each part of Problem 5–9.

5–18   Minimize the gates required to implement the functions in each part of Problem 5–10.

5–19   Minimize the gates required to implement the functions in each part of Problem 5–11.

Sections 5–3 and 5–4

5–20   Implement the logic circuit in Figure 5–45 using only NAND gates.

5–21   Implement the logic circuit in Figure 5–47 using only NAND gates.

5–22   Repeat Problem 5–20 using only NOR gates.

5–23   Repeat Problem 5–21 using only NOR gates.

Section 5–5

5–24   Given the logic circuit and the input waveforms in Figure 5–48, sketch the output waveform.

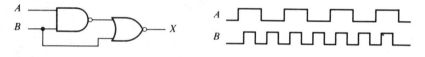

FIGURE 5–48

5–25   For the logic circuit in Figure 5–49, sketch the output waveform in proper relation to the inputs.

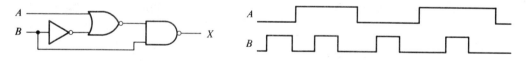

FIGURE 5–49

5–26   For the input waveforms in Figure 5–50, what logic circuit would generate the output waveform shown?

FIGURE 5–50

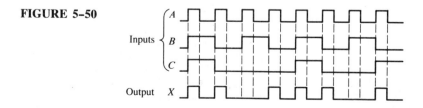

Inputs { A B C

Output   X

5-27 Repeat Problem 5-26 for the waveforms in Figure 5-51.

FIGURE 5-51

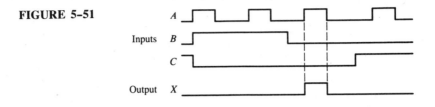

5-28 For the circuit in Figure 5-52, sketch the waveforms at each point indicated, in the proper relationship to each other.

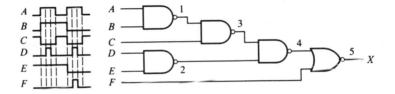

FIGURE 5-52

5-29 Assuming a propagation delay through each gate of 10 nanoseconds (ns), determine if the *desired* output waveform in Figure 5-53 (a pulse with a minimum $t_W = 25$ ns positioned as shown) will be generated properly with the given inputs.

FIGURE 5-53

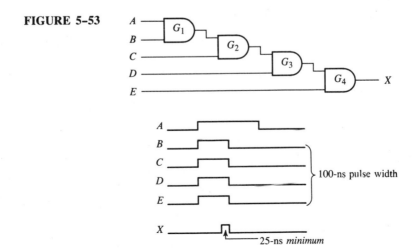

Section 5-6

5-30 For the logic circuit and the input waveforms in Figure 5-54, the indicated output waveform is observed. Determine if this is the correct output waveform.

FIGURE 5-54

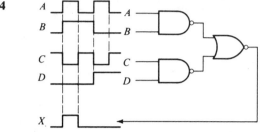

5-31 The output waveform in Figure 5-55 is incorrect for the inputs that are applied to the circuit. Assuming that one gate in the circuit has failed with its output either a constant HIGH or a constant LOW, determine the faulty gate and the failure (output open or shorted) mode that would cause the erroneous output.

FIGURE 5-55

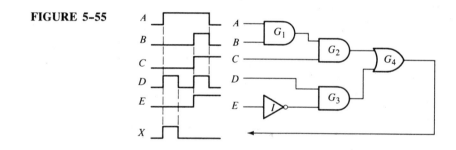

5-32 Repeat Problem 5-31 for the circuit in Figure 5-56, with input and output waveforms as shown.

FIGURE 5-56

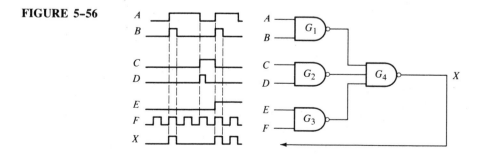

5-33 The node in Figure 5-57 has been found to be stuck in the LOW state. A pulser and current tracer yield the indications shown in the diagram. From this, what would you conclude?

FIGURE 5-57

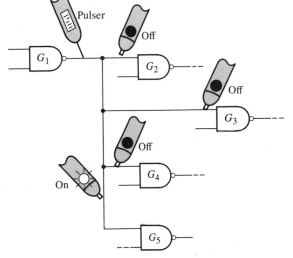

5-34 Figure 5-58(a) is a logic network under test. Figure 5-58(b) shows the waveforms as observed on a logic analyzer. The output waveform is incorrect for the inputs that are applied to the circuit. Assuming that one gate in the network has failed with its output either a constant HIGH or a constant LOW, determine the faulty gate and the failure mode that would cause the erroneous output.

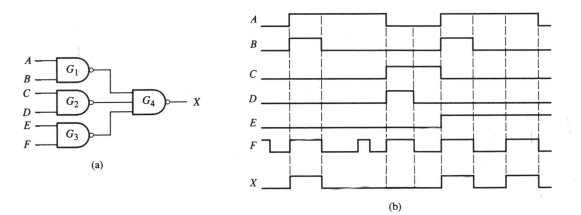

(a)

(b)

FIGURE 5-58

ANSWERS TO SECTION REVIEWS

Section 5–1
1. (a) 1   (b) 0   (c) 0
2. (a) 1   (b) 0   (c) 1   (d) 0
3. $X = 1$ when $ABC = 000, 011, 101, 110,$ and 111,
   $X = 0$ when $ABC = 001, 010,$ and 100.
4. $X = AB + \overline{A}\,\overline{B}$; the circuit consists of two AND gates, one OR gate, and two inverters.

Section 5–2
1. (a) Three AND gates, one OR gate   (b) Three AND gates, one OR gate
2. $X = ABC + \overline{A}\,\overline{B}\,\overline{C}$; two AND gates, one OR gate, and three inverters.

Section 5–3
1. (a) A two-input NAND gate with A and $\overline{B}$ on its inputs.
   (b) A two-input NAND with A and $\overline{B}$ on its inputs, followed by one NAND used as an inverter.
2. A three-input NAND with inputs A, B, and C, with its output connected to a second three-input NAND with two other inputs, D and E.

Section 5–4
1. (a) A two-input NOR with inputs $\overline{A}$ and B, followed by one NOR used as an inverter.
   (b) A two-input NOR with $\overline{A}$ and B on its inputs.
2. A three-input NOR with inputs A, B, and C, with its output connected to a second three-input NOR with two other inputs, D and E.

Section 5–5
1. A 15-μs pulse followed by a 25-μs pulse, with a separation of 10 μs between the pulses.
2. The output is HIGH when both inputs are HIGH or when both inputs are LOW.

Section 5–6
1. Open input or output; shorted input or output.
2. Output is always LOW.

In this chapter, several types of MSI combinational logic functions are studied, including adders, comparators, decoders, encoders, code converters, multiplexers (data selectors), demultiplexers, and parity generators/checkers.

Examples of applications of many of these devices are presented to provide a basic idea of how the devices can be used in practical situations. Also, a simple application of the logic analyzer and oscilloscope in troubleshooting digital systems is introduced.

Specific devices introduced in this chapter are as follows:

1. 7482 two-bit binary adder
2. 7483A four-bit parallel binary adder
3. 7485 four-bit magnitude comparator
4. 74154 four-line-to-sixteen-line decoder
5. 7442A BCD-to-decimal decoder
6. 7449 BCD-to-seven-segment decoder/driver
7. 74147 decimal-to-BCD priority encoder
8. 74157 quadruple two-input data selector/multiplexer
9. 74151A eight-input data selector/multiplexer
10. 74150 sixteen-input data selector/multiplexer
11. 74139 dual two-line-to-four-line decoder/demultiplexer
12. 74180 nine-bit parity generator/checker
13. 74138 three-line-to-eight-line decoder/demultiplexer

In this chapter, you will learn

☐ How half-adders and full-adders work.
☐ How full-adders are used to implement multi-bit parallel binary adders.
☐ What limits the add times of parallel adders and how the addition process can be speeded up with the *look-ahead-carry* method.
☐ The purpose of the magnitude comparator and how it works.
☐ How to implement a basic binary decoder.
☐ How binary, BCD-to-decimal, and BCD-to-seven-segment decoders work.
☐ The principles of LED and LCD displays.
☐ How to use BCD-to-seven-segment decoders in display systems.
☐ How the decimal-to-BCD priority encoder works and how it can be applied in a simple keyboard application.
☐ Methods for converting from BCD to binary, binary to Gray code, and Gray code to binary.
☐ How multiplexers work and how they are applied in data selection, multiplexed displays, logic function generation, and simple communications systems.
☐ How decoders are used as demultiplexers.
☐ What *parity* means.
☐ How errors are detected in digital systems using parity generators and checkers.
☐ How a simple data communication system can be implemented.
☐ What *glitches* are and how their causes can be identified.

6

Functions of Combinational Logic

6–1   ADDERS

Half-Adder

Recall the basic rules for binary addition as stated in Chapter 2.

$$0 + 0 = 0$$
$$0 + 1 = 1$$
$$1 + 0 = 1$$
$$1 + 1 = 10_2$$

These operations are performed by a logic circuit called a *half-adder*. The half-adder accepts two binary digits on its inputs and produces two binary digits on its outputs, a *sum* bit and a *carry* bit, as shown by the ANSI/IEEE Std. 91–1984 logic symbol in Figure 6–1.

FIGURE 6–1 *Logic symbol for a half-adder.*

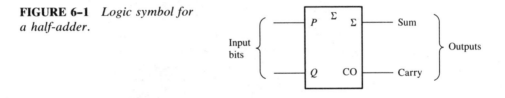

From the logical operation of the half-adder as expressed in Table 6–1, expressions can be derived for the sum and carry outputs as functions of the inputs. Notice that the carry output CO is a 1 only when both P and Q are 1s; therefore, CO can be expressed as the AND of the input variables.

$$CO = PQ \tag{6–1}$$

Now observe that the sum output (Σ) is a 1 only if the input variables are not equal. The sum can therefore be expressed as the exclusive-OR of the input variables.

$$\Sigma = P \oplus Q \tag{6–2}$$

From these two expressions, the implementation required for the half-adder function is apparent. The carry output is produced with an AND gate with P and Q

TABLE 6–1 *Half-adder truth table.*

P	Q	CO	Σ
0	0	0	0
0	1	0	1
1	0	0	1
1	1	1	0

Σ = sum
CO = carry output
P and Q = input variables (operands)

on the inputs, and the sum output is generated with an exclusive-OR gate as shown in Figure 6–2.

FIGURE 6–2 *Half-adder logic diagram.*

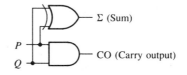

Full-Adder

The second basic category of adder is the *full-adder*. The full-adder accepts *three* inputs and generates a sum output and a carry output. So, the basic difference between a full-adder and a half-adder is that the full-adder accepts an additional input that allows for handling input carries. An ANSI/IEEE logic symbol for a full-adder is shown in Figure 6–3, and the truth table in Table 6–2 shows its operation.

FIGURE 6–3 *Logic symbol for a full-adder.*

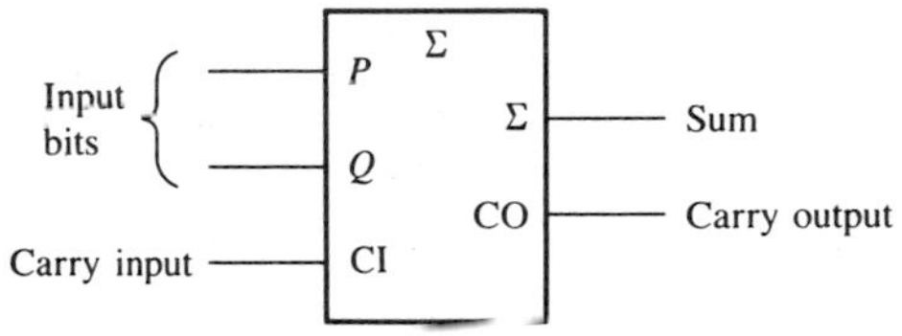

TABLE 6–2 *Truth table for a full adder.*

P	Q	CI	CO	Σ
0	0	0	0	0
0	0	1	0	1
0	1	0	0	1
0	1	1	1	0
1	0	0	0	1
1	0	1	1	0
1	1	0	1	0
1	1	1	1	1

CI = input carry
CO = output carry
Σ = sum
P and Q = input variables (operands)

The full-adder must sum the two input bits and the input carry bit. From the half-adder we know that the sum of the input bits P and Q is the exclusive-OR of those two variables, $P \oplus Q$. In order to add the input carry CI to the input bits, it must be exclusive-ORed with $P \oplus Q$, yielding the equation for the *sum* output of the full-adder.

$$\Sigma = (P \oplus Q) \oplus CI \qquad (6–3)$$

This means that to implement the full-adder sum function, two exclusive-OR gates can be used. The first must generate the term $P \oplus Q$, and the second has as its inputs the output of the first gate and the input carry as illustrated in Figure 6–4(a).

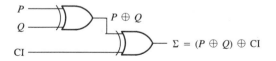

(a) Logic required to form the sum of three bits

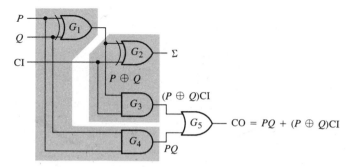

(b) Complete logic circuitry for a full-adder (each half-adder is shown in a shaded block).

FIGURE 6–4 *Full-adder logic.*

The output *carry* is a 1 when both inputs to the first exclusive-OR gate are 1s or when both inputs to the second exclusive-OR gate are 1s. You can verify this by studying Table 6–2. The output carry of the full-adder is therefore P ANDed

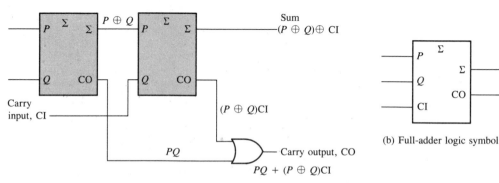

(a) Arrangement of two half-adders to form a full-adder

(b) Full-adder logic symbol

FIGURE 6–5 *Full-adder.*

with Q, $P \oplus Q$ ANDed with CI, and these two terms ORed as expressed in Equation (6–4). This function is implemented and combined with the sum logic to form a complete full-adder circuit, as shown in Figure 6–4(b).

$$CO = PQ + (P \oplus Q)CI \qquad \qquad \textbf{(6–4)}$$

Note that in Figure 6–4(b) there are two half-adders, connected as shown in the block diagram of Figure 6–5(a), with their output carries ORed. The logic symbol as shown in Figure 6–5(b) will normally be used to represent the full-adder.

EXAMPLE 6-1

Determine an alternate method for implementing the full-adder.

Solution Going back to Table 6-2, we can write sum-of-products expressions for both Σ and CO by observing the input conditions that make them 1s. The expressions are as follows:

$$\Sigma = \overline{P}\overline{Q}CI + \overline{P}Q\overline{CI} + P\overline{Q}\overline{CI} + PQCI$$
$$CO = \overline{P}QCI + P\overline{Q}CI + PQ\overline{CI} + PQCI$$

These two functions are implemented with AND/OR logic as shown in Figure 6–6 without simplification. The CO function can be simplified; see if you can do it.

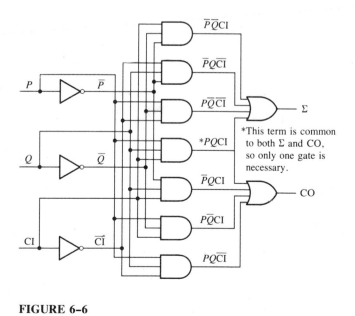

FIGURE 6–6

1. Determine the sum (Σ) and the carry output (CO) of a half-adder for each set of input bits.
    (a) 01   (b) 00   (c) 10   (d) 11
2. A full-adder has CI $= 1$. What are the sum (Σ) and the carry output (CO) when $P = 1$ and $Q = 1$?

6–2   PARALLEL BINARY ADDERS

As we have seen, a single full-adder is capable of adding two one-bit numbers and an input carry. In order to add binary numbers with more than one bit, additional full-adders must be employed. When one binary number is added to another, each column generates a sum and a 1 or 0 carry to the next higher-order column, as illustrated with two-bit numbers:

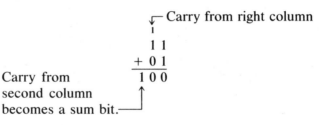

To implement the addition of binary numbers with logic circuits, a full-adder is required for each column. So, for two-bit numbers, two adders are needed; for three-bit numbers, three adders are used; and so on. The carry output of each adder is connected to the carry input of the next higher-order adder, as shown in Figure 6–7 for a two-bit adder. It should be pointed out that either a half-adder can be used for the least significant position or the carry input of a full-adder is made 0 because there is no carry into the least significant bit position.

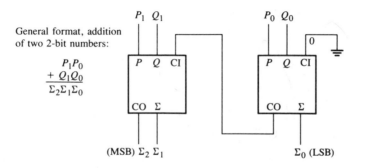

FIGURE 6–7 *Block diagram of a two-bit parallel adder.*

In Figure 6–7, the least significant bits of the two numbers are represented by P_0 and Q_0. The next higher-order bits are represented by P_1 and Q_1. The three complete sum bits are Σ_0, Σ_1, and Σ_2. Notice that the output carry from the leftmost full-adder becomes the most significant sum bit.

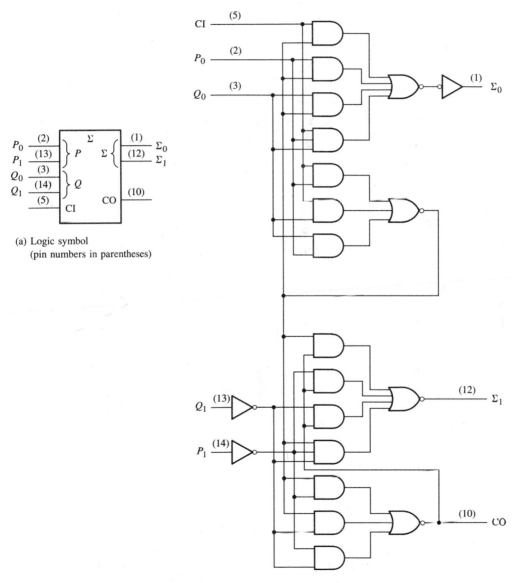

(a) Logic symbol
(pin numbers in parentheses)

(b) Logic diagram

FIGURE 6–8 *7482 two-bit binary full-adder.*

The 7482 Two-Bit Binary Full-Adder

The 7482 is an example of an integrated circuit two-bit adder. It is a TTL medium-scale integrated (MSI) circuit with two interconnected full-adders in one package. A logic symbol with pin numbers and a logic diagram are shown in Figure 6–8 (p. 205). Table 6–3 is the truth table for this device.

Notice the input and output labels on this device: P_0 and Q_0 are the LSB inputs, and P_1 and Q_1 are the MSB inputs. CI is the carry input to the least significant bit adder; CO is the carry output of the most significant bit adder; and Σ_0 (LSB) and Σ_1 (MSB) are the sum outputs. (Labels may vary from one manufacturer to another.)

TABLE 6–3 *Truth table for a 7482 two-bit binary full-adder.*

				When CI = 0			When CI = 1		
P_0	Q_0	P_1	Q_1	Σ_0	Σ_1	CO	Σ_0	Σ_1	CO
0	0	0	0	0	0	0	1	0	0
1	0	0	0	1	0	0	0	1	0
0	1	0	0	1	0	0	0	1	0
1	1	0	0	0	1	0	1	1	0
0	0	1	0	0	1	0	1	1	0
1	0	1	0	1	1	0	0	0	1
0	1	1	0	1	1	0	0	0	1
1	1	1	0	0	0	1	1	0	1
0	0	0	1	0	1	0	1	1	0
1	0	0	1	1	1	0	0	0	1
0	1	0	1	1	1	0	0	0	1
1	1	0	1	0	0	1	1	0	1
0	0	1	1	0	0	1	1	0	1
1	0	1	1	1	0	1	0	1	1
0	1	1	1	1	0	1	0	1	1
1	1	1	1	0	1	1	1	1	1

The Inputs columns are P_0, Q_0, P_1, Q_1. The Outputs section is divided into When CI = 0 and When CI = 1.

EXAMPLE 6–2 Verify that the two-bit parallel adder in Figure 6–9 properly performs the following addition:

$$\begin{array}{r} 11 \\ + 10 \\ \hline 101 \end{array}$$

Solution The logic levels at each point in the circuit for the given input numbers are determined from the truth table of each gate. By following these levels through the circuit as indicated on the logic diagram, we find that the proper levels appear on the sum outputs.

FIGURE 6–9

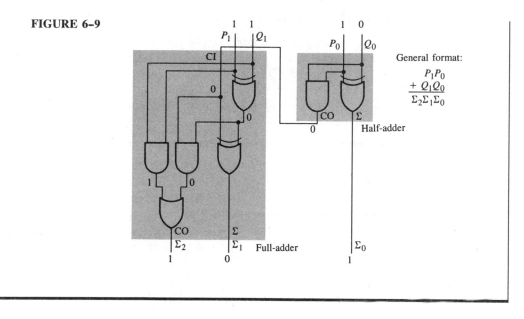

General format:

$$P_1P_0$$
$$+ Q_1Q_0$$
$$\overline{\Sigma_2\Sigma_1\Sigma_0}$$

Half-adder

Full-adder

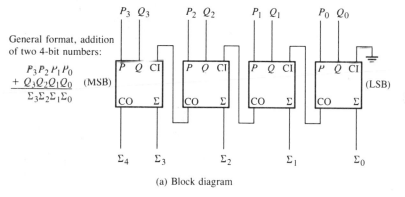

General format, addition
of two 4-bit numbers:

$$P_3P_2P_1P_0$$
$$+ Q_3Q_2Q_1Q_0$$
$$\overline{\Sigma_3\Sigma_2\Sigma_1\Sigma_0}$$

(a) Block diagram

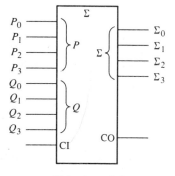

(b) Logic symbol

FIGURE 6–10 *Four-bit parallel binary adder.*

Four-Bit Parallel Adders

A four-bit binary parallel adder is shown in Figure 6–10 (p. 207). Again, the least significant bits in each number being added (operands) go into the right-most full-adder; the higher-order bits are applied as shown to the successively higher-order adders, with the most significant bits in each number being applied to the left-most full-adder. The carry output of each adder is connected to the carry input of the next higher-order adder.

EXAMPLE 6–3

Show how to connect two 7482 adders to form a four-bit adder.

Solution   Figure 6–11 shows two 7482s used as a four-bit adder. Additional bits can be added by connecting the CO of the most significant adder to the CI of the next adder, and so on. Note that the CI of the least significant adder is grounded (0) because there is no carry input to this stage.

FIGURE 6–11   *A four-bit adder using two 7482s.*

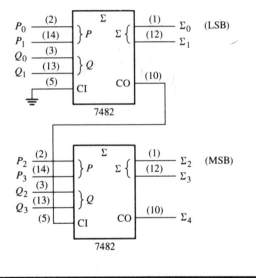

EXAMPLE 6–4

Show how two 7483A four-bit parallel adders can be connected to form an eight-bit parallel adder. Show outputs for

$$P_7P_6P_5P_4P_3P_2P_1P_0 = 10111001$$

and

$$Q_7Q_6Q_5Q_4Q_3Q_2Q_1Q_0 = 10011110$$

Solution   The only connection between the two 7483As is the carry output (CO) of the lower-order adder to the carry input (CI) of the higher-order adder, as shown in Figure 6–12. Pin 13 of the least significant 7483A is grounded (no input carry).

FIGURE 6-12 *Two 7483A four-bit parallel binary adders connected to form an eight-bit parallel adder (pin numbers in parentheses).*

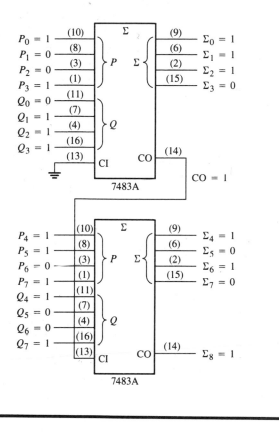

SECTION REVIEW 6-2

1. Two 2-bit binary numbers (11_2 and 10_2) are applied to a two-bit parallel adder. The input carry is 1. Determine the sum (Σ) and the output carry.
2. Draw a diagram showing 7482s connected to form an eight-bit parallel adder.

6-3   A METHOD OF SPEEDING ADDITION

The parallel adders covered so far are *ripple carry* types in which the carry output of each full-adder stage is connected to the carry input of the next higher-order stage. The sum and carry outputs of any stage cannot be produced until the input carry occurs; this leads to a time delay in the addition process, as illustrated in Figure 6–13. The carry propagation delay for each full-adder is the time from the

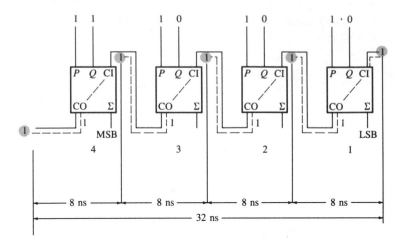

FIGURE 6–13   *A four-bit parallel binary ripple carry adder showing typical carry propagation delays.*

application of the input carry until the output carry occurs, assuming that the *P* and *Q* inputs are present.

Full-adder 1 cannot produce a potential carry output until a carry input is applied. Full-adder 2 cannot produce a potential carry output until full-adder 1 produces a carry output. Full-adder 3 cannot produce a potential carry output until a carry output is produced by full-adder 1 followed by a carry output from full-adder 2, and so on. As you can see, the input carry to the least significant stage has to "ripple" through all of the adders before a final sum is produced. A cumulative delay through all of the adder stages is a "worst-case" addition time. The total delay can vary, depending on the carries produced by each stage. If two numbers are added such that no carries occur between stages, the add time is simply the propagation time through a single full-adder from the application of the data bits on the inputs to the occurrence of a sum output.

The Look-Ahead-Carry Adder

As you have seen in the discussion of the parallel adder, the speed with which an addition can be performed is limited by the time required for the carries to propagate or ripple through all of the stages of the adder. One method of speeding up this process by eliminating this ripple carry delay is called *look-ahead-carry* addition; this method is based on two functions of the full-adder, called the *carry generate* and the *carry propagate* functions.

The *carry generate* (CG) function indicates when an output carry is produced (generated) by the full-adder. A carry is *generated* only when *both input bits are 1s*. This condition is expressed as the AND function of the two input bits, *P* and *Q*:

$$CG = PQ$$

A carry input may be *propagated* by the full-adder when either or both of the input bits are 1s. This condition is expressed as the OR function of the input bits:

$$CP = P + Q$$

The *carry generate* and *carry propagate* conditions are illustrated in Figure 6–14.

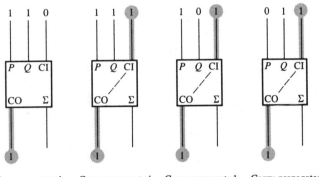

Carry generated   Carry propagated   Carry propagated   Carry propagated

FIGURE 6–14 *Illustration of carry generate and carry propagate conditions.*

How can the carry output of a full-adder be expressed in terms of the carry generate (CG) and the carry propagate (CP)? The output carry (CO) is a 1 if the carry generate is a 1 OR if the carry propagate is a 1 AND the input carry (CI) is a 1. In other words, we get an output carry of 1 if it is *generated* by the full-adder ($P = 1$ AND $Q = 1$) or if the adder can *propagate* the input carry ($P = 1$ OR $Q = 1$) AND CI = 1. This relationship is expressed as

$$CO = CG + CP \cdot CI$$

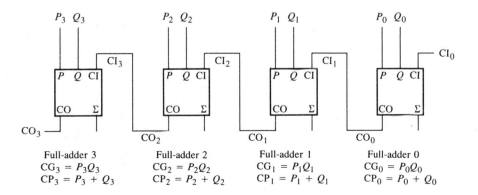

Full-adder 3	Full-adder 2	Full-adder 1	Full-adder 0
$CG_3 = P_3 Q_3$	$CG_2 = P_2 Q_2$	$CG_1 = P_1 Q_1$	$CG_0 = P_0 Q_0$
$CP_3 = P_3 + Q_3$	$CP_2 = P_2 + Q_2$	$CP_1 = P_1 + Q_1$	$CP_0 = P_0 + Q_0$

FIGURE 6–15 *Carry generate and carry propagate functions in terms of the input bits to a four-bit adder.*

Now we will see how this concept can be applied to a parallel adder whose individual stages are shown in Figure 6–15 (p. 211) for a four-bit example. For each full-adder, the output carry is dependent on its carry generate (CG), its carry propagate (CP), and its carry input (CI). The CG and CP functions for each stage are immediately available as soon as the input bits P and Q and the input carry to the LSB adder are applied, because they are dependent only on these bits. The carry input to each stage is the carry output of the previous stage. Based on this, we will now develop expressions for the carry out, CO, of each full-adder stage for the four-bit example.

Full-adder 0:

$$CO_0 = CG_0 + CP_0CI_0 \tag{6–5}$$

Full-adder 1:

$$CI_1 = CO_0$$
$$CO_1 = CG_1 + CP_1CI_1$$
$$= CG_1 + CP_1CO_0$$
$$= CG_1 + CP_1(CG_0 + CP_0CI_0)$$
$$CO_1 = CG_1 + CP_1CG_0 + CP_1CP_0CI_0 \tag{6–6}$$

Full-adder 2:

$$CI_2 = CO_1$$
$$CO_2 = CG_2 + CP_2CO_1$$
$$= CG_2 + CP_2(CG_1 + CP_1G_0 + CP_1CP_0CI_0)$$
$$CO_2 = CG_2 + CP_2CG_1 + CP_2CP_1CG_0 + CP_2CP_1CP_0CI_0 \tag{6–7}$$

Full-adder 3:

$$CI_3 = CO_2$$
$$CO_3 = CG_3 + CP_3CI_3$$
$$= CG_3 + CP_3CO_2$$
$$= CG_3 + CP_3(CG_2 + CP_2CG_1 + CP_2CP_1CG_0 + CP_2CP_1CP_0CI_0)$$
$$CO_3 = CG_3 + CP_3CG_2 + CP_3CP_2CG_1 + CP_3CP_2CP_1CG_0 + CP_3CP_2CP_1CP_0CI_0 \tag{6–8}$$

Notice that in each of these expressions, the carry output for each full-adder stage is dependent only on the initial input carry (CI_0), its CG and CP functions, and the CG and CP functions of the preceding stages. Since each of the CG and CP functions can be expressed in terms of the P and Q inputs to the full-adders, all of the output carries are immediately available (except for gate delays), and we do not have to wait for a carry to ripple through all of the stages before a final result is achieved. Thus the look-ahead-carry technique speeds up the addition process.

Equations (6–5) through (6–8) can be implemented with logic gates and connected to the full-adders to create a look-ahead-carry adder, as shown in Figure 6–16.

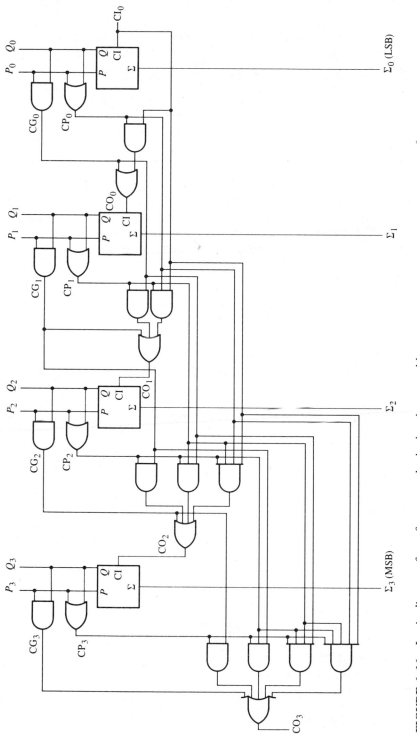

FIGURE 6-16 *Logic diagram for a four-stage look-ahead-carry adder.*

1. The input bits to a full-adder are $P = 1$ and $Q = 0$. Determine CG and CP.
2. Determine the carry output of a full-adder when CI $= 1$, CG $= 0$, and CP $= 1$.

6-4   MAGNITUDE COMPARATORS

The basic function of a *magnitude comparator* is to *compare* the magnitudes of two quantities in order to determine the relationship of those quantities. In its simplest form, a comparator circuit determines whether two numbers are equal.

The exclusive-OR gate is a basic comparator because its output is a 1 if its two input bits are not equal and a 0 if the inputs are equal. Figure 6–17 shows the exclusive-OR as a two-bit comparator.

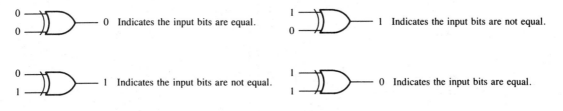

FIGURE 6–17   *Basic comparator operation.*

In order to compare binary numbers containing two bits each, an additional exclusive-OR gate is necessary. The two least significant bits (LSBs) of the two numbers are compared by gate G_1, and the two most significant bits (MSBs) are compared by gate G_2, as shown in Figure 6–18. If the two numbers are equal, their corresponding bits are the same and the output of each exclusive-OR gate is a 0. If the corresponding sets of bits are not equal, a 1 occurs on that exclusive-OR gate output. In order to produce a *single* output indicating an equality or inequality of

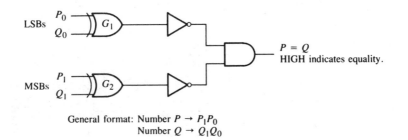

FIGURE 6–18   *Logic diagram for comparison of two 2-bit binary numbers.*

two numbers, two inverters and an AND gate can be employed, as shown in Figure 6–18. The output of each exclusive-OR gate is inverted and applied to the AND gate input. When the two input bits for each exclusive-OR are equal, the corresponding bits of each number are equal, producing a 1 on both inputs to the AND gate and thus a 1 on the output. When the two numbers are not equal, one or both sets of corresponding bits are unequal, and a 0 appears on at least one input to the AND gate to produce a 0 on its output. Thus the output of the AND gate indicates equality (1) or inequality (0) of the two numbers. Example 6–5 illustrates this operation for two specific cases. The exclusive-OR gate and inverter are replaced by an exclusive-NOR symbol.

EXAMPLE 6–5

Apply each of the following sets of binary numbers to comparator inputs, and determine the output by following the logic levels through the circuit:
(a) 10 and 10 **(b)** 11 and 10

Solutions
(a) The output is 1 for inputs 10 and 10, as shown in Figure 6–19(a).
(b) The output is 0 for inputs 11 and 10, as shown in Figure 6–19(b).

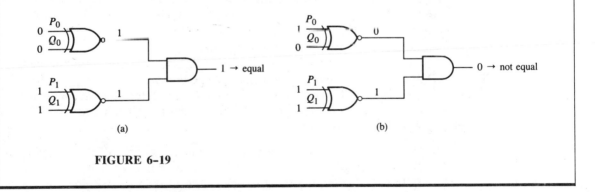

FIGURE 6–19

The basic comparator circuit can be expanded to any number of bits, as illustrated in Figure 6–20 for two 4-bit numbers. The AND gate sets the condition

FIGURE 6–20 *Logic diagram for the comparison of two 4-bit binary numbers, $P_3P_2P_1P_0$ and $Q_3Q_2Q_1Q_0$.*

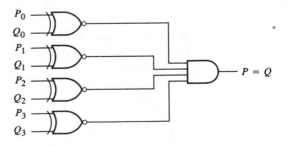

that all corresponding bits of the two numbers must be equal if the two numbers themselves are equal.

Some integrated circuit comparators provide additional outputs that indicate which of the two numbers being compared is the larger. That is, there is an output that indicates when number P is greater than number Q ($P > Q$) and an output that indicates when number P is less than number Q ($P < Q$), as shown in the logic symbol for a four-bit comparator in Figure 6–21.

FIGURE 6–21 *Logic symbol for a four-bit comparator with inequality indication.*

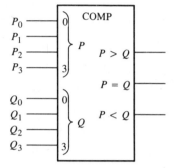

A general method of implementing these two additional output functions is shown in Figure 6–22. In order to understand the logic circuitry required for the $P > Q$ and $P < Q$ functions, let us examine two binary numbers and determine what characterizes an inequality of the numbers.

For our purposes we will use two 4-bit binary numbers with the general format $P_3P_2P_1P_0$ for one number, which we will call number P, and $Q_3Q_2Q_1Q_0$ for the other number, which we will call number Q. To determine an inequality of numbers P and Q, we first examine the highest-order bit in each number. The following conditions are possible:

1. $P_3 = 1$ and $Q_3 = 0$ indicates that number P is greater than number Q.
2. $P_3 = 0$ and $Q_3 = 1$ indicates that number P is less than number Q.
3. If $P_3 = Q_3$, then we must examine the next lower bit position for an inequality.

The three observations are valid for each bit position in the numbers. The general procedure is to check for an inequality in a bit position, starting with the highest order. When such an inequality is found, the relationship of the two numbers is established and any other inequalities in lower-order bit positions *must be ignored* because it is possible for an opposite indication to occur—the highest-order indication *must take precedence*. To illustrate, let us assume that number P is 0111 and number Q is 1000. Comparison of bits P_3 and Q_3 indicates that $P < Q$ because $P_3 = 0$ and $Q_3 = 1$. However, comparison of bits P_2 and Q_2 indicates $P > Q$ because $P_2 = 1$ and $Q_2 = 0$. The same is true for the remaining lower-order

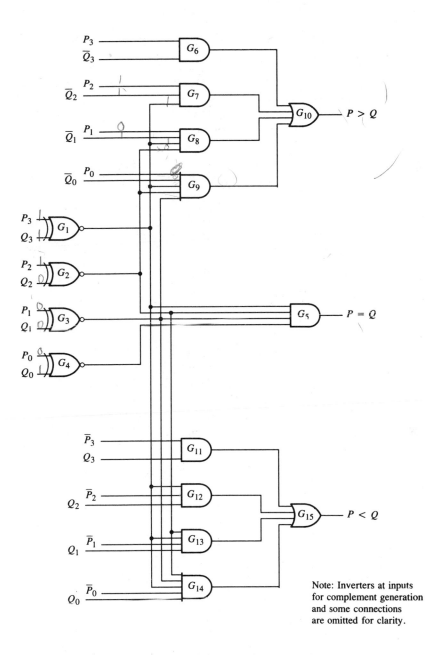

FIGURE 6–22 *Logic diagram for a four-bit comparator.*

EXAMPLE 6–6 Analyze the comparator operation for the numbers $P = 1010$ and $Q = 1001$.

Solution Figure 6-23 shows all logic levels within the comparator for the specified inputs.

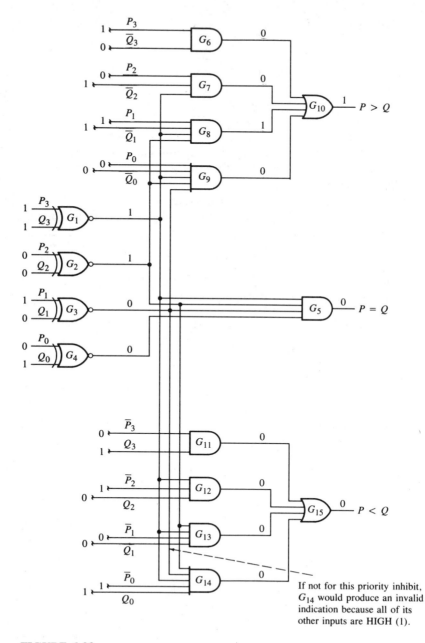

If not for this priority inhibit, G_{14} would produce an invalid indication because all of its other inputs are HIGH (1).

FIGURE 6-23

bits. In this case, priority must be given to P_3 and Q_3 because they determine the proper inequality condition.

Figure 6–22 shows a method of comparing two 4-bit numbers and generating a $P > Q$, or a $P < Q$, or a $P = Q$ output. The $P > Q$ condition is determined by gates G_6 through G_{10}. Gate G_6 checks for $P_3 = 1$ and $Q_3 = 0$, and its function is expressed as $P_3\overline{Q_3}$; gate G_7 checks for $P_2 = 1$ and $Q_2 = 0$ ($P_2\overline{Q_2}$); gate G_8 checks for $P_1 = 1$ and $Q_1 = 0$ ($P_1\overline{Q_1}$); gate G_9 checks for $P_0 = 1$ and $Q_0 = 0$ ($P_0\overline{Q_0}$). These conditions all indicate that number P is greater than number Q. The output of each of these gates is ORed by gate G_{10} to produce the $P > Q$ output.

Notice that the output of gate G_1 is connected to inputs of gates G_7, G_8, and G_9. This provides a *priority inhibit* so that if the proper inequality occurs in bits P_3 and Q_3 ($P_3 < Q_3$), the lower-order bit checks will be inhibited. A priority inhibit is also provided by gate G_2 to gates G_8 and G_9, and by gate G_3 to gate G_9.

Gates G_{11} through G_{15} check for a $P < Q$ condition. Each AND gate checks a given bit position for the occurrence of a 0 in the number P and a 1 in number Q. Each AND gate output is ORed by gate G_{15} to provide the $P < Q$ output. Priority inhibiting is provided as previously discussed. When all four bits in P equal the bits in Q, the output of each exclusive-OR gate is 1. This enables G_5 to produce a 1 on the $P = Q$ output.

Example 6–6 shows the comparison of specific numbers and indicates the logic levels throughout the circuitry. You should also go through the analysis with numbers of your own choosing to verify the operation.

The 7485 Four-Bit Magnitude Comparator

The 7485 is a representative integrated circuit comparator in the 54/74 family. The logic symbol is shown in Figure 6–24.

Notice that this device has all of the inputs and outputs of the generalized comparator just discussed, and in addition it has three *cascading inputs*. These inputs allow several comparators to be cascaded for comparison of any number of bits. To expand the comparator, the $P < Q$, $P = Q$, $P > Q$ outputs of the less

FIGURE 6–24 *Logic symbol for the 7485 four-bit magnitude comparator (pin numbers are in parentheses).*

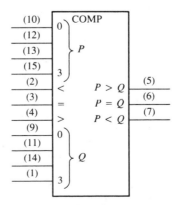

significant comparator are connected to the corresponding cascading inputs of the next higher comparator. The least significant comparator must have a HIGH on the = input and LOWs on the < and > inputs.

EXAMPLE 6–7

Use 7485 comparators to compare the magnitudes of two 8-bit binary numbers. Show the comparators with proper interconnections.

Solution   Two 7485s are required for eight bits. They are connected as shown in Figure 6–25 in a cascaded arrangement.

FIGURE 6–25   *An eight-bit magnitude comparator using two 7485s.*

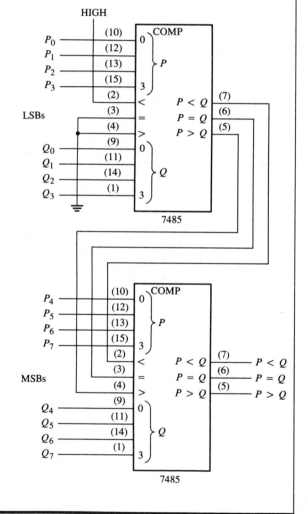

1. $P = 1011_2$ and $Q = 1010_2$ are applied to the operand inputs of a 7485. Determine the outputs.
2. $P = 11001011_2$ and $Q = 11010100_2$ are applied to the eight-bit comparator in Figure 6-25. Determine the states of pins 5, 6, and 7 on each 7485.

6-5   DECODERS

The basic function of a decoder is to detect the presence of a specified combination of bits (code) on its inputs and to indicate that presence by a specified output level. In its general form, a decoder has n input lines to handle n bits and from one to 2^n output lines to indicate the presence of one or more n-bit combinations.

The Basic Binary Decoder

Suppose we wish to determine when a binary 1001 occurs on the inputs of a digital circuit. An AND gate can be used as the basic decoding element because it produces a HIGH output only when all of its inputs are HIGH. Therefore, we must make sure that all of the inputs to the AND gate are HIGH when the binary number 1001 occurs; this can be done by inverting the two middle bits (the 0s), as shown in Figure 6-26.

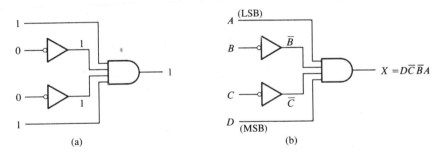

FIGURE 6-26   *Decoding logic for 1001 with an active-HIGH output.*

The logic equation for the decoder of Figure 6-26(a) is developed as illustrated in Figure 6-26(b). You should verify that the output function is 0 except when $A = 1$, $B = 0$, $C = 0$, and $D = 1$ are applied to the inputs. A is the LSB and D is the MSB. In the representation of a binary number or other weighted code in this book, the LSB is always the right-most bit in a horizontal arrangement and the top-most bit in a vertical arrangement, unless specified otherwise.

If a NAND gate is used in place of the AND gate, as shown in Figure 6–27, a LOW output will indicate the presence of the proper binary code.

FIGURE 6-27　*Decoding logic for 1001 with an active-LOW output.*

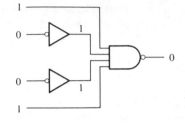

EXAMPLE 6-8

Determine the logic required to decode the binary number 1011_2 by producing a HIGH indication on the output.

Solution　The decoding function can be formed by complementing only the variables that appear as 0 in the binary number as follows:

$$X = D\overline{C}BA$$

This function can be implemented by connecting the true (uncomplemented) variables A, B, and D directly to the inputs of an AND gate, and inverting the variable C before applying it to the AND gate input. The decoding logic is shown in Figure 6–28.

FIGURE 6-28　*Decoding logic for producing a HIGH output when 1011_2 is on the inputs.*

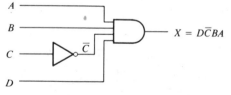

The Four-Bit Binary Decoder

In order to decode all possible combinations of four bits, sixteen decoding gates are required ($2^4 = 16$). This type of decoder is commonly called a *4-line-to-16-line* decoder because there are four inputs and sixteen outputs. A list of the sixteen binary code words and their corresponding decoding functions is given in Table 6–4.

　　If an active-LOW output is desired for each decoded number, the entire decoder can be implemented with NAND gates and inverters as follows:

First, since each variable and its complement are required in the decoder as seen from Table 6–4, the complements can be generated once and then used for all decoding gates as required, rather than duplicating an inverter each place a complement is used. This arrangement is shown in Figure 6–29.

FIGURE 6–29 *Logic for a 4-line-to-16-line decoder.*

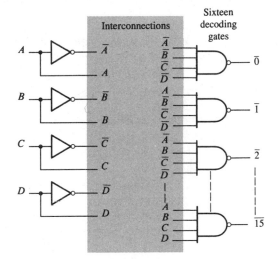

TABLE 6–4 *Decoding functions and truth table for a 4-line-to-16-line decoder.*

Decimal Digit	Binary Inputs				Logic Function	Outputs															
	D	C	B	A		0	1	2	3	4	5	6	7	8	9	10	11	12	13	14	15
0	0	0	0	0	$\overline{D}\,\overline{C}\,\overline{B}\,\overline{A}$	0	1	1	1	1	1	1	1	1	1	1	1	1	1	1	1
1	0	0	0	1	$\overline{D}\,\overline{C}\,\overline{B}A$	1	0	1	1	1	1	1	1	1	1	1	1	1	1	1	1
2	0	0	1	0	$\overline{D}\,\overline{C}B\overline{A}$	1	1	0	1	1	1	1	1	1	1	1	1	1	1	1	1
3	0	0	1	1	$\overline{D}\,\overline{C}BA$	1	1	1	0	1	1	1	1	1	1	1	1	1	1	1	1
4	0	1	0	0	$\overline{D}C\overline{B}\,\overline{A}$	1	1	1	1	0	1	1	1	1	1	1	1	1	1	1	1
5	0	1	0	1	$\overline{D}C\overline{B}A$	1	1	1	1	1	0	1	1	1	1	1	1	1	1	1	1
6	0	1	1	0	$\overline{D}CB\overline{A}$	1	1	1	1	1	1	0	1	1	1	1	1	1	1	1	1
7	0	1	1	1	$\overline{D}CBA$	1	1	1	1	1	1	1	0	1	1	1	1	1	1	1	1
8	1	0	0	0	$D\overline{C}\,\overline{B}\,\overline{A}$	1	1	1	1	1	1	1	1	0	1	1	1	1	1	1	1
9	1	0	0	1	$D\overline{C}\,\overline{B}A$	1	1	1	1	1	1	1	1	1	0	1	1	1	1	1	1
10	1	0	1	0	$D\overline{C}B\overline{A}$	1	1	1	1	1	1	1	1	1	1	0	1	1	1	1	1
11	1	0	1	1	$D\overline{C}BA$	1	1	1	1	1	1	1	1	1	1	1	0	1	1	1	1
12	1	1	0	0	$DC\overline{B}\,\overline{A}$	1	1	1	1	1	1	1	1	1	1	1	1	0	1	1	1
13	1	1	0	1	$DC\overline{B}A$	1	1	1	1	1	1	1	1	1	1	1	1	1	0	1	1
14	1	1	1	0	$DCB\overline{A}$	1	1	1	1	1	1	1	1	1	1	1	1	1	1	0	1
15	1	1	1	1	$DCBA$	1	1	1	1	1	1	1	1	1	1	1	1	1	1	1	0

In order to decode each of the sixteen binary code words, sixteen NAND gates are required (AND gates can be used to produce active-HIGH outputs). The decoding gate arrangement is illustrated in Figure 6–29.

Rather than reproduce the complex logic diagram for the decoder each time it is required in a schematic, a simpler representation is normally used. A logic symbol for a 4-line-to-16-line decoder is shown in Figure 6–30.

FIGURE 6–30   *Logic symbol for a 4-line-to-16-line decoder.*

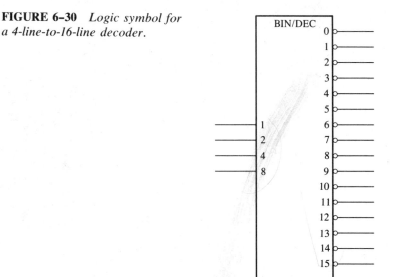

The 74154 4-Line-to-16-Line Decoder

The 74154 is a good example of a TTL MSI decoder. Its logic diagram is shown in Figure 6–31(a), and the logic symbol appears in Figure 6–31(b). The additional inverters on the inputs are required to prevent excessive loading of the driving source(s). Each input is connected to the input of only one inverter, rather than to the inputs of several NAND gates as in Figure 6–29. There is also an *enable* function provided on this particular device, which is implemented with a NOR gate used as a negative AND. A LOW level on each input, $\overline{G}_1$ and $\overline{G}_2$, is required in order to make the enable gate output (G) HIGH. The enable gate output is connected to an input of *each* NAND gate, so it must be HIGH for the gates to be enabled. If the enable gate is not activated (LOW on both inputs), then all sixteen decoder outputs will be HIGH *regardless* of the states of the four input variables, A, B, C, and D.

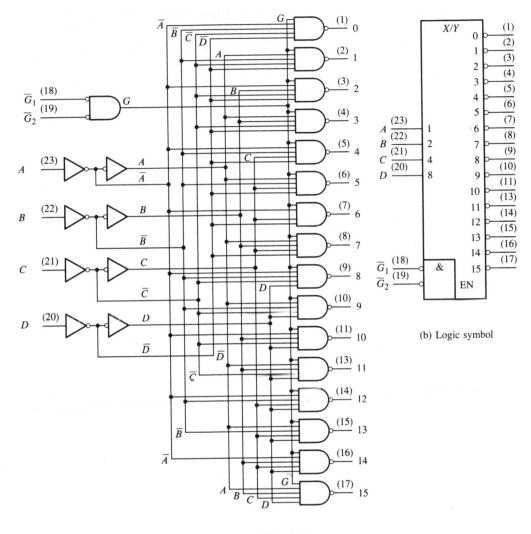

(a) Logic diagram

(b) Logic symbol

FIGURE 6–31 *The 74154 4-line-to-16-line decoder.*

EXAMPLE 6-9 A certain application requires that a five-bit binary number be decoded. Use 74154 decoders to implement the logic. The binary number is represented as $A_4A_3A_2A_1A_0$.

Solution Since the 74154 can handle only four bits, two decoders must be used to decode five bits. The fifth bit, A_4, is connected to the enable inputs, $\overline{G}_1$ and $\overline{G}_2$, of one decoder, and $\overline{A}_4$ is connected to the enable inputs of the other

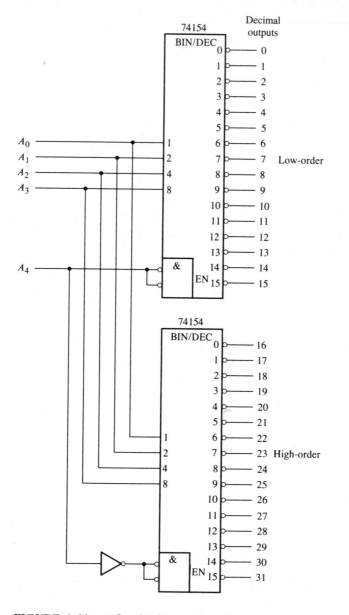

FIGURE 6–32 *A five-bit binary decoder using 74154s.*

decoder, as shown in Figure 6–32. When the binary number is 15 or less, $A_4 = 0$, and the low-order decoder is enabled and the high-order decoder is disabled. When the binary number is greater than 15, $A_4 = 1$, and the high-order decoder is enabled and the low-order decoder is disabled.

The BCD-to-Decimal Decoder

The BCD-to-decimal decoder converts each BCD code word (8421 code) into one of ten possible decimal digit indications. It is frequently referred to as a *4-line-to-10-line* decoder, although other types of decoders also fall into this category (such as an Excess-3 decoder).

The method of implementation is essentially the same as for the 4-line-to-16-line decoder previously discussed, except that only *ten* decoding gates are required because the BCD code represents only the ten decimal digits 0 through 9. A list of the ten BCD code words and their corresponding decoding functions is given in Table 6–5. Each of these decoding functions is implemented with NAND gates to provide active-LOW outputs. If an active-HIGH output is required, AND gates are used for decoding. The logic is identical to that of the first ten decoding gates in the 1-of-16 decoder.

TABLE 6–5 *BCD decoding functions.*

Decimal Digit	BCD Code				Logic Function
	D	C	B	A	
0	0	0	0	0	$\overline{D}\,\overline{C}\,\overline{B}\,\overline{A}$
1	0	0	0	1	$\overline{D}\,\overline{C}\,\overline{B}\,A$
2	0	0	1	0	$\overline{D}\,\overline{C}\,B\,\overline{A}$
3	0	0	1	1	$\overline{D}\,\overline{C}\,B\,A$
4	0	1	0	0	$\overline{D}\,C\,\overline{B}\,\overline{A}$
5	0	1	0	1	$\overline{D}\,C\,\overline{B}\,A$
6	0	1	1	0	$\overline{D}\,C\,B\,\overline{A}$
7	0	1	1	1	$\overline{D}\,C\,B\,A$
8	1	0	0	0	$D\,\overline{C}\,\overline{B}\,\overline{A}$
9	1	0	0	1	$D\,\overline{C}\,\overline{B}\,A$

EXAMPLE 6–10

The 7442A is an integrated circuit BCD-to-decimal decoder. The logic symbol is shown in Figure 6–33.

If the input waveforms in Figure 6–34 are applied to the inputs of the 7442A, sketch the output waveforms.

FIGURE 6–33 *The 7442A BCD-to-decimal decoder.*

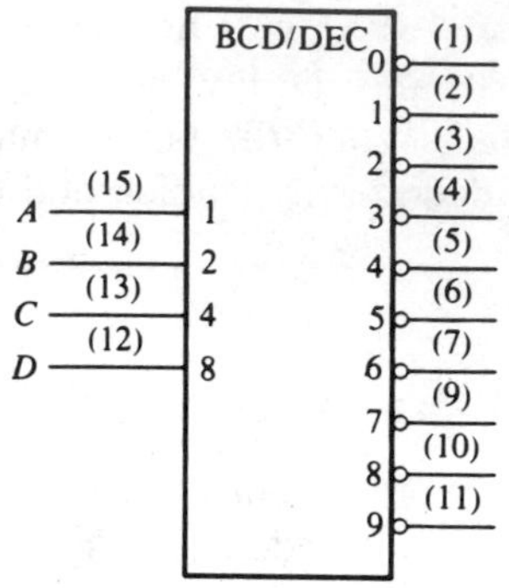

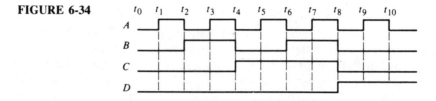

FIGURE 6-34

Solution   The output waveforms are shown in Figure 6–35. As you can see, the inputs are sequenced through the BCD for digits 0 through 9. The output waveforms indicate that sequence.

FIGURE 6–35

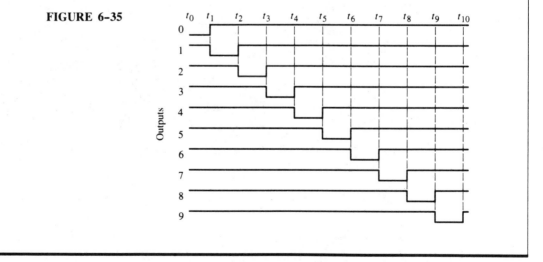

The BCD-to-Seven-Segment Decoder/Driver

This type of decoder accepts the BCD code on its inputs and provides outputs to energize seven-segment display devices in order to produce a decimal readout.

FIGURE 6–36 *Seven-segment display format showing arrangement of segments.*

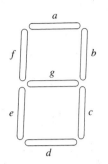

Before proceeding with a discussion of this decoder, let us examine the basics of a seven-segment display device.

Figure 6–36 shows a common display format composed of seven elements or segments. By energizing certain combinations of these segments, each of the ten decimal digits can be produced. Figure 6–37 illustrates this method of digital display for each of the ten digits by using a darker segment to represent one that is energized. To produce a 1, segments *b* and *c* are energized; to produce a 2, segments *a*, *b*, *g*, *e*, and *d* are used; and so on.

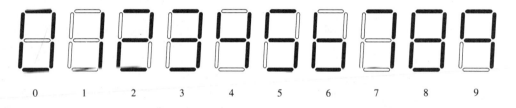

FIGURE 6–37 *Display of decimal digits with a seven-segment device.*

LED displays One common type of seven-segment display consists of light-emitting diodes (LEDs) arranged as shown in Figure 6–38. Each segment is an LED that emits light when current flows through it. In Figure 6–38(a), the common-anode arrangement requires the driving circuit to provide a LOW level voltage in order to activate a given segment. When a LOW is applied to a segment input, the LED is forward biased and current flows through it. In Figure 6–38(b), the common-cathode arrangement requires the driver to provide a HIGH level voltage to activate a segment. When a HIGH is applied to a segment input, the LED is forward biased and current flows through it.

The activated segments for each of the ten decimal digits are listed in Table 6–6.

We will now examine the decoding logic required to produce the format for a seven-segment display with a BCD input. Notice that segment *a* is activated for digits 0, 2, 3, 5, 7, 8, and 9. Segment *b* is activated for digits 0, 1, 2, 3, 4, 7, 8, and 9, and so on. If we let the BCD inputs to the decoder be represented by the general form *DCBA*, a Boolean expression can be found for each segment in the display,

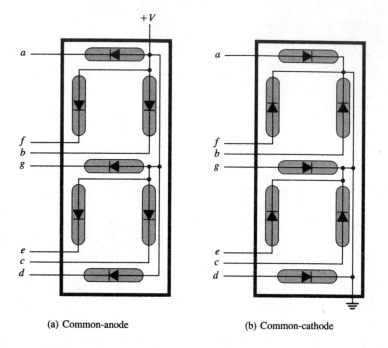

(a) Common-anode   (b) Common-cathode

FIGURE 6–38   *Arrangements of seven-segment LED displays.*

and this will tell us the logic circuitry required to drive or activate each segment. For example, the equation for segment *a* is as follows:

$$a = \overline{D}\,\overline{C}\,\overline{B}\,\overline{A} + \overline{D}\,\overline{C}B\overline{A} + \overline{D}\,\overline{C}BA + \overline{D}C\overline{B}A + \overline{D}CBA + D\overline{C}\,\overline{B}\,\overline{A} + D\overline{C}\,\overline{B}A$$

This equation says that segment *a* is activated or "true" if the BCD code is "0 OR 2 OR 3 OR 5 OR 7 OR 8 OR 9." Table 6–7 lists the logic function for each of the seven segments.

TABLE 6–6   *Seven-segment display format.*

Digit	Segments Activated
0	*a, b, c, d, e, f*
1	*b, c*
2	*a, b, g, e, d*
3	*a, b, c, d, g*
4	*b, c, f, g*
5	*a, c, d, f, g*
6	*c, d, e, f, g*
7	*a, b, c*
8	*a, b, c, d, e, f, g*
9	*a, b, c, f, g*

TABLE 6-7 *Seven-segment decoding functions.*

Segment	Used in Digits	Logic Function
a	0, 2, 3, 5, 7, 8, 9	$\overline{D}\,\overline{C}\,\overline{B}\,\overline{A} + \overline{D}\,\overline{C}B\overline{A} + \overline{D}\,\overline{C}BA + \overline{D}C\overline{B}A + \overline{D}CBA$ $+ D\overline{C}\,\overline{B}\,\overline{A} + D\overline{C}\,\overline{B}A$
b	0, 1, 2, 3, 4, 7, 8, 9	$\overline{D}\,\overline{C}\,\overline{B}\,\overline{A} + \overline{D}\,\overline{C}\,\overline{B}A + \overline{D}\,\overline{C}B\overline{A} + \overline{D}\,\overline{C}BA + \overline{D}C\overline{B}\,\overline{A} + \overline{D}CBA$ $+ D\overline{C}\,\overline{B}\,\overline{A} + D\overline{C}\,\overline{B}A$
c	0, 1, 3, 4, 5, 6, 7, 8, 9	$\overline{D}\,\overline{C}\,\overline{B}\,\overline{A} + \overline{D}\,\overline{C}\,\overline{B}A + \overline{D}\,\overline{C}BA + \overline{D}C\overline{B}\,\overline{A} + \overline{D}C\overline{B}A + \overline{D}CB\overline{A} + \overline{D}CBA$ $+ D\overline{C}\,\overline{B}\,\overline{A} + D\overline{C}\,\overline{B}A$
d	0, 2, 3, 5, 6, 8	$\overline{D}\,\overline{C}\,\overline{B}\,\overline{A} + \overline{D}\,\overline{C}B\overline{A} + \overline{D}\,\overline{C}BA + \overline{D}C\overline{B}A + \overline{D}CB\overline{A} + D\overline{C}\,\overline{B}\,\overline{A}$
e	0, 2, 6, 8	$\overline{D}\,\overline{C}\,\overline{B}\,\overline{A} + \overline{D}\,\overline{C}B\overline{A} + \overline{D}CB\overline{A} + D\overline{C}\,\overline{B}\,\overline{A}$
f	0, 4, 5, 6, 8, 9	$\overline{D}\,\overline{C}\,\overline{B}\,\overline{A} + \overline{D}C\overline{B}\,\overline{A} + \overline{D}C\overline{B}A + \overline{D}CB\overline{A} + D\overline{C}\,\overline{B}\,\overline{A} + D\overline{C}\,\overline{B}A$
g	2, 3, 4, 5, 6, 8, 9	$\overline{D}\,\overline{C}B\overline{A} + \overline{D}\,\overline{C}BA + \overline{D}C\overline{B}\,\overline{A} + \overline{D}C\overline{B}A + \overline{D}CB\overline{A}$ $+ D\overline{C}\,\overline{B}\,\overline{A} + D\overline{C}\,\overline{B}A$

From the expressions in Table 6–7, the logic for the BCD-to-seven-segment decoder can be implemented. Each of the ten BCD code words is decoded, and then the decoding gates are ORed as dictated by the logic expression for each segment. For instance, segment a requires that the decoded BCD digits 0, 2, 3, 5, 7, 8, and 9 be ORed, as shown in Figure 6–39.

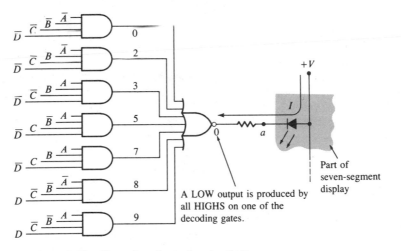

FIGURE 6-39 *Decoding logic for the "a" segment.*

Segment b requires that the decoded BCD digits 0, 1, 2, 3, 4, 7, 8, and 9 be ORed. Segment c is activated by ORing the outputs of the 0, 1, 3, 4, 5, 6, 7, 8, and 9 decode gates, and so on. A logic symbol for a seven-segment decoder is shown in Figure 6–40. The bubbles indicate active-0 (LOW) outputs for compatibility with the common-anode type display.

FIGURE 6–40 *Logic symbol for a seven-segment decoder/driver.*

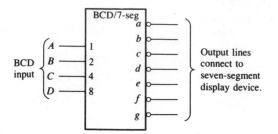

LCD displays   Another common type of seven-segment display is the *liquid crystal display*. LCDs operate by polarizing light so that a nonactivated segment reflects incident light and thus appears invisible against its background. An activated segment does not reflect incident light and thus appears dark. LCDs consume much less power than LEDs but cannot be seen in the dark, while LEDs can.

LCDs operate from a low-frequency signal voltage (30 Hz to 60 Hz) applied between the segment and a common element called the *backplane* (bp). The basic operation is as follows: Figure 6–41 shows a square wave used as the source signal. Each segment in the display is driven by an exclusive-OR gate with one input connected to an output of the seven-segment decoder/driver and the other input connected to the signal source. When the decoder/driver output is HIGH (1), the exclusive-OR output is a square wave that is 180° out-of-phase with the

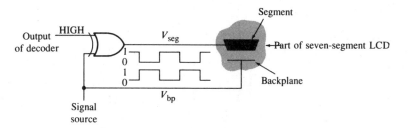

(a) Segment on

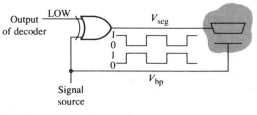

(b) Segment off

FIGURE 6–41 *Basic operation of an LCD.*

signal source, as shown in Figure 6–41(a). You can verify this by reviewing the truth table operation of the exclusive-OR. The resulting voltage between the LCD segment and the backplane is also a square wave because when $V_{seg} = 1$, $V_{bp} = 0$, and vice versa. This turns the segment on.

When the decoder/driver output is LOW (0), the exclusive-OR output is a square wave that is in-phase with the signal source, as shown in Figure 6–41(b). The resulting voltage difference between the segment and the backplane is *zero* because $V_{seg} = V_{bp}$. This turns the segment off.

TTL is not recommended for driving LCDs because its LOW level voltage is typically a few tenths of a volt, thus creating a dc component across the LCD which degrades its performance. CMOS is therefore used in LCD applications.

Zero Suppression An additional feature found on many seven-segment decoders is the *zero suppression logic*. This extra function is useful in multidigit displays because it is used to blank out unnecessary zeros in the display. For instance, the number 0006.400 would be displayed as 6.4, which is read more easily. Blanking of the zeros on the front of the number is called *leading zero suppression,* and blanking of the zeros after the number is called *trailing zero suppression*.

A block diagram of a four-digit display is shown in Figure 6–42. It will be used to illustrate the requirements for leading zero suppression. Notice that two additional functions have been added to each BCD-to-seven-segment decoder, a *ripple blanking input* (RBI) and a *ripple blanking output* (RBO). The highest-order digit position is always blanked if a 0 code appears on its BCD inputs *and* the blanking input is HIGH. Each lower-order digit position is blanked if a 0 code appears on its BCD inputs, *and* the next higher-order digit is a 0 as indicated by a HIGH on its blanking output. The ripple blanking output of any decoder indicates that it has a BCD 0 on its inputs, and *all* higher-order digits are also 0. The blanking output of each stage is connected to the blanking input of the next lower-order stage, as shown in the diagram. As an example, in Figure 6–42 the highest-order digit is 0, which is therefore blanked. Also, the next digit is a 0, and because the highest-order digit is 0, it is also blanked. The remaining two digits are displayed.

For the fractional portion of a display (the digits to the right of the decimal point), trailing zero suppression is used. That is, the lowest-order digit is blanked if it is 0, and each digit that is 0 *and* is followed by 0s in all the lower-order positions is also blanked. To illustrate, Figure 6–43 shows a block diagram in which there are three digits to the right of the decimal point. In this example, the lowest-order digit is blanked because it is a 0. The next digit is also 0, *and* its blanking input is HIGH, so it is blanked. The highest-order digit is a 5, and it is displayed. Notice that the blanking output of each decoder stage is connected to the blanking input of the next higher-order stage.

In this section, several decoders have been introduced. The basic principles can be extended to other types, such as the 2-line-to-4-line decoder, the 3-line-to-8-line decoder, and the Excess-3-to-decimal 4-line-to-10-line decoder, all of which are available in MSI.

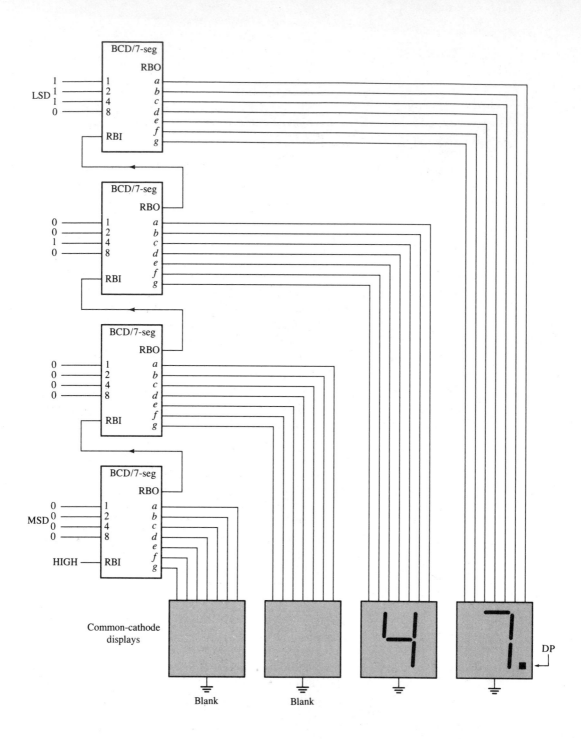

FIGURE 6–42 *Illustration of leading zero suppression.*

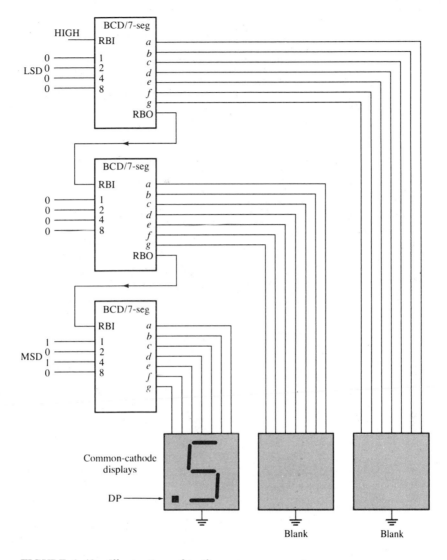

FIGURE 6–43 *Illustration of trailing zero suppression.*

SECTION
REVIEW
6–5

1. A 3-line-to-8-line decoder can be used for octal-to-decimal decoding. When a 101_2 is on the inputs, which output line is activated?
2. How many 74154 4-line-to-16-line decoders are necessary to decode a six-bit binary number?
3. Would you select a decoder/driver with active-HIGH or active-LOW outputs to drive a common-anode seven-segment LED display?

6–6   ENCODERS

An *encoder* is a combinational logic circuit that essentially performs a "reverse" decoder function. An encoder accepts an active level on one of its inputs representing a digit, such as a decimal or octal digit, and converts it to a coded output, such as binary or BCD. Encoders can also be devised to encode various symbols and alphabetic characters. This process of converting from familiar symbols or numbers to a coded format is called *encoding*.

The Decimal-to-BCD Encoder

This type of encoder has ten inputs—one for each decimal digit—and four outputs corresponding to the BCD code, as shown in Figure 6–44. This is a basic 10-line-to-4-line encoder.

FIGURE 6–44 *Logic symbol for a decimal-to-BCD encoder.*

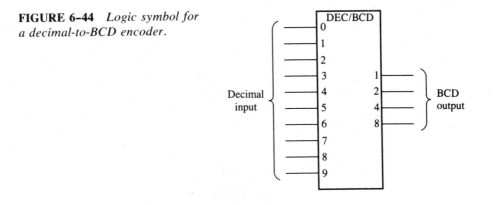

The BCD (8421) code is listed in Table 6–8, and from this we can determine the relationship between each BCD bit and the decimal digits. For instance, the most significant bit of the BCD code, D, is a 1 for decimal digit 8 or 9. The OR expression for bit D in terms of the decimal digits can therefore be written

$$D = 8 + 9$$

Bit C is a 1 for decimal digits 4, 5, 6, or 7 and can be expressed as an OR function as follows:

$$C = 4 + 5 + 6 + 7$$

Bit B is a 1 for decimal digits 2, 3, 6, or 7 and can be expressed as

$$B = 2 + 3 + 6 + 7$$

Finally, A is a 1 for digits 1, 3, 5, 7, or 9. The expression for A is

$$A = 1 + 3 + 5 + 7 + 9$$

TABLE 6-8

Decimal Digit	BCD Code			
	D	C	B	A
0	0	0	0	0
1	0	0	0	1
2	0	0	1	0
3	0	0	1	1
4	0	1	0	0
5	0	1	0	1
6	0	1	1	0
7	0	1	1	1
8	1	0	0	0
9	1	0	0	1

Now we can implement the logic circuitry required for encoding each decimal digit to a BCD code by using the logic expressions just developed. It is simply a matter of ORing the appropriate decimal digit input lines to form each BCD output. The basic encoder logic resulting from these expressions is shown in Figure 6–45.

FIGURE 6–45 *Basic logic diagram of a decimal-to-BCD encoder. A 0 digit input is not needed because the BCD outputs are all LOW when there are no HIGH inputs.*

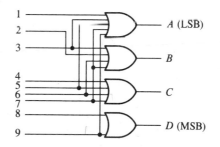

The basic operation is as follows: When a HIGH appears on one of the decimal digit input lines, the appropriate levels occur on the four BCD output lines. For instance, if input line 9 is HIGH (assuming all other input lines are LOW), this will produce a HIGH on outputs A and D and LOWs on outputs B and C, which is the BCD code (1001) for decimal 9.

The Decimal-to-BCD Priority Encoder

This type of encoder performs the same basic encoding function as that previously discussed. It also offers additional flexibility in that it can be used in applications requiring priority detection. The *priority* function means that the encoder will produce a BCD output corresponding to the *highest-order decimal digit* appearing on the inputs, and will ignore all others. For instance, if the 6 and the 3 inputs are both HIGH, the BCD output is 0110 (which represents decimal 6).

Now let us look at the requirements for the priority detection logic. The purpose of this logic circuitry is to prevent a lower-order digit input from disrupting the encoding of a higher-order digit; this is accomplished by using *inhibit gates*. We will start by examining each BCD output (beginning with output A). Referring to Figure 6–45, note that A is HIGH when 1, 3, 5, 7, or 9 is HIGH. Digit input 1 must be allowed to activate the A output only if no higher-order digits *other than those that also activate A* are HIGH. This can be stated as follows:

1. A is HIGH if 1 is HIGH and 2, 4, 6, and 8 are LOW.
2. A is HIGH if 3 is HIGH and 4, 6, and 8 are LOW.
3. A is HIGH if 5 is HIGH and 6 and 8 are LOW.
4. A is HIGH if 7 is HIGH and 8 is LOW.
5. A is HIGH if 9 is HIGH.

These five statements describe the priority of encoding for the BCD bit A. The A output is HIGH if any of the conditions listed occur; that is, A is true if statement 1, statement 2, statement 3, statement 4, or statement 5 is true. This can be expressed in the form of the following logic equation:

$$A = 1 \cdot \overline{2} \cdot \overline{4} \cdot \overline{6} \cdot \overline{8} + 3 \cdot \overline{4} \cdot \overline{6} \cdot \overline{8} + 5 \cdot \overline{6} \cdot \overline{8} + 7 \cdot \overline{8} + 9$$

From this expression the logic circuitry required for the A output with priority inhibits can be readily implemented, as shown in Figure 6–46.

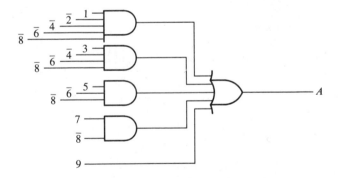

FIGURE 6–46 *Logic for the A-bit output of a decimal-to-BCD priority encoder.*

The same reasoning process can be applied to output B, and the following logical statements can be made:

1. B is HIGH if 2 is HIGH and 4, 5, 8, and 9 are LOW.
2. B is HIGH if 3 is HIGH and 4, 5, 8, and 9 are LOW.
3. B is HIGH if 6 is HIGH and 8 and 9 are LOW.
4. B is HIGH if 7 is HIGH and 8 and 9 are LOW.

These statements are summarized in the following equation, and the logic implementation is shown in Figure 6–47:

$$B = 2 \cdot \overline{4} \cdot \overline{5} \cdot \overline{8} \cdot \overline{9} + 3 \cdot \overline{4} \cdot \overline{5} \cdot \overline{8} \cdot \overline{9} + 6 \cdot \overline{8} \cdot \overline{9} + 7 \cdot \overline{8} \cdot \overline{9}$$

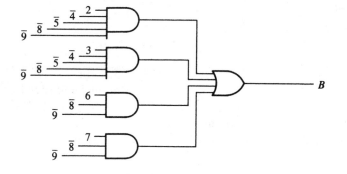

FIGURE 6-47   *Logic for the B-bit output of a decimal-to-BCD priority encoder.*

Output *C* can be described as follows:

1.   *C* is HIGH if 4 is HIGH and 8 and 9 are LOW.
2.   *C* is HIGH if 5 is HIGH and 8 and 9 are LOW.
3.   *C* is HIGH if 6 is HIGH and 8 and 9 are LOW.
4.   *C* is HIGH if 7 is HIGH and 8 and 9 are LOW.

In equation form, output *C* is

$$C = 4 \cdot \overline{8} \cdot \overline{9} + 5 \cdot \overline{8} \cdot \overline{9} + 6 \cdot \overline{8} \cdot \overline{9} + 7 \cdot \overline{8} \cdot \overline{9}$$

The logic circuitry for the *C* output appears in Figure 6-48.

FIGURE 6-48   *Logic for the C-bit output of a decimal-to-BCD priority encoder.*

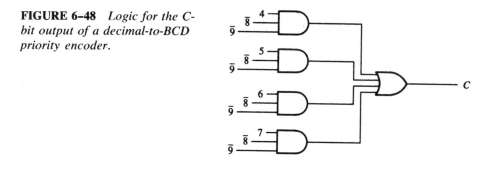

Finally, for the *D* output,

D is HIGH if 8 is HIGH or if 9 is HIGH.

This statement appears in equation form as follows:

$$D = 8 + 9$$

The logic for this output is shown in Figure 6-49. No inhibits are required. We now have developed the basic logic for the decimal-to-BCD priority encoder. All of the complements of the input digits are realized by inverting the inputs.

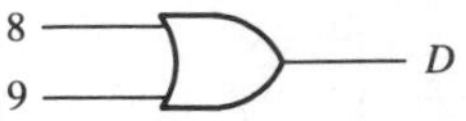

FIGURE 6-49 *Logic for the D-bit output of a decimal-to-BCD priority encoder.*

EXAMPLE 6-11

The 74147 is a decimal-to-BCD priority encoder. As indicated on the logic symbol in Figure 6–50, its inputs and outputs are all active-LOW.

If LOW levels appear on pins 1, 4, and 13, indicate the state of the four outputs. All other inputs are HIGH. Pin numbers are in parentheses.

FIGURE 6-50 *Logic symbol for the 74147 decimal-to-BCD priority encoder (HPRI means highest priority).*

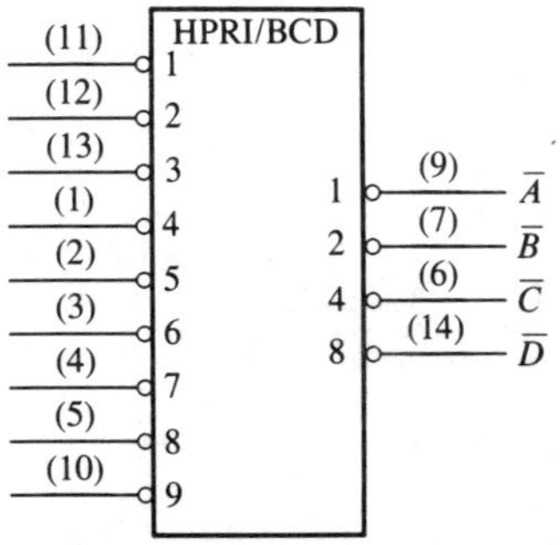

Solution   Pin 4 is the highest-order decimal input having a LOW level and represents decimal 7. Therefore, the output levels indicate the BCD code for decimal 7 where $\overline{A}$ is the LSB and $\overline{D}$ is the MSB. $\overline{A}$ is LOW, $\overline{B}$ is LOW, $\overline{C}$ is LOW, and $\overline{D}$ is HIGH.

An Encoder Application

A classic application example is a keyboard encoder. The ten decimal digits on the keyboard of a calculator, for example, must be encoded for processing by the logic circuitry. When one of the keypads is pressed, the decimal digit is encoded to the corresponding BCD code. Figure 6–51 shows a simple keyboard encoder arrangement using a 74147 priority encoder. Notice that there are ten push-button switches, each with a *pull-up* resistor to $+V_{CC}$. The pull-up resistor ensures that the line is HIGH when a switch is not depressed. When a switch is depressed, the line is connected to ground and a LOW is applied to the corresponding encoder input. The zero key is not connected because the BCD output represents zero when none of the other keys are depressed.

The BCD output of the encoder goes into a storage device, and each successive BCD code is stored until the entire number has been entered. Methods of storing BCD numbers and binary data are covered in the following chapters.

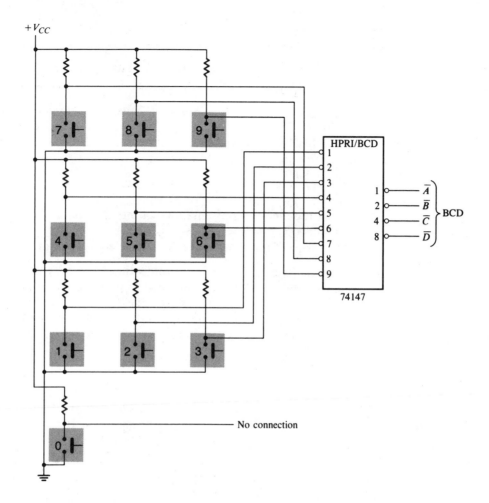

FIGURE 6–51 *A simplified keyboard encoder.*

**SECTION
REVIEW
6–6**

1. Suppose that HIGH levels are applied to the 2 input and the 9 input of the circuit in Figure 6–45.
   (a) What are the states of the output lines?
   (b) Does this represent a valid BCD code?
   (c) What is the restriction on the encoder logic in Figure 6–45?
2. (a) What is the $\overline{D}\,\overline{C}\,\overline{B}\,\overline{A}$ output when LOWs are applied to pins 1 and 5 of the 74147 in Figure 6–50?
   (b) What does this output represent?

6-7   CODE CONVERSION

In this section we will examine some methods of using combinational logic circuits to convert from one code to another.

BCD-to-Binary Conversion

One method of BCD-to-binary code conversion involves the use of adder circuits. The basic conversion process is as follows:

1. The *value* of *each* bit in the BCD number is represented by a binary number.
2. All of the binary representations of bits that are 1s in the BCD number are added.
3. The result of this addition is the binary equivalent of the BCD number.

A more concise statement of this operation is that the binary numbers representing the weights of the BCD bits are summed to produce the total binary number.

We will examine an eight-bit BCD code (one that represents a two-digit decimal number) in order to understand the relationship between BCD and binary. For instance, we already know the decimal number 87 can be expressed in BCD as

$$\underbrace{1000}_{8}\,\underbrace{0111}_{7}$$

The left-most four-bit group represents 80, and the right-most four-bit group represents 7. That is, the left-most group has a weight of 10, and the right-most has a weight of 1. Within each group, the binary weight of each bit is as follows:

	Tens Digit				*Units Digit*			
Weight:	80	40	20	10	8	4	2	1
Bit designation:	D_1	C_1	B_1	A_1	D_0	C_0	B_0	A_0

The binary equivalent of each BCD bit is a binary number representing the *weight* of that bit within the total BCD number. This representation is given in Table 6-9.

If the binary representations for the weight of each 1 in the BCD number are added, the result is the binary number corresponding to the BCD number. Example 6-12 illustrates this.

With this basic procedure in mind, let us determine how the process can be implemented with logic circuits. Once the binary representation for each 1 in the BCD number is determined, adder circuits can be used to add the 1s in each column of the binary representation. The 1s occur in a given column only when the corresponding BCD bit is a 1. The occurrence of a BCD 1 can therefore be used to generate the proper binary 1 in the appropriate column of the adder structure. To handle a two-decimal digit (two-decade) BCD code, eight BCD input lines and seven binary outputs are required. (It takes 7 binary bits to represent numbers up through 99.)

TABLE 6-9 *Binary representations of BCD bit weights.*

BCD Bit	BCD Weight	Binary Representation						
		64	32	16	8	4	2	1
A_0	1	0	0	0	0	0	0	1
B_0	2	0	0	0	0	0	1	0
C_0	4	0	0	0	0	1	0	0
D_0	8	0	0	0	1	0	0	0
A_1	10	0	0	0	1	0	1	0
B_1	20	0	0	1	0	1	0	0
C_1	40	0	1	0	1	0	0	0
D_1	80	1	0	1	0	0	0	0

EXAMPLE 6-12

Convert the BCD numbers 00100111 (decimal 27) and 10011000 (decimal 98) to binary.

Solutions Write the binary for the weights of all 1s appearing in the numbers, and then add them together.

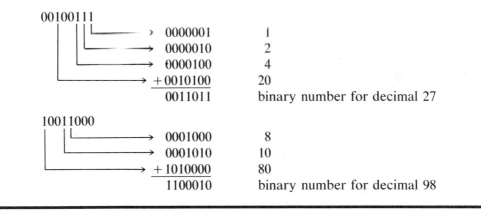

Referring to Table 6-9, notice that the "1" (LSB) column of the binary representation has only a single 1 and no possibility of an input carry, so that a straight connection from the A_0 bit of the BCD input to the least significant binary output is sufficient. In the "2" column of the binary representation, the possible occurrence of the two 1s can be accommodated by adding the B_0 bit and the A_1 bit of the BCD number. In the "4" column of the binary representation, the possible occurrence of the two 1s is handled by adding the C_0 bit and the B_1 bit of the BCD number. In the "8" column of the binary representation, the possibility of the three 1s is handled by adding the D_0, A_1, and C_1 bits of the BCD number. In the "16" column, the B_1 and the D_1 bits are added. In the "32" column, only a single

1 is possible, so the C_1 bit is added to the carry from the "16" column. In the "64" column, only a single 1 can occur, so the D_1 bit is added only to the carry from the "32" column. A method of implementing these requirements with full-adders is shown in Figure 6–52.

Most MSI code converters are implemented with preprogrammed read only memories (ROMs). These are covered in a later chapter.

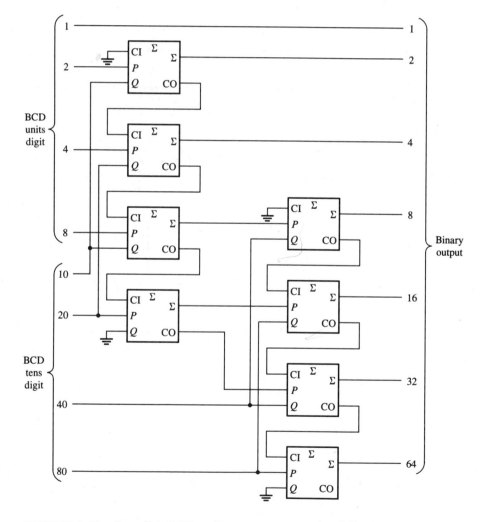

FIGURE 6–52 *Two-digit BCD-to-binary converter using full-adders.*

Binary-to-Gray and Gray-to-Binary Conversion

The basic process for Gray/binary conversions was covered in Chapter 2. We will now see how exclusive-OR gates can be used for these conversions. Figure 6–53 shows a four-bit binary-to-Gray code converter, and Figure 6–54 illustrates a four-bit Gray-to-binary converter.

FIGURE 6–53 *Four-bit binary-to-Gray conversion logic.*

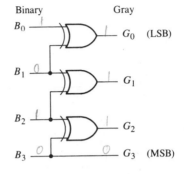

FIGURE 6–54 *Four-bit Gray-to-binary conversion logic.*

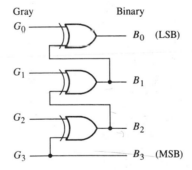

EXAMPLE 6–13 Convert the following binary codes to Gray code using exclusive-OR gates:
(a) 0101_2 (b) 00111_2 (c) 101011_2

Solutions
(a) Binary 0101_2 is 0111 Gray.
(b) 00111_2 is 00100 Gray.
(c) 101011_2 is 111110 Gray.
See Figure 6–55.

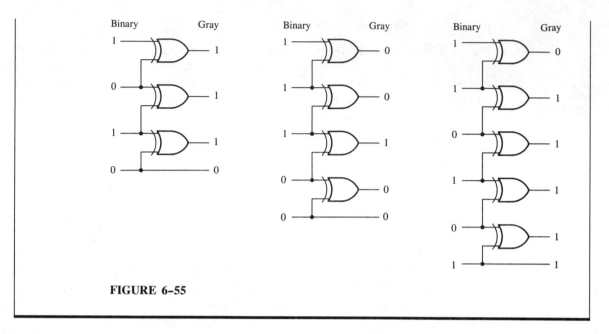

FIGURE 6–55

EXAMPLE 6–14   Convert the following Gray codes to binary using exclusive-OR gates:
(a) 1011   (b) 11000   (c) 1001011

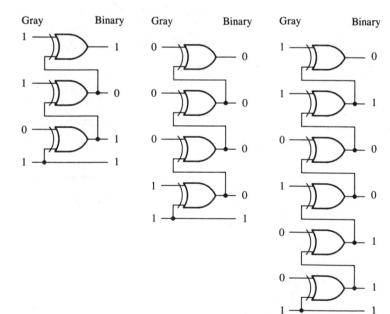

FIGURE 6–56

Solutions
(a) Gray code 1011 is 1101_2.
(b) 11000 is 10000_2.
(c) 1001011 is 1110010_2.
See Figure 6–56.

SECTION REVIEW 6-7

1. Convert the BCD number 10000101 to binary.
2. Draw the logic diagram for converting an eight-bit binary number to Gray code.

6-8 MULTIPLEXERS (DATA SELECTORS)

A multiplexer (MUX) is a device that allows digital information from several sources to be routed onto a single line for transmission over that line to a common destination. The basic multiplexer, then, has several data-input lines and a single output line. It also has *data-select* inputs that permit digital data on any one of the inputs to be switched to the output line. A logic symbol for a four-input multiplexer is shown in Figure 6–57. Notice that there are two data-select lines because with two select bits, each of the four *data-input* lines can be selected.

FIGURE 6-57 *Logic symbol for a 1-of-4 data selector/multiplexer.*

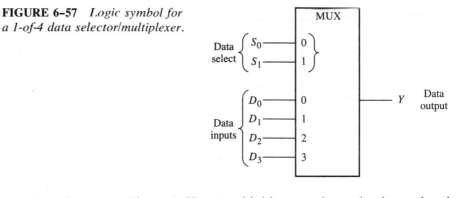

In reference to Figure 6–57, a two-bit binary code on the data-select inputs will allow the data on the corresponding data input to pass through to the data output. If a binary 0 ($S_1 = 0$ and $S_0 = 0$) is applied to the data-select lines, the data on input D_0 appear on the data-output line. If a binary 1 ($S_1 = 0$ and $S_0 = 1$) is applied to the data-select lines, the data on input D_1 appear on the data output. If a binary 2 ($S_1 = 1$ and $S_0 = 0$) is applied, the data on D_2 appear on the output. If a binary 3 ($S_1 = 1$ and $S_0 = 1$) is applied, the data on D_3 are switched to the output line. A summary of this operation is given in Table 6–10.

Now let us look at the logic circuitry required to perform this multiplexing operation. The data output is equal to the state of the *selected* data input. We

TABLE 6-10 *Data selection for a four-input multiplexer.*

Data-Select Inputs		Input Selected
S_1	S_0	
0	0	D_0
0	1	D_1
1	0	D_2
1	1	D_3

should, therefore, be able to derive a logical expression for the output in terms of the *data input* and the *select inputs*. This can be done as follows:

The data output Y is equal to the data input D_0 if and only if $S_1 = 0$ and $S_0 = 0$:

$$Y = D_0 \overline{S}_1 \overline{S}_0$$

The data output is equal to D_1 if and only if $S_1 = 0$ and $S_0 = 1$:

$$Y = D_1 \overline{S}_1 S_0$$

The data output is equal to D_2 if and only if $S_1 = 1$ and $S_0 = 0$:

$$Y = D_2 S_1 \overline{S}_0$$

The data output is equal to D_3 if and only if $S_1 = 1$ and $S_0 = 1$:

$$Y = D_3 S_1 S_0$$

If these terms are ORed, the total expression for the data output is

$$Y = D_0 \overline{S}_1 \overline{S}_0 + D_1 \overline{S}_1 S_0 + D_2 S_1 \overline{S}_0 + D_3 S_1 S_0$$

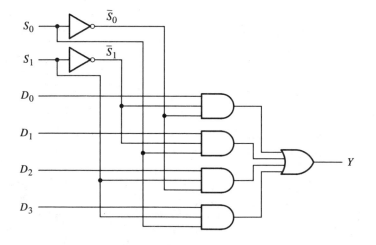

FIGURE 6-58 *Logic diagram for a four-input multiplexer.*

The implementation of this equation requires four 3-input AND gates, a four-input OR gate, and two inverters to generate the complements of S_1 and S_0, as shown in Figure 6–58.

Because data can be selected from any of the input lines, this circuit is also referred to as a *data selector*. As you can see, the logic diagram is equivalent to the AND-OR logic described in Chapter 5.

EXAMPLE 6–15   The *data-input* and *data-select* waveforms in Figure 6–59 are applied to the multiplexer in Figure 6–58. Determine the output waveform in relation to the inputs.

FIGURE 6–59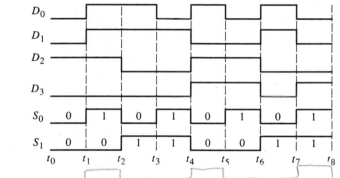

Solution   The binary state of the data-select inputs during each interval determines which data input is selected. Notice that the data-select inputs go through a binary sequence 00, 01, 10, 11, 00, 01, 10, 11, and so on. The resulting output waveform is shown in Figure 6–60.

FIGURE 6–60

Three representative MSI multiplexers are now introduced:

1. The 74157 quadruple two-input data selector/multiplexer.
2. The 74151A eight-input data selector/multiplexer.
3. The 74150 sixteen-input data selector/multiplexer.

The 74157 Quadruple Two-Input Data Selector/Multiplexer

The 74157 consists of four separate two-input multiplexers on a single chip. Each of the four multiplexers shares a common *data-select* line and a common $\overline{enable}$, as shown in Figure 6–61(a). Because there are only two inputs to be selected in each multiplexer, a single data-select input is sufficient ($2^1 = 2$).

As you can see in the logic diagram, the data-select input is ANDed with the *B* input of each two-input multiplexer, and the complement of data select is ANDed with each *A* input.

A LOW on the $\overline{\text{enable}}$ input allows the selected input data to pass through to the output. A HIGH on the $\overline{\text{enable}}$ input prevents data from going through to the output; that is, it *disables* the multiplexers. The ANSI/IEEE logic symbol for this device is shown in Figure 6–61(b). Notice that the four multiplexers are indicated by the partitioned outline and that the inputs common to all four multiplexers are indicated as inputs to the "notched" block at the top, which is called the *common control block*. All labels within the upper MUX block apply to the other blocks below it.

Notice the 1 and $\overline{1}$ labels in the MUX blocks and the G1 label in the common control block. These labels are an example of the *dependency notation* system specified in the ANSI/IEEE Std. 91–1984. In this case, G1 indicates an AND relationship between the data-select input and the data inputs with 1 or $\overline{1}$ labels. (The $\overline{1}$ means that the AND relationship applies to the complement of the G1 input.) In other words, when the data-select input is HIGH, the *B* inputs of the multiplexers are selected, and when the data-select input is LOW, the *A* inputs are selected. G is always used to denote AND dependency.

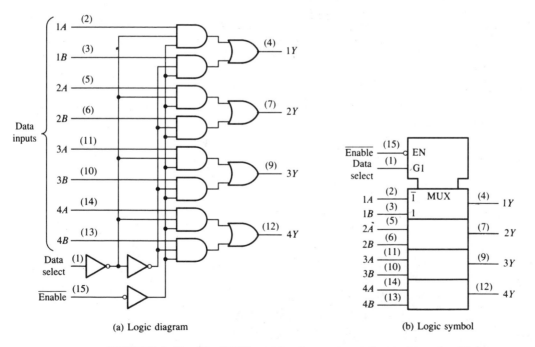

(a) Logic diagram   (b) Logic symbol

FIGURE 6–61 *The 74157 quadruple two-input data selector/multiplexer.*

Other aspects of dependency notation are introduced as appropriate throughout the book.

The 74151A Eight-Input Data Selector/Multiplexer

The 74151A has eight data inputs and, therefore, three data-select input lines. Three bits are required to select any one of the eight data inputs ($2^3 = 8$). A LOW on the $\overline{\text{enable}}$ input allows the selected input data to pass through to the output. Notice that the data output and its complement are both available. The logic diagram is shown in Figure 6–62(a), and the ANSI/IEEE logic symbol is shown in part (b). In this case, there is no need for a common control block on the logic symbol because there is only *one* multiplexer to be controlled, not four as in the

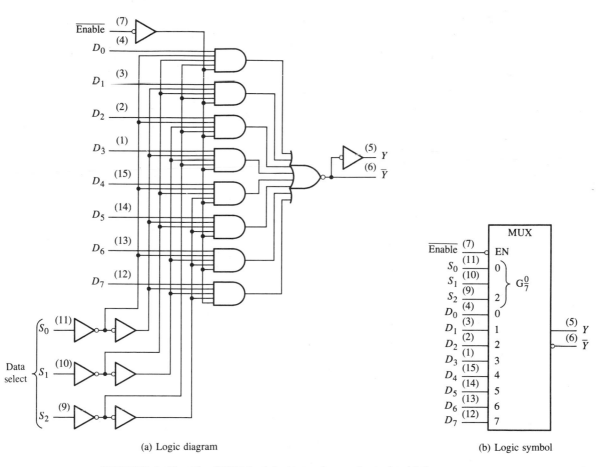

(a) Logic diagram   (b) Logic symbol

FIGURE 6–62 *The 74151A eight-input data selector/multiplexer.*

74157. The G_7^0 label within the logic symbol indicates the AND relationship between the data-select inputs and each of the data inputs 0 through 7.

The 74150 Sixteen-Input Data Selector/Multiplexer

The 74150 has sixteen data inputs and four data-select lines. In this case, four bits are required to select any one of the sixteen data inputs ($2^4 = 16$). There is also an active-LOW enable input. On this particular device, only the complement of the output data is available. The logic symbol is shown in Figure 6–63.

FIGURE 6–63 *The 74150 sixteen-input data selector/multiplexer.*

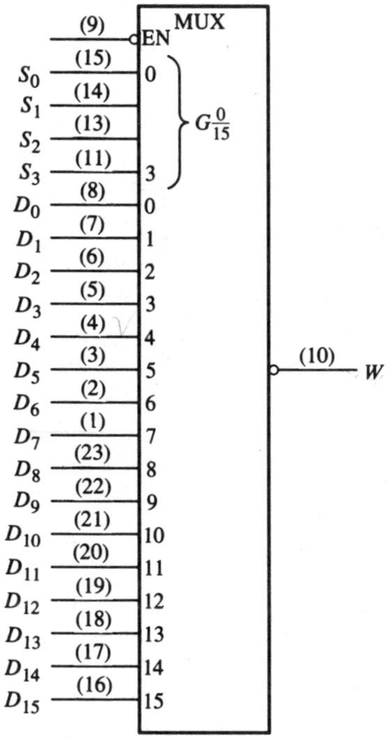

EXAMPLE 6–16   Use 74150s and any other logic necessary to multiplex 32 data lines onto a single data-output line.

Solution   An implementation of this system is shown in Figure 6–64. Five bits are required to select one of 32 data inputs ($2^5 = 32$). In this application, the *enable* input is used as the most significant data-select bit. When the MSB in the data-select code is LOW, the upper 74150 is enabled, and one of the data inputs (D_0 through D_{15}) is selected by the other four data-select bits. When the data-select MSB is HIGH, the lower 74150 is enabled, and one of the data inputs (D_{16} through D_{31}) is selected. The selected input data is then passed through to the negative-OR gate and onto the single output line.

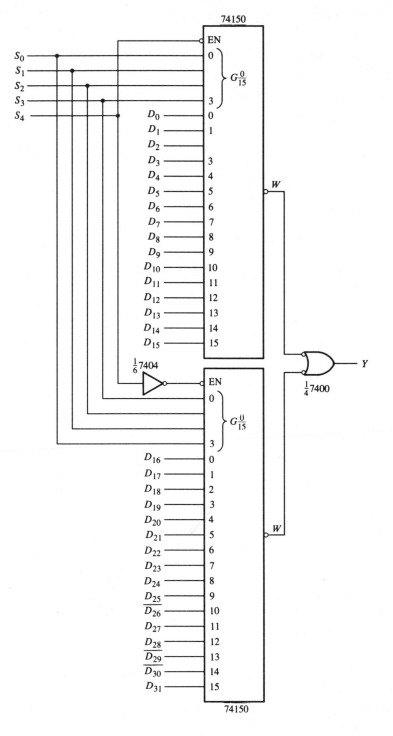

FIGURE 6-64 *A 32-input multiplexer.*

Now two examples of data selector/multiplexer applications are presented:

1. The seven-segment display multiplexer.
2. The logic function generator.

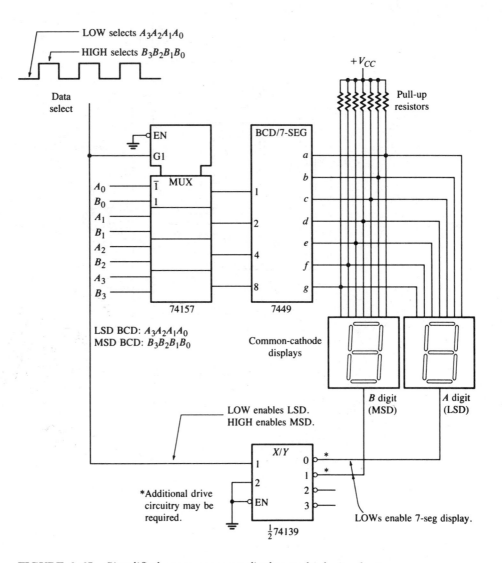

FIGURE 6–65 *Simplified seven-segment display multiplexing logic.*

The Seven-Segment Display Multiplexer

Figure 6–65 shows a simplified method of multiplexing BCD numbers to a seven-segment display. In this example, two-digit numbers are displayed on the seven-segment readout using a single BCD-to-seven-segment decoder. This basic method of display multiplexing can be extended to displays with any number of digits. The basic operation is as follows:

Two BCD digits (A and B) are applied to the multiplexer inputs. A square wave is applied to the data-select line, and when it is LOW, the A bits ($A_3 A_2 A_1 A_0$) are passed through to the inputs of the 7449 BCD-to-seven-segment decoder. The LOW on the data-select also puts a LOW on the 1 input of the 74139 2-line-to-4-line decoder, thus activating its 0 output and enabling the A-digit display by effectively connecting its common terminal to ground. The A digit is now on and the B digit is off.

When the data-select line goes HIGH, the B bits ($B_3 B_2 B_1 B_0$) are passed through to the inputs of the BCD-to-seven-segment decoder. Also, the 74139 decoder's 1 output is activated, thus enabling the B-digit display. The B digit is now on and the A digit is off. The cycle repeats at the frequency of the data-select square wave. This frequency must be high enough (about 30 Hz) to prevent visual flicker as the digit displays are multiplexed.

The Logic Function Generator

A very useful application of the data selector/multiplexer is in the generation of combinational logic functions in sum-of-products form. When used in this way, the device can replace discrete gates and often save many ICs and make design changes much easier.

To illustrate, a 74151A eight-input data selector/multiplexer can be used to implement any specified three-variable logic function by connecting the variables to the data-select inputs and setting each data input to the logic level required in the truth table for that function. For example, if the function is a 1 when the variable combination is $\overline{C}\,\overline{B}\,A$, the D_2 input (selected by 010) is connected to a HIGH. This HIGH is passed through to the output when this particular combination of variables occurs on the data-select lines. An example will help clarify this procedure.

EXAMPLE 6–17 Implement the logic function as specified in Table 6–11 using a 74151A eight-input data selector/multiplexer. Compare this method with a discrete logic gate implementation.

TABLE 6–11

Inputs			Output
C	B	A	Y
0	0	0	0
0	0	1	1
0	1	0	0
0	1	1	1
1	0	0	0
1	0	1	1
1	1	0	1
1	1	1	0

Solution   Notice from the truth table that Y is a 1 for the following input variable combinations: 001, 011, 101, and 110. Y is 0 for all other combinations. To implement this function with the data selector, the data input selected by each of the above-mentioned combinations must be connected to a HIGH. All the other data inputs must be connected to a LOW, as shown in Figure 6–66.

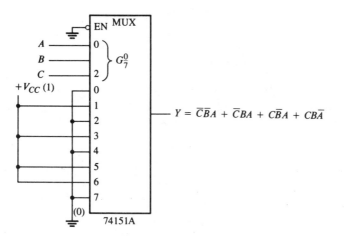

FIGURE 6–66   *Data selector as a three-variable logic function generator.*

The implementation of this function with discrete logic gates would require as many as four 3-input AND gates, one 4-input OR gate, and three inverters. Notice also that no time is spent in simplifying the logic.

Example 6–17 illustrated how the eight-input data selector can be used as a logic function generator for three variables. Actually, this device can be also used as a four-variable logic function generator by utilizing the data inputs in conjunction with the fourth variable (the MSB).

A four-variable truth table has sixteen combinations of input variables. Using an eight-bit data selector, each input is selected twice for each combination of the three lower-order bits, C, B, and A: the first time when the fourth bit, D, is 0 and the second time when the fourth bit is 1. With this in mind, the following rules can be applied (Y is the output and D is the fourth and most significant bit):

1.   If $Y = 0$ both times a given input is selected by CBA, connect that input to ground (0).
2.   If $Y = 1$ both times a given input is selected by CBA, connect that input to V_{CC} (1).
3.   If Y is different both times a given input is selected by CBA and if $Y = D$, connect that input to D.
4.   If Y is different both times a given input is selected by CBA and if $Y = \overline{D}$, connect that input to $\overline{D}$.

The following example illustrates this method.

EXAMPLE 6–18   Implement the logic function in Table 6–12 using a 74151A eight-input data selector/multiplexer. Compare this method with a discrete logic gate implementation.

TABLE 6–12

Inputs				Output
D	C	B	A	Y
0	0	0	0	0
0	0	0	1	1
0	0	1	0	1
0	0	1	1	0
0	1	0	0	0
0	1	0	1	1
0	1	1	0	1
0	1	1	1	1
1	0	0	0	1
1	0	0	1	0
1	0	1	0	1
1	0	1	1	0
1	1	0	0	1
1	1	0	1	1
1	1	1	0	0
1	1	1	1	1

Solution   The data-select inputs are CBA. In the first row of the table, $CBA = 000$ and $Y = D = 0$. In the ninth row where CBA again is 000, $Y = D = 1$. Thus, D is connected to the 0 input. In the second row of the table, $CBA = 001$ and $Y = \overline{D} = 1$. Also, in the tenth row, when CBA again is 001, $Y = \overline{D} = 0$. Thus, D is inverted and connected to the 1 input. This analysis is continued until each input is properly connected according to the specified rules. The implementation is shown in Figure 6–67.

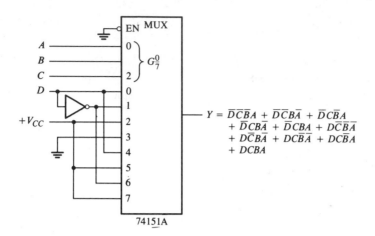

$$Y = \overline{D}\,\overline{C}\,\overline{B}A + \overline{D}\,\overline{C}B\overline{A} + \overline{D}C\overline{B}A$$
$$+ \overline{D}CB\overline{A} + \overline{D}CBA + D\overline{C}\,\overline{B}A$$
$$+ D\overline{C}B\overline{A} + DC\overline{B}\,\overline{A} + DC\overline{B}A$$
$$+ DCBA$$

FIGURE 6-67 *Data selector as a four-variable logic function generator.*

If implemented with discrete logic, the function would require as many as ten 4-input AND gates, one 10-input OR gate, and four inverters.

SECTION REVIEW 6-8

1. In Figure 6-58, $D_0 = 1$, $D_1 = 0$, $D_2 = 1$, $D_3 = 0$, $S_0 = 1$, and $S_1 = 0$. What is the output?
2. Identify each MSI device.
   (a) 74157   (b) 74151A   (c) 74150
3. A 74151A has alternate LOW and HIGH levels on its data inputs beginning with $D_0 = 0$. The data-select lines are sequenced through a binary count (000, 001, 010, and so on) at a frequency of 1 kHz. The enable input is LOW. Draw the timing diagram for this operation.
4. Briefly describe the purpose of each of the following devices in Figure 6-65:
   (a) 74157   (b) 7449   (c) 74139

6-9   DEMULTIPLEXERS

A demultiplexer (DMUX) basically reverses the multiplexing function. It takes data from one line and distributes them to a given number of output lines. Figure 6-68 shows a 1-line-to-4-line demultiplexer circuit. The input data line goes to all of the AND gates. The two select lines enable only one gate at a time, and the data appearing on the input line will pass through the selected gate to the associated output line.

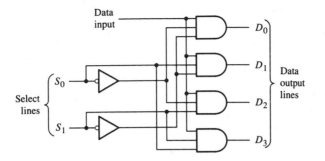

FIGURE 6-68 *A 1-line-to-4-line demultiplexer.*

EXAMPLE 6-19   The serial data-input waveform and select inputs are shown in Figure 6–69. Determine the data-output waveforms for the demultiplexer in Figure 6–68.

FIGURE 6-69

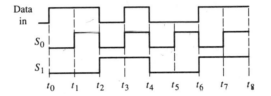

Solution   Notice that the select lines go through a binary sequence so that each successive input bit is routed to D_0, D_1, D_2, and D_3 in sequence, as shown in Figure 6–70.

FIGURE 6-70

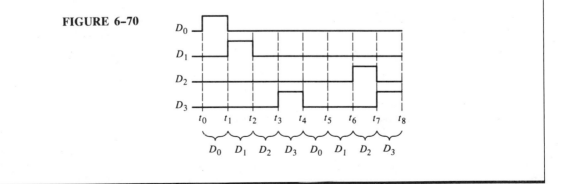

The 74154 as a Demultiplexer

We have already discussed the 74154 in its application as a 1-of-16 or as a binary-to-decimal decoder (Section 6–5). This device and other decoders can also be used in demultiplexing applications. The logic symbol for this device used as a de-multiplexer is shown in Figure 6–71.

In demultiplexer applications the input lines are used as the data-select lines. One of the enable inputs is used as the data-input line with the other enable input held LOW to enable the internal negative-AND gate.

FIGURE 6–71 *The 74154 used as a demultiplexer.*

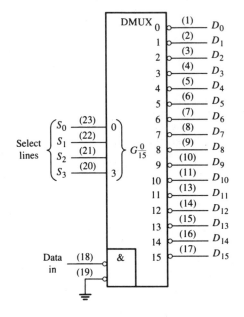

1. Generally, how can an MSI decoder be used as a demultiplexer?
2. The 74154 demultiplexer in Figure 6–71 has a binary code of 1010 on the select lines, and the data-in line is HIGH. What are the states of the output lines?

6–10   PARITY GENERATORS/CHECKERS

Parity

Errors can occur as digital codes are being transferred from one point to another within a digital system or while codes are being transmitted from one system to another. The errors take the form of undesired changes in the bits that make up the coded information; that is, a 1 can change to a 0, or a 0 to 1, due to component malfunctions or electrical noise. In most digital systems, the probability that even a single bit error will occur is very small, and the likelihood that more than one

TABLE 6-13 *8421 BCD code with parity bits.*

Even Parity		Odd Parity	
P	8421	P	8421
0	0000	1	0000
1	0001	0	0001
1	0010	0	0010
0	0011	1	0011
1	0100	0	0100
0	0101	1	0101
0	0110	1	0110
1	0111	0	0111
1	1000	0	1000
0	1001	1	1001

will occur is even smaller. Many systems, however, employ a *parity bit* as a means of detecting a bit error. Binary information is normally handled by a digital system in groups of bits called *words*. A word always contains either an even or an odd number of 1s. A parity bit is attached to the group of information bits in order to make the *total* number of 1s *always even* or *always odd*. An even parity bit makes the total number of 1s even, and an odd parity bit makes the total odd.

A given system operates with even or odd parity, but not both. For instance, if a system operates with even parity, a check is made on each group of bits received to make sure the total number of 1s in that group is even. If there is an odd number of 1s, an error has occurred.

As an illustration of how parity bits are attached to a code word, Table 6-13 lists the parity bits for each 8421 BCD code number for both even and odd parity. The parity bit for each BCD number is in the *P* column.

The parity bit can be attached to the code group at either the beginning or the end, depending on system design. Notice that the total number of 1s, *including the parity bit*, is always even for even parity and always odd for odd parity.

Detecting an Error

A parity bit provides for the detection of a *single* error (or any odd number of errors, which is very unlikely) but cannot check for two errors. For instance, let us assume that we wish to transmit the BCD code 0101. (Parity can be used with any number of bits; we are using four for illustration.) The total code transmitted, including the even parity bit, is

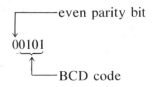

Now, let us assume that an error occurs in the third bit from the left (the 1 becomes a 0), as follows:

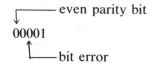

When this code is received, the parity check circuitry determines that there is only a single 1 (odd number), when there should be an even number of 1s. Because an even number of 1s does not appear in the code when it is received, an error is indicated.

Let us now consider what happens if two bit errors occur as follows:

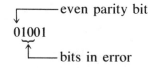

When a check is made, an even number of 1s appears, and although there are two errors, the parity check indicates a *correct* code.

An odd parity bit also provides in a similar manner for the detection of a single error in a given group of bits.

EXAMPLE 6–20

Assign the proper even parity bit to the following code words:

(a) 1010   (b) 111000   (c) 101101
(d) 100011100101   (e) 101101011111

Solutions Make the parity bit either 1 or 0 to make the total number of 1s even. The parity will be the left-most bit (shaded).

(a) 0   1010
(b) 1   111000
(c) 0   101101
(d) 0   100011100101
(e) 1   101101011111

EXAMPLE 6–21

An odd parity system receives the following code words: 10110, 11010, 110011, 110101110100, and 1100010101010. Determine which ones, if any, are in error.

Solutions Since odd parity is required, any code with an even number of 1s is incorrect. The following codes are in error: 110011 and 1100010101010.

Several specific codes also provide inherent error detection; a few of the most important, as well as an *error-correcting* code, are discussed in Appendix C.

Parity Logic

In order to check for or generate the proper parity in a given code word, a very basic principle can be used: *The sum* (disregarding carries) *of an even number of 1s is always 0, and the sum of an odd number of 1s is always 1.* Therefore, in order

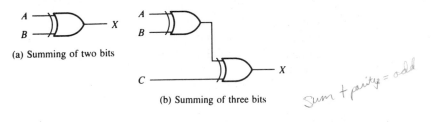

(a) Summing of two bits

(b) Summing of three bits

Sum + parity = odd

FIGURE 6–72

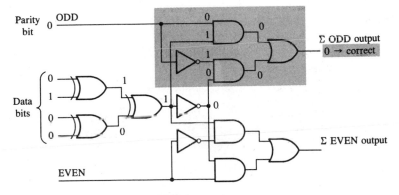

(a) Code correct

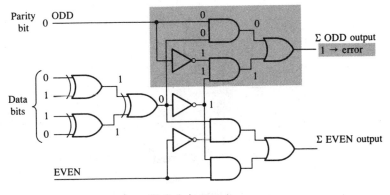

(b) Code in error

FIGURE 6–73 *Examples of odd parity detection.*

to determine if a given code word is even or odd parity, all of the bits in that code word are summed. As we know, the sum of two bits can be generated by an exclusive-OR gate, as shown in Figure 6–72(a) (p. 263); the sum of three bits can be formed by two exclusive-OR gates connected as shown in Figure 6–72(b); and so on.

A typical five-bit generator/checker circuit is shown in Figure 6–73 (p. 263). It can be used for either odd or even parity. When used as an *odd parity checker* as shown, the operation is as follows:

A five-bit code (four data bits and one parity bit) is applied to the inputs. The four data bits are on the exclusive-OR inputs, and the parity bit is applied to the ODD input line. When the number of 1s in the five-bit code is odd, the Σ ODD output is LOW, indicating proper parity. When there is an even number of 1s, the

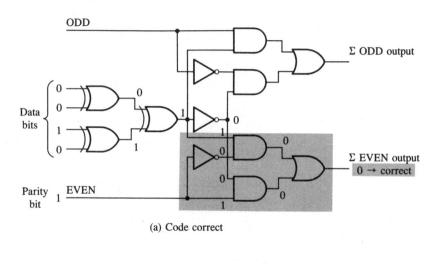

(a) Code correct

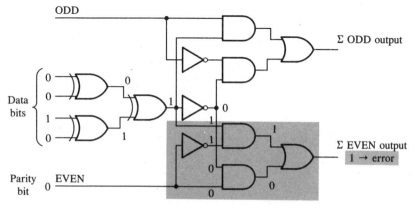

(b) Code in error

FIGURE 6–74 *Examples of even parity detection.*

Σ ODD output is HIGH, indicating incorrect parity. Both of these conditions are illustrated in Figure 6–73. Similarly, even parity checks are illustrated for both nonerror and error conditions in Figure 6–74.

The same circuit can be used as a *parity generator*, as shown in Figure 6–75. The operation for odd parity generation is illustrated in part (a). A four-bit code is applied to the inputs, and the ODD line is held LOW (by grounding in this case).

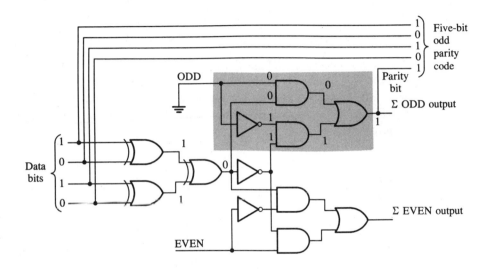

(a) Odd parity generation

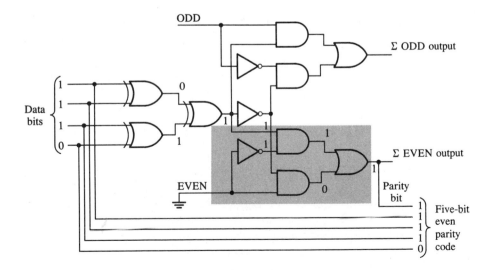

(b) Even parity generation

FIGURE 6–75 *Examples of parity generation.*

When the four-bit code has an even number of 1s as shown, the Σ ODD output is HIGH (1). This 1 output is the odd parity bit and is combined with the four-bit code to form a five-bit odd parity code as shown. Similarly, a 0 parity bit is produced when there is an odd number of 1s in the input code.

Figure 6–75(b) shows an example of the same circuit used in an even parity system. Notice that the EVEN line is grounded in this case.

This basic logic can be expanded to accommodate any number of input bits by adding more exclusive-OR gates.

The 74180 Nine-Bit Parity Generator/Checker

The 74180 is represented in Figure 6–76. This particular MSI device can be used to check for odd or even parity on a nine-bit code (eight data bits and one parity bit), or it can be used to generate a nine-bit odd or even parity code. The truth table operation varies slightly from the simpler basic circuits just discussed, but the principle is the same.

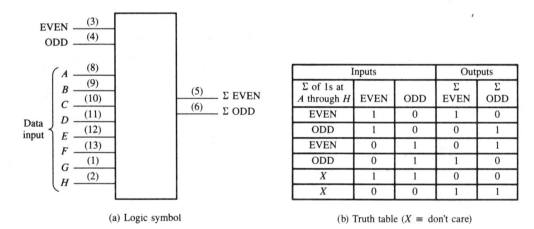

Inputs			Outputs	
Σ of 1s at A through H	EVEN	ODD	Σ EVEN	Σ ODD
EVEN	1	0	1	0
ODD	1	0	0	1
EVEN	0	1	0	1
ODD	0	1	1	0
X	1	1	0	0
X	0	0	1	1

(a) Logic symbol   (b) Truth table ($X \equiv$ don't care)

FIGURE 6–76   *The 74180 nine-bit parity generator/checker.*

A Data Transmission System with Error Detection

A simplified data transmission system is shown in Figure 6–77 to illustrate an application of parity generators/checkers as well as multiplexers and demultiplexers and to illustrate the need for *data storage* in some applications.

In this application, digital data from seven sources are multiplexed onto a single line for transmission to a distant point. The seven data bits (D_0 through D_6) are applied to the multiplexer data inputs and, at the same time, to the even parity generator inputs. The Σ ODD output of the parity generator is used as the even parity bit. This bit is 0 if the number of 1s in the inputs A through H is even and is a 1 if the number of 1s in A through H is odd. This bit is D_7 of the transmitted code.

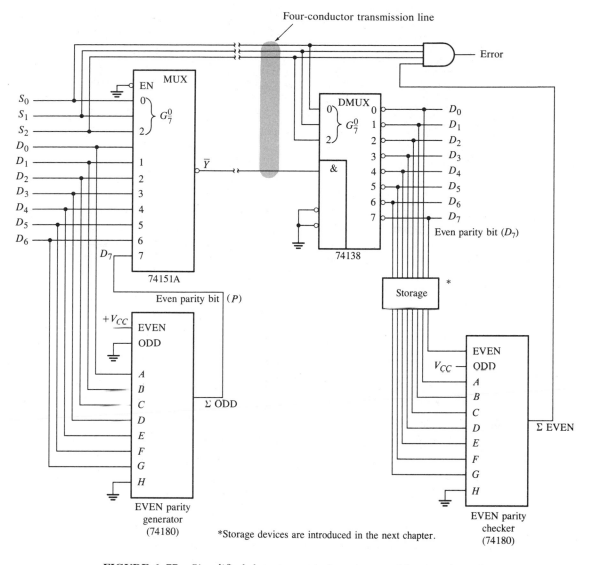

FIGURE 6–77 *Simplified data transmission system with error detection.*

The data-select inputs are repeatedly cycled through a binary sequence, and each data bit, beginning with D_0, is serially passed through and onto the transmission line ($\overline{Y}$). In this example, the transmission line consists of four conductors: One carries the serial data and three carry the timing signals (data selects). There are more sophisticated ways of sending the timing information, but we are using this direct method to illustrate a basic application.

At the demultiplexer end of the system, the data-select signals and the serial data stream are applied to the demultiplexer. The data bits are distributed by the demultiplexer onto the output lines in the same order in which they occurred on the multiplexer inputs. That is, D_0 comes out on the D_0 output, D_1 comes out on the D_1 output, and so on. The parity bit comes out on the D_7 output. These eight bits are temporarily stored and applied to the even parity checker. All of the bits are not present on the parity checker inputs until the parity bit D_7 comes out and is stored. At this time, the error gate is enabled by the 111 data-select code. If the parity is correct, a 0 appears on the Σ EVEN output, keeping the ERROR output at 0. If the parity is incorrect, all 1s appear on the error gate inputs, and a 1 on the ERROR output results.

This particular application has demonstrated the need for data storage so that you can better appreciate the usefulness of the storage devices introduced in the next chapter.

The timing diagram in Figure 6–78 illustrates a specific case in which two eight-bit words are transmitted, one with correct parity and one with an error.

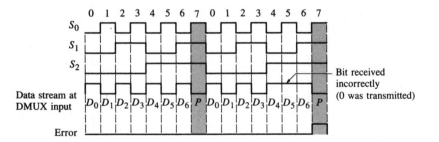

FIGURE 6–78 *Example of data transmission with and without error for the system in Figure 6–77.*

SECTION REVIEW 6–10

1. Add an even parity bit to each of the following codes:
    (a)   110100   (b)   01100011
2. Add an odd parity bit to each of the following codes:
    (a)   1010101   (b)   1000001
3. Check each of the even parity codes for an error.
    (a)   100010101   (b)   1110111001

6–11   TROUBLESHOOTING

In this section, the problem of *decoder glitches* is introduced and examined from a troubleshooting standpoint. A *glitch* is any undesired voltage spike (pulse) of very short duration. A glitch can be interpreted as a valid signal by a logic circuit and may cause improper operation.

A 3-line-to-8-line decoder in Figure 6–79 is used to illustrate how glitches occur and how to identify their cause. The *CBA* inputs of the decoder are sequenced through a binary count, and the resulting waveforms of the inputs and outputs can be displayed on the screen of a logic analyzer as shown in Figure 6–79.

The output waveforms are correct except for the glitches that occur on some of the output signals. An oscilloscope can be used to examine the critical timing details of the input waveforms in an effort to pinpoint the cause of the glitches. The oscilloscope is useful in this case because the waveforms can be "magnified" so that very small time differences between waveform transitions are observable.

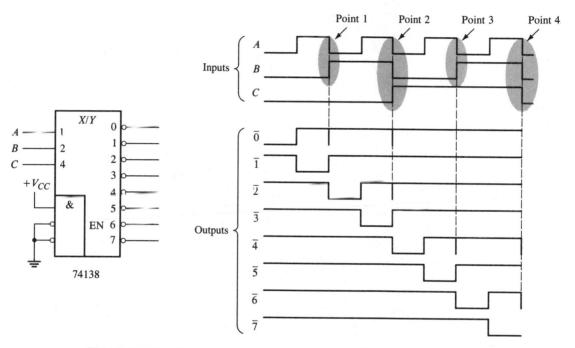

FIGURE 6–79 *Decoder waveforms with output glitches.*

The points of interest indicated by the shaded areas on the input waveforms in Figure 6–79 are displayed on an oscilloscope as shown in Figure 6–80. At point 1, there is a transitional state of 000 due to delay differences in the waveforms. This causes the first glitch on the $\overline{0}$ output of the decoder. At point 2, there are two transitional states, 010 and 000. These cause the glitch on the $\overline{2}$ output of the decoder and the second glitch on the $\overline{0}$ output, respectively. At point 3, the transitional state is 100 which causes the first glitch on the $\overline{4}$ output of the decoder. At point 4, the two transitional states, 110 and 100, result in the glitch on the $\overline{6}$ output and the second glitch on the $\overline{4}$ output, respectively.

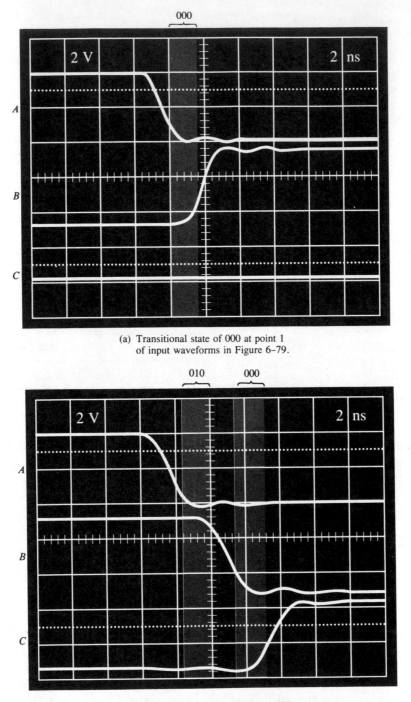

(a) Transitional state of 000 at point 1
of input waveforms in Figure 6–79.

(b) Transitional states of 010 and 000
at point 2 of input waveforms in
Figure 6–79.

FIGURE 6–80 *Oscilloscope displays of transitions of decoder input waveforms for glitch analysis.*

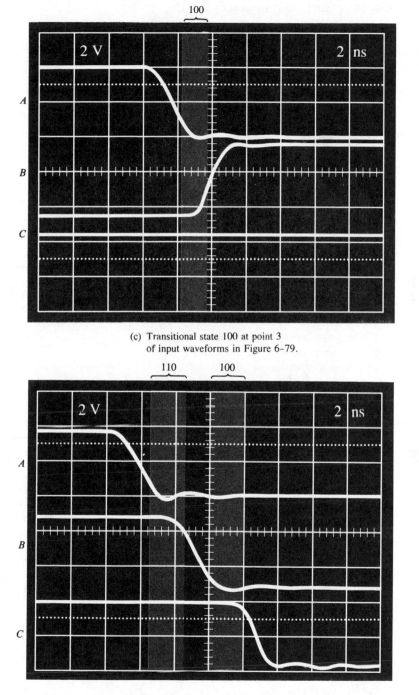

(c) Transitional state 100 at point 3
of input waveforms in Figure 6–79.

(d) Transitional states of 110 and 100
at point 4 of input waveforms
in Figure 6–79.

FIGURE 6–80 *(Continued)*

One way to eliminate the glitch problem is a method called *strobing*, in which the decoder is enabled only during the times when the waveforms are not in transition. This method is discussed in a later chapter.

SECTION REVIEW 6–11

1. Define the term *glitch*.
2. Explain the basic cause of glitches in decoder logic.
3. Define the term *strobing*.

SUMMARY

- ☐ The difference between a half-adder and a full-adder is that the half-adder does not have a carry input.
- ☐ Σ stands for sum, CO stands for output carry, and CI stands for input carry. P and Q are the input variables or operands in an adder.
- ☐ The sum bit is formed by the exclusive-OR of the two operand bits. The carry output is formed by the AND of the two operand bits.
- ☐ Look-ahead carry is a method of decreasing the time required to add two multibit numbers.
- ☐ A carry is generated from a full-adder when both input bits are 1s.
- ☐ A carry is propagated by a full-adder when either or both input bits are 1s.
- ☐ A magnitude comparator indicates when two numbers are equal or which number is greater than the other.
- ☐ A decoder detects the presence of a specified combination of bits (code) on its inputs and activates the proper output. For example, it selects a decimal digit for a given binary code.
- ☐ LEDs are light-emitting diodes.
- ☐ LCDs are liquid crystal displays.
- ☐ The encoder operates as a "decoder in reverse." For example, it can produce a binary code representing a selected decimal digit.
- ☐ In a priority encoder, the highest-value input digit is encoded and any other active input is ignored.
- ☐ A code converter converts one type of code to another, for example, BCD to binary.
- ☐ Multiplexers take data in parallel from several sources and send the data out serially on one line in a time sequence.
- ☐ Multiplexers (MUX) are also known as *data selectors*.
- ☐ Demultiplexers (DMUX) take serial data from one line and distribute them onto several lines in parallel.
- ☐ Parity bits are used for error detection in digital codes.
- ☐ Even parity means that there is an even number of 1s in a code. Odd parity means that there is an odd number of 1s in a code.

SELF-TEST

1. Describe the difference between a half-adder and a full-adder.
2. The following input bits are applied to a half-adder. Determine the Σ and CO for each.
   (a) $P = 1, Q = 0$ (b) $P = 1, Q = 1$ (c) $P = 0, Q = 1$
   (d) $P = 0, Q = 0$
3. The following input bits are applied to a full-adder. Determine the Σ and CO for each.
   (a) $P = 1, Q = 0, CI = 0$ (b) $P = 1, Q = 0, CI = 1$
   (c) $P = 0, Q = 1, CI = 1$ (d) $P = 1, Q = 1, CI = 1$
4. A 7482 adder has the following inputs: $P_0 = 1, P_1 = 0, Q_0 = 1, Q_1 = 1$, and $CI = 0$. Determine Σ_0, Σ_1, and CO.
5. Show how to connect 7482 two-bit adders to form an eight-bit parallel adder.
6. Show how to connect 7483A four-bit adders to form a sixteen-bit adder.
7. Determine the carry generate (CG) and the carry propagate (CP) functions for a full-adder when the input bits are as follows:
   (a) $P = 1, Q = 0$ (b) $P = 0, Q = 0$ (c) $P = 1, Q = 1$
8. A 7485 four-bit magnitude comparator has $P = 1011$ and $Q = 1001$.
   (a) Determine the outputs.
   (b) Show how to connect the $<$, $=$, and $>$ inputs if this is to be the least significant stage.
9. Use 7485 comparators to form a twelve-bit comparator. Show the diagram.
10. Determine which output is LOW for each set of inputs to a 74154 decoder.
    (a) $DCBA = 1100$ (b) $DCBA = 1000$ (c) $DCBA = 0010$
11. Determine which output is LOW for each set of inputs to a 7442A decoder
    (a) $DCBA = 0101$ (h) $DCBA = 1001$ (c) $D\check{C}BA = 1100$
12. A BCD-to-seven-segment decoder/driver is connected to an LED display. What segments are illuminated for each of the following input codes?
    (a) $DCBA = 0001$ (b) $DCBA = 0111$ (c) $DCBA = 0011$
13. A 74147 decimal-to-BCD priority encoder has the following input pins LOW: 3, 4, and 12. The other inputs are HIGH. What BCD code is on the outputs?
14. (a) Convert BCD 10010011 to binary.
    (b) Convert BCD 01100111 to binary.
15. Draw the logic diagram for a three-bit binary-to-Gray converter.
16. Draw the logic diagram for a nine-bit Gray-to-binary converter.
17. A 74157 quadruple two-input data selector/multiplexer has the following bits on its inputs:

$$
\begin{array}{ll}
1A = 1 & 1B = 0 \\
2A = 0 & 2B = 0 \\
3A = 1 & 3B = 0 \\
4A = 0 & 4B = 1
\end{array}
$$

The enable is LOW. Determine the outputs when (a) the data-select input is 0 and (b) the data-select input is 1.
18. The data-select inputs of a 74151A eight-input multiplexer are repetitively sequenced through a binary count from 0 to 7. If there is a 1 on each odd-numbered data input and a 0 on all the others, sketch the output waveform for one cycle through the binary count.

19. Show how to connect three 74151A multiplexers to form a 24-input multiplexer.
20. Implement the logic equation $Y = \overline{CB}\overline{A} + C\overline{B}A + CBA$ with a 74151A data selector/multiplexer.
21. The 74154 is used as a demultiplexer. If the select lines are $S_3S_2S_1S_0 = 1011$ and a 1 is applied to the data input (pin 18), on which output line does the data bit appear?
22. Attach the proper even parity bit to the following codes:
    (a) 11010 **(b)** 1001 **(c)** 0111101
23. Repeat Problem 22 for odd parity.

PROBLEMS

Section 6–1

6–1 For the full-adder of Figure 6–4, determine the logic state (1 or 0) at each gate output for the following inputs:
   (a) $P = 1, Q = 1, CI = 1$ **(b)** $P = 0, Q = 1, CI = 1$
   (c) $P = 0, Q = 1, CI = 0$

6–2 Repeat Problem 6–1 for the following inputs:
   (a) $P = 0, Q = 0, CI = 0$ **(b)** $P = 1, Q = 0, CI = 0$
   (c) $P = 1, Q = 0, CI = 1$

6–3 Simplify, if possible, the full-adder circuit of Example 6–1 (Figure 6–6) using the Karnaugh map method.

Section 6–2

6–4 For the parallel adder in Figure 6–81, determine the sum by analysis of the logical operation of the circuit. Verify your result by long-hand addition of the two input numbers.

FIGURE 6–81

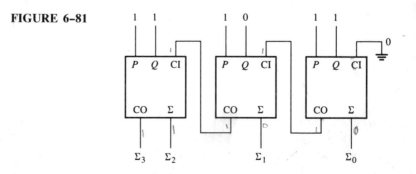

6–5 Repeat Problem 6–4 for the circuit and input conditions in Figure 6–82.

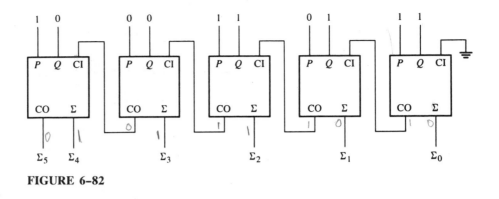

FIGURE 6–82

6–6 The input waveforms in Figure 6–83 are applied to a 7482 two-bit adder. Determine the waveforms for the sum and carry outputs in relation to the inputs.

FIGURE 6–83

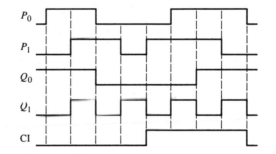

6–7 The following sequences of bits (right-most bit first) appear on the inputs to the adder developed in Example 6–3 (Figure 6–11). Determine the resulting sequence of bits on each sum output.

$$
\begin{array}{ll}
P_0 & 1001011010110100 \\
P_1 & 1110100010110010 \\
Q_0 & 0000101011110010 \\
Q_1 & 1011101011111001 \\
P_2 & 1111100001111000 \\
P_3 & 1100110011001100 \\
Q_2 & 1010101010101010 \\
Q_3 & 0010010010010010
\end{array}
$$

6–8 In the process of checking a 7483A four-bit full-adder, the following voltage levels are observed on its pins: 1-LOW, 2-HIGH, 3-HIGH, 4-HIGH, 6-HIGH, 7-HIGH, 8-LOW, 9-LOW, 10-LOW, 11-LOW, 13-LOW, 14-HIGH, 15-LOW, and 16-HIGH. Determine if the 1C is functioning properly.

Section 6–3

6–9 Each of eight full-adders in an eight-bit parallel ripple carry adder exhibits the following propagation delays:

$$P \text{ to } \Sigma \text{ and CO:} \quad 40 \text{ ns}$$
$$Q \text{ to } \Sigma \text{ and CO:} \quad 40 \text{ ns}$$
$$\text{CI to } \Sigma: \quad 35 \text{ ns}$$
$$\text{CI to CO:} \quad 25 \text{ ns}$$

Determine the maximum total time for the addition of two eight-bit numbers.

6–10 Show the additional logic circuitry necessary to make the four-bit look-ahead-carry adder in Figure 6–16 into a five-bit adder.

Section 6–4

6–11 The waveforms in Figure 6–84 are applied to the comparator as shown. Determine the output $(P = Q)$ waveform.

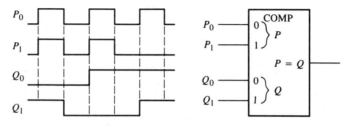

FIGURE 6–84

6–12 For the four-bit comparator in Figure 6–85, plot each output waveform for the inputs shown. The outputs are active-HIGH.

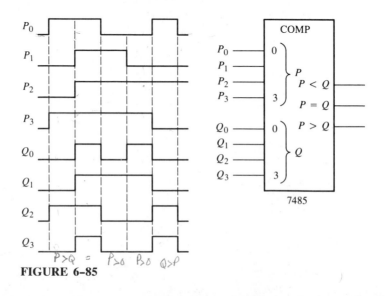

FIGURE 6–85

6–13 For each following set of binary numbers, determine the logic states at each point in the comparator circuit of Figure 6–22, and verify that the output indications are correct:

(a) $P_3P_2P_1P_0 = 1100$ (b) $P_3P_2P_1P_0 = 1000$
      $Q_3Q_2Q_1Q_0 = 1001$ $Q_3Q_2Q_1Q_0 = 1011$

(c) $P_3P_2P_1P_0 = 0100$
      $Q_3Q_2Q_1Q_0 = 0100$

Section 6–5

6–14 When a HIGH is on the output of each of the decoding gates in Figure 6–86, what is the binary code appearing on the inputs? D is the MSB.

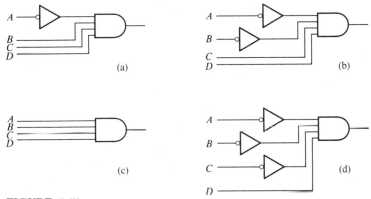

FIGURE 6–86

6–15 Show the decoding logic for each of the following codes if an active-HIGH (1) indication is required:

(a) 1101   (b) 1000   (c) 11011   (d) 11100
(e) 101010   (f) 111110   (g) 000101   (h) 1110110

6–16 Repeat Problem 6–15 given that an active-LOW (0) output is required.

6–17 We wish to detect only the presence of the codes 1010, 1100, 0001, and 1011. An active-HIGH output is required to indicate their presence. Develop the complete decoding logic with a single output that will tell us when any one of these codes is on the inputs. For any other code, the output must be LOW.

6–18 If the input waveforms are applied to the decoding logic as indicated in Figure 6–87, sketch the output waveform in proper relation to the inputs.

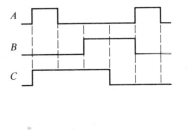 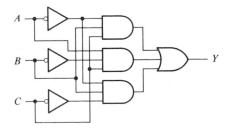

FIGURE 6–87

6–19   BCD numbers are applied sequentially to the BCD-to-decimal decoder in Figure 6–88. Draw the ten output waveforms, showing each in the proper relationship to the others and to the inputs.

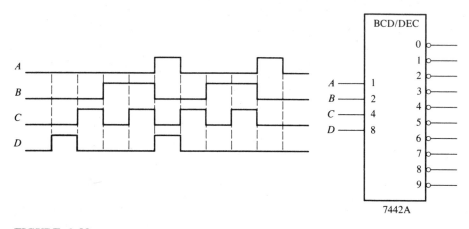

FIGURE 6–88

6–20   A seven-segment decoder/driver drives the display in Figure 6–89. If the waveforms are applied as indicated, determine the sequence of digits that appears on the display.

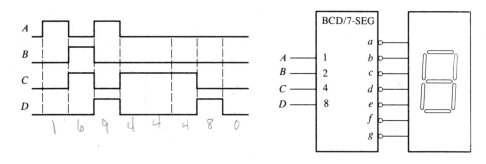

FIGURE 6–89

Section 6–6

6–21   In the decimal-to-BCD encoder of Figure 6–45, assume that the 9 input and the 3 input are both HIGH. What is the output code? Is it a valid BCD (8421) code?

6–22   A 74147 encoder has LOW levels on pins 2, 5, and 12. What BCD code appears on the outputs if all the other inputs are HIGH?

Section 6-7

6-23 Show the logic required to convert a ten-bit binary number to Gray code, and use that logic to convert the following binary code words to Gray code:
(a) 1010101010 (b) 1111100000
(c) 0000001110 (d) 1111111111

6-24 Show the logic required to convert a ten-bit Gray code to binary, and use that logic to convert the following Gray code words to binary:
(a) 1010000000 (b) 0011001100
(c) 1111000111 (d) 0000000001

6-25 Convert the following decimal numbers first to BCD. Then, using the logical operation of the BCD-to-binary converter of Figure 6-52, convert the BCD to binary. Verify the result in each case.
(a) 2 (b) 8 (c) 13 (d) 26 (e) 33

Section 6-8

6-26 For the multiplexer in Figure 6-90, determine the output for the following input states: $D_0 = 0, D_1 = 1, D_2 = 1, D_3 = 0, S_0 = 1, S_1 = 0$.

6-27 If the data-select inputs to the multiplexer in Figure 6-90 are sequenced as shown by the waveforms in Figure 6-91, determine the output waveform with the data inputs as specified in Problem 6-26.

FIGURE 6-90

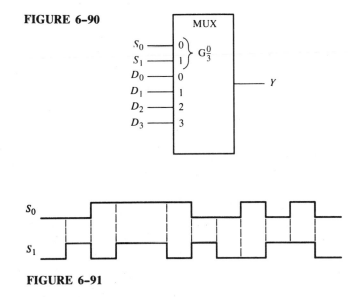

FIGURE 6-91

6-28 The waveforms in Figure 6-92 are observed on the inputs of a 74151A eight-input multiplexer. Sketch the Y output waveform.

FIGURE 6–92

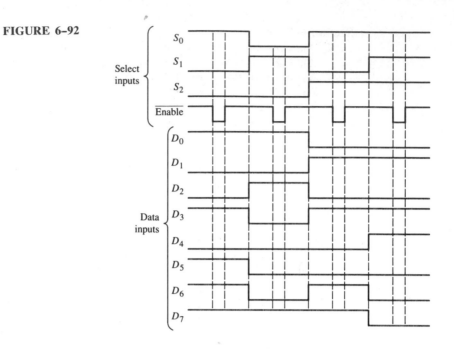

6–29 Modify the design of the seven-segment display multiplexing system in Figure 6–65 to accommodate two additional digits.

6–30 Implement the sum logic in Figure 6–6 using a 74151A data selector. The three input variables are P, Q, and CI.

6–31 Implement the logic function specified in Table 6–14 using a 74151A data selector.

TABLE 6–14

Inputs				Output
D	C	B	A	Y
0	0	0	0	0
0	0	0	1	0
0	0	1	0	1
0	0	1	1	1
0	1	0	0	0
0	1	0	1	0
0	1	1	0	1
0	1	1	1	1
1	0	0	0	1
1	0	0	1	0
1	0	1	0	1
1	0	1	1	1
1	1	0	0	0
1	1	0	1	1
1	1	1	0	0
1	1	1	1	1

Section 6–9

6–32 Develop the total timing diagram (inputs and outputs) for a 74154 used in a de-multiplexing application in which the inputs are as follows: The select inputs are repetitively sequenced through a straight binary count beginning with 0000, and the data input is a serial data stream carrying BCD data representing the decimal number 2468. The least significant digit (8) is first in the sequence with its LSB first, and it should appear in the first four-bit positions of the output.

Section 6–10

6–33 The waveforms in Figure 6–93 are applied to the four-bit parity logic. Determine the output waveform in proper relation to the inputs. How many times does even parity occur, and how is it indicated?

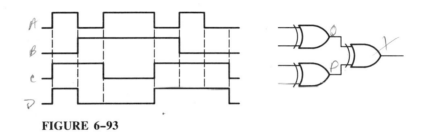

FIGURE 6–93

6–34 Determine the Σ EVEN and the Σ ODD outputs of a 74180 nine-bit parity generator/checker for the inputs in Figure 6–94. Refer to the truth table in Figure 6–76.

FIGURE 6–94

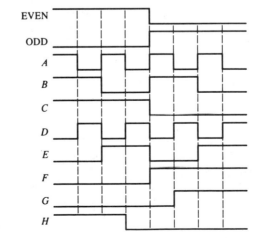

ANSWERS TO SECTION REVIEWS

Section 6–1

1.   (a) $\Sigma = 1, CO = 0$ (b) $\Sigma = 0, CO = 0$ (c) $\Sigma = 1, CO = 0$
     (d) $\Sigma = 0, CO = 1$
2. $\Sigma = 1, CO = 1$

Section 6–2

1. $CO\Sigma_1\Sigma_0 = 110$
2.   Four 7482s are connected as indicated in Figure 6–11 for two 7482s.

Section 6–3

1. $CG = 0, CP = 1$ 2. $CO = 1$

Section 6–4

1. $P > Q = 1, P < Q = 0, P = Q = 0$
2.   Pin 7: $P < Q = 1$. Pin 6: $P = Q = 0$. Pin 5: $P > Q = 0$

Section 6–5

1.   The 5 input   2.   Two   3.   Active-LOW

Section 6–6

1.   (a) $A = 1, B = 1, C = 0, D = 1$ (b) Only one input can be active for valid
     output.
2.   (a) 1010   (b) The complement of 0101 (binary 5).

Section 6–7

1. 1010101_2
2.   Seven exclusive-OR gates in an arrangement such as that in Figure 6–54.

Section 6–8

1.   0
2.   (a) Quad two-input data selector   (b) Eight-input data selector
     (c) Sixteen-input data selector
3.   The data output alternates between LOW and HIGH as the select inputs sequence
     through the binary states.
4.   (a) It multiplexes the two BCD codes to the seven-segment decoder.
     (b) It decodes the BCD to energize the display.
     (c) It alternately enables the seven-segment displays.

Section 6–9

1.   By using the input lines for data selection and an enable line for data input.
2.   All HIGH except D_{10}, which is LOW.

Section 6–10

1.   (a) 1110100   (b) 001100011
2.   (a) 11010101   (b) 11000001
3.   (a) Correct, four 1s   (b) Error, seven 1s

Section 6–11
1. A very short duration voltage spike (usually unwanted).
2. Caused by transition states.
3. Enabling a device for a specified period of time when the device is not in transition.

The flip-flop is a *bistable* logic circuit which is a type of multivibrator. There are three types of multivibrators: the *bistable*, the *monostable*, and the *astable*. In this chapter, the bistable elements are emphasized, but a coverage of the other two types is also provided.

There are four basic categories of bistable elements: the *latch*, the *edge-triggered flip-flop*, the *pulse-triggered (master-slave) flip-flop*, and the *data lock-out flip-flop*.

Bistable elements (latches and flip-flops) exhibit two stable states. They are capable of residing in either of these two states indefinitely. These two states are called SET and RESET. Because of their ability to retain a given state, bistable elements are useful as storage (memory) devices. The flip-flop is a basic building-block for registers, memories, counters, control logic, and other functions in digital systems.

The monostable multivibrator, commonly called a *one-shot*, has only one stable state. It produces a pulse in response to a triggering input. The astable multivibrator has no stable states and is used primarily as an *oscillator* to generate periodic pulse waveforms for timing purposes.

Specific devices introduced in this chapter are as follows:

1. 7475 four-bit latch
2. 7474 dual edge-triggered D flip-flop
3. 74LS76A dual edge-triggered J-K flip-flop
4. 74L71 master-slave S-R flip-flop
5. 74107 dual master-slave J-K flip-flop
6. 74111 dual data lock-out J-K flip-flop
7. 74122 and 74123 one-shots
8. 555 timer

The devices studied in this chapter are identified by standard ANSI/IEEE logic symbols.

In this chapter, you will learn

☐ The operation of S-R and D latches.
☐ The operation of the S-R, D, and J-K edge-triggered flip-flops.
☐ The operation of the S-R, D, and J-K pulse-triggered flip-flops and why they are commonly called *master-slave* flip-flops.
☐ The operation of a data lock-out flip-flop and how this device compares to both the edge-triggered and pulse-triggered devices.
☐ The advantages of a master-slave flip-flop over an edge-triggered type in certain applications.
☐ The significance of propagation delays, set-up time, hold time, maximum operating frequency, minimum clock pulse widths, and power dissipation in the application of flip-flops.
☐ How several specific TTL flip-flops compare to each other and to a specific CMOS device in operating characteristics and parameters.
☐ How flip-flops are used in several important applications.
☐ The operation of the one-shot (monostable multivibrator).
☐ The operation of the astable multivibrator using the 555 timer.
☐ How to approach debugging a new design and how to correct design faults as illustrated by a specific example.

7

Flip-Flops and Other Multivibrators

7–1   LATCHES

The latch is a type of bistable device that is normally placed in a category separate from that of flip-flops. Latches are basically similar to flip-flops because they are bistable devices that can reside in either of two states by virtue of a feedback arrangement. The main difference between latches and flip-flops is in the method used for changing their state.

The S-R Latch

An active-HIGH input S-R (SET-RESET) latch is formed with two cross-coupled NOR gates as shown in Figure 7–1(a), or an active-LOW input S-R latch is formed with two cross-coupled NAND gates as shown in Figure 7–1(b). Notice that the output of each gate is connected to an input of the opposite gate. This produces the *feedback* that is characteristic of all multivibrators.

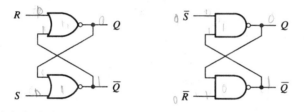

(a) Active-HIGH input S-R latch   (b) Active-LOW input S-R latch

FIGURE 7–1   *Two versions of SET-RESET latches.*

To understand the operation of the latch, we will use the NAND gate S-R latch in Figure 7–1(b). This latch is redrawn in Figure 7–2 using the negative-OR gate equivalents in place of the NAND gates. This is done because LOWs on the $\overline{S}$ and $\overline{R}$ lines are the activating inputs.

FIGURE 7–2
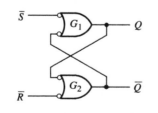

The latch has two inputs, $\overline{S}$ and $\overline{R}$, and two outputs, Q and $\overline{Q}$. We will start by assuming that both inputs and the Q output are HIGH. Since the Q output is connected back to an input of gate G_2, and the $\overline{R}$ input is HIGH, the output of G_2 must be LOW. This LOW output is coupled back to an input of gate G_1, insuring that its output is HIGH.

When the Q output is HIGH, the latch is in the SET state. It will remain in this state indefinitely until a LOW is temporarily applied to the $\overline{R}$ input. With a LOW on the $\overline{R}$ input and a HIGH on $\overline{S}$, the output of gate G_2 is forced HIGH. This HIGH on the $\overline{Q}$ output is coupled back to an input of G_1, and since the $\overline{S}$ input is

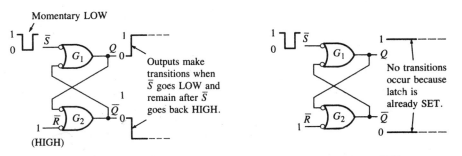

(a) Two possibilities for the SET operation

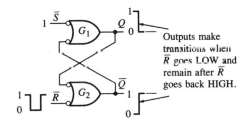
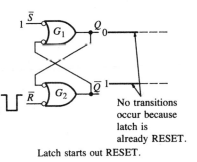

(b) Two possibilities for the RESET operation

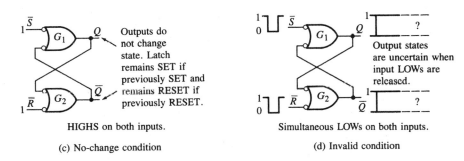

(c) No-change condition

(d) Invalid condition

FIGURE 7–3 *The four modes of basic latch operation.*

HIGH, the output of G_1 goes LOW. This LOW on the Q output is coupled back to an input of G_2, ensuring that the $\overline{Q}$ output remains HIGH even when the LOW on the $\overline{R}$ input is removed. When the Q output is LOW, the latch is in the RESET state. Now the latch remains indefinitely in the RESET state until a LOW is applied to the $\overline{S}$ input.

Notice that in the operation just described, the outputs are always complements of each other: When Q is HIGH, $\overline{Q}$ is LOW, and when Q is LOW, $\overline{Q}$ is HIGH.

A condition that is not allowed in the operation of an active-LOW input S-R latch occurs when LOWs are applied to both $\overline{S}$ and $\overline{R}$ at the same time. As long as the LOW levels are simultaneously held on the inputs, both the Q and $\overline{Q}$ outputs are forced HIGH, thus violating the basic complementary operation of the outputs. Also, if the LOWs are released simultaneously, both outputs will attempt to go LOW. Since there is always some small difference in the propagation delay of the gates, one of the gates will dominate in its transition to the LOW output state. This, in turn, forces the output of the slower gate to remain HIGH. This creates an unpredictable situation because we cannot reliably predict the next state of the latch.

Figure 7–3 (p. 287) illustrates the active-LOW input S-R latch operation for each of the four possible combinations of levels on the inputs. (The first three combinations are valid, but the last is not.) Table 7–1 summarizes the logical operation in truth table form. Operation of the active-HIGH input NOR gate latch in Figure 7–1(a) is similar but requires the use of opposite logic levels. You should analyze its operation as an exercise.

Logic symbols for the active-HIGH input and the active-LOW input S-R latches are shown in Figure 7–4.

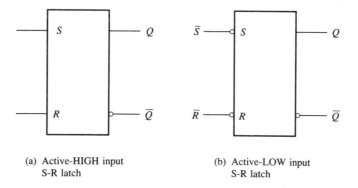

(a) Active-HIGH input
    S-R latch

(b) Active-LOW input
    S-R latch

FIGURE 7–4   *Logic symbols for the S-R latch.*

The following example illustrates how an active-LOW input S-R latch responds to conditions on its inputs. We pulse LOW levels on each input in a certain sequence and observe the resulting Q output waveform. The $\overline{S} = 0$, $\overline{R} = 0$

condition is avoided because it results in an invalid mode of operation and is a major drawback of any SET-RESET type of latch.

EXAMPLE 7–1

If the $\overline{S}$ and $\overline{R}$ waveforms in Figure 7–5(a) are applied to the inputs of the latch of Figure 7–4(b), determine the waveform that would be observed on the Q output. Assume that Q is initially LOW.

Solution   See Figure 7–5(b).

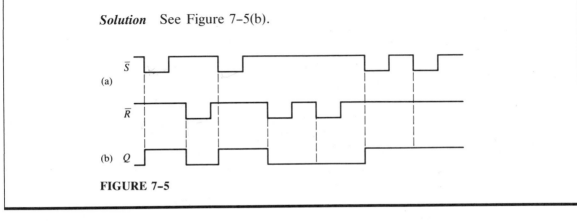

FIGURE 7–5

TABLE 7–1   *Truth table for an active-LOW input S-R latch.*

Inputs		Outputs		
$\overline{S}$	$\overline{R}$	Q	$\overline{Q}$	Comments
1	1	NC	NC	No change. Latch remains in present state.
0	1	1	0	Latch SETS.
1	0	0	1	Latch RESETS.
0	0	1	1	Invalid condition.

The Latch as a Contact-Bounce Eliminator

A good example of an application for an S-R latch is in the elimination of mechanical switch contact "bounce." When the pole of a switch strikes the contact upon switch closure, it physically vibrates or bounces several times before finally making a solid contact. Although these bounces are minute, they produce voltage spikes which are often not acceptable in a digital system. This situation is illustrated in Figure 7–6(a).

An S-R latch can be used to eliminate the effects of switch bounce as shown in Figure 7–6(b). The switch is normally in position 1, keeping the $\overline{R}$ input LOW and the latch RESET. When the switch is thrown to position 2, $\overline{R}$ goes HIGH

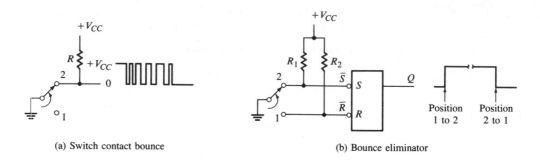

(a) Switch contact bounce   (b) Bounce eliminator

FIGURE 7-6   *The S-R latch used to eliminate switch contact bounce.*

because of the pull-up resistor to V_{CC}, and S goes LOW on the first contact. Although $\overline{S}$ remains LOW for only a very short time before the switch bounces, this is sufficient to SET the latch. Any further voltage spikes on the $\overline{S}$ input due to switch bounce do not affect the latch, and it remains SET. Notice that the Q output of the latch provides a clean transition from LOW to HIGH, thus eliminating the voltage spikes caused by contact bounce. Similarly, a clean transition from HIGH to LOW is made when the switch is thrown back to position 1.

The Gated S-R Latch

A gated latch requires an enable input, EN. The logic diagram and logic symbol for a gated S-R latch are shown in Figure 7–7. The S and R inputs control the state to which the latch will go upon application of a HIGH level on the EN input. The latch will not change until the EN input is HIGH, but as long as it remains HIGH, the output is determined by the state of the S and R inputs. In this circuit, the invalid state occurs when both S and R are simultaneously HIGH.

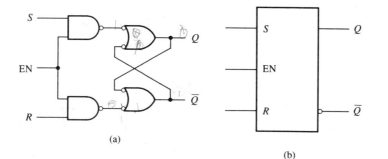

FIGURE 7-7   *A gated S-R latch.*

EXAMPLE Determine the Q output waveform if the inputs shown in Figure 7–8(a) are
7–2   applied to a gated S-R latch that is initially RESET.

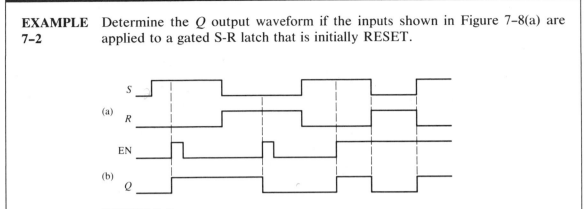

FIGURE 7–8

Solution The Q waveform is shown in Figure 7–8(b). Anytime S is HIGH and
R is LOW, a HIGH on the EN input SETS the latch. Anytime S is LOW and R
is HIGH, a HIGH on the EN input RESETS the latch.

The Gated D Latch

Another type of gated latch is called the D latch. It differs from the S-R latch in
that it has only one input in addition to EN. This input is called the D input. Figure
7–9 contains a logic diagram and logic symbol of a D latch. When the D input is
HIGH and the EN input is HIGH, the latch will SET. When the D input is LOW
and the EN is HIGH, the latch will RESET. Stated another way, the output Q
follows the input D when EN is HIGH.

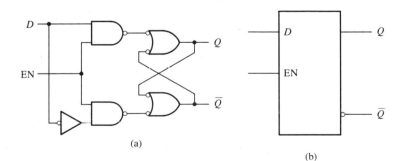

FIGURE 7–9 *A gated D latch.*

**EXAMPLE
7–3**

Determine the Q output waveform if the inputs shown in Figure 7–10(a) are applied to the gated D latch which is initially RESET.

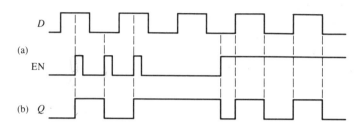

FIGURE 7–10

Solution   The Q waveform is shown in Figure 7–10(b). Anytime D is HIGH and the EN input is HIGH, Q goes HIGH. Anytime D is LOW and the EN input is HIGH, Q goes LOW. When the EN input is LOW, the state of the latch is not affected by the D input.

The 7475 Four-Bit Latches

An example of an IC D-type latch is the 7475 device represented by the logic symbol in Figure 7–11(a). This device has four latches on a single chip. Notice that each EN input is shared by two latches and is designated as a control input (C). The truth table for each latch is shown in Figure 7–11(b). The X in the truth table represents a "don't care" or irrelevant condition. In this case, when the EN input

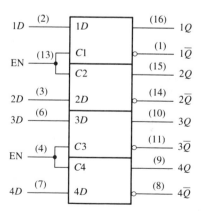

(a) Logic symbol

Inputs		Outputs		Comment
D	EN	Q	$\overline{Q}$	
0	1	0	1	RESET
1	1	1	0	SET
X	0	Q_0	$\overline{Q}_0$	No change

Q_0 is the prior output level.

(b) Truth table (each latch)

FIGURE 7–11 *The 7475 four-bit latches.*

is LOW, it does not matter what the D input is because the outputs are unaffected and remain in their prior states.

SECTION	**1.**	List three types of latches.
REVIEW	**2.**	Develop the truth table for the active-HIGH input S-R latch in Figure 7–1(a).
7–1	**3.**	What is the Q output of a D latch when EN $= 1$ and $D = 1$?

7–2   EDGE-TRIGGERED FLIP-FLOPS

Flip-flops are synchronous bistable devices. In this case, the term *synchronous* means that the output changes state *only* at a specified point on a triggering input called the *clock* (designated as a control input, C); that is, changes in the output occur in synchronization with the clock.

The term *edge-triggered* means that the flip-flop changes state either at the positive edge (rising edge) or at the negative edge (falling edge) of the clock pulse and is sensitive to its inputs only at this transition of the clock. Three basic types of edge-triggered flip-flops are covered in this section: S-R, D, and J-K. The logic symbols for all of these are shown in Figure 7–12. Notice that each type can be either positive edge-triggered (no bubble at C input) or negative edge-triggered (bubble at C input). The key to identifying an edge-triggered flip-flop by its logic symbol is the small "triangle" at the clock (C) input. This is called the *dynamic input indicator*.

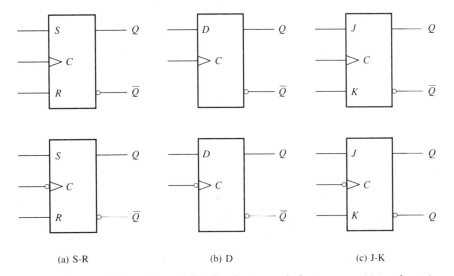

(a) S-R   (b) D   (c) J-K

FIGURE 7–12 *Edge-triggered flip-flop logic symbols (top—positive edge-triggered; bottom—negative edge-triggered).*

The Edge-Triggered S-R Flip-Flop

The S and R inputs of the S-R flip-flop are called the *synchronous inputs* because data on these inputs are transferred to the flip-flop's output only on the triggering edge of the clock pulse. When S is HIGH and R is LOW, the Q output goes HIGH on the triggering edge of the clock pulse and the flip-flop is SET. When S is LOW and R is HIGH, the Q output goes LOW on the triggering edge of the clock pulse and the flip-flop is RESET. When both S and R are LOW, the output does not change from its prior state. An *invalid* condition exists when both S and R are HIGH.

This basic operation is illustrated in Figure 7–13, and Table 7–2 is the truth table for this type of flip-flop. Remember, *the flip-flop cannot change state except on the triggering edge of a clock pulse*. The S and R inputs can be changed anytime when the clock input is LOW or HIGH (except for a very short interval around the triggering transition of the clock) without affecting the output.

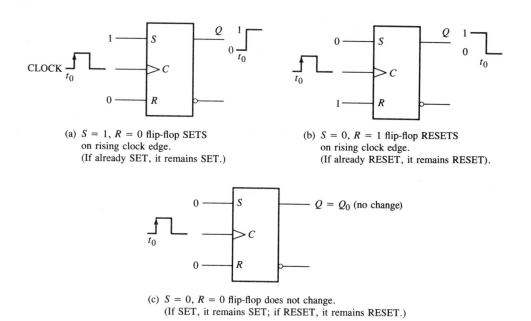

(a) $S = 1$, $R = 0$ flip-flop SETS on rising clock edge. (If already SET, it remains SET.)

(b) $S = 0$, $R = 1$ flip-flop RESETS on rising clock edge. (If already RESET, it remains RESET).

(c) $S = 0$, $R = 0$ flip-flop does not change. (If SET, it remains SET; if RESET, it remains RESET.)

FIGURE 7–13   *Operation of a positive edge-triggered S-R flip-flop.*

The operation and truth table for a negative edge-triggered S-R flip-flop are the same as those for a positive except that the falling edge of the clock pulse is the triggering edge.

TABLE 7-2 *Truth table for a positive edge-triggered S-R flip-flop.*

Inputs			Outputs		
S	R	C	Q	$\overline{Q}$	Comments
0	0	X	Q_0	$\overline{Q}_0$	No change
0	1	↑	0	1	RESET
1	0	↑	1	0	SET
1	1	↑	?	?	Invalid

↑ = clock transition LOW to HIGH.
X = irrelevant (don't care).
Q_0 = prior output level.

EXAMPLE 7-4

Given the waveforms for the S, R, and C inputs in Figure 7–15(a), to the flip-flop of Figure 7–14, determine the Q and $\overline{Q}$ output waveforms. Assume that the positive edge-triggered flip-flop is initially RESET.

FIGURE 7-14

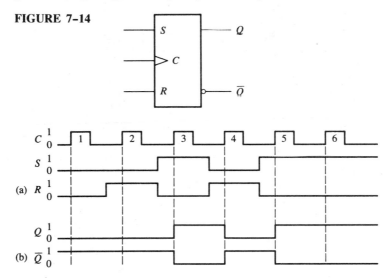

FIGURE 7-15

Solutions

1. At clock pulse 1, S is LOW and R is LOW. Q does not change.
2. At clock pulse 2, S is LOW and R is HIGH. Q remains LOW (RESET).
3. At clock pulse 3, S is HIGH and R is LOW. Q goes HIGH (SET).
4. At clock pulse 4, S is LOW and R is HIGH. Q goes LOW (RESET).
5. At clock pulse 5, S is HIGH and R is LOW. Q goes HIGH (SET).
6. At clock pulse 6, S is HIGH and R is LOW. Q stays HIGH.

Once Q is determined, $\overline{Q}$ is easily found since it is simply the complement of Q. The resulting waveforms for Q and $\overline{Q}$ are shown in Figure 7–15(b) for the input waveforms in (a).

The Method of Edge-Triggering

Although the internal circuitry of the flip-flops is not of primary concern to us, a simplified discussion of the basic operation of edge-triggering is presented here. A simplified implementation of an edge-triggered S-R flip-flop is illustrated in Figure 7–16(a) and is used to demonstrate the concept of edge-triggering. This coverage of the S-R flip-flop does not imply that it is the most important type. Actually, the D flip-flop and the J-K flip-flop are much more widely used and more available in IC form than is the S-R type. However, the S-R is a good base upon which to build because both the D and the J-K flip-flops are derived from the S-R.

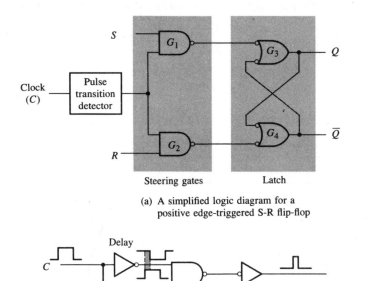

(a) A simplified logic diagram for a positive edge-triggered S-R flip-flop

(b) A type of pulse transition detector

FIGURE 7–16   *Edge triggering.*

Notice that the S-R flip-flop differs from the gated S-R latch only in that it has a pulse transition detector. The purpose of this circuit is to produce a very short duration spike on the positive-going transition of the clock pulse. One basic type of pulse transition detector is shown in Figure 7–16(b). As you can see, there is a small delay on one input to the NAND gate so that the inverted clock pulse arrives at the gate input a few nanoseconds after the true clock pulse. This produces an output spike with a time duration of only a few nanoseconds. A negative edge-triggered flip-flop inverts the clock pulse, thus producing a narrow spike on the negative-going edge.

Notice that the circuit in Figure 7–16 is partitioned into two sections, one labeled *steering gates,* and the other *latch.* The steering gates direct or "steer" the clock spike either to the input to gate G_3 or to the input to gate G_4, depending on the state of the S and R inputs. In order to understand the operation of this flip-flop, let us begin with the assumptions that it is in the RESET state ($Q = 0$) and that the S, R, and clock inputs are all LOW. For this condition, the outputs of gate G_1 and gate G_2 are both HIGH. The LOW on the Q output is coupled back into one input of gate G_4, making the $\overline{Q}$ output HIGH. Because $\overline{Q}$ is HIGH, both inputs to gate G_3 are HIGH (remember, the output of gate G_1 is HIGH), holding the Q output LOW. If a pulse is applied to the clock input, the outputs of gates G_1 and G_2 remain HIGH because they are disabled by the LOWs on the S input and the R input; therefore, there is no change in the state of the flip-flop—it remains RESET.

We will now make S HIGH, leave R LOW, and apply a clock pulse. Because the S input to gate G_1 is now HIGH, the output of gate G_1 goes LOW for a very short time (spike) when the clock input goes HIGH, causing the Q output to go HIGH. Both inputs to gate G_4 are now HIGH (remember, gate G_2 output is HIGH because R is LOW), forcing the $\overline{Q}$ output LOW. This LOW on $\overline{Q}$ is coupled back into one input of gate G_3, ensuring that the Q output will remain HIGH. The flip-flop is now in the SET state. Figure 7–17 illustrates the logic level transitions that take place within the flip-flop for this condition.

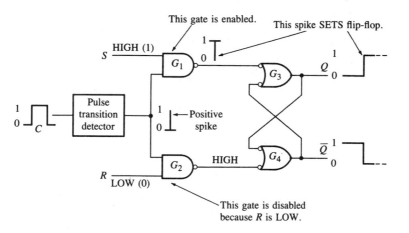

FIGURE 7–17 *Flip-flop making transition from the RESET state to the SET state on the positive leading edge of the clock pulse.*

Next, we will make S LOW and R HIGH, and apply a clock pulse. Because the R input is now HIGH, the positive-going edge of the clock produces a negative-going spike on the output of gate G_2, causing the $\overline{Q}$ output to go HIGH.

Because of this HIGH on $\overline{Q}$, both inputs to gate G_3 are now HIGH (remember, the output of gate G_1 is HIGH because of the LOW on S), forcing the Q output to go LOW. This LOW on Q is coupled back into one input of gate G_4, ensuring that $\overline{Q}$ will remain HIGH. The flip-flop is now in the RESET state. Figure 7–18 illustrates the logic level transitions that occur within the flip-flop for this condition.

As with the gated latch, an invalid condition exists when both S and R are HIGH at the same time.

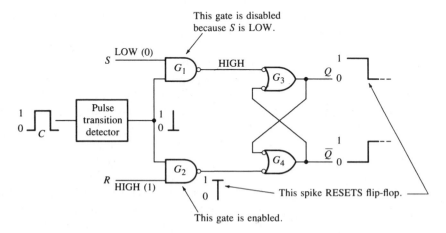

FIGURE 7–18   *Flip-flop making transition from the SET state to the RESET state on the positive leading edge of the clock pulse.*

The Edge-Triggered D Flip-Flop

The D flip-flop is very useful when a single data bit (1 or 0) is to be stored. The simple addition of an inverter to an S-R flip-flop creates a basic D flip-flop, as shown in Figure 7–19, which is a positive edge-triggered type.

FIGURE 7–19   *A positive edge-triggered D flip-flop formed with an S-R flip-flop and inverter.*

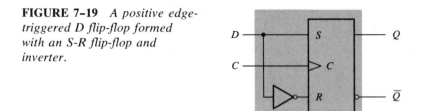

Notice that this flip-flop has only one input in addition to the clock. This is called the D input. If there is a HIGH on the D input when a clock pulse is applied, the flip-flop will SET, and the HIGH on the D input is "stored" by the flip-flop on the positive leading edge of the clock pulse. If there is a LOW on the D input when the clock pulse is applied, the flip-flop will RESET, and the LOW on the D input is

TABLE 7–3 *Truth table for a positive edge-triggered D flip-flop.*

Inputs		Outputs		
D	C	Q	$\overline{Q}$	Comments
1	↑	1	0	SET (stores a 1)
0	↑	0	1	RESET (stores a 0)

↑ = clock transition LOW to HIGH.

thus stored by the flip-flop on the leading edge of the clock pulse. In the SET state, the flip-flop is storing a logic 1, and in the RESET state it is storing a logic 0.

The operation of the positive edge-triggered D flip-flop is summarized in Table 7–3. The operation of a negative edge-triggered device is, of course, the same, except that triggering occurs on the falling edge of the clock pulse. Remember, *Q* follows *D* at the clock edge.

EXAMPLE 7–5

Given the waveforms in Figure 7–20(a) for the *D* input and the clock, determine the *Q* output waveform if the flip-flop starts out RESET.

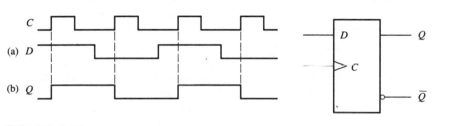

FIGURE 7–20

Solution The *Q* output assumes the state of the *D* input at the time of the positive leading clock edge. The resultant output is shown in Figure 7–20(b).

The Edge-Triggered J-K Flip-Flop

The J-K flip-flop is very versatile and is perhaps the most widely used type of flip-flop. The *J* and *K* designations for the inputs have no known significance except that they are adjacent letters in the alphabet.

The functioning of the J-K flip-flop is identical to that of the S-R flip-flop in SET, RESET, and *no-change* conditions of operation. The difference is that the J-K flip-flop has no invalid state as does the S-R flip-flop. Therefore, the J-K flip-flop is a very versatile device that finds wide application in digital systems.

Figure 7–21 shows a basic positive edge-triggered J-K flip-flop. Notice that it differs from the S-R edge-triggered flip-flop in that *the Q output is connected back to the input of gate G_2, and the $\overline{Q}$ output is connected back to the input of gate G_1.*

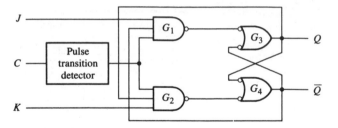

FIGURE 7–21 *A simplified logic diagram for a positive edge-triggered J-K flip-flop.*

The two control inputs are labeled J and K. A J-K flip-flop can also be of the negative edge-triggered type, in which case the clock input is inverted.

We will start by assuming that the flip-flop is RESET and that the J input is HIGH and the K input is LOW. When a clock pulse occurs, a leading-edge spike is passed through gate G_1 because $\overline{Q}$ is HIGH and J is HIGH. This will cause the latch portion of the flip-flop to change to the SET state.

The flip-flop is now SET. If we now make J LOW and K HIGH, the next clock spike will pass through gate G_2 because Q is HIGH and K is HIGH. This will cause the latch portion of the flip-flop to change to the RESET state.

Now if a LOW is applied to both the J and K inputs, the flip-flop will stay in its present state when a clock pulse occurs. So, a LOW on both J and K results in a *no-change* condition.

So far, the logical operation of the J-K flip-flop is the same as that of the S-R type in the SET, RESET, and *no-change* modes. The difference in operation occurs when both the J and K inputs are HIGH. To see this, assume that the flip-flop is RESET. The HIGH on the $\overline{Q}$ enables gate G_1 so that the clock spike passes through to SET the flip-flop. Now, there is a HIGH on Q which allows the next clock spike to pass through gate G_2 and RESET the flip-flop. As you can see, on each successive clock, the flip-flop changes to the opposite state. This is called *toggle* operation. Figure 7–22 illustrates the transitions when the flip-flop is in the toggle mode.

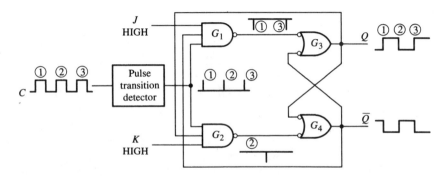

FIGURE 7–22 *Transitions illustrating the toggle operation when J = 1 and K = 1.*

TABLE 7–4 *Truth table for a positive edge-triggered J-K flip-flop.*

Inputs			Outputs		
J	K	C	Q	$\overline{Q}$	Comments
0	0	↑	Q_0	$\overline{Q}_0$	No change
0	1	↑	0	1	RESET
1	0	↑	1	0	SET
1	1	↑	$\overline{Q}_0$	Q_0	Toggle

↑ = clock transition LOW to HIGH.
Q_0 = output level prior to clock transition.

Table 7–4 summarizes the operation of the edge-triggered J-K flip-flop in truth table form. Notice that there is no invalid state as there is with an S-R flip-flop. The truth table for a negative edge-triggered device is identical except that it is triggered on the falling edge of the clock pulse.

EXAMPLE 7–6
The waveforms in Figure 7–23(a) are applied to the *J*, *K*, and clock inputs as indicated. Determine the *Q* output, assuming that the flip-flop is initially RESET.

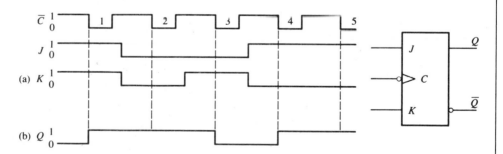

FIGURE 7–23

Solution

1. First, since this is a negative edge-triggered flip-flop, the *Q* output will change only on the negative-going edge of the clock pulse.
2. At the first clock pulse, both *J* and *K* are HIGH, and because this is a toggle condition, *Q* goes HIGH.
3. At clock pulse 2, a *no-change* condition exists on the inputs, keeping *Q* at a HIGH level.
4. When clock pulse 3 occurs, *J* is LOW and *K* is HIGH, resulting in a RESET condition. *Q* goes LOW.
5. At clock pulse 4, *J* is HIGH and *K* is LOW, resulting in a SET condition. *Q* goes HIGH.

6. A SET condition still exists on *J* and *K* when clock pulse 5 occurs. *Q* will remain HIGH.

The resulting *Q* waveform is indicated in Figure 7–23(b).

Asynchronous Inputs

For the flip-flops just discussed, the *S-R, D,* and *J-K* inputs are called *synchronous inputs* because data on these inputs are transferred to the flip-flop's output only on the triggering edge of the clock pulse; that is, the data are transferred synchronously with the clock.

Most integrated circuit flip-flops also have *asynchronous inputs*. These are inputs that affect the state of the flip-flop independent of the clock. They are normally labeled *preset* (PRE) and *clear* (CLR), or *direct set* (S_D) and *direct reset* (R_D) by some manufacturers. An active level on the preset input will SET the flip-flop, and an active level on the clear input will RESET it. A logic symbol for a J-K flip-flop with preset and clear is shown in Figure 7–24. These inputs are active-LOW as indicated by the bubbles. These preset and clear inputs must both be kept HIGH for synchronous operation.

FIGURE 7–24 *Logic symbol for a J-K flip-flop with active-LOW preset and clear inputs.*

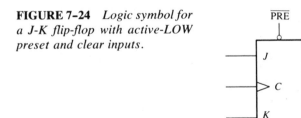

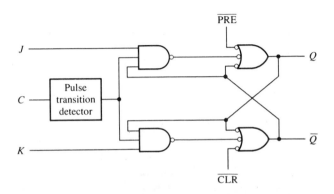

FIGURE 7–25 *Logic diagram for a basic J-K flip-flop with active-LOW preset and clear.*

Figure 7–25 shows the logic diagram for an edge-triggered J-K flip-flop with *preset* ($\overline{\text{PRE}}$) and *clear* ($\overline{\text{CLR}}$) inputs. This figure illustrates basically how these inputs work. As you can see, they are connected directly into the latch portion of the flip-flop so that they override the effect of the synchronous inputs, *J*, *K*, and the clock.

EXAMPLE 7–7   For the positive edge-triggered J-K flip-flop with preset and clear inputs in Figure 7–26(a), determine the *Q* output for the inputs shown in the timing diagram. *Q* is initially LOW.

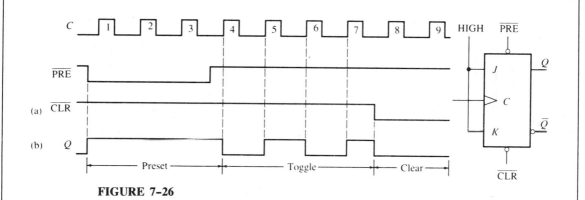

FIGURE 7–26

Solution

1. During clock pulses 1, 2, and 3, the preset ($\overline{\text{PRE}}$) is LOW, keeping the flip-flop SET regardless of the synchronous inputs.
2. For clock pulses 4, 5, 6, and 7, toggle operation occurs because *J* is HIGH, *K* is HIGH, and both $\overline{\text{PRE}}$ and $\overline{\text{CLR}}$ are HIGH.
3. For clock pulses 8 and 9, the clear ($\overline{\text{CLR}}$) input is LOW, keeping the flip-flop RESET regardless of the synchronous inputs. The resulting *Q* output is shown in Figure 7–26(b).

Specific Devices

Two specific IC edge-triggered flip-flops are now examined. They are representative of the various types available in IC form.

7474 dual D flip-flops   This TTL device has two identical flip-flops on a single chip. The flip-flops are independent of each other except for sharing V_{CC} and ground. The flip-flops are positive edge-triggered and have active-LOW asynchronous preset and clear inputs. The logic symbols for the individual flip-flops within the package are shown in Figure 7–27(a), and an ANSI/IEEE stan-

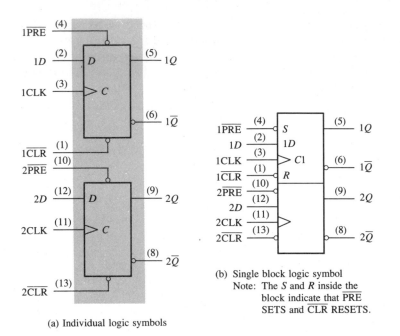

(a) Individual logic symbols

(b) Single block logic symbol
Note: The S and R inside the block indicate that $\overline{PRE}$ SETS and $\overline{CLR}$ RESETS.

FIGURE 7–27 *Logic symbols for the 7474 dual edge-triggered D flip-flops.*

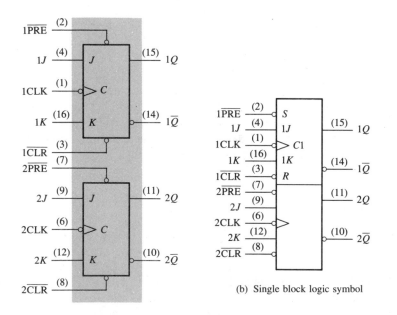

(a) Individual logic symbols

(b) Single block logic symbol

FIGURE 7–28 *Logic symbols for the 74LS76A dual edge-triggered J-K flip-flops.*

dard single block symbol representing the entire device is shown in part (b) of the figure. The pin numbers are shown in parentheses.

74LS76A dual J-K flip-flops This TTL device also has two identical flip-flops which are negative edge-triggered and have asynchronous preset and clear inputs. The logic symbols are shown in Figure 7–28.

EXAMPLE 7–8

The J, K, C, $\overline{PRE}$, and $\overline{CLR}$ waveforms in Figure 7–29(a) are applied to one of the flip-flops in a 74LS76A package. Determine the Q output waveform.

FIGURE 7–29

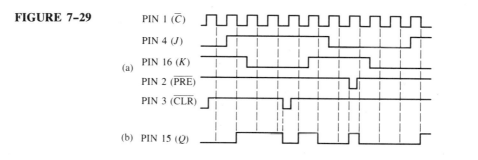

Solution The resulting Q waveform is shown in Figure 7–29(b). Notice that each time a LOW is applied to the $\overline{PRE}$ or $\overline{CLR}$, the flip-flop is SET or RESET regardless of the states of the other inputs. Note that LOWs on both $\overline{PRE}$ and $\overline{CLR}$ at the same time causes Q and $\overline{Q}$ both to go HIGH, which is a logically invalid state.

SECTION REVIEW 7–2

1. Describe the main difference between a gated S-R latch and an edge-triggered S-R flip-flop.
2. How does a J-K flip-flop differ from an S-R flip-flop in its basic operation?
3. Assume that the flip-flop in Figure 7–20 is negative edge-triggered. Sketch the output waveform for the same C and D waveforms.

7–3 PULSE-TRIGGERED (MASTER-SLAVE) FLIP-FLOPS

The second class of flip-flop is the *pulse-triggered* or *master-slave*. The term *pulse-triggered* means that data are entered into the flip-flop on the leading edge of the clock pulse, but the output does not reflect the input state until the trailing edge of the clock pulse. The inputs must be set up prior to the clock pulse's leading edge, but the output is *postponed* until the trailing edge of the clock. A major restriction of the pulse-triggered flip-flop is that the data inputs must not change while the clock pulse is HIGH because the flip-flop is sensitive to any change of input levels during this time.

As with the edge-triggered flip-flops, there are three basic types of pulse-triggered flip-flops: S-R, D, and J-K. The J-K is by far the most commonly available in integrated circuit form. The logic symbols for all of these are shown in Figure 7–30. The key to identifying a pulse-triggered (master-slave) flip-flop by its logic symbol is the ANSI/IEEE *postponed output* symbol (⌐) at the outputs. Notice that there is no dynamic input indicator (▷) at the clock input.

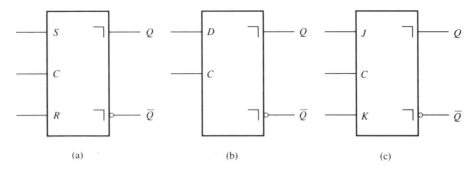

(a)   (b)   (c)

FIGURE 7–30   *Pulse-triggered (master-slave) flip-flop logic symbols.*

The Pulse-Triggered (Master-Slave) S-R Flip-Flop

A basic master-slave S-R flip-flop is shown in Figure 7–31. The truth table operation is the same as that for the edge-triggered S-R flip-flop except for the way it is clocked. Internally, though, it is quite different.

This type of flip-flop is composed of two sections, the *master* section and the *slave* section. The master section is basically a gated S-R latch, and the slave section is the same except that it is clocked on the inverted clock pulse and is controlled by the outputs of the master section rather than by the external *S-R* inputs.

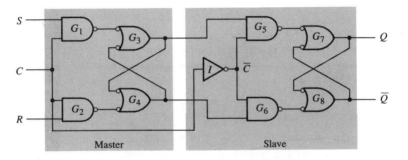

FIGURE 7–31   *Logic diagram for a basic master-slave S-R flip-flop.*

The master section will assume the state determined by the S and R inputs at the positive-going edge of the clock pulse. The state of the master section is then transferred into the slave section on the negative-going edge of the clock pulse because the outputs of the master are applied to the inputs of the slave, and the clock pulse to the slave is inverted. The state of the slave then immediately appears on the Q and $\overline{Q}$ outputs.

The timing diagram in Figure 7–32 illustrates the master-slave operation in reference to Figure 7–31. To begin, the flip-flop is RESET, S is HIGH, and R is LOW. The following statements describe what happens on the *leading* edge of the first clock pulse:

1. The output of gate G_1 goes from a HIGH to a LOW because both of its inputs are HIGH.
2. The output of gate G_3 goes from a LOW to a HIGH, and the output of gate G_4 goes from a HIGH to a LOW.
3. The inverted clock ($\overline{C}$) input to gates G_5 and G_6 goes LOW; this disables the G_5 and G_6 gates and forces their outputs HIGH.

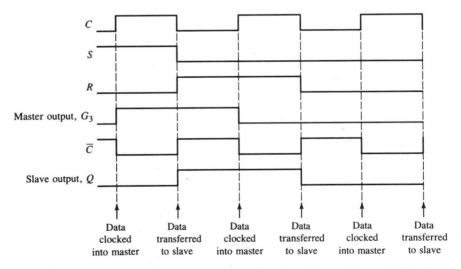

FIGURE 7–32 *Timing diagram for master-slave flip-flop in Figure 7–31, showing SET, RESET, and no-change conditions.*

Let us review what has taken place on the leading edge of the clock pulse. The master section has been SET because the S input is HIGH and the R input is LOW. The slave section has not changed state, and therefore the Q and $\overline{Q}$ outputs remain in the initial RESET state.

Now let us look at what happens on the *trailing* edge of the first clock pulse.

1. The master section remains in the SET condition.
2. The output of gate G_5 goes from a HIGH to a LOW because both of its inputs are now HIGH.

3. The output of gate G_7 goes from a LOW to a HIGH, and the output of gate G_8 goes from a HIGH to a LOW.

The flip-flop is now in the SET state, and therefore the Q output is HIGH. It did not become SET until the trailing edge of the clock pulse, although the master section was SET on the leading edge. You should verify the operations for the RESET and no-change conditions on the second and third clock pulses, respectively.

The operation of the S-R master-slave flip-flop is summarized in Table 7–5. As mentioned, it is the same as for an edge-triggered S-R except for the manner in which it is triggered. This is indicated on the truth table by the pulse symbols in the C column, which mean that it takes the *complete pulse* to get data from the input to the output rather than just a single transition as with the edge-triggered type.

TABLE 7–5 *Truth table for the S-R master-slave flip-flop.*

Inputs			Outputs		
S	R	C	Q	$\overline{Q}$	Comments
0	0	⊓	Q_0	$\overline{Q}_0$	No change
0	1	⊓	0	1	RESET
1	0	⊓	1	0	SET
1	1	⊓	?	?	Invalid

⊓ = clock pulse.
Q_0 = output level prior to clock pulse.

The Pulse-Triggered (Master-Slave) D Flip-Flop

The truth table operation of the master-slave D flip-flop is shown in Table 7–6. Notice that it is the same as for the edge-triggered D flip-flop except for the way it is triggered. The basic method of master-slave triggering discussed in relation to the S-R flip-flop also applies to the D flip-flop as well as to the J-K flip-flop.

EXAMPLE 7-9 Determine the Q output for the master-slave D flip-flop with the D and clock (C) inputs shown in Figure 7–33(a). The flip-flop is initially RESET.

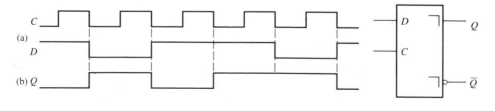

FIGURE 7–33

Solution   The data are clocked into the master portion on the leading edges of the clock pulses, and the output changes accordingly on the trailing edges as shown in Figure 7–33(b).

TABLE 7–6 *Truth table for the master-slave D flip-flop.*

Inputs		Outputs		
D	C	Q	$\overline{Q}$	Comments
0	⎍	0	1	RESET
1	⎍	1	0	SET

It was mentioned in the beginning of this section that a major restriction of the master-slave flip-flop is that the input data must be held constant while the clock is HIGH. If the *D* input changes while the clock pulse is HIGH, the new data bit is immediately stored by the master section, thus losing the data bit that was on the input at the start of the clock pulse. That is, the flip-flop responds to any change in the input level when the clock is HIGH. The last input level seen while the clock pulse is HIGH is then triggered into the slave portion and appears on the output. This is illustrated in Figure 7–34.

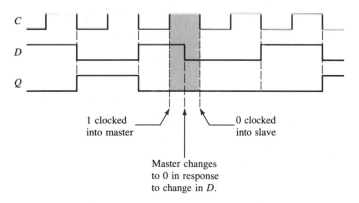

FIGURE 7–34 *Example of the effect of a change in input level while clock is HIGH in a master-slave flip-flop.*

Pulse-Triggered (Master-Slave) J-K Flip-Flop

Figure 7–35 is the basic logic diagram for a master-slave J-K flip-flop. A brief inspection of this diagram shows that this circuit is very similar to the master-slave S-R flip-flop. Notice the two main differences: First, the *Q* output is connected back into the input of gate G_2, and the $\overline{Q}$ output is connected back into

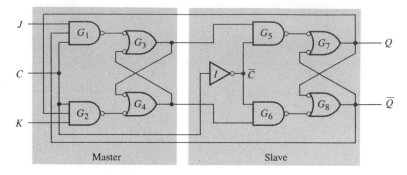

FIGURE 7–35 *Logic diagram for a master-slave J-K flip-flop.*

the input of gate G_1. Second, one input is now designated J and the other input is now designated K. The logical operation is summarized in Table 7–7.

EXAMPLE 7–10

Determine the Q output of the master-slave J-K flip-flop for the input waveforms shown in Figure 7–36(a). The flip-flop starts out RESET.

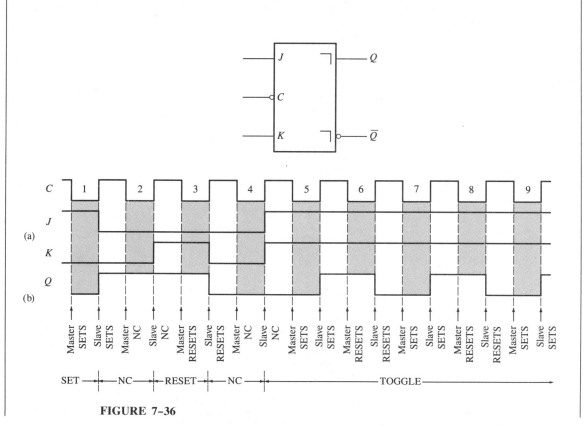

FIGURE 7–36

Solution   The Q waveform is shown in Figure 7–36(b). The input states and events at each clock pulse are labeled to demonstrate the operation.

TABLE 7–7   *Truth table for the master-slave J-K flip-flop.*

Inputs			Outputs		Comments
J	K	C	Q	$\overline{Q}$	
0	0	⎍	Q_0	$\overline{Q}_0$	No change
0	1	⎍	0	1	RESET
1	0	⎍	1	0	SET
1	1	⎍	$\overline{Q}_0$	Q_0	Toggle

⎍ = clock pulse.
Q_0 = output level before clock pulse.

Specific Devices

Two specific TTL pulse-triggered flip-flops are now presented, and examples of their operation are given.

The 74L71 master-slave S-R flip-flop   Two versions of the logic symbol for this flip-flop are shown in Figure 7–37. The more traditional symbol is in part (a), and the ANSI/IEEE Std. 91–1984 recommended symbol is shown in part (b). Notice that this particular device has AND gates for the S and R inputs. The three SET inputs, $S1$, $S2$, and $S3$, must all be HIGH to produce an $S = 1$ condition within the flip-flop. Likewise, $R1$, $R2$, and $R3$ must all be HIGH to produce an $R = 1$ condition. These AND-gated inputs simply provide additional flexibility in controlling the flip-flop. This device also has active-LOW preset ($\overline{\text{PRE}}$) and clear ($\overline{\text{CLR}}$) inputs.

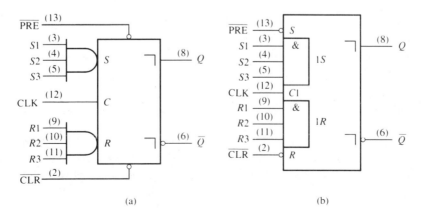

FIGURE 7–37   *Logic symbols for the 74L71 master-slave S-R flip-flop.*

In the symbol in part (b), the preset is labeled S inside the block because a LOW on the $\overline{\text{PRE}}$ line SETS the flip-flop. The clear is labeled R inside the block because a LOW on the $\overline{\text{CLR}}$ line RESETS the flip-flop. These two inputs are *asynchronous* and not dependent on the clock. The $1S$ and $1R$ labels inside the block are the *synchronous* SET and RESET functions. The 1 suffix after C and the 1 prefixes before S and R are examples of *control dependency notation*. The 1s indicate that $1S$ and $1R$ are dependent on the clock, $C1$.

EXAMPLE 7–11   From the input waveforms in Figure 7–38(a), determine the Q and $\overline{Q}$ outputs for a 74L71 master-slave S-R flip-flop. Assume that $Q = 0$ initially.

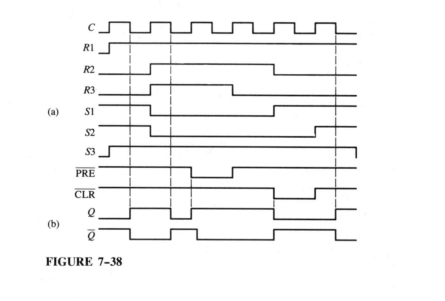

FIGURE 7–38

Solution   The resulting output waveforms are shown in Figure 7–38(b). $\overline{\text{PRE}}$ and $\overline{\text{CLR}}$ override synchronous operation.

The 74107 dual master-slave J-K flip-flop   Figure 7–39(a) shows the individual logic symbols for each flip-flop in the 74107 package. Part (b) represents the device with a single logic symbol, including dependency notation. Notice that this device has a $\overline{\text{CLR}}$ input but not a $\overline{\text{PRE}}$. This allows it to be asynchronously RESET for purposes of initialization. The labels inside the upper block in part (b) apply also to the lower block.

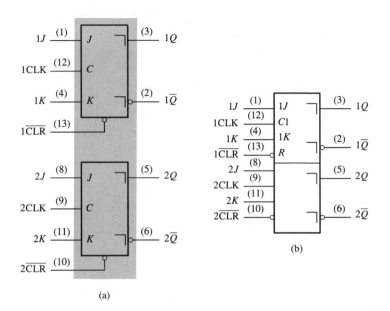

FIGURE 7-39 *Logic symbols for the 74107 dual master-slave J-K flip-flops.*

EXAMPLE 7-12

A clock pulse waveform with a frequency of 1 kHz is applied to a 74107 master-slave J-K flip-flop. J and K are both connected to V_{CC}. The $\overline{CLR}$ line is held LOW for the three initial clock pulses, and then it goes HIGH. Draw the timing diagram for ten clock pulses.

Solution The timing diagram is shown in Figure 7-40. Since J and K are connected to V_{CC}, the flip-flop is in the toggle condition. As long as the $\overline{CLR}$ is LOW, toggle operation is overridden and the flip-flop is held in RESET. As soon as $\overline{CLR}$ is released from its LOW state, the flip-flop begins toggling on the fourth pulse.

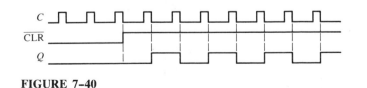

FIGURE 7-40

1. Describe the basic difference between pulse-triggered and edge-triggered flip-flops.
2. Suppose that the D input of a flip-flop changes from LOW to HIGH in the middle of a clock pulse.
   (a) Describe what happens if the flip-flop is a positive edge-triggered type.
   (b) Describe what happens if the flip-flop is a pulse-triggered (master-slave) type.
3. What is the advantage of a J-K flip-flop over an S-R flip-flop?

7-4 DATA LOCK-OUT FLIP-FLOPS

The data lock-out flip-flop is similar to the pulse-triggered (master-slave) flip-flop except that it has a *dynamic* clock input so that it is sensitive to the data inputs

FIGURE 7–41 *Logic symbol for a data lock-out J-K flip-flop.*

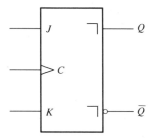

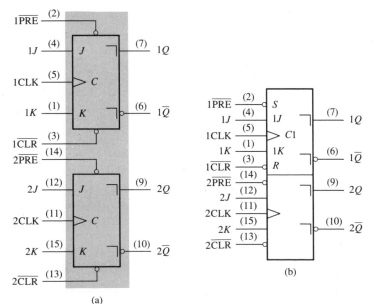

(a)

(b)

FIGURE 7–42 *Logic symbols for the 74111 dual J-K master-slave flip-flop with data lock-out.*

only during a clock transition. After the leading-edge clock transition, the data inputs are disabled and do not have to be held constant while the clock pulse is HIGH. In essence, the master portion of this flip-flop is like an edge-triggered device, and the slave portion performs as in a pulse-triggered device to produce a postponed output.

A logic symbol for a data lock-out J-K flip-flop is shown in Figure 7–41. Notice that this symbol has both the dynamic input indicator for the clock and the postponed output indicators. This type of flip-flop is actually classified by most manufacturers as a master-slave with a special lock-out feature.

The 74111 is an example of this type of device. It is a dual J-K master-slave flip-flop with data lock-out. It has both preset and clear inputs. The logic symbols are shown in Figure 7–42. Again, part (a) is the more traditional symbol, and part (b) is a single block symbol as specified by ANSI/IEEE Std. 91–1984.

SECTION REVIEW 7–4

1. Describe how a data lock-out flip-flop differs from a pulse-triggered type.

7–5 OPERATING CHARACTERISTICS

Several operating characteristics or parameters important in the application of flip-flops specify the performance, operating requirements, and operating limitations of the circuit. They are typically found in the data sheets for integrated circuits, and they are applicable to all flip-flops regardless of the particular form of the circuit.

Propagation Delay Time

A propagation delay is the interval of time required after the input signal has been applied for the resulting output change to occur. Several categories of propagation delay are important in the operation of a flip-flop:

1. *Propagation delay* t_{PLH} measured from the triggering edge of the clock pulse to the *LOW-to-HIGH transition* of the output. This delay is illustrated in Figure 7–43(a).

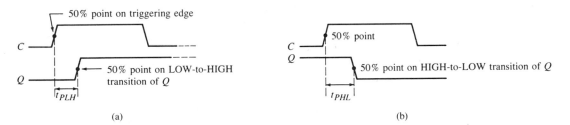

(a)   (b)

FIGURE 7–43 *Propagation delays, clock to output.*

2. *Propagation delay* t_{PHL} measured from the triggering edge of the clock pulse to the *HIGH-to-LOW transition* of the output. This delay is illustrated in Figure 7–43(b).

3. *Propagation delay* t_{PLH} measured from the *preset* input to the *LOW-to-HIGH transition* of the output. This delay is illustrated in Figure 7–44(a) for an active-LOW *preset*.

FIGURE 7–44 *Propagation delays, preset and clear to output.*

4. *Propagation delay* t_{PHL} measured from the *clear* input to the *HIGH-to-LOW transition* of the output. This delay is illustrated in Figure 7–44(b) for an active-LOW *clear*.

Set-Up Time

The *set-up time* (t_s) is the minimum interval required for the control levels to be maintained constantly on the inputs (*J* and *K*, or *S* and *R*, or *D*) *prior* to the triggering edge of the clock pulse in order for the levels to be reliably clocked into the flip-flop. This is illustrated in Figure 7–45 for a D flip-flop.

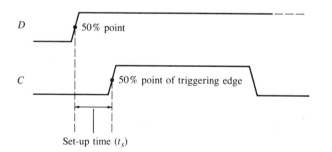

FIGURE 7–45 *Set-up time* (t_s).

Hold Time

The *hold time* (t_h) is the minimum interval required for the control levels to remain on the inputs *after* the triggering edge of the clock pulse in order for the levels to be reliably clocked into the flip-flop. This is illustrated in Figure 7–46 for a D flip-flop.

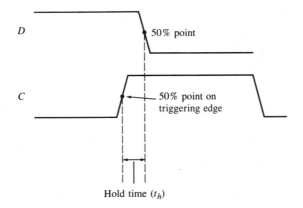

FIGURE 7–46 *Hold time (t_h).*

Maximum Clock Frequency

The *maximum clock frequency* ($f_{\max}$) is the highest rate at which a flip-flop can be reliably triggered. At clock frequencies above the maximum, the flip-flop would be unable to respond quickly enough and its operation would be impaired.

Pulse Widths

Minimum pulse widths (t_W) for reliable operation are usually specified by the manufacturer for the clock, preset, and clear inputs. Typically, the clock is specified by its minimum HIGH time and its minimum LOW time.

Power Dissipation

The power dissipation of a flip-flop is the total power consumption of the device. For example, if the flip-flop operates on a +5-V dc source and draws 50 mA of current, the power dissipation is

$$P = V_{CC} \times I_{CC} = 5 \text{ V} \times 50 \text{ mA} = 250 \text{ mW}$$

The power dissipation is very important in most applications in which the capacity of the dc supply is a concern. As an example, let us assume that we have a digital system that requires a total of ten flip-flops, and each flip-flop dissipates 250 mW of power. The total power requirement is

$$P_{\text{TOT}} = 10 \times 250 \text{ mW} = 2500 \text{ mW} = 2.5 \text{ W}$$

This tells us the type of dc supply that is required as far as output capacity is concerned. If the flip-flops operate on $+5$ V dc, then the amount of current that the supply must provide is as follows:

$$I = \frac{2.5 \text{ W}}{5 \text{ V}} = 0.5 \text{ A}$$

We must use a $+5$-V dc supply that is capable of providing at least 0.5 A of current.

Other Characteristics

Many characteristics discussed in Chapter 3 in relation to gates—such as fan-out, input voltages, output voltages, and noise margin—apply to flip-flops and are not repeated here.

Comparison of Specific Flip-Flops

Table 7–8 provides a comparison of several TTL devices that have been studied in this chapter and a CMOS device in terms of the operating parameters discussed in this section.

TABLE 7–8 *Comparison of operating parameters of flip-flops.*

Parameter (Times in ns)	TTL					CMOS
	7474	74LS76A	74L71	74107	74111	74HC112
t_{PHL} (CLK to Q)	40	20	150	40	30	31
t_{PLH} (CLK to Q)	25	20	75	25	17	31
t_{PHL} ($\overline{\text{CLR}}$ to Q)	40	20	200	40	30	41
t_{PLH} ($\overline{\text{PRE}}$ to Q)	25	20	75	25	18	41
t_s (set-up)	20	20	0	0	0	25
t_h (hold)	5	0	0	0	30	0
t_W (CLK HI)	30	20	200	20	25	25
t_W ($\overline{\text{CLK LO}}$)	37	—	200	47	25	25
t_W ($\overline{\text{CLR/PRE}}$)	30	25	100	25	25	25
f_{max} (MHz)	15	45	3	20	25	20
Power (mW/F-F)	43	10	3.8	50	70	0.12

SECTION REVIEW 7–5

1. Define the following:
   (a) *set-up time* (b) *hold time*
2. Which specific flip-flop can be operated at the highest frequency?

7–6   BASIC FLIP-FLOP APPLICATIONS

In this section, several basic applications are presented to give you a better feel for how flip-flops can be applied in digital systems.

Parallel Data Storage

A common requirement in digital systems is to take several bits of data on parallel lines and store them simultaneously in a group of flip-flops. This operation is illustrated in Figure 7–47(a). Each of the four parallel data lines is connected to the D input of a flip-flop. The clock inputs of all the flip-flops are connected to a common clock input, so that each flip-flop is triggered at the same time. In this

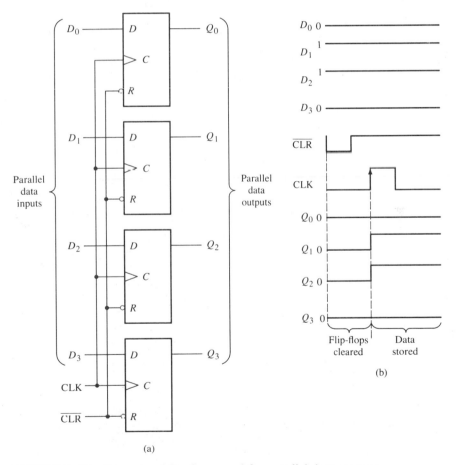

FIGURE 7–47   *Example of flip-flops used for parallel data storage.*

example, positive edge-triggered flip-flops are used, so the data on the D inputs are stored simultaneously by the flip-flops on the positive edge of the clock as indicated in the timing diagram in Figure 7–47(b). Also, the asynchronous *clear* inputs are connected to a common $\overline{CLR}$ line which resets all the flip-flops. In this application, resetting clears out all data bits previously stored.

Groups of flip-flops used for data storage are called *registers*. In digital systems, data are normally stored in groups of bits that represent numbers, codes, or other information.

Data Transfer

It is often necessary to transfer a data bit from one flip-flop to another, as illustrated in Figure 7–48 using pulse-triggered (master-slave) flip-flops. A data bit is clocked into flip-flop A on the leading edge of the first clock pulse. (The 1 in this case is indicated by the shaded area in the timing diagram.) The Q_A output goes HIGH on the trailing edge of the clock pulse. Since the Q output of flip-flop A is connected to the D input of flip-flop B, the 1 is clocked into flip-flop B on the leading edge of the second clock pulse and appears on the Q_B output on the trailing edge. Thus, the transfer of the data bit from flip-flop A to flip-flop B has been achieved. This concept is important in certain types of registers that are covered in a later chapter.

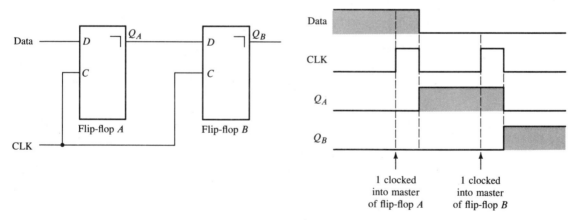

FIGURE 7–48   *Data bit (1 indicated by shaded area) being transferred into flip-flop A and then to flip-flop B.*

This example is a good way to demonstrate the advantage of master-slave flip-flops over edge-triggered flip-flops in certain situations. Figure 7–49 shows the circuit of Figure 7–48 implemented with negative edge-triggered D flip-flops. The data bit is clocked into flip-flop A on the trailing edge of the first clock pulse, and Q_A goes HIGH at this time (actually t_{PLH} later). At the trailing edge of the second clock pulse, Q_A goes LOW a short time (t_{PHL}) after the 1 is being clocked

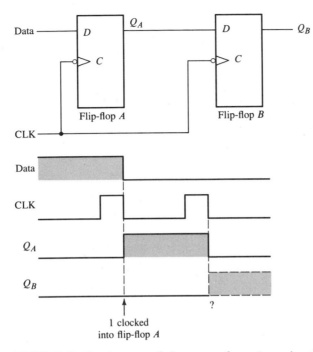

FIGURE 7–49 *A case of data transfer using edge-triggered flip-flops. An unreliable transfer is possible due to a race condition.*

into flip-flop *B*, creating what is called a *race* condition. This produces a marginal condition because if Q_A (which is also the *D* input of flip-flop *B*) does not remain HIGH long enough after the triggering edge of the clock, the 1 may not be transferred. Q_A must remain HIGH for a time equal to the hold time (t_h) in order to be reliably clocked into flip-flop *B*. The 1 is properly transferred only if $t_h <$ t_{PHL} (CLK to *Q*). So, a potentially unreliable data-transfer situation is created. The very low hold times on some newer edge-triggered flip-flops eliminate the need for master-slave flip-flops in many applications in which they previously had an advantage.

Frequency Division

Another basic application of a flip-flop is dividing (reducing) the frequency of a periodic waveform. When a pulse waveform is applied to the clock input of a J-K flip-flop that is connected to toggle, the *Q* output is a square wave with one-half the frequency of the clock input. Thus, a single flip-flop is a divide-by-2 device, as is illustrated in Figure 7–50. As you can see, the flip-flop changes state on each triggering clock edge (positive edge-triggered in this case). This results in an output that changes at half the frequency of the clock waveform.

FIGURE 7–50 *The J-K flip-flop as a divide-by-2 device.*

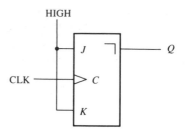

Further division of a clock frequency can be achieved by using the output of one flip-flop as the clock input to a second flip-flop, as shown in Figure 7–51. The frequency of the Q_A output is divided by 2 by flip-flop B. The Q_B output is, therefore, one-fourth the frequency of the original clock input.

By connecting flip-flops in this way, a frequency division of 2^n is achieved, where n is the number of flip-flops. For example, three flip-flops divide the clock frequency by $2^3 = 8$; four flip-flops divide the clock frequency by $2^4 = 16$; and so on.

FIGURE 7–51 *Example of two J-K flip-flops used to divide the clock frequency by 4.*

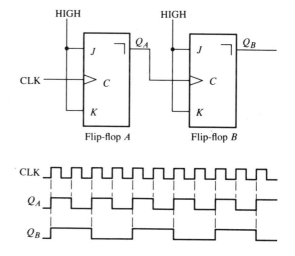

Counting

Another very important application of flip-flops is in *digital counters*, which are covered in detail in the next chapter. The concept is illustrated in Figure 7–52. The flip-flops are negative edge-triggered J-Ks. Both flip-flops are initially reset. Flip-flop A toggles on the negative-going transition of each clock pulse. The Q output of flip-flop A clocks flip-flop B so that each time Q_A makes a HIGH-to-LOW transition, flip-flop B toggles. The resulting Q_A and Q_B waveforms are shown in the figure.

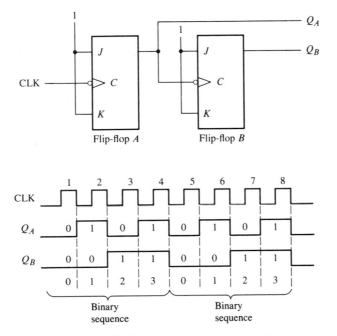

FIGURE 7–52   *Flip-flops used to generate a binary count sequence.*

Observe the sequence of Q_A and Q_B. Prior to clock pulse 1, $Q_A = 0$ and $Q_B = 0$; after clock pulse 1, $Q_A = 1$ and $Q_B = 0$; after clock pulse 2, $Q_A = 0$ and $Q_B = 1$; and after clock pulse 3, $Q_A = 1$ and $Q_B = 1$. If we take Q_A as the least significant bit, a two-bit binary sequence is produced as the flip-flops are clocked. This binary sequence repeats every four clock pulses, as shown in the timing diagram of Figure 7–52. Thus, the flip-flops are counting in sequence from 0 to 3 (00, 01, 10, 11) and then recycling back to 0 to begin the sequence again.

SECTION REVIEW 7–6

1. A group of flip-flops used for data storage is called a _____.
2. How must a J-K flip-flop be connected to function as a divide-by-2 element?
3. How many flip-flops are required to produce a divide-by-32 device?

7-7   ONE-SHOTS

A second type of multivibrator is the *one-shot* (*monostable*) multivibrator. This device has only *one stable state*. When triggered, it changes from its stable to its unstable state and remains there for a specified length of time before returning automatically to its stable state.

Figure 7–53 shows a basic one-shot circuit composed of a logic gate and an inverter. When a pulse is applied to the *trigger* input, the output of gate G_1 goes LOW. This HIGH-to-LOW transition is coupled through the capacitor to the input to inverter, G_2. The apparent LOW on G_2 makes its output go HIGH. This HIGH is connected back into G_1, keeping its output LOW. Up to this point, the trigger pulse has caused the output of the one-shot, Q, to go HIGH.

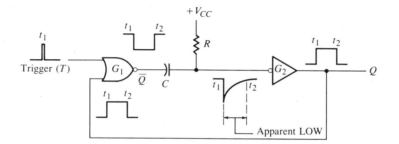

FIGURE 7–53   *A simple one-shot circuit.*

The capacitor immediately begins to charge through R toward the high voltage level. The rate at which it charges is determined by the RC time constant. When the capacitor charges to a certain level which appears as a HIGH to G_2, the output goes back LOW.

To summarize, the output of inverter G_2 goes HIGH in response to the trigger input. It remains HIGH for a time set by the RC time constant. At the end of this time, it goes LOW. So, a single narrow trigger pulse produces a *single* output pulse whose time duration is controlled by the RC time constant. This operation is illustrated in Figure 7–53. A typical one-shot logic symbol is shown in Figure 7–54(a), and the same symbol with an external R and C is shown in Figure 7–54(b).

Integrated Circuit One-Shots

The two basic types of IC one-shots are nonretriggerable and retriggerable. A nonretriggerable type will not respond to any additional trigger pulses once it is

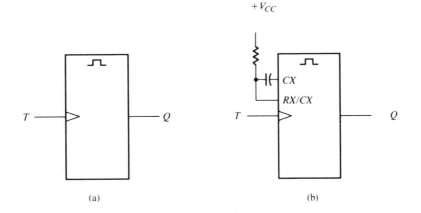

FIGURE 7-54 *One-shot logic symbols (ANSI/IEEE Std.).*

triggered into its unstable state (fired) until it returns to its stable state (times out). In other words, the period of the trigger pulses must be greater than the time the one-shot remains in its unstable state. The time that it remains in its unstable state is the *pulse width* of the output.

Figure 7–55 shows the one-shot being triggered at intervals greater than its pulse width and at intervals less than the pulse width. Notice that in the second case, the additional pulses are ignored.

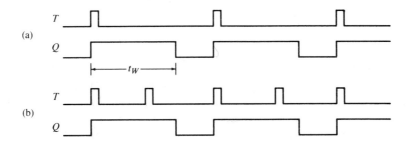

FIGURE 7-55 *Nonretriggerable one-shot action.*

A retriggerable one-shot can be triggered before it times out. The result of retriggering is an extension of the pulse width as illustrated in Figure 7–56.

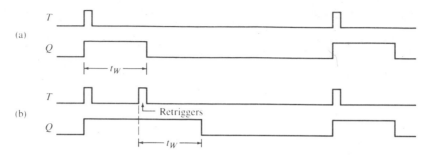

FIGURE 7–56   *Retriggerable one-shot action.*

The 74122 One-Shot

The 74122 is an example of an IC one-shot with a gated trigger and a clear input. It has provisions for external R and C, as shown in Figure 7–57. $A1$, $A2$, $B1$, and $B2$ are the gated trigger inputs. R_{INT} is an internal timing resistor.

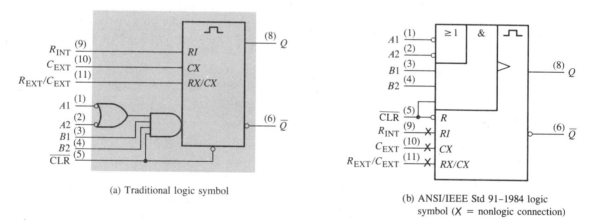

(a) Traditional logic symbol

(b) ANSI/IEEE Std 91–1984 logic symbol (X = nonlogic connection)

FIGURE 7–57   *Logic symbol for the 74122 retriggerable one-shot (monostable multivibrator).*

The pulse width is set by the external resistor and capacitor connected to pins 10 and 11. A general formula for calculating the values of these components for a specified pulse width (t_W) is

$$t_W = KR_{EXT}C_{EXT}\left(1 + \frac{0.7}{R_{EXT}}\right) \qquad (7\text{–}1)$$

where K is a constant determined by the particular type of one-shot and is usually given on manufacturers' data sheets. For example, K is 0.32 for the 74122.

The 74123 One-Shot

Another example of an integrated circuit one-shot is the 74123, which is a dual retriggerable one-shot as shown in Figure 7–58.

FIGURE 7–58 *The 74123 dual retriggerable one-shot.*

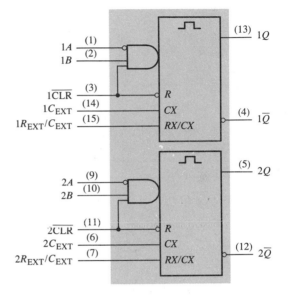

EXAMPLE 7–13

Determine the values of R_{EXT} and C_{EXT} that will produce a pulse width of 1 μs when connected to a 74123. Show the connections.

Solution The manufacturer's data sheet gives $K = 0.28$ for this device. R_{EXT} is in kilohms (kΩ), C_{EXT} is in picofarads (pF), and t_W is in nanoseconds (ns).

$$t_W = K R_{EXT} C_{EXT} \left(1 + \frac{0.7}{R_{EXT}} \right)$$

Assume a value of $C_{EXT} = 500$ pF and then solve for R_{EXT}.

$$t_W = K R_{EXT} C_{EXT} + 0.7 \left(\frac{K R_{EXT} C_{EXT}}{R_{EXT}} \right)$$

$$= K R_{EXT} C_{EXT} + 0.7 K C_{EXT}$$

$$R_{EXT} = \frac{t_W - 0.7 K C_{EXT}}{K C_{EXT}} = \frac{t_W}{K C_{EXT}} - 0.7$$

$$= \frac{1000 \text{ ns}}{(0.28)500 \text{ pF}} - 0.7$$

$$= 7.14 \text{ k}\Omega$$

Connections are shown in Figure 7–59.

FIGURE 7–59

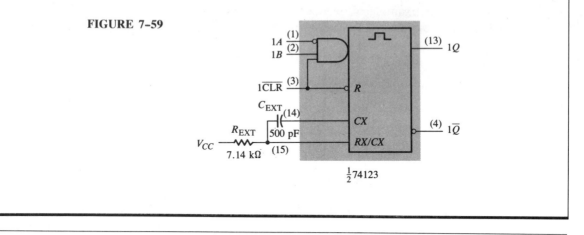

SECTION REVIEW 7–7

1. Describe the difference between a nonretriggerable and a retriggerable one-shot.
2. How is the output pulse width adjusted in most IC one-shots?

7–8   ASTABLE MULTIVIBRATORS

A third type of multivibrator is the *astable* (free-running) type. This device has no stable states, but it switches back and forth between two unstable states. The astable multivibrator is used as an oscillator to provide clock signals for timing purposes.

Figure 7–60 shows a basic astable using two inverters. Notice the capacitive coupling from the output of each inverter to the input of the other. The operation is similar to that of the one-shot in its unstable state, but the two capacitive coupling networks prevent either inverter from having a stable state. If the circuit is designed properly, it will start oscillating on its own and requires no initial input trigger.

FIGURE 7–60 *An example of an astable multivibrator.*

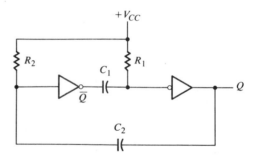

The 555 Timer

The 555 timer is an integrated circuit that can be used as an astable or monostable multivibrator and for many other applications. Figure 7–61 shows a functional diagram of a 555 timer.

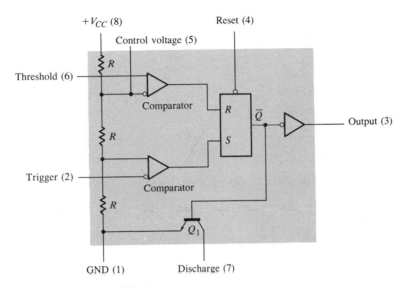

FIGURE 7–61 *A 555 timer.*

The *threshold* and *trigger* levels are normally $\frac{2}{3}V_{CC}$ and $\frac{1}{3}V_{CC}$, respectively. These levels can be altered with external connections to the *control voltage* terminal. When the *trigger* input goes below the trigger level, the internal latch is SET and the output goes HIGH. When the *threshold* input goes above the threshold level, the latch is RESET and the output goes LOW.

The RESET inputs can override the other inputs. Here is what happens when the RESET is LOW: The latch is RESET, causing the output to go LOW. This turns on Q_1, providing a low impedance path from the *discharge* terminal to ground.

Astable operation A 555 timer is shown in Figure 7–62 connected as a *free-running multivibrator*. Notice that the trigger input and threshold inputs are connected together. The capacitor C charges through R_1 and R_2 and discharges through R_2. Therefore, the frequency and duty cycle of the output waveform can be set by selecting proper values for these two resistors. Figure 7–63 shows typical waveforms produced by this device. It can be used as a pulse signal source in various applications.

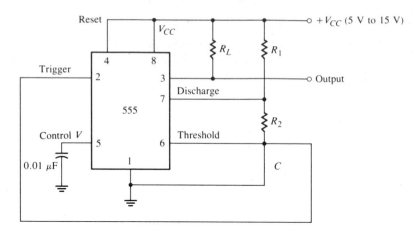

FIGURE 7–62 *A 555 timer connected for astable operation.*

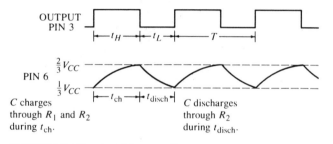

FIGURE 7–63 *Astable waveforms.*

The following formula can be used to calculate the frequency of the 555 timer for astable operation:

$$f = \frac{1}{T} = \frac{1.44}{(R_1 + 2R_2)C} \tag{7-2}$$

The time that the output is in the HIGH state is

$$t_H = 0.693(R_1 + R_2)C \tag{7-3}$$

The time that the output is in the LOW state is

$$t_L = 0.693R_2C \tag{7-4}$$

The period of the output waveform is the sum of the HIGH time and the LOW time.

$$T = t_H + t_L = 0.693(R_1 + 2R_2)C \tag{7-5}$$

Finally, the duty cycle is developed as follows:

$$\text{Duty cycle} = \frac{t_H}{T} = \frac{t_H}{t_H + t_L} = \frac{R_1 + R_2}{R_1 + 2R_2} \tag{7-6}$$

EXAMPLE 7-14

A 555 timer is to be used as an astable multivibrator. Determine the values for R_1 and R_2 if the output frequency is to be 10 kHz with a duty cycle of approximately 50%. The capacitor C is 0.001 μF.

Solution For a 50% duty cycle, $t_H = t_L$; therefore $R_1 + R_2 \cong R_2$. To satisfy this requirement, $R_2 >> R_1$.

$$f = \frac{1.44}{(R_1 + 2R_2)C} \cong \frac{1.44}{2R_2C}$$

$$R_2 = \frac{1.44}{2fC} = \frac{1.44}{2(10 \text{ kHz})(0.001 \ \mu\text{F})}$$

$$= 72 \text{ k}\Omega$$

Make $R_1 = 1$ kΩ.

SECTION REVIEW 7-8

1. Explain the difference in operation between an astable multivibrator and a monostable multivibrator.
2. For a certain astable multivibrator, $t_H = 15$ ms and $T = 20$ ms. What is the duty cycle of the output?

7-9 TROUBLESHOOTING

In industry, it is standard practice to test a new circuit design to be sure that it is operating as specified. New designs are usually "breadboarded" and tested before the design is finalized. The term *breadboard* refers to a method of temporarily hooking up a circuit so that its operation can be verified and any faults (bugs) worked out before a prototype unit is built. We will now consider an example case.

The purpose of the circuit in Figure 7-64(a) is to generate two clock waveforms (CLK A and CLK B), having an alternating occurrence of pulses. Each waveform is to be one-half the frequency of the original clock (CLK), as shown in the ideal timing diagram in part (b).

When the circuit is tested, the CLK A and CLK B waveforms appear on the oscilloscope as shown in Figure 7-65(a). Since glitches are observed on both waveforms, something is wrong with the circuit either in its basic design or in the way it is connected. Further investigation reveals that the glitches are caused by a race condition between the CLK signal and the Q and $\overline{Q}$ signals at the inputs of the AND gates. As displayed in Figure 7-65(b), the propagation delays between CLK and Q and $\overline{Q}$ create a short-duration coincidence of HIGH levels at the leading edges of alternate clock pulses. Thus, there is a basic design flaw.

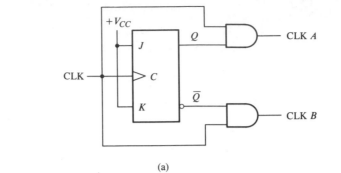

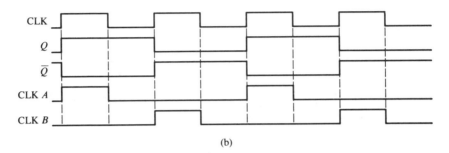

(a)

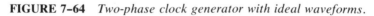

(b)

FIGURE 7–64   *Two-phase clock generator with ideal waveforms.*

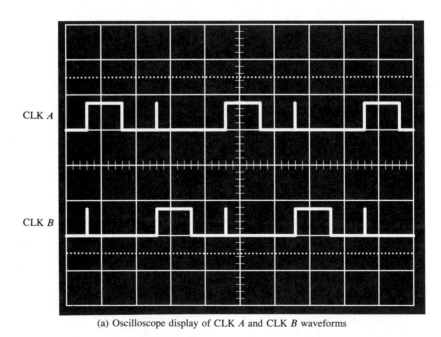

(a) Oscilloscope display of CLK *A* and CLK *B* waveforms

FIGURE 7–65   *Oscilloscope displays for the circuit in Figure 7–64.*

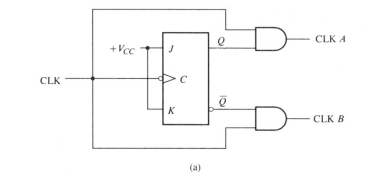

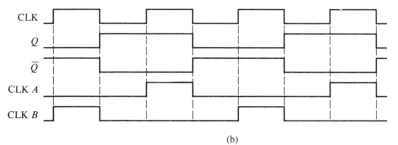

(b)

FIGURE 7–66 *Two-phase clock generator using negative edge-triggered flip-flop to eliminate glitches.*

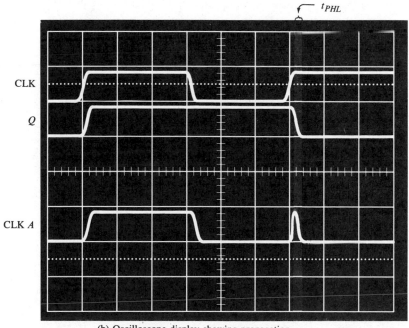

(b) Oscilloscope display showing propagation
delay that creates glitch on CLK *A* waveform

FIGURE 7–65 *(Continued)*

The problem can be corrected by using a negative edge-triggered flip-flop in place of the positive edge-triggered device as shown in Figure 7–66(a) (p. 333). Although the propagation delays between CLK and Q and $\overline{Q}$ still exist, they are initiated on the trailing edges of the clock (CLK), thus eliminating the glitches as shown in the timing diagram of Figure 7–66(b).

SECTION REVIEW 7–9

1. Can a master-slave J-K flip-flop be used in the circuit of Figure 7–66 as effectively as the negative edge-triggered flip-flop?

SUMMARY

□ The symbols for all devices in this summary conform with ANSI/IEEE Std. 91–1984. (Preset and clear inputs are not shown on devices.)

□ Latches are bistable elements whose state normally depends on asynchronous inputs. Figure 7–67 shows symbols and truth tables for latches.

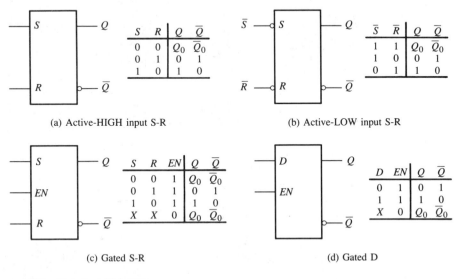

S	R	Q	$\overline{Q}$
0	0	Q_0	$\overline{Q}_0$
0	1	0	1
1	0	1	0

(a) Active-HIGH input S-R

$\overline{S}$	$\overline{R}$	Q	$\overline{Q}$
1	1	Q_0	$\overline{Q}_0$
1	0	0	1
0	1	1	0

(b) Active-LOW input S-R

S	R	EN	Q	$\overline{Q}$
0	0	1	Q_0	$\overline{Q}_0$
0	1	1	0	1
1	0	1	1	0
X	X	0	Q_0	$\overline{Q}_0$

(c) Gated S-R

D	EN	Q	$\overline{Q}$
0	1	0	1
1	1	1	0
X	0	Q_0	$\overline{Q}_0$

(d) Gated D

Note: Q_0 is the initial state.

FIGURE 7–67 *Latches.*

☐ Edge-triggered flip-flops are bistable elements with synchronous inputs whose state depends on the inputs only at the triggering transition of a clock pulse. Changes in the outputs occur at the triggering transition of the clock. Figure 7–68 shows symbols and truth tables for edge-triggered flip-flops.

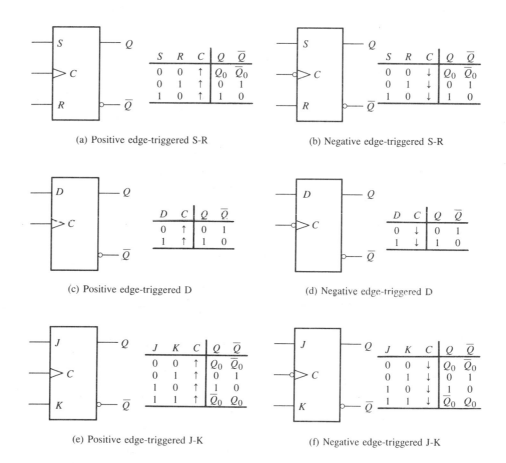

(a) Positive edge-triggered S-R

(b) Negative edge-triggered S-R

(c) Positive edge-triggered D

(d) Negative edge-triggered D

(e) Positive edge-triggered J-K

(f) Negative edge-triggered J-K

FIGURE 7–68 *Edge-triggered flip-flops.*

☐ Pulse-triggered or master-slave flip-flops are bistable elements with synchronous inputs whose state depends on the inputs at the leading edge of the clock pulse, but whose output is postponed and does not reflect the internal state until the trailing clock edge. The synchronous inputs should not be allowed to change while the clock is HIGH. Figure 7–69 shows symbols and truth tables for master-slave flip-flops.

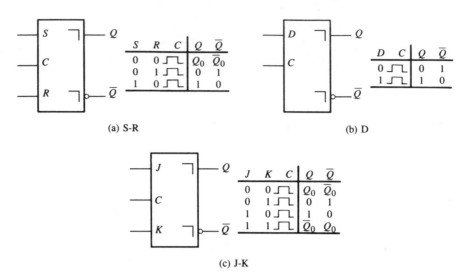

(a) S-R   (b) D

(c) J-K

FIGURE 7–69   *Pulse-triggered (master-slave) flip-flops.*

☐   Data lock-out flip-flops are similar to master-slave types in that they are sensitive to the inputs only at the leading clock edge, but the output is postponed until the trailing clock edge. However, inputs can be changed while clock is HIGH without affecting the operation. Figure 7–70 shows symbols and truth tables for data lock-out flip-flops.

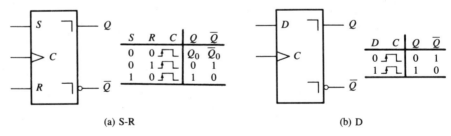

(a) S-R   (b) D

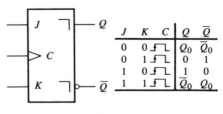

(c) J-K

FIGURE 7–70   *Data lock-out flip-flops.*

☐ Monostable multivibrators (one-shots) have one stable state. When the one-shot is triggered, the output goes to its unstable state for a time determined by an *RC* circuit. Figure 7–71 shows the symbol for a one-shot.

FIGURE 7–71 *Monostable multivibrator (one-shot).*

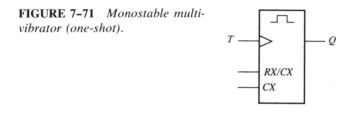

☐ Astable multivibrators have no stable states and are used as oscillators to generate timing waveforms in digital systems.

SELF-TEST

1. An active-HIGH input S-R latch has a 1 on the *S* input and a 0 on the *R* input. What state is the latch in?
2. Why is the $S = 1$, $R = 1$ input state invalid in an active-HIGH S-R latch?
3. The following sequence of levels is applied to the inputs of a gated D latch: *D* goes HIGH, then EN goes HIGH, then *D* goes LOW, then EN goes LOW and back HIGH. What is the sequence of levels on the *Q* output?
4. List the three basic types of edge-triggered flip-flops classified by inputs.
5. Name the two types of edge-triggered flip-flops classified by the method of triggering, and explain the difference.
6. What advantage does a J-K flip-flop have over an S-R?
7. For a negative edge-triggered J-K flip-flop, $J = 0$, $K = 1$, and $Q = 0$. When the clock input goes HIGH, what happens to Q? When the clock input goes LOW, what happens to Q?
8. Describe basically how data are clocked into a pulse-triggered (master-slave) flip-flop.
9. What is the major restriction when operating a pulse-triggered flip-flop?
10. Symbolically, how can a data lock-out flip-flop be distinguished from a pulse-triggered flip-flop?
11. Typically, a manufacturer's data sheet specifies four different propagation delay times associated with a flip-flop. Name and describe each one.
12. The data sheet of a certain flip-flop specifies that the minimum HIGH time for the clock pulse is 30 ns and the minimum LOW time is 37 ns. What is the maximum operating frequency?
13. List four basic flip-flop applications.
14. Sketch the output of a one-shot with an output pulse width of 50 μs that is triggered with a 10-kHz signal.
15. How are astable multivibrators normally used in digital systems?

PROBLEMS

Section 7-1

7-1   If the waveforms in Figure 7-72 are applied to an active-LOW input S-R latch, sketch the resulting Q output waveform in relation to the inputs. Assume that Q starts LOW.

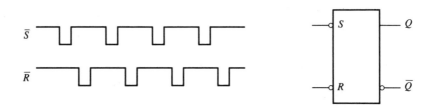

FIGURE 7-72

7-2   Repeat Problem 7-1 for the input waveforms in Figure 7-73 applied to an active-HIGH S-R latch.

FIGURE 7-73

7-3   Repeat Problem 7-1 for the input waveforms in Figure 7-74.

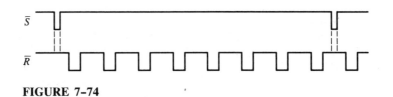

FIGURE 7-74

7-4   For a gated S-R latch, determine the Q and $\overline{Q}$ outputs for the given inputs in Figure 7-75. Show them in proper relation to the enable. Assume that Q starts LOW.

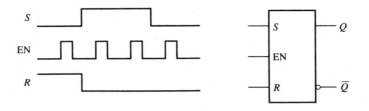

FIGURE 7-75

7-5 Repeat Problem 7–4 for the inputs in Figure 7–76.

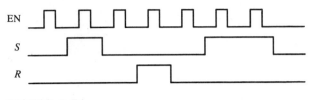

FIGURE 7–76

7-6 Repeat Problem 7–4 for the inputs in Figure 7–77.

FIGURE 7–77

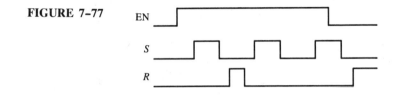

7-7 For a gated D latch, the waveforms shown in Figure 7–78 are observed on its inputs. Sketch the output waveform you would expect to see at Q if the latch is initially RESET.

FIGURE 7–78

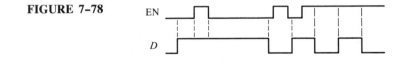

Section 7-2

7-8 Two edge-triggered S-R flip-flops are shown in Figure 7–79. If the inputs are as shown, sketch the Q output of each flip-flop, indicating the relative difference between the two. The flip-flops are initially RESET.

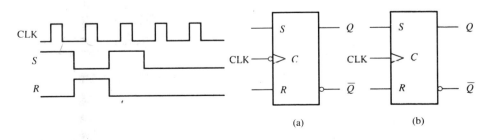

FIGURE 7–79

7–9 The Q output of the edge-triggered S-R flip-flop in Figure 7–80 is shown in relation to the clock signal. Determine the input waveforms on the S and R inputs that are required to produce this output if the flip-flop is a positive edge-triggered type.

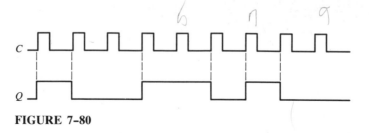

FIGURE 7–80

7–10 Sketch the Q output for a D flip-flop with the inputs as shown in Figure 7–81. Assume positive edge-triggering and Q initially LOW.

FIGURE 7–81

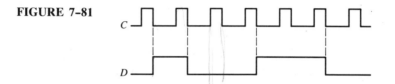

7–11 Repeat Problem 7–10 for the inputs in Figure 7–82.

FIGURE 7–82

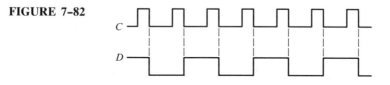

7–12 For a positive edge-triggered J-K flip-flop with inputs as shown in Figure 7–83, determine the Q output. Assume that Q starts LOW.

FIGURE 7–83

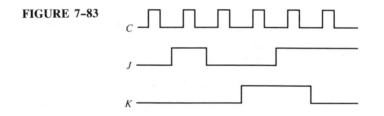

7–13 Repeat Problem 7–12 for the inputs in Figure 7–84.

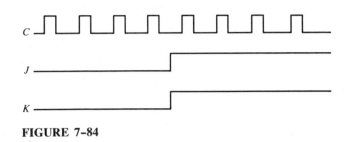

FIGURE 7–84

7–14 Determine the Q waveform if the signals shown in Figure 7–85 are applied to the inputs of the J-K flip-flop. Q is initially LOW.

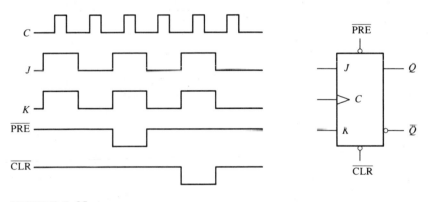

FIGURE 7–85

Section 7–3

7–15 For a pulse-triggered (master-slave) J-K flip-flop with the inputs in Figure 7–86, sketch the Q output waveform. Assume that Q is initially LOW.

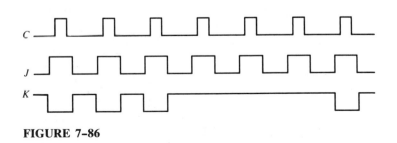

FIGURE 7–86

7-16 The following serial data are applied to the flip-flop as indicated in Figure 7–87. Determine the resulting serial data that appear on the Q output. There is one clock pulse for each bit time. Assume that Q is initially 0. Right-most bits are applied first.

$$J_1:\ 1010011$$
$$J_2:\ 0111010$$
$$J_3:\ 1111000$$
$$K_1:\ 0001110$$
$$K_2:\ 1101100$$
$$K_3:\ 1010101$$

FIGURE 7–87

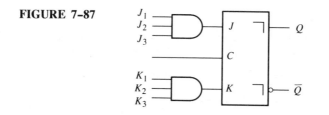

7-17 The 7472 is an AND-gated master-slave J-K flip-flop as shown in Figure 7–88(a). Complete the timing diagram in Figure 7–88(b) by sketching the Q output (which is initially LOW).

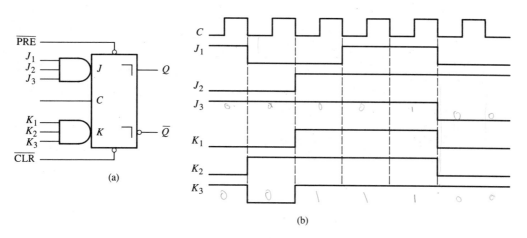

(a)

(b)

FIGURE 7–88

7-18 Sketch the Q output of flip-flop B in Figure 7–89 in proper relation to the clock. The flip-flops are initially RESET.

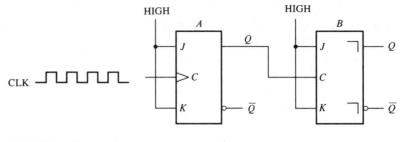

FIGURE 7-89

7-19 Sketch the Q output of flip-flop B in Figure 7-90 in proper relation to the clock. The flip-flops are initially RESET.

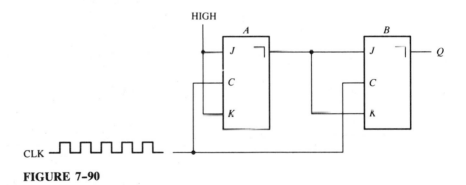

FIGURE 7-90

Section 7-4

7-20 For a data lock-out flip-flop with the inputs in Figure 7-86, draw the timing diagram showing Q starting in the LOW state.

7-21 The D input and a single clock pulse are shown in Figure 7-91. Compare the resulting Q outputs for positive edge-triggered, negative edge-triggered, pulse-triggered, and data lock-out flip-flops. The flip-flops are initially RESET.

FIGURE 7-91

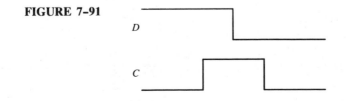

Section 7–5

7–22   The flip-flop in Figure 7–92 is initially RESET. Show the relation between the Q output and the clock pulse if the propagation delay t_{PLH} (clock to Q) is 8 ns.

FIGURE 7–92

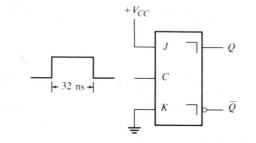

7–23   The direct current required by a particular flip-flop that operates on a $+5$ V dc source is found to be 10 mA. A certain digital device uses 15 of these flip-flops. Determine the current capacity required for the $+5$ V dc supply and the total power dissipation of the system.

7–24   For the circuit in Figure 7–93, determine the maximum frequency of the clock signal for reliable operation if the set-up time for each flip-flop is 20 ns and the propagation delays (t_{PLH} and t_{PHL}) from clock to output are 50 ns for each flip-flop.

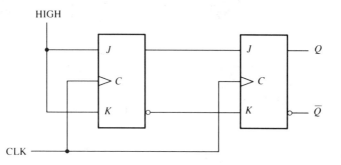

FIGURE 7–93

Section 7–6

7–25   An S-R flip-flop is connected as shown in Figure 7–94. Determine the Q output in relation to the clock. What specific function does this device perform?

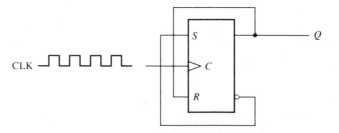

FIGURE 7–94

7–26 Draw the logic diagram for a circuit that will divide the input clock frequency by 16. Use negative edge-triggered J-K flip-flops.

7–27 Devise a basic counter circuit that produces a binary sequence from 0 through 7 using negative edge-triggered J-K flip-flops.

7–28 In the shipping department of a softball factory, the balls roll down a conveyor and through a chute single-file into boxes for shipment. Each ball passing through the chute activates a switch circuit that produces an electrical pulse. The capacity of each box is 32 balls. Design a logic circuit to indicate when a box is full so that an empty box can be moved into position.

Section 7–7

7–29 Determine the pulse width of a 74123 one-shot if the external resistor is 3.3 kΩ and the external capacitor is 2000 pF. Assume that the data sheet gives $K = 0.28$.

7–30 An output pulse of 5 μs duration is to be generated by a 74123 one-shot. Using a capacitor of 10,000 pF and $K = 0.28$, determine the value of external resistance required.

Section 7–8

7–31 A 555 timer is configured to run as an astable multivibrator as shown in Figure 7–95. Determine its frequency.

FIGURE 7–95

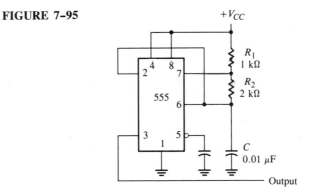

7–32 Determine the value of R_2 for a 555 timer used as an astable multivibrator with an output frequency of 20 kHz, if the external capacitor C is 0.002 μF and the duty cycle is to be approximately 50%. Make $R_1 = 1$ kΩ.

ANSWERS TO SECTION REVIEWS

Section 7–1

1. S-R, D, gated S-R, gated D
2. $SR = 00$, NC; $SR = 01$, $Q = 0$; $SR = 10$, $Q = 1$; $SR = 11$, invalid.
3. 1

Section 7–2

1. The output of a gated S-R latch can change anytime the gate enable (EN) input is active. The output of an edge-triggered S-R flip-flop can change only on the triggering edge of a clock pulse.
2. The J-K flip-flop does not have the invalid state as does the S-R flip-flop.
3. Q goes HIGH on the trailing edge of the first clock pulse, LOW on the trailing edge of the second pulse, HIGH on the trailing edge of the third pulse, and LOW on the trailing edge of the fourth pulse.

Section 7–3

1. In the pulse-triggered flip-flop (master-slave), a data bit goes into the master section on the leading edge of a clock pulse and is transferred to the output (slave) on the trailing edge. In the edge-triggered flip-flop, a data bit goes into the flip-flop and appears on the output on a single clock edge.
2. (a) Nothing.   (b) The output will change.
3. The J-K does not have an invalid state.

Section 7–4

1. A data lock-out is sensitive to the input data only during a clock transition and not while the clock is in its active state.

Section 7–5

1. (a) Time required for input data to be present before triggering edge of clock.
   (b) Time required for data to remain on input after triggering edge of clock.
2. 74LS76A in Table 7–8.

Section 7–6

1. Register   2. Toggle ($J = 1$, $K = 1$)   3. 5

Section 7–7

1. Nonretriggerable times out before it can respond to another trigger input. Retriggerable responds to each trigger input.
2. With external R and C.

Section 7–8

1. Astable has no stable state. Monostable has one stable state.
2. 75%

Section 7–9

1. Yes.

As you saw in the last chapter, flip-flops can be connected together to perform counting operations. Such a group of flip-flops is called a *counter*. The number of flip-flops used and the way in which they are connected determine the number of states (called the *modulus*) and the sequence of states that the counter goes through in each complete cycle.

Counters are classified into two broad categories according to the way they are clocked: *asynchronous* and *synchronous*. In asynchronous counters, commonly called *ripple counters,* the first flip-flop is clocked by the external clock pulse, and then each successive flip-flop is clocked by the Q or $\overline{Q}$ output of the previous flip-flop. Therefore, in asynchronous counters, the flip-flops are not clocked simultaneously. In synchronous counters, the clock input is connected to all of the flip-flops, and thus they are clocked simultaneously.

Within each of these two categories, counters are classified primarily by the sequence of states, by the number of states, or by the number of flip-flops (stages) within the counter.

Specific devices introduced in this chapter are the following:

1. 7493A four-bit asynchronous binary counter
2. 74LS163A four-bit synchronous binary counter
3. 74LS160A synchronous decade counter
4. 74190 up/down decade counter

In this chapter, you will learn

- ☐ The difference between asynchronous and synchronous counters.
- ☐ How to analyze counter timing diagrams.
- ☐ How asynchronous binary counters operate.
- ☐ The meaning of *ripple clocking* and its major drawback.
- ☐ How propagation delays affect the operation of a counter.
- ☐ How asynchronous decade counters operate.
- ☐ How the number of states (modulus) of a counter is determined.
- ☐ How to change the modulus of a counter.
- ☐ How synchronous binary and decade counters operate.
- ☐ How an up/down counter operates.
- ☐ How to design counters with any specified sequence.
- ☐ How to determine the sequence of a counter from its logical operation.
- ☐ How to increase the overall modulus to achieve greater divide-by factors by connecting counters in cascade.
- ☐ How to use logic gates to decode a counter's states.
- ☐ Why glitches are often produced by the decoding process and how to eliminate them.
- ☐ How counters are used in practical applications by using specific examples.
- ☐ The meaning of *dependency notation* as specified by ANSI/IEEE Std. 91–1984.

8
Counters

8-1   ASYNCHRONOUS COUNTERS

The term *asynchronous* refers to events that do not occur at the same time. With respect to counter operation, *asynchronous* means that the flip-flops within the counter are not made to change states at exactly the same time; they do not because the clock pulses are not connected directly to the C input of each flip-flop in the counter.

Figure 8–1 shows a two-bit counter connected for asynchronous operation. Notice that the clock line (CLK) is connected to the clock input (C) of only the first stage, FFA. The second stage, FFB, is triggered by the $\overline{Q}_A$ output of FFA. FFA changes state at the positive-going edge of each clock pulse, but FFB changes only when triggered by a positive-going transition of the $\overline{Q}_A$ output of FFA. Because of the inherent propagation delay time through a flip-flop, a transition of the input clock pulse and a transition of the $\overline{Q}_A$ output of FFA can never occur at exactly the same time. Therefore, the two flip-flops are *never simultaneously triggered*, which results in *asynchronous* counter operation.

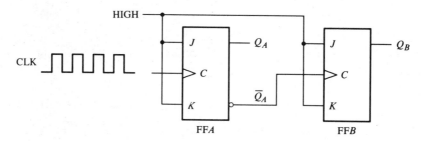

FIGURE 8–1   *A two-bit asynchronous binary counter.*

Let us examine the basic operation of the counter of Figure 8–1 by applying four clock pulses to FFA and observing the Q output of each flip-flop; Figure 8–2 illustrates the changes in the state of the flip-flop outputs in response to the clock

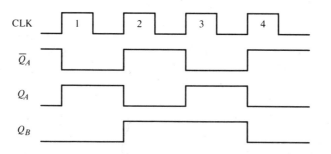

FIGURE 8–2   *Timing diagram for the counter of Figure 8–1.*

pulses. Also, both flip-flops are connected for toggle operation ($J = 1$, $K = 1$) and are assumed to be initially RESET.

The positive-going edge of CLK_1 (clock pulse 1) causes the Q_A output of FFA to go HIGH. The $\overline{Q}_A$ output at the same time goes LOW, but it has no effect on FFB because a *positive-going* transition must occur to trigger the flip-flop. After the leading edge of CLK_1, $Q_A = 1$ and $Q_B = 0$. The positive-going edge of CLK_2 causes Q_A to go LOW. $\overline{Q}_A$ goes HIGH and triggers FFB, causing Q_B to go HIGH. After the leading edge of CLK_2, $Q_A = 0$ and $Q_B = 1$. The positive-going edge of CLK_3 causes Q_A to go HIGH again. $\overline{Q}_A$ goes LOW and has no effect on FFB. Thus, after the leading edge of CLK_3, $Q_A = 1$ and $Q_B = 1$. The positive-going edge of CLK_4 causes Q_A to go LOW. $\overline{Q}_A$ goes HIGH and triggers FFB, causing Q_B to go LOW. After the leading edge of CLK_4, $Q_A = 0$ and $Q_B = 0$. The counter has now recycled back to its original state (both flip-flops are RESET).

The waveforms of the Q_A and Q_B outputs are shown relative to the clock pulses in Figure 8-2. This waveform relationship is called a *timing diagram*. It should be pointed out that for simplicity, the transitions of Q_A, Q_B, and the clock pulses are shown simultaneous even though this is an asynchronous counter. There is, of course, some small delay between the CLK, Q_A, and Q_B transitions.

Notice that the two-bit counter exhibits four different states, as you would expect with two flip-flops ($2^2 - 4$). Also, notice that if Q_A represents the least significant bit (LSB) and Q_B represents the most significant bit (MSB), the sequence of counter states is actually a sequence of binary numbers as shown in Table 8-1. In this book, Q_A is always the LSB unless otherwise specified.

Since it goes through a binary sequence, the counter in Figure 8-1 is a *binary counter*. It actually counts the number of clock pulses up to three, and on the fourth pulse it recycles to its original state ($Q_A = 0$, $Q_B = 0$). The term *recycle* is commonly applied to counter operation; it refers to the transition of the counter from its final state back to its original state.

TABLE 8-1

Clock Pulse	Q_B	Q_A
0	0	0
1	0	1
2	1	0
3	1	1

A Three-Bit Asynchronous Binary Counter

A three-bit asynchronous binary counter is shown in Figure 8-3(a). The basic operation, of course, is the same as that of the two-stage counter just discussed, except that it has *eight* states due to its three stages. A timing diagram appears in Figure 8-3(b) for eight clock pulses.

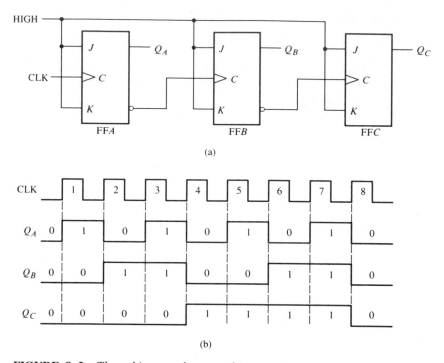

(a)

(b)

FIGURE 8-3 *Three-bit asynchronous binary counter and its timing diagram for one cycle.*

Notice that the counter progresses through a binary count of 0 to 7 and then recycles to the 0 state. This counter sequence is presented in Table 8–2.

TABLE 8-2 *State sequence for a three-stage binary counter.*

Clock Pulse	Q_C	Q_B	Q_A
0	0	0	0
1	0	0	1
2	0	1	0
3	0	1	1
4	1	0	0
5	1	0	1
6	1	1	0
7	1	1	1

Asynchronous counters are commonly referred to as *ripple* counters for the following reason. The effect of the input clock pulse is first "felt" by FFA. This effect cannot get to FFB immediately due to the propagation delay through FFA.

Then there is the propagation delay through FF*B* before FF*C* can be triggered. Thus, the effect of an input clock pulse "ripples" through the counter, taking some time, due to propagation delays, to reach the last flip-flop. To illustrate, notice that all three flip-flops in the counter of Figure 8–3 change state as a result of CLK_4. The HIGH-to-LOW transition of Q_A occurs one delay time after the positive-going transition of the clock pulse. The HIGH-to-LOW transition of Q_B occurs one delay time after the positive-going transition of $\overline{Q}_A$. The LOW-to-HIGH transition of Q_C occurs one delay time after the positive-going transition of $\overline{Q}_B$. As you can see, FF*C* is not triggered until two delay times after the positive-going edge of the clock pulse, CLK_4. Thus, it takes three propagation delay times for the effect of the clock pulse (CLK_4) to "ripple" through the counter and change Q_C from LOW to HIGH. This ripple clocking effect is illustrated in Figure 8–4 for the first four clock pulses with the propagation delays shown.

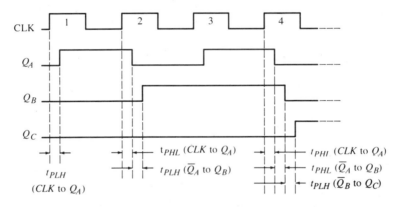

FIGURE 8-4 *Propagation delays in a ripple clocked binary counter.*

This cumulative delay of an asynchronous counter is a major disadvantage in many applications because it limits the rate at which the counter can be clocked and creates decoding problems.

EXAMPLE 8-1

A four-stage asynchronous binary counter is shown in Figure 8–5(a). Each flip-flop is negative edge-triggered and has a propagation delay of 10 nanoseconds (ns). Draw a timing diagram showing the Q output of each stage, and determine the total delay time from the triggering edge of a clock pulse until a corresponding change can occur in the state of Q_D.

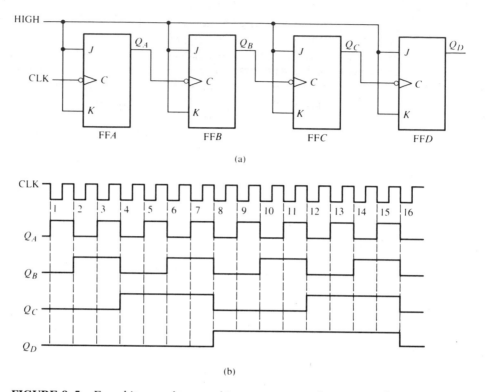

FIGURE 8–5 *Four-bit asynchronous binary counter and its timing diagram.*

Solutions The timing diagram with delays omitted is as shown in Figure 8–5(b). To determine the total delay time, the effect of CLK_8 or CLK_{16} must propagate through four flip-flops before Q_D changes.

$$t_p = 4 \times 10 \text{ ns} = 40 \text{ ns total delay}$$

Asynchronous Decade Counters

Regular binary counters, such as those previously introduced, have a maximum modulus; that is, they progress through all of their possible states. Recall that the maximum possible number of states (*maximum modulus*) of a counter is 2^n, where *n* is the number of flip-flops in the counter.

Counters can also be designed to have a number of states in their sequence that is less than 2^n. The resulting sequence is called a *truncated* sequence.

One common modulus for counters with truncated sequences is ten. Counters with ten states in their sequence are called *decade* counters. A decade counter with a count sequence of 0 (0000) through 9 (1001) is a *BCD decade counter* because its ten-state sequence is the BCD code. This type of counter is very useful in display applications in which BCD is required for conversion to a decimal readout.

To obtain a truncated sequence, it is necessary to force the counter to recycle before going through all of its normal states. For example, the BCD decade counter must recycle back to the 0000 state after the 1001 state.

A decade counter requires four flip-flops (three flip-flops are insufficient because $2^3 = 8$). We will now take a four-bit asynchronous counter such as the one in Figure 8–5(a) and *modify* its sequence in order to understand the principle of truncated counters. One method of achieving this recycling after the count of 9 (1001) is to *decode* count 10_{10} (1010) with a NAND gate and connect the output of the NAND gate to the clear (CLR) inputs of the flip-flops, as shown in Figure 8–6(a).

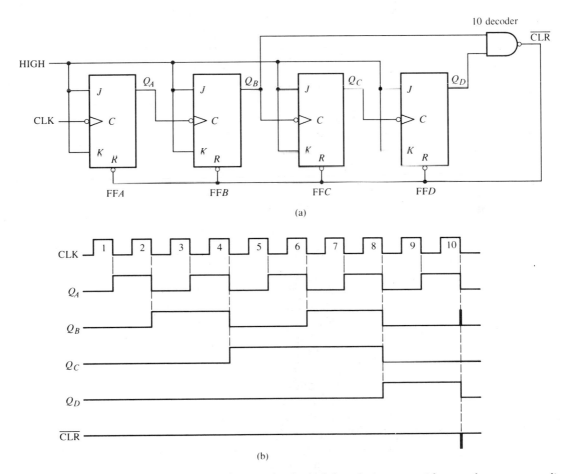

(a)

(b)

FIGURE 8-6 *An asynchronously clocked decade counter with asynchronous recycling.*

Notice that only Q_B and Q_D are connected to the NAND gate inputs. This is an example of *partial decoding*, in which the two unique states ($Q_B = 1$ and $Q_D = 1$) are sufficient to decode the count of 10_{10} because none of the other states (0 through 9) have both Q_B and Q_D HIGH at the same time. When the counter goes into count 10_{10} (1010_2), the decoding gate output goes LOW and asynchronously RESETS all of the flip-flops.

The resulting timing diagram is shown in Figure 8–6(b). Notice that there is a *glitch* on the Q_B waveform. The reason for this glitch is that Q_B must first go HIGH before the count of 10_{10} can be decoded. Not until several nanoseconds after the counter goes to the count of 10_{10} does the output of the decoding gate go LOW (both inputs are HIGH). Thus, the counter is in the 1010 state for a short time before it is RESET back to 0000, thus producing the glitch on Q_B.

Other truncated sequences can be implemented in a similar way, as the next example shows.

EXAMPLE 8–2

Show how an asynchronous counter can be implemented with a modulus of twelve with a straight binary sequence from 0000 through 1011.

Solution   Since three flip-flops can produce a maximum of eight states, four flip-flops are required to produce any modulus greater than eight but less than or equal to sixteen.

When the counter gets to its last state, 1011, it must recycle back to 0000 rather than going to its normal next state of 1100, as illustrated in the sequence chart below.

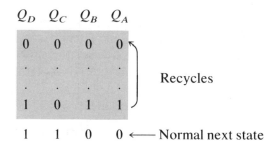

Observe that Q_A and Q_B both go to 0 anyway, but Q_C and Q_D must be forced to 0 on the twelfth clock pulse. Figure 8–7(a) shows the modulus-12 counter. The NAND gate partially decodes count 12 (1100) and RESETS flip-flop C and flip-flop D. Thus, on the twelfth clock pulse, the counter is made to recycle from count 11_2 to count 0, as shown in the timing diagram of Figure 8–7(b). (It is in count 12 for only a few nanoseconds before it is RESET.)

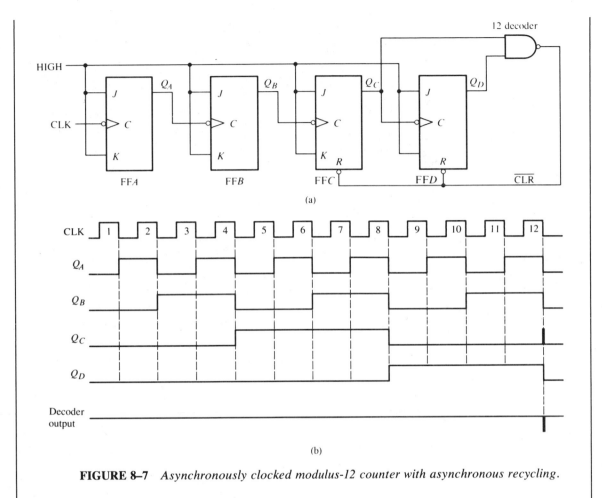

(a)

(b)

FIGURE 8–7 *Asynchronously clocked modulus-12 counter with asynchronous recycling.*

The 7493A Four-Bit Binary Counter

The 7493A is presented as an example of a specific integrated circuit asynchronous counter. As the logic diagram in Figure 8–8 shows, this device actually consists of a single flip-flop and a three-bit asynchronous counter. This arrangement is for flexibility. It can be used as a divide-by-2 device using only the single flip-flop, or it can be used as a modulus-8 counter using only the three-bit counter portion. This device also provides gated reset inputs, $R0(1)$ and $R0(2)$. When both of these inputs are HIGH, the counter is RESET to the 0000 state by $\overline{CLR}$.

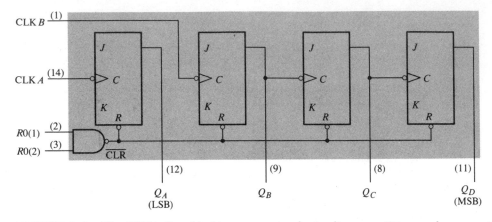

FIGURE 8-8 *The 7493A four-bit binary counter logic diagram. (Pin numbers are in parentheses, and all J-K inputs are internally connected HIGH.)*

Additionally, the 7493A can be used as a four-bit modulus-16 counter (counts 0 through 15) by connecting the Q_A output to the CLKB input as shown in Figure 8–9(a). It can also be configured as a decade counter (counts 0 through 9) with asynchronous recycling by using the gated reset inputs for partial decoding of count 10_{10}, as shown in Figure 8–9(b).

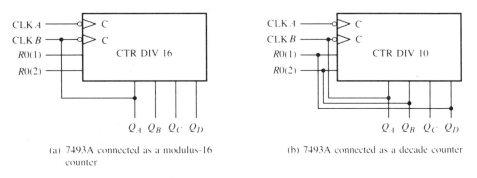

(a) 7493A connected as a modulus-16 counter

(b) 7493A connected as a decade counter

FIGURE 8-9 *Two configurations of the 7493A asynchronous counter. (The qualifying lable, CTR DIVn, indicates a counter with n states.)*

EXAMPLE 8–3

Show how the 7493A can be used as a modulus-12 counter.

Solution Use the gated reset inputs, $R0(1)$ and $R0(2)$, to partially decode count 12 (remember, there is an internal NAND gate associated with these inputs). The count-12 decoding is accomplished by connecting Q_D to $R0(1)$ and Q_C to $R0(2)$, as shown in Figure 8–10. Q_A is connected to CLKB to create a four-bit counter.

FIGURE 8–10 *7493A connected as a modulus-12 counter.*

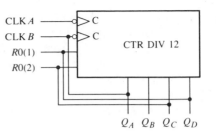

Immediately after the counter goes to count 12 (1100), it is reset to 0000. This, however, results in a glitch on Q_C because the counter must reside in the 1100 state for several nanoseconds before recycling.

SECTION REVIEW 8–1

1. What does the term *asynchronous* mean in relation to counters?
2. How many states does a modulus-14 counter have?

8–2 SYNCHRONOUS COUNTERS

The term *synchronous* as applied to counter operation means that the counter is clocked such that each flip-flop in the counter is *triggered at the same time*. This is accomplished by connecting the clock line to *each* stage of the counter, as shown in Figure 8–11 for a two-stage counter. Notice that an arrangement different from that for the asynchronous counter must be used for the *J* and *K* inputs of FF*B* in order to achieve a binary sequence.

The operation of this counter is as follows: First, we will assume that the counter is initially in the binary 0 state; that is, both flip-flops are RESET. When the positive edge of the first clock pulse is applied, FF*A* will toggle, and Q_A will therefore go HIGH. What happens to FF*B* at the positive-going edge of CLK_1? To

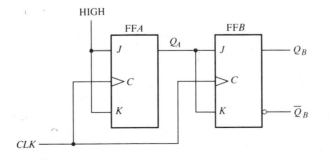

FIGURE 8–11 *A two-bit synchronous binary counter.*

find out, let us look at the input conditions of FF*B*. *J* and *K* are both LOW because Q_A, to which they are connected, has not yet gone HIGH. Remember, there is a propagation delay from the triggering edge of the clock pulse until the *Q* output actually makes a transition. So, $J = 0$ and $K = 0$ when the leading edge of the first clock pulse is applied. This is a *no-change* condition, and therefore FF*B* does not change state. A timing detail of this portion of the counter operation is given in Figure 8–12(a).

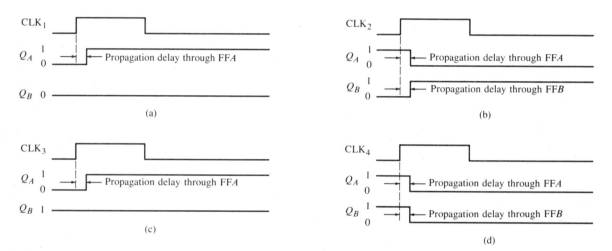

FIGURE 8–12   *Timing details for the two-bit synchronous counter operation.*

After CLK_1, $Q_A = 1$ and $Q_B = 0$ (which is the binary 1 state). At the leading edge of CLK_2, FF*A* will toggle, and Q_A will go LOW. Since FF*B* "sees" a HIGH on its *J* and *K* inputs when the triggering edge of this clock pulse occurs, the flip-flop toggles and Q_B goes HIGH. Thus, after CLK_2, $Q_A = 0$ and $Q_B = 1$ (which is a binary 2 state). The timing detail for this condition is given in Figure 8–12(b). At the leading edge of CLK_3, FF*A* again toggles to the SET state ($Q_A = 1$), and FF*B* remains SET ($Q_B = 1$) because its *J* and *K* inputs are both LOW. After this triggering edge, $Q_A = 1$ and $Q_B = 1$ (which is a binary 3 state). The timing detail is shown in Figure 8–12(c).

Finally, at the leading edge of CLK_4, Q_A and Q_B go LOW because they both have a toggle condition on their *J* and *K* inputs. The timing detail is shown in Figure 8–12(d). The counter has now recycled back to its original state, binary 0. The complete timing diagram is shown in Figure 8–13.

Notice that all of the waveform transitions appear coincident; that is, the delays are not indicated. Although the delays are a very important factor in the counter operation, as we have seen in the preceding discussion, in an overall timing diagram they are normally omitted for simplicity. Major waveform relationships resulting from the logical operation of a circuit can be conveyed completely without showing small delay and timing differences.

FIGURE 8-13 *Timing diagram for the counter of Figure 8-11.*

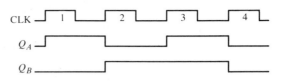

A Three-Bit Synchronous Binary Counter

A three-bit synchronous binary counter is shown in Figure 8–14 and its timing diagram in Figure 8–15. An understanding of this counter can be achieved by a careful examination of its sequence of states as shown in Table 8–3.

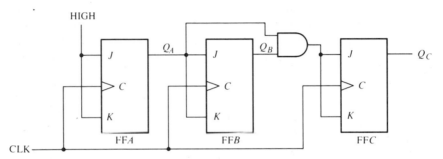

FIGURE 8-14 *A three-bit synchronous binary counter.*

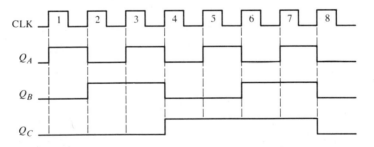

FIGURE 8-15 *Timing diagram for the counter of Figure 8-14.*

TABLE 8-3 *State sequence for a three-stage binary counter.*

Clock Pulse	Q_C	Q_B	Q_A
0	0	0	0
1	0	0	1
2	0	1	0
3	0	1	1
4	1	0	0
5	1	0	1
6	1	1	0
7	1	1	1

First, let us look at Q_A. Notice that Q_A changes on each clock pulse as we progress from its original state to its final state and then back to its original state. To produce this operation, FFA must be held in the toggle mode by constant HIGHs on its J and K inputs. Now let us see what Q_B does. Notice that it goes to the opposite state following each time Q_A is a 1. This occurs at CLK_2, CLK_4, CLK_6, and CLK_8. CLK_8 causes the counter to recycle. To produce this operation, Q_A is connected to the J and K inputs of FFB. When Q_A is a 1 and a clock pulse occurs, FFB is in the toggle mode and will change state. The other times when Q_A is a 0, FFB is in the *no-change* mode and remains in its present state.

Next, let us see how FFC is made to change at the proper times according to the binary sequence. Notice that both times Q_C changes state, it is preceded by the unique condition of both Q_A and Q_B being HIGH. This condition is detected by the AND gate and applied to the J and K inputs of FFC. Whenever both Q_A and Q_B are HIGH, the output of the AND gate makes the J and K inputs of FFC HIGH, and FFC toggles on the following clock pulse. At all other times, the J and K inputs of FFC are held LOW by the AND gate output, and FFC does not change state.

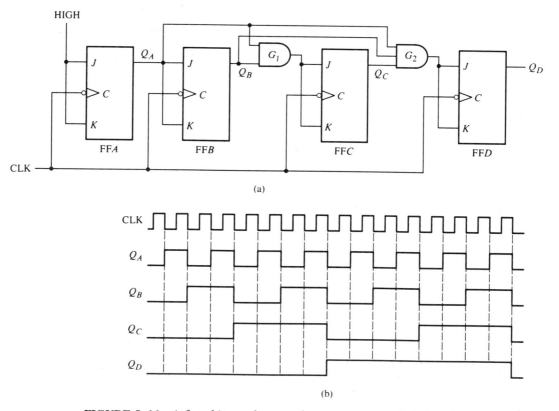

FIGURE 8–16 *A four-bit synchronous binary counter and timing diagram.*

A Four-Bit Synchronous Binary Counter

Figure 8–16(a) shows a four-bit binary counter, and Figure 8–16(b) shows its timing diagram. This particular counter is implemented with negative edge-triggered flip-flops. The reasoning behind the J and K input control for the first three flip-flops is the same as presented previously for the three-stage counter. The fourth stage, FFD, changes only twice in the sequence. Notice that both of these transitions occur following the times Q_A, Q_B, and Q_C are all HIGH. This condition is decoded by AND gate G_2, so that when a clock pulse occurs, FFD will change state. For all other times the J and K inputs of FFD are LOW, and it is in a *no-change* condition.

Synchronous Decade Counters

As you know, the BCD decade counter exhibits a truncated sequence and goes through a straight binary sequence through the 1001 state. Rather than going to the 1010 state, it recycles to the 0000 state. A synchronous BCD decade counter is shown in Figure 8–17.

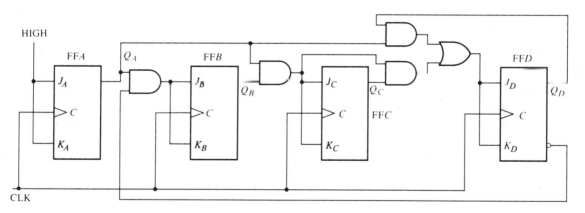

FIGURE 8–17 *A synchronous BCD decade counter.*

The counter operation can be understood by examining the sequence of states in Table 8–4. First, notice that FFA toggles on each clock pulse, so the logic equation for its J and K inputs is

$$J_A = K_A = 1$$

This is implemented by connecting these inputs to a constant HIGH level. Next notice that FFB changes on the next clock pulse each time $Q_A = 1$ and $Q_D = 0$, so the logic equation for its J and K inputs is

$$J_B = K_B = Q_A\overline{Q}_D$$

TABLE 8-4 *States of a BCD decade counter.*

Clock Pulse	Q_D	Q_C	Q_B	Q_A
0	0	0	0	0
1	0	0	0	1
2	0	0	1	0
3	0	0	1	1
4	0	1	0	0
5	0	1	0	1
6	0	1	1	0
7	0	1	1	1
8	1	0	0	0
9	1	0	0	1

This is implemented by ANDing Q_A and $\overline{Q_D}$ and connecting the gate output to the J and K inputs of FFB. FFC changes on the next clock pulse each time both $Q_A = 1$ and $Q_B = 1$. This requires an input logic equation as follows:

$$J_C = K_C = Q_A Q_B$$

This is implemented by ANDing Q_A and Q_B and connecting the gate output to the J and K inputs of FFC. Finally, FFD changes to the opposite state on the next clock pulse each time $Q_A = 1$, $Q_B = 1$, and $Q_C = 1$ (count 7), or when $Q_A = 1$ and $Q_D = 1$ (count 9). The equation for this is as follows:

$$J_D = K_D = Q_A Q_B Q_C + Q_A Q_D$$

This function is implemented with the AND/OR logic connected to FFD as shown in the logic diagram in Figure 8-17. Notice that the only difference between this decade counter and a modulus-16 binary counter is the $Q_A Q_D$ AND gate and the OR gate; this essentially detects the occurrence of the 1001 state and causes the counter to recycle properly on the next clock pulse. The timing diagram for the decade counter is given in Figure 8-18.

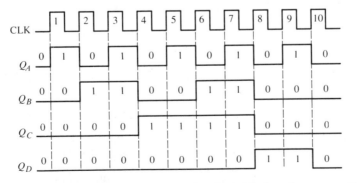

FIGURE 8-18 *Timing diagram for the BCD decade counter (Q_A is the LSB).*

The 74LS163A Synchronous Four-Bit Binary Counter

The 74LS163A is an example of an integrated circuit synchronous binary counter. A logic symbol is shown in Figure 8–19. This counter has several features in addition to the basic functions previously discussed for the general synchronous binary counter.

FIGURE 8–19 *The 74LS163A four-bit synchronous binary counter. (The qualifying label CTR DIV 16 indicates a counter with sixteen states.)*

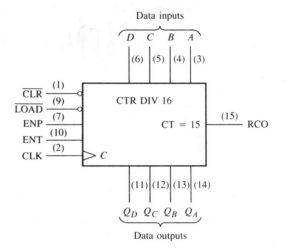

First, the counter can be synchronously *preset* to any four-bit binary number by applying the proper levels to the *data inputs*. When a LOW is applied to the LOAD input, the counter will assume the state of the data inputs on the next clock pulse. This, of course, allows the counter sequence to be started with any four-bit binary number.

Also, there is an active-LOW *clear* input ($\overline{CLR}$) which synchronously resets all four flip-flops in the counter. There are two *enable inputs*, ENP and ENT. These inputs must both be high for the counter to sequence through its binary states. When at least one is LOW, the counter is disabled. The *ripple carry output* (RCO) goes HIGH when the counter reaches the last state in the sequence, 15 (1111). This output, in conjunction with the enable inputs, allows these counters to be cascaded for higher count sequences, as will be discussed later.

Figure 8–20 shows a timing diagram of this counter being preset to 12 (1100) and then counting up to its terminal count, 15 (1111). A and Q_A are the least significant input and output bits, respectively.

Let us examine this timing diagram in detail. This will aid you in interpreting timing diagrams found later in this chapter or in manufacturers' data sheets.

To begin, the LOW level pulse on the $\overline{CLR}$ input causes all the *outputs* (Q_A, Q_B, Q_C, and Q_D) to go LOW.

Next, the LOW level pulse on the $\overline{LOAD}$ input enters the data on the *data inputs* (A, B, C, and D) into the counter. These data appear on the Q outputs at the

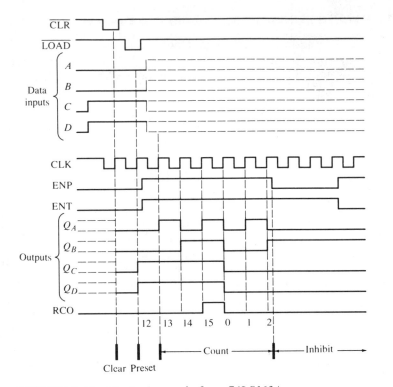

FIGURE 8-20 *Timing example for a 74LS163A.*

time of the first positive-going *clock* edge after $\overline{\text{LOAD}}$ goes LOW. This is the *preset* operation. In this particular example, Q_A is LOW, Q_B is LOW, Q_C is HIGH, and Q_D is HIGH. This, of course, is a binary 12 (Q_A is the LSB).

The counter now advances through binary counts 13, 14, 15 on the next three positive-going clock edges. It then recycles to 0, 1, 2 on the following clock pulses. Notice that both ENP and ENT inputs are HIGH during the count sequence. When ENP goes LOW, the count is inhibited, and the counter remains in the binary 2 state.

The 74LS160A Synchronous Decade Counter

This device has the same inputs and outputs as the 74LS163A binary counter previously discussed. It may be preset to any BCD count using the data inputs and a LOW on the $\overline{\text{LOAD}}$ input. A LOW on the $\overline{\text{CLR}}$ will RESET the counter. The count enable inputs ENP and ENT must both be high for the counter to advance through its sequence of states in response to a positive transition on the CLK input. As in the 74LS163A, the enable inputs in conjunction with the carry out (RCO; terminal count of 1001) provide for cascading several decade counters. Cascaded counters will be discussed later in this chapter.

Figure 8–21 shows a logic symbol, and Figure 8–22 is a timing diagram showing the counter being preset to count 7 (0111).

FIGURE 8–21 *The 74LS160A synchronous decade (BCD) counter. (The qualifying label CTR DIV 10 indicates a counter with ten states.)*

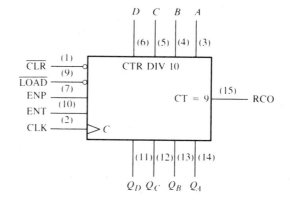

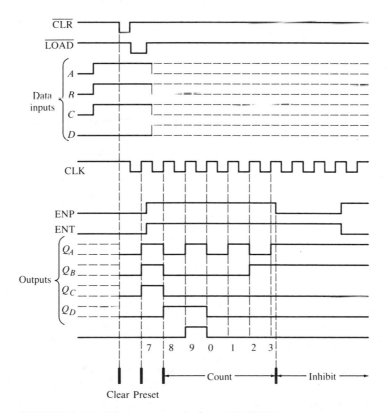

FIGURE 8–22 *Timing example for a 74LS160A.*

SECTION
REVIEW
8–2

1. How does a synchronous counter differ from an asynchronous counter?
2. Explain the function of the presettable feature of counters such as the 74LS160A and the 74LS163A.
3. Describe the purpose of the ENP and ENT inputs and the RCO output for the two specific counters introduced in this section.

8–3   UP/DOWN SYNCHRONOUS COUNTERS

An up/down counter is one that is capable of progressing in *either direction* through a certain sequence. An up/down counter, sometimes called a *bidirectional* counter, can have any specified sequence of states. For example, a three-bit binary counter that advances upward through its sequence (0, 1, 2, 3, 4, 5, 6, 7) and then can be reversed so that it goes through the sequence in the opposite direction (7, 6, 5, 4, 3, 2, 1, 0) is an illustration of up/down sequential operation.

In general, most up/down counters can be reversed at any point in their sequence. For instance, the three-bit binary counter mentioned can be made to go through the following sequence:

$$\overbrace{0, 1, 2, 3, 4, 5,}^{\text{up}} \underbrace{4, 3, 2,}_{\text{down}} \overbrace{3, 4, 5, 6, 7,}^{\text{up}} \underbrace{6, 5,}_{\text{down}} \text{ etc.}$$

TABLE 8–5 *Up/down sequence for a three-bit binary counter.*

Clock Pulse	UP	Q_C	Q_B	Q_A	DOWN
0		0	0	0	
1		0	0	1	
2		0	1	0	
3		0	1	1	
4		1	0	0	
5		1	0	1	
6		1	1	0	
7		1	1	1	

Table 8–5 shows the complete up/down sequence for a three-bit binary counter. The arrows indicate the state-to-state movement of the counter for both its UP and its DOWN modes of operation. An examination of Q_A for both the UP

and DOWN sequences shows that FFA toggles on each clock pulse. So, the J and K inputs of FFA are

$$J_A = K_A = 1$$

For the UP sequence, Q_B changes state on the next clock pulse when $Q_A = 1$. For the DOWN sequence, Q_B changes on the next clock pulse when $Q_A = 0$. Thus, the J and K inputs of FFB must equal 1 under the conditions expressed by the following equation:

$$J_B = K_B = Q_A \cdot \text{UP} + \overline{Q}_A \cdot \text{DOWN}$$

For the UP sequence, Q_C changes state on the next clock pulse when $Q_A = Q_B = 1$. For the DOWN sequence, Q_C changes on the next clock pulse when $Q_A = Q_B = 0$. Thus, the J and K inputs of FFC must equal 1 under the conditions expressed by the following equation:

$$J_C = K_C = Q_A \cdot Q_B \cdot \text{UP} + \overline{Q}_A \cdot \overline{Q}_B \cdot \text{DOWN}$$

Each of the conditions for the J and K inputs of each flip-flop produces a toggle at the appropriate point in the counter sequence.

Figure 8–23 shows a basic implementation of a three-bit up/down binary counter using the logic equations just developed for the J and K inputs of each flip-flop. Notice that the UP/$\overline{\text{DOWN}}$ control input is HIGH for UP and LOW for DOWN.

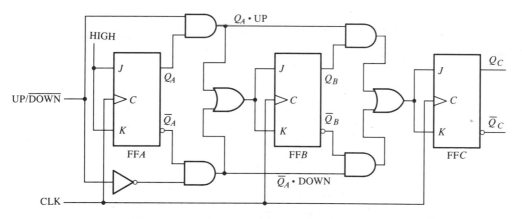

FIGURE 8–23 *A basic three-bit up/down synchronous counter.*

EXAMPLE 8–4   Determine the sequence of a synchronous four-bit *binary* up/down counter if the clock and up/down control inputs have waveforms as shown in Figure 8–24(a). The counter starts in the all 0s state and is positive edge-triggered.

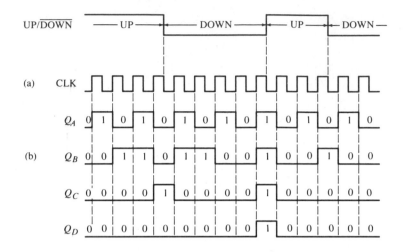

FIGURE 8–24

Solution From the waveforms in Figure 8–24(b), the counter sequence is as follows:

Q_D	Q_C	Q_B	Q_A	
0	0	0	0	
0	0	0	1	
0	0	1	0	up
0	0	1	1	
0	1	0	0	
0	0	1	1	
0	0	1	0	
0	0	0	1	down
0	0	0	0	
1	1	1	1	
0	0	0	0	
0	0	0	1	up
0	0	1	0	
0	0	0	1	down
0	0	0	0	

The 74190 Up/Down Decade Counter

Figure 8–25 shows a logic diagram for the 74190, a good example of an integrated circuit up/down counter. The direction of the count is determined by the level of the *up/down input* ($D/\overline{U}$). When this input is HIGH, the counter counts down; and when it is LOW, the counter counts up. Also, this device can be preset to any

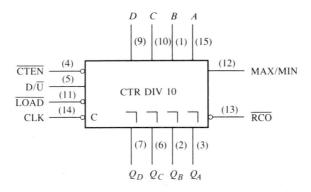

FIGURE 8–25 *The 74190 up/down synchronous decade counter.*

desired BCD digit as determined by the states of the *data inputs* when the *load input* is LOW.

The *MAX-MIN output* produces a HIGH pulse when the terminal count 9 (1001) is reached in the up mode or when the terminal count 0 (0000) is reached in the down mode. This MAX-MIN output along with the *ripple clock output* ($\overline{RCO}$)

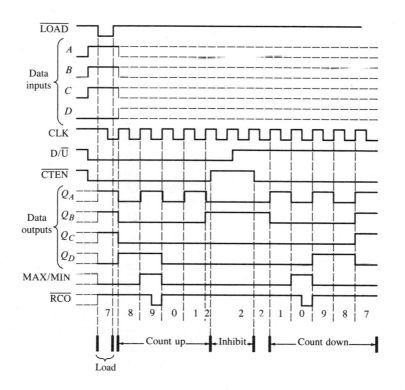

FIGURE 8–26 *Timing example for a 74190.*

and the *count enable input* ($\overline{\text{CTEN}}$) are used when cascading counters. (The topic of cascading counters is discussed later in this chapter.) Notice that this counter is implemented with master-slave flip-flops that are negative pulse-triggered.

Figure 8–26 (p. 371) is an example timing diagram showing the 74190 counter preset to 7 (0111) and then going through a count-up sequence followed by a count-down sequence. The MAX/MIN output is HIGH when the counter is in either the all 0s state (MIN) or the 1001 state (MAX).

SECTION REVIEW 8–3

1. A four-bit up/down binary counter is in the down mode and in the 1010 state. On the next clock pulse, to what state does the counter go?
2. What is the terminal count of a four-bit binary counter in the up mode? In the down mode? What is the next state after the terminal count in the down mode?

8–4   A PROCEDURE FOR THE DESIGN OF SEQUENTIAL CIRCUITS

This section may be omitted without affecting the flow of material in the remainder of the book. Inclusion of this material is recommended for those who need an introduction to counter design or state machine design in general.

Before proceeding with a specific design technique, a general definition of a sequential circuit (state machine) is given: A general sequential circuit consists of a *combinational logic* section and a *memory* section (flip-flops) as shown in Figure 8–27. In a clocked sequential circuit, there is a clock input to the memory section as indicated.

The information stored in the memory section, as well as the inputs to the combinational logic ($I_1, I_2, \ldots, I_m$), is required for proper operation of the circuit. At any given time, the memory is in a state called the *present state* and will advance to a *next state* on a clock pulse as determined by conditions on the excitation lines ($Y_1, Y_2, \ldots, Y_p$). The present state of the memory is represented by the *state variables* ($Q_1, Q_2, \ldots, Q_x$). These state variables, along with the inputs ($I_1, I_2, \ldots, I_m$), determine the system outputs ($O_1, O_2, \ldots, O_n$).

Not all sequential circuits have input and output variables as in the general model just discussed. However, all have excitation variables and state variables. Counters are a special case of clocked sequential circuits. In this section, a general design procedure for sequential circuits as applied to counters is presented.

In this procedure, a clocked sequential circuit is first described by a *state diagram* which shows the progression of states and input and output conditions, if any. Figure 8–28 is an example of a state diagram for a three-bit Gray code counter. The circuit has no inputs other than the clock and no outputs other than its internal state (outputs are taken off each flip-flop in the counter).

Once the sequential circuit is defined by a state diagram, the second step is to derive a *next-state table* which lists each state of the counter (present state)

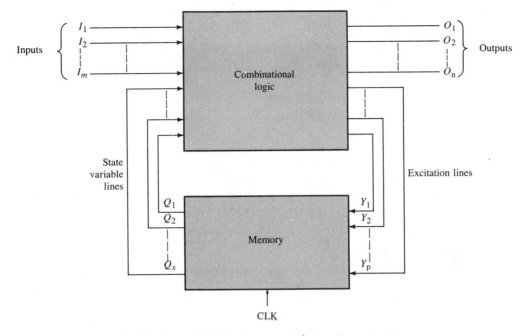

FIGURE 8–27 *General clocked sequential circuit.*

FIGURE 8–28 *State diagram for a three-bit Gray code counter.*

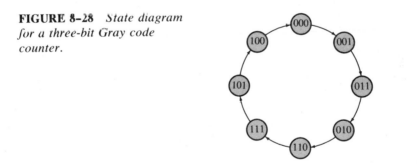

along with the corresponding *next state*. The *next state* is the state that the counter goes to from its present state upon application of a clock pulse. The *next-state table* is derived from the state diagram and is shown in Table 8–6 for the three-bit Gray code counter example.

The third step is to develop a *truth table* that shows the flip-flop inputs required to make the counter go to its proper next state from each of its present states. If a flip-flop is RESET in its present state and must remain RESET in its next state, it can be in either a *no-change* or a RESET condition at the next clock pulse. This is indicated by NC/R (no-change or RESET). If a flip-flop is SET and must remain SET in its next state, it can be in either a *no-change* or a SET

TABLE 8-6 *Next-state table for three-bit Gray code counter.*

Present State			Next State		
Q_C	Q_B	Q_A	Q_C	Q_B	Q_A
0	0	0	0	0	1
0	0	1	0	1	1
0	1	1	0	1	0
0	1	0	1	1	0
1	1	0	1	1	1
1	1	1	1	0	1
1	0	1	1	0	0
1	0	0	0	0	0

condition on the next clock pulse. This is indicated by NC/S (no-change or SET). If a flip-flop is RESET and must go to the SET state, it can be in either a SET or a toggle condition on the next clock pulse. This is indicated by S/T (SET or toggle). If a flip-flop is SET and must go to the RESET state, it can be in either a RESET or a toggle condition on the next clock pulse. This is indicated by R/T (RESET or toggle).

The truth table for the three-bit Gray code counter is shown in Table 8-7 using J-K flip-flops. The X's represent *don't care* conditions (the variable can be either a 1 or a 0).

Now let's see how the flip-flop inputs in Table 8-7 were determined. In the first row, $Q_C = 0$ in the present state and remains 0 in the next state. Therefore,

TABLE 8-7 *Truth table for the three-bit Gray code counter.*

Present State			Next State			Flip-Flop Input Conditions			Flip-Flop Inputs (present state)		
Q_C	Q_B	Q_A	Q_C	Q_B	Q_A	FFC	FFB	FFA	$J_C K_C$	$J_B K_B$	$J_A K_A$
0	0	0	0	0	1	NC/R	NC/R	S/T	0 X	0 X	1 X
0	0	1	0	1	1	NC/R	S/T	NC/S	0 X	1 X	X 0
0	1	1	0	1	0	NC/R	NC/S	R/T	0 X	X 0	X 1
0	1	0	1	1	0	S/T	NC/S	NC/R	1 X	X 0	0 X
1	1	0	1	1	1	NC/S	NC/S	S/T	X 0	X 0	1 X
1	1	1	1	0	1	NC/S	R/T	NC/S	X 0	X 1	X 0
1	0	1	1	0	0	NC/S	NC/R	R/T	X 0	0 X	X 1
1	0	0	0	0	0	R/T	NC/R	NC/R	X 1	0 X	0 X

X = don't care (can be either 1 or 0).
NC/R = no-change or RESET.
NC/S = no-change or SET.
R/T = RESET or toggle.
S/T = SET or toggle.

FFC must be either in a *no-change* or a RESET condition. Thus, J_C must be 0, and K_C can be either a 0 or a 1 ("don't care"); it does not matter because FFC will remain RESET ($Q_C = 0$) in either case. The same explanation applies to FFB. Q_A must go from a 0 in its present state to a 1 in its next state. Therefore, FFA must be made either to SET or to toggle by making $J_A = 1$ and K_A either a 0 or a 1. Similar reasoning applies to the development of the rest of the table.

The fourth step is to transfer the information from the truth table to Karnaugh maps in order to derive a simplified Boolean expression for each flip-flop input. This procedure is shown in Figure 8–29, which shows a present-state Karnaugh map for the J and the K inputs of each flip-flop. Zeros are not transferred because only 1s are grouped with appropriate "don't cares" (X's) to obtain each expression.

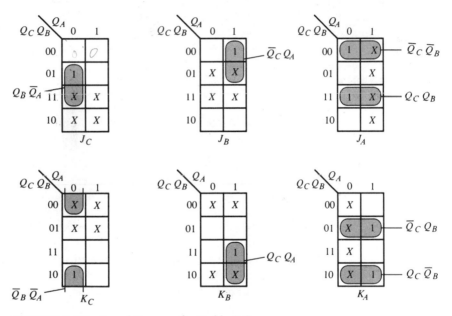

FIGURE 8–29 *J and K maps for Table 8–7.*

In the fifth step, the 1s in the Karnaugh maps of Figure 8–29 are grouped with "don't cares," and the following expressions for the J and K inputs of each flip-flop are obtained:

$$J_A = Q_C Q_B + \overline{Q}_C \overline{Q}_B$$
$$K_A = Q_C \overline{Q}_B + \overline{Q}_C Q_B$$

$$J_B = \overline{Q}_C Q_A$$
$$K_B = Q_C Q_A$$

$$J_C = Q_B \overline{Q}_A$$
$$K_C = \overline{Q}_B \overline{Q}_A$$

The final step is to implement the combinational logic from the equations and connect the flip-flops to form the sequential circuit. The complete three-bit Gray code counter is shown in Figure 8–30.

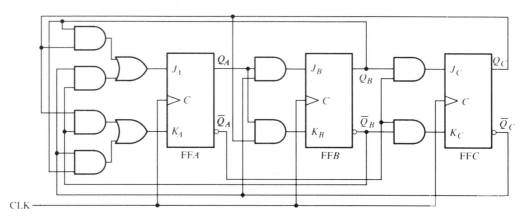

FIGURE 8–30   *Three-bit Gray code counter.*

A summary of steps used in the design of this counter is listed below. In general, these steps can be applied to any sequential circuit.

1. Specify the sequential circuit and draw a state diagram.
2. Derive a next-state table from the state diagram.
3. Develop a truth table for the combinational logic required for the flip-flop inputs to make each flip-flop go to its proper next state from each of its present states.
4. Transfer the information from the truth table to Karnaugh maps. There is a Karnaugh map for each input of each flip-flop.
5. Factor the maps to generate an expression for each flip-flop input.
6. Implement the expressions with combinational logic and combine with the flip-flops.

This procedure is now applied to the design of another counter.

EXAMPLE 8–5

Develop a synchronous three-bit up/down counter with a Gray code sequence. The counter should count up when an up/down control input is 1 and count down when the control input is 0.

Solution
Step 1   The state diagram is shown in Figure 8–31. The 1 or 0 beside each arrow indicates the state of the up/down control input, Y.

FIGURE 8–31 *State diagram for a three-bit up/down Gray code counter.*

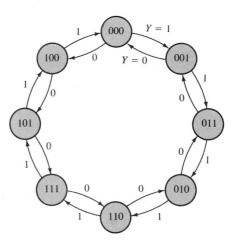

Step 2 The next-state table is derived from the state diagram and is shown in Table 8–8.

TABLE 8–8 *Next-state table for three-bit up/down Gray code counter.*

Present State			Next State					
			$Y = 0$ (down)			$Y = 1$ (up)		
Q_C	Q_B	Q_A	Q_C	Q_B	Q_A	Q_C	Q_B	Q_A
0	0	0	1	0	0	0	0	1
0	0	1	0	0	0	0	1	1
0	1	1	0	0	1	0	1	0
0	1	0	0	1	1	1	1	0
1	1	0	0	1	0	1	1	1
1	1	1	1	1	0	1	0	1
1	0	1	1	1	1	1	0	0
1	0	0	1	0	1	0	0	0

Y = up/down control input.

Step 3 The truth table for the flip-flop inputs is shown in Table 8–9. The control input, Y, is treated as a fourth variable along with the three present-state variables, Q_A, Q_B, and Q_C.

TABLE 8–9 *Truth table for three-bit up/down Gray code counter.*

Present State			Input	Next State			Flip-Flop Input Conditions*			Flip-Flop Input (present state)		
Q_C	Q_B	Q_A	Y	Q_C	Q_B	Q_A	FFC	FFB	FFA	$J_C K_C$	$J_B K_B$	$J_A K_A$
0	0	0	0	1	0	0	S/T	NC/R	NC/R	1 X	0 X	0 X
0	0	0	1	0	0	1	NC/R	NC/R	S/T	0 X	0 X	1 X
0	0	1	0	0	0	0	NC/R	NC/R	R/T	0 X	0 X	X 1
0	0	1	1	0	1	1	NC/R	S/T	NC/S	0 X	1 X	X 0
0	1	1	0	0	0	1	NC/R	R/T	NC/S	0 X	X 1	X 0
0	1	1	1	0	1	0	NC/R	NC/S	R/T	0 X	X 0	X 1
0	1	0	0	0	1	1	NC/R	NC/S	S/T	0 X	X 0	1 X
0	1	0	1	1	1	0	S/T	NC/S	NC/R	1 X	X 0	0 X
1	1	0	0	0	1	0	R/T	NC/S	NC/R	X 1	X 0	0 X
1	1	0	1	1	1	1	NC/S	NC/S	S/T	X 0	X 0	1 X
1	1	1	0	1	1	0	NC/S	NC/S	R/T	X 0	X 0	X 1
1	1	1	1	1	0	1	NC/S	R/T	NC/S	X 0	X 1	X 0
1	0	1	0	1	1	1	NC/S	S/T	NC/S	X 0	1 X	X 0
1	0	1	1	1	0	0	NC/S	NC/R	R/T	X 0	0 X	X 1
1	0	0	0	1	0	1	NC/S	NC/R	S/T	X 0	0 X	1 X
1	0	0	1	0	0	0	R/T	NC/R	NC/R	X 1	0 X	0 X

* These conditions permit the flip-flop to go to the specified next state on the next clock pulse.
NC/R = no-change or RESET.
NC/S = no-change or SET.
R/T = RESET or toggle.
S/T = SET or toggle.

Step 4 The Karnaugh maps for the *J* and *K* inputs of the flip-flops are shown in Figure 8–32. As mentioned before, the control input *Y* is considered one of the variables. The information in the "Flip-Flop Input" columns of Table 8–9 is transferred onto the maps as indicated.

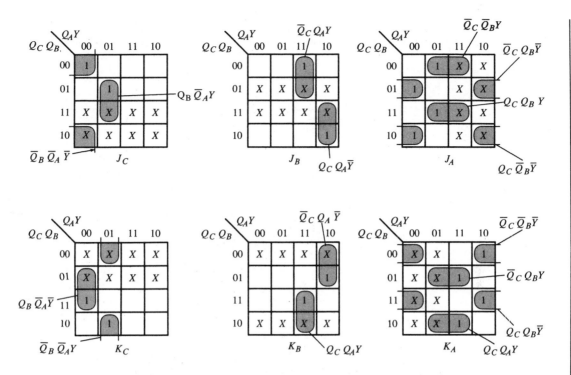

FIGURE 8-32 *J and K maps for Table 8-9.*

Step 5 The 1s are combined in the largest possible groupings using "don't cares" (X's) where possible. The groups are factored, and the expressions for each J and K input are as follows:

$$J_A = Q_C\overline{Q}_B Y + Q_C\overline{Q}_B\overline{Y} + \overline{Q}_C Q_B\overline{Y} + \overline{Q}_C\overline{Q}_B Y$$
$$K_A = \overline{Q}_C\overline{Q}_B\overline{Y} + \overline{Q}_C Q_B Y + Q_C Q_B\overline{Y} + Q_C Q_A Y$$

$$J_B = \overline{Q}_C Q_A Y + Q_C Q_A\overline{Y}$$
$$K_B = \overline{Q}_C Q_A\overline{Y} + Q_C Q_A Y$$

$$J_C = Q_B\overline{Q}_A Y + \overline{Q}_B\overline{Q}_A\overline{Y}$$
$$K_C = Q_B\overline{Q}_A\overline{Y} + \overline{Q}_B\overline{Q}_A Y$$

Step 6 The J and K equations are implemented with combinational logic and the complete counter is shown in Figure 8-33.

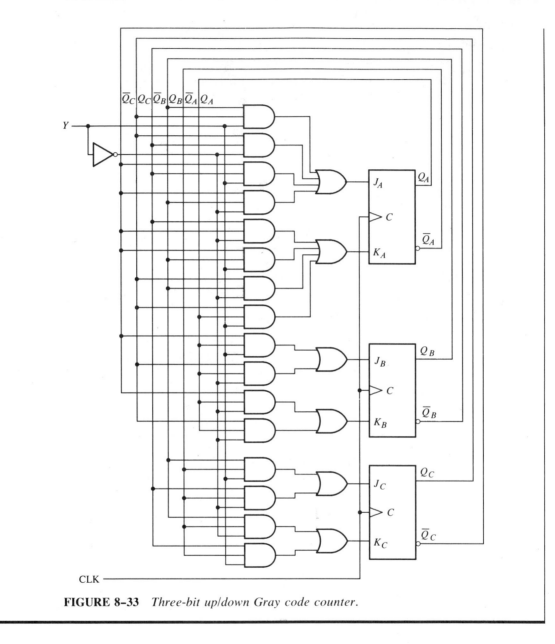

FIGURE 8–33 *Three-bit up/down Gray code counter.*

You have seen how sequential circuit design techniques can be applied specifically to counter design. In general, sequential circuits can be classified into two types: (1) those in which the output or outputs depend only on the present internal state (called *Moore* circuits) and (2) those in which the output or outputs depend on both the present state and the input or inputs (called *Mealy* circuits).

SECTION
REVIEW
8–4
1. A flip-flop is presently in the RESET state and must go to the SET state on the next clock pulse. What must J and K be?
2. A flip-flop is presently in the SET state and must remain SET on the next clock pulse. What must J and K be?
3. A binary counter is in the 1010_2 state.
   (a) What is its next state?
   (b) What condition must exist on each flip-flop input to insure that it goes to the proper next state on the clock pulse?

8–5   CASCADED COUNTERS

Counters can be connected in *cascade* in order to achieve higher modulus operation. In essence, *cascading* means that the last stage output of one counter drives the input of the next counter. An example of two counters connected in cascade is shown for a two-bit and a three-bit ripple counter in Figure 8–34. The timing diagram is in Figure 8–35.

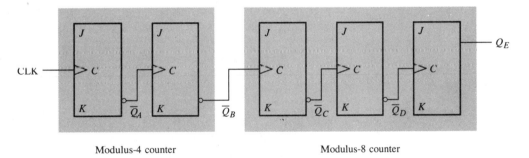

Modulus-4 counter   Modulus-8 counter

FIGURE 8–34 *Two cascaded counters (all J, K inputs are HIGH).*

Notice that in the timing diagram of Figure 8–35, the final output of the modulus-8 counter, Q_E, occurs once for every 32 input clock pulses. The overall modulus of the cascaded counters is 32; that is, they act as a divide-by-32 counter.

In general, the overall modulus of cascaded counters is equal to the product of each individual modulus. For instance, for the counter in Figure 8–34, $4 \times 8 = 32$.

When operating synchronous counters in a cascaded configuration, it is necessary to use the *count enable* and the *terminal count* functions to achieve higher modulus operation. On some devices, the count enable is labeled simply CTEN or some other similar designation, and terminal count (TC) is analogous to *ripple clock* or *ripple carry output* (RCO) on some IC counters.

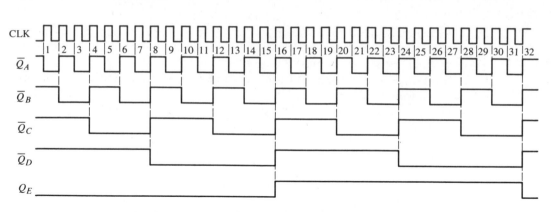

FIGURE 8–35 *Timing diagram for the counter configuration of Figure 8–34.*

Figure 8–36 shows two decade counters connected in cascade. The terminal count (TC) output of counter 1 is connected to the count enable (CTEN) input of counter 2. Counter 2 is inhibited by the LOW on its CTEN input until counter 1 reaches its last or terminal state and its terminal count output goes HIGH. This HIGH now enables counter 2, so that when the first clock pulse after counter 1 reaches its terminal count (CLK_{10}), counter 2 goes from its initial state to its second state. Upon completion of the entire second cycle of counter 1 (when counter 1 reaches terminal count the second time), counter 2 is again enabled and advances to its next state. This sequence continues. Since these are decade counters, counter 1 must go through ten complete cycles before counter 2 completes its first cycle. In other words, for every ten cycles of counter 1, counter 2 goes through one cycle. Thus, counter 2 will complete one cycle after 100 clock pulses. The overall modulus of these two cascaded counters is $10 \times 10 = 100$.

When viewed as a frequency divider, the circuit of Figure 8–36 divides the input clock frequency by 100. Cascaded counters are often used to divide a high-

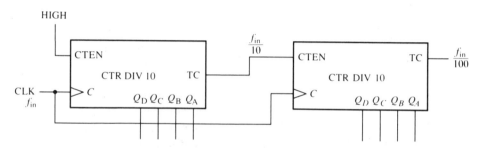

FIGURE 8–36 *A modulus-100 cascaded counter using two decade counters.*

frequency clock signal to obtain highly accurate pulse frequencies. Cascaded counter configurations used for such purposes are sometimes called *countdown chains*.

For example, suppose that we have a basic clock frequency of 1 MHz and we wish to obtain 100 kHz, 10 kHz, and 1 kHz; a series of cascaded counters can be used. If the 1-MHz signal is divided by 10, we get 100 kHz. Then if the 100-kHz signal is divided by 10, we get 10 kHz. Another division by 10 yields the 1-kHz frequency. The general implementation of this countdown chain is shown in Figure 8–37.

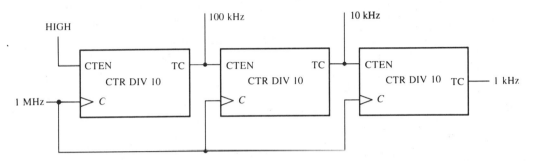

FIGURE 8–37   *Three cascaded decade counters forming a divide-by-1000.*

EXAMPLE 8–6   Determine the overall modulus of the two cascaded counter configurations in Figure 8–38.

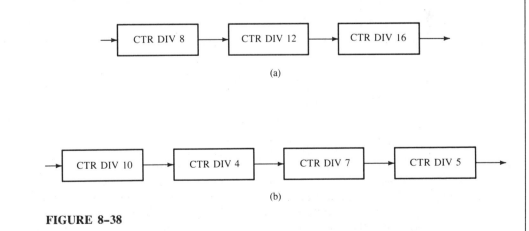

FIGURE 8–38

Solutions

(a)   The overall modulus for the three-counter configuration is

$$8 \times 12 \times 16 = 1536$$

(b) The modulus for the four counters is

$$10 \times 4 \times 7 \times 5 = 1400$$

EXAMPLE 8-7   Use 74LS160A counters to obtain a 10-kHz waveform from a 1-MHz clock. Show the logic diagram.

Solution   To obtain 10 kHz from a 1-MHz clock requires a division factor of 100. Two 74LS160A counters must be cascaded as shown in Figure 8–39. The left counter produces an RCO pulse for every 10 clock pulses. The right counter produces an RCO pulse for every 100 clock pulses.

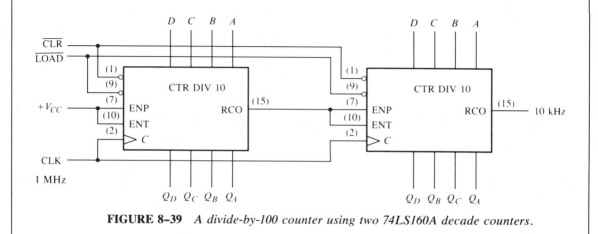

FIGURE 8–39   *A divide-by-100 counter using two 74LS160A decade counters.*

Cascaded IC Counters with Truncated Sequences

The previous material has shown how to achieve an overall modulus (*divide-by* factor) that is the product of the individual moduli of all the cascaded counters. This can be considered *full-modulus cascading*.

Often, an application requires an overall modulus that is *less* than that achieved by full-modulus cascading. That is, a truncated sequence must be implemented with cascaded counters.

To illustrate this method, the cascaded counter configuration in Figure 8–40 is used as an example. This particular circuit uses four 74LS161A four-bit binary counters. If these four counters (sixteen bits total) were cascaded in a full-modulus arrangement, the modulus would be

$$2^{16} = 65,536$$

Let's assume that a certain application requires a divide-by-40,000 counter (modulus 40,000). The difference between 65,536 and 40,000 is 25,536, which is

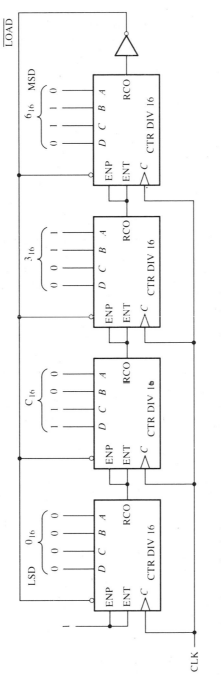

FIGURE 8–40 *A divide-by-40,000 counter using 74LS161A four-bit binary counters.*

the number of states that must be deleted from the full-modulus sequence. The technique used in the circuit of Figure 8–40 is to *preset* the cascaded counter to 25,536 (63C0 in hexadecimal) each time that it recycles so that it counts from 25,536 up to 65,536 on each full cycle. Therefore, each full cycle of the counter consists of 40,000 states.

Notice that the RCO output of the right-most counter is inverted and applied to the $\overline{\text{LOAD}}$ input of each four-bit counter. Each time the count reaches its terminal value of 65,536, RCO goes HIGH and causes the number on the data inputs ($63C0_{16}$) to be preset into the counter. Thus, there is one RCO pulse from the right-most four-bit counter for every 40,000 clock pulses.

With this technique, any modulus can be achieved by simply presetting the counter to the appropriate initial state on each cycle.

SECTION REVIEW 8–5

1. How many decade counters are necessary to implement a divide-by-1000 (modulus-1000) counter? A divide-by-10,000?
2. Show with general block diagrams how to achieve each of the following using a flip-flop, a decade counter, and a four-bit binary counter, or any combination of these:
   - **(a)** Divide-by-20 counter
   - **(b)** Divide-by-32 counter
   - **(c)** Divide-by-160 counter
   - **(d)** Divide-by-320 counter

8–6   COUNTER DECODING

In many digital applications it is necessary that some or all of the counter states be decoded. The decoding of a counter involves using decoders or logic gates to determine when the counter is in a certain state or states in its sequence. For instance, the terminal count function previously discussed is a single state (the last state) in the counter sequence.

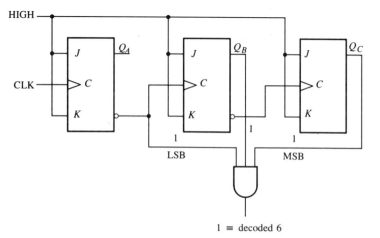

FIGURE 8–41 *Decoding of count 6.*

For example, suppose that we wish to decode count 6 (110) of a three-bit binary counter. This can be done as shown in Figure 8–41. When $Q_C = 1$, $Q_B = 1$, and $Q_A = 0$, a HIGH appears on the output of the decoding gate, indicating that the counter is at count 6. This is called *active-HIGH decoding*. Replacing the AND gate with a NAND gate provides active-LOW decoding.

EXAMPLE 8–8

Implement the decoding of state 2 and state 7 of a three-bit synchronous counter. Show the entire counter timing diagram and the output waveforms of the decoding gates. Binary $2 = \overline{Q}_C Q_B \overline{Q}_A$, and binary $7 = Q_C Q_B Q_A$.

Solution   See Figure 8–42.

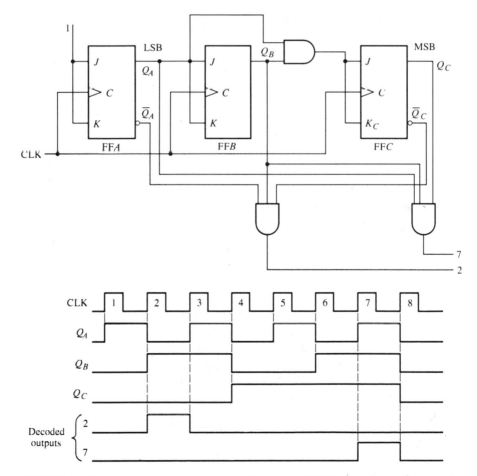

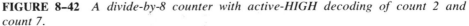

FIGURE 8–42 *A divide-by-8 counter with active-HIGH decoding of count 2 and count 7.*

Decoding Glitches

The problem of glitches produced by the decoding process was introduced in Chapter 6. As you have learned, the propagation delays due to the ripple effect in asynchronous counters create transitional states in which the counter outputs are changing at slightly different times. These transitional states produce undesired voltage spikes of short duration (glitches) on the outputs of a decoder. The glitch problem can also occur to some degree with synchronous counters because the propagation delays from clock to Q outputs of each flip-flop in a counter can vary slightly. As an example of the glitch problem in counter decoding, Figure 8–43 shows a basic asynchronous BCD counter connected to a BCD-to-decimal decoder.

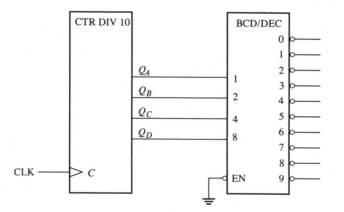

FIGURE 8–43   *A basic decade (BCD) counter and decoder.*

To see what happens in this case, we will look at a timing diagram in which the propagation delays are taken into account, as in Figure 8–44. Notice how these delays cause false states of short duration. The decimal value of the false binary state at each critical transition is indicated on the diagram. The resulting glitches can be seen on the decoder outputs.

One way to eliminate the glitches is to enable the decoded outputs at a time *after* the glitches have had time to disappear. This method is known as *strobing* and can be accomplished in this case by using the LOW level of the clock to enable the decoder, as shown in Figure 8–45. The resulting improved timing diagram is shown in Figure 8–46.

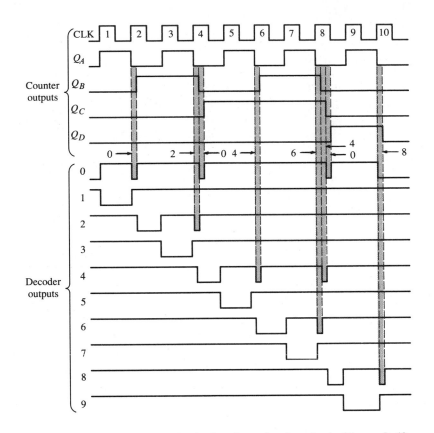

FIGURE 8–44 *Outputs with glitches from the decoder in Figure 8–43.*

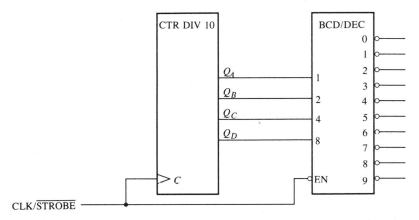

FIGURE 8–45 *The basic decade counter and decoder with strobing to eliminate glitches.*

389

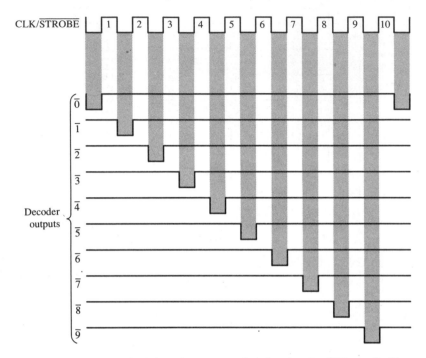

FIGURE 8–46 *Strobed decoder outputs for the circuit of Figure 8–45.*

SECTION REVIEW 8–6

1. What transitional states are possible when an asynchronous four-bit binary counter changes from
   (a) Count 2 to count 3
   (b) Count 3 to count 4
   (c) Count 10_{10} to count 11_{10}
   (d) Count 15 to count 0

8–7   COUNTER APPLICATIONS

The digital counter is a very useful and versatile device that is found in many applications. In this section, some representative counter applications are presented.

The Digital Clock

A very common example of a counter application is in timekeeping systems. Figure 8–47 is a simplified logic diagram of a digital clock that displays seconds, minutes, and hours.

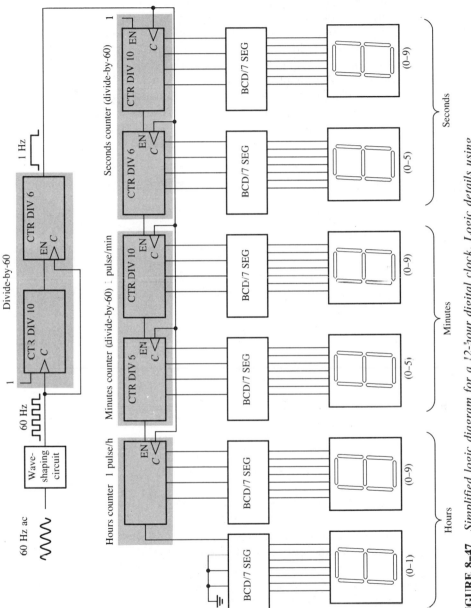

FIGURE 8-47 *Simplified logic diagram for a 12-hour digital clock. Logic details using specific devices are shown in the following diagrams.*

First, a 60-Hz sinusoidal ac voltage is converted to a 60-Hz pulse waveform and divided down to a 1-Hz pulse waveform by a divide-by-60 counter formed by a divide-by-10 counter followed by a divide-by-6 counter. Both the *seconds* and *minutes* counts are also produced by divide-by-60 counters, the details of which are shown in Figure 8–48. These counters count from 0 to 59 and then recycle to 0; 74LS160A synchronous decade counters are used in this particular implementation. Notice that the divide-by-6 portion is formed with a decade counter with a truncated sequence achieved by using the decoded count 6 to asynchronously clear the counter. The terminal count of 5 is also decoded to enable the next counter in the chain.

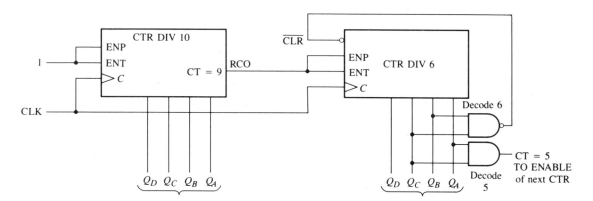

FIGURE 8–48 *Logic diagram of typical divide-by-60 counter using 74LS160A synchronous decade counters.*

The *hours* counter is implemented with a decade counter and a flip-flop as shown in Figure 8–49. Consider that initially both the decade counter and the flip-flop are RESET, and the decode-12 gate output is HIGH. The decade counter advances from 0 to 9, and as it recycles from 9 back to 0, the flip-flop is toggled to the SET state by the HIGH-to-LOW transition of Q_D. This illuminates a 1 on the tens-of-hours display. The total count is now 10_{10} (the decade counter is in the 0 state and the flip-flop is SET).

Next, the total count advances to 11_{10} and then to 12. In state 12, the Q_B output of the decade counter is HIGH, the flip-flop is still SET, and thus the decode gate output is LOW. This activates the LOAD input of the decade counter. On the next clock pulse, the decade counter is preset to state 1 by the data inputs, and the flip-flop is RESET ($J = 0$, $K = 1$). As you can see, this logic always causes the counter to recycle from 12 back to 1 rather than back to 0.

Auto Parking Control

Now a simple application example illustrates the use of an up/down counter to solve an everyday problem. The problem is to devise a means of monitoring

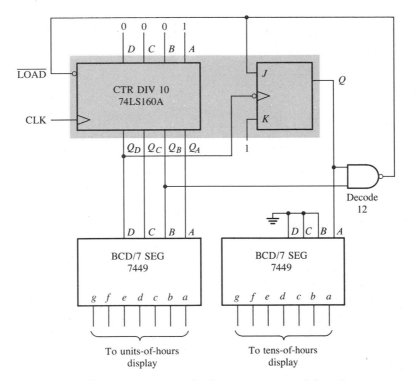

FIGURE 8-49 *Logic diagram for hours counter and decoders.*

available spaces in a 100-space parking garage and provide for an indication of a *full* condition by illuminating a display sign and lowering a gate bar at the entrance.

A system that solves this problem consists of (1) optoelectronic sensors at the entrance and exit of the garage, (2) an up/down counter and associated circuitry, and (3) an interface circuit that takes the counter output to turn the *full* sign on or off as required and lower or raise the gate bar at the entrance. A general block diagram of this system is shown in Figure 8-50.

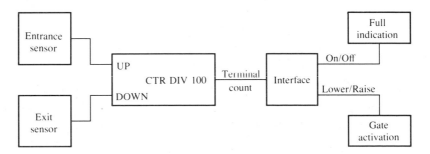

FIGURE 8-50 *Functional block diagram for parking garage control.*

A logic diagram of the up/down counter is shown in Figure 8–51. It consists of two cascaded 74190 up/down decade counters. The operation is as follows:

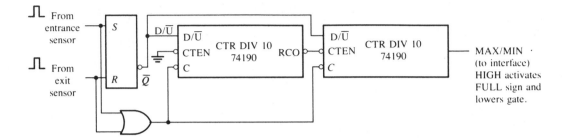

FIGURE 8–51 *Logic diagram for modulus-100 up/down counter for auto parking control.*

The counter is initially preset to 0. Each automobile entering the garage breaks a light beam, activating a sensor that produces an electrical pulse. This positive pulse SETS the S-R latch on its leading edge. The LOW on the $\overline{Q}$ output of the latch puts the counter in the UP mode. Also, the sensor pulse goes through the OR gate and clocks the counter on the HIGH-to-LOW transition of its trailing edge. Each time an automobile enters the garage, the counter is advanced by 1 (incremented). When the one-hundredth automobile enters, the counter goes to its last state (100_{10}). The MAX/MIN output goes HIGH and activates the interface circuit (no detail), which lights the *full* sign and lowers the gate bar to prevent further entry.

When an automobile exits, an optoelectronic sensor produces a positive pulse which RESETS the S-R latch and puts the counter in the DOWN mode. The trailing edge of the clock decreases the count by 1 (decremented).

If the garage is full and an automobile leaves, the MAX/MIN output of the counter goes LOW, turning off the *full* sign and raising the gate.

Parallel-to-Serial Data Conversion (Multiplexing)

A simplified example of data transmission using multiplexing and demultiplexing techniques was introduced in Chapter 6. Essentially, the parallel data bits on the multiplexer inputs were converted to serial data bits on the single transmission line. A group of bits appearing simultaneously on parallel lines is called *parallel data*. A group of bits appearing on a single line in a time sequence is called *serial data*.

Parallel-to-serial conversion is normally accomplished using a counter to provide a binary sequence for the data-select inputs of a data selector/multiplexer as illustrated in Figure 8–52. The *Q* outputs of the modulus-8 counter are connected to the data-select inputs of an eight-bit multiplexer. Figure 8–53 is a timing diagram illustrating the operation of this circuit.

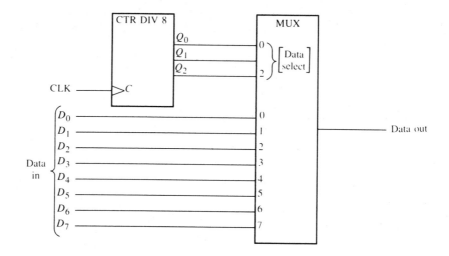

FIGURE 8–52 *Parallel-to-serial data conversion logic.*

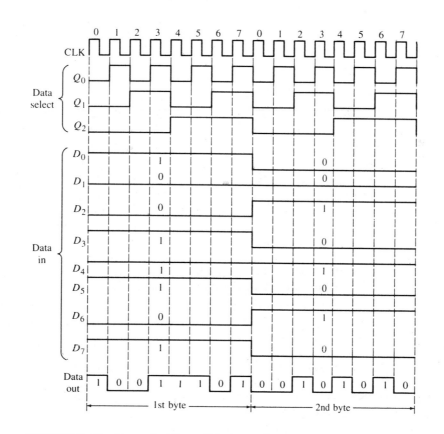

FIGURE 8–53 *Example of parallel-to-serial conversion timing.*

The first byte of parallel data is applied to the multiplexer inputs. (A *byte* is an eight-bit group.) As the counter goes through a binary sequence from 0 to 7, each bit beginning with D_0 is sequentially selected and passed through the multiplexer to the output line. After eight clock pulses, the data byte has been converted to a serial format and sent out on the transmission line. When the counter recycles back to 0, the next byte is applied to the data inputs and is sequentially converted to serial form as the counter cycles through its eight states. This process continues repeatedly as each parallel byte is converted to a serial byte. An end-of-chapter problem requires that this circuit be implemented with specific devices.

SECTION REVIEW 8–7

1. Explain the purpose of the NAND gate in Figure 8–49.
2. Identify the two recycle conditions for the hours counter in Figure 8–47, and explain the reason for each.

8–8   LOGIC SYMBOLS WITH DEPENDENCY NOTATION

Up to this point, the logic symbols with dependency notation as specified in ANSI/IEEE Std. 91–1984 have been introduced on a limited basis. In many cases, the new symbols do not deviate greatly from the traditional symbols. A significant departure from what we are accustomed to does occur, however, for some devices, including counters and other more complex devices.

Dependency notation is fundamental to the new ANSI/IEEE standard. Dependency notation is used in conjunction with the logic symbols to specify the relationships of inputs and outputs so that the logical operation of a given device can be determined entirely from its logic symbol without a prior knowledge of the details of its internal structure and without a detailed logic diagram for reference. Dependency notation was introduced in relation to specific flip-flops in Chapter 7.

Although we will continue to use primarily the more traditional and familiar symbols throughout this book, a brief coverage of logic symbols with dependency notation is provided in this section to prepare you for a projected increase in usage of these symbols in the future. This section can be treated as optional, and its omission will not affect the remainder of the book.

The 74LS163A four-bit binary counter is used for illustration. For comparison, Figure 8–54 shows a traditional block symbol and the ANSI/IEEE symbol with dependency notation. Basic descriptions of the symbol and the dependency notation follow.

Common Control Block

The upper block with notched corners in Figure 8–54(b) has inputs and an output that are considered *common* to all elements in the device and not unique to any one of the elements.

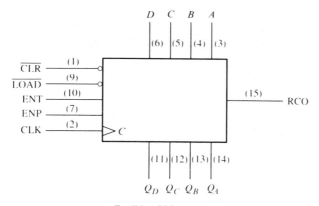

(a) Traditional block symbol

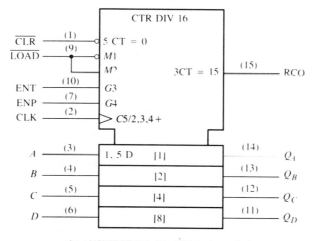

(b) ANSI/IEEE Std. 91–1984 logic symbol

FIGURE 8–54 *The 74LS163A four-bit synchronous counter.*

Individual Elements

The lower block, which is partitioned into four abutted sections, represents the four storage elements (D flip-flops) in the counter with inputs A, B, C, and D and outputs Q_A, Q_B, Q_C, and Q_D.

Qualifying Symbol

"CTR DIV 16" identifies the device as a counter (CTR) with sixteen states (DIV 16).

C (Control) Dependency

The letter *C* denotes control dependency. Control inputs usually enable or disable the data inputs (*D, J, K, S,* and *R*) of a storage element. The *C* input is usually the clock input. In this case, the digit 5 following *C* (*C*5/2,3,4+) indicates that the inputs labeled with a 5 prefix are dependent on the clock (synchronous with the clock). For example, 5CT = 0 on the CLR input indicates that the clear function is dependent on the clock; that is, it is a *synchronous* clear. When the $\overline{\text{CLR}}$ input is LOW (0), the counter is reset to 0 (CT = 0) on the triggering edge of the clock pulse. Also, the 5*D* label at the input of storage element [1] indicates that the data storage is dependent on (synchronous with) the clock. All labels in the [1] storage element apply to the [2], [4], and [8] elements below it, since they are not labeled differently.

M (Mode) Dependency

The letter *M* denotes mode dependency. This is used to indicate how the functions of various inputs or outputs depend on the mode in which the device is operating. In this case, the device has two modes of operation. When the $\overline{\text{LOAD}}$ input is LOW (0), as indicated by the bubble input, the counter is in a preset mode (*M*1) in which the input data (*A, B, C,* and *D*) are synchronously loaded into the four flip-flops. The digit 1 following *M* in *M*1 and the 1 in 1,5*D* shows a dependency relationship and indicates that input data are stored only when the device is in the preset mode (*M*1) in which $\overline{\text{LOAD}}$ = 0. When the $\overline{\text{LOAD}}$ input is HIGH (1), as indicated by the nonbubble input, the counter advances through its normal binary sequence as indicated by *M*2 and the 2 in *C*5/2,3,4+.

G (AND) Dependency

The letter *G* denotes AND dependency, indicating that an input designated with *G* followed by a digit is ANDed with any other input or output having the same digit as a prefix in its label. In this particular example, the *G*3 at the ENT input and the 3CT = 15 at the RCO output are related as indicated by the 3, and that relationship is an AND dependency as indicated by the *G*. This tells us that ENT must be HIGH (no bubble on the input) AND the count must be fifteen (CT = 15) for the RCO output to be HIGH (no bubble on output).

Also, the digits 2, 3, and 4 in the label *C*5/2,3,4+ indicate that the counter advances through its states when $\overline{\text{LOAD}}$ = 1, as indicated by the mode dependency label *M*2, and when ENT = 1 AND ENP = 1, as indicated by the AND dependency labels *G*3 and *G*4. The + indicates that the counter advances by one count when these conditions exist.

This coverage of a specific logic symbol with dependency notation is intended to aid in the interpretation of other such symbols that you may encounter in the future.

| SECTION REVIEW 8–8 | 1. | In dependency notation, what do the letters *C*, *M*, and *G* stand for? |
| | 2. | By what letter is data storage denoted? |

SUMMARY

☐ An asynchronous counter is one in which the flip-flops are not simultaneously triggered. Each flip-flop (after the least significant stage) is clocked by the output of the preceding one.

☐ Asynchronous counters are also called *ripple counters*.

☐ A synchronous counter is one in which all the flip-flops are simultaneously triggered from the same clock input.

☐ Asynchronous counters are slower (have a lower maximum frequency) than synchronous counters because of the ripple clocking delay. Any individual flip-flop must wait for all the preceding flip-flops to change before it can change.

☐ The modulus of a counter is the number of states in its sequence. The "divide-by" factor is the same as the modulus.

☐ A full-modulus counter has 2^n states, where *n* is the number of flip-flops.

☐ Counters with truncated sequences have modulus numbers less than 2^n.

☐ A decade counter has ten states.

☐ The total modulus of a cascaded counter arrangement is the *product* of the individual moduli of all the cascaded counters.

☐ AND gates and NAND gates can be used to decode each state of a counter.

☐ Glitches at the outputs of decoders are caused by the counter's transitional states.

SELF-TEST

1. What is the reason for the difference between asynchronous and synchronous counter performance?
2. What is the primary disadvantage of an asynchronous counter?
3. What is the maximum modulus for a counter with each of the following numbers of flip-flops?
   (a) 2   **(b)** 4   **(c)** 5   **(d)** 6   **(e)** 7   **(f)** 8
4. Determine the number of flip-flops in each of the following counters:
   (a) Modulus-3   **(b)** Modulus-8   **(c)** Modulus-12
   (d) Modulus-16   **(e)** Modulus-18   **(f)** Modulus-36
   (g) Modulus-64   **(h)** Modulus-144
5. A four-bit asynchronous (ripple) counter consists of flip-flops each of which has a clock to *Q* propagation delay of 12 ns. How long does it take the counter to recycle from 1111 to 0000 after the triggering edge of the clock pulse?
6. Repeat Problem 5 for a synchronous counter.

7. Explain how a BCD decade counter can be used as a divide-by-10 device, and illustrate it with a timing diagram.

8. A decade counter does not use its maximum possible modulus, so there are several invalid states. List these states.

9. A 74LS163A four-bit synchronous counter is in the 1010 state, and the following input conditions exist: $\overline{CLR} = 1$, $\overline{LOAD} = 0$, $A = 1$, $B = 0$, $C = 1$, $D = 0$, ENT = 1, and ENP = 1. To what state does the counter go on the next clock pulse?

10. A four-bit binary up/down counter (modulus-16) is in the binary state of 0. What is its next state in the up mode? In the down mode?

11. A flip-flop is in the RESET state and must stay RESET at the next clock pulse. What must J and K be?

12. With general block diagrams, show how to obtain the following frequencies from a 10-MHz clock using single flip-flops, modulus-5 counters, and decade counters:
   (a) 5 MHz   **(b)** 2.5 MHz   **(c)** 2 MHz   **(d)** 1 MHz   **(e)** 500 kHz
   (f) 250 kHz   **(g)** 62.5 kHz   **(h)** 40 kHz   **(i)** 10 kHz   **(j)** 1 kHz

13. Show how to decode the following states of a four-bit binary counter using active-LOW decoding:
   (a) 1   **(b)** 5   **(c)** 10   **(d)** 14

14. In Figure 8–49, explain how the recycle to count 1 is accomplished.

15. If the clock frequency is 10 kHz in Figure 8–52, how long does it take for two bytes to be transmitted in serial form?

PROBLEMS

Section 8–1

8–1   For the ripple counter shown in Figure 8–55, draw the complete timing diagram for eight clock pulses showing the clock, Q_A, and Q_B waveforms.

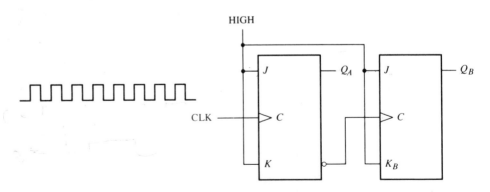

FIGURE 8–55

8–2   For the ripple counter in Figure 8–56, draw the complete timing diagram for sixteen clock pulses. Show the clock, Q_A, Q_B, and Q_C waveforms.

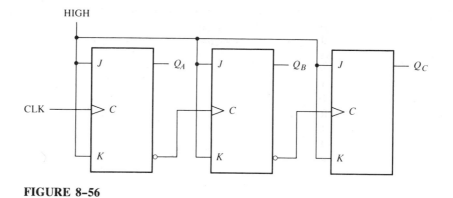

FIGURE 8-56

8-3 In the counter of Problem 8-2, assume that each flip-flop has a propagation delay from the triggering edge of the clock to a change in the Q output of 8 ns. Determine the worst-case (longest) delay time from a clock pulse to the arrival of the counter in a given state. Specify the state or states for which this worst-case delay occurs.

8-4 Show how to connect a 7493A four-bit asynchronous counter for each modulus below:

    (a) 9   **(b)** 11   **(c)** 13   **(d)** 14   **(e)** 15

Section 8-2

8-5 If the counter of Problem 8-3 were synchronous rather than asynchronous, what would be the longest delay time?

8-6 Draw the complete timing diagram for the five-stage synchronous binary counter in Figure 8-57. Verify that the waveforms of the Q outputs represent the proper binary number after each clock pulse.

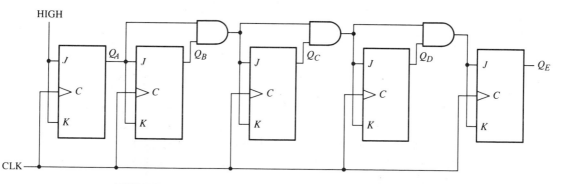

FIGURE 8-57

8-7 By analyzing the J and K inputs to each flip-flop prior to each clock pulse, prove that the decade counter in Figure 8–58 progresses through a BCD sequence. Explain how these conditions in each case cause the counter to go to the next proper state.

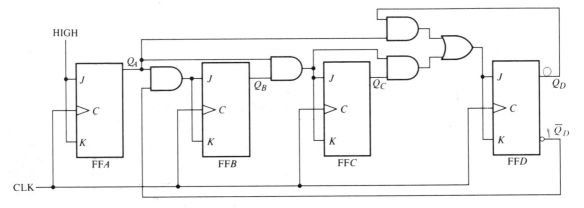

FIGURE 8–58

8-8 The waveforms in Figure 8–59 are applied to the count enable, clear, and clock inputs as indicated. Sketch the counter output waveforms in proper relation to these inputs. The clear input is asynchronous.

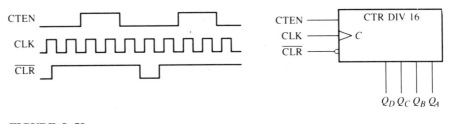

FIGURE 8–59

8-9 A BCD decade counter is shown in Figure 8–60. The waveforms are applied to the clock and clear inputs as indicated. Determine the waveforms for each of the counter outputs (Q_A, Q_B, Q_C, and Q_D). The clear is synchronous, and the counter is initially in the 1000_2 state.

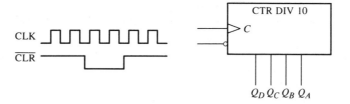

FIGURE 8-60

8-10 The waveforms in Figure 8-61 are applied to a 74LS163A counter. Determine the Q outputs and RCO. The inputs are $A = 1$, $B = 1$, $C = 0$, and $D = 1$.

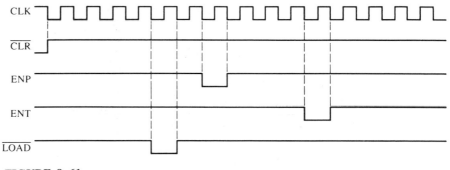

FIGURE 8-61

8-11 The waveforms in Figure 8-61 are applied to a 74LS160A counter. Determine the Q outputs and RCO. The inputs are $A = 1$, $B = 0$, $C = 0$, and $D = 1$.

Section 8-3

8-12 Draw a complete timing diagram for a three-bit up/down counter that goes through the following sequence. Indicate when the counter is in the up mode and when it is in the down mode. Assume positive edge-triggering.

0, 1, 2, 3, 2, 1, 2, 3, 4, 5, 6, 5, 4, 3, 2, 1, 0

8-13 Sketch the Q output waveforms for a 74190 up/down counter with the input waveforms shown in Figure 8-62. A binary 0 is on the data inputs.

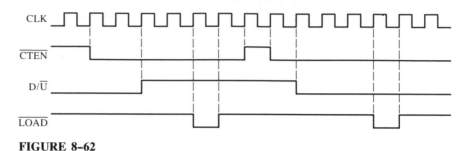

FIGURE 8-62

Section 8-4

8-14 Determine the sequence of the counter in Figure 8-63.

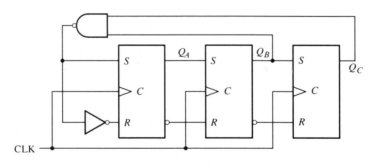

FIGURE 8-63

8-15 Determine the sequence of the counter in Figure 8-64. Begin with the counter cleared.

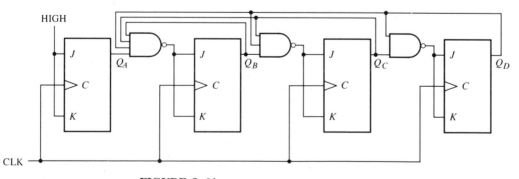

FIGURE 8-64

8-16 Design a counter to produce the following sequence. Use J-K flip-flops.

0, 2, 1, 3, 0

Clock Pulse	Q_B	Q_A
0	0	0
1	1	0
2	0	1
3	1	1

8-17 Design a counter to produce the following sequence. Use J-K flip-flops.

Clock Pulse	Q_C	Q_B	Q_A
0	0	0	1
1	1	0	0
2	0	1	1
3	1	0	1
4	1	1	1
5	1	1	0
6	0	1	0

8-18 Design a counter to produce the following sequence. Use J-K flip-flops.

Clock Pulse	Q_D	Q_C	Q_B	Q_A
0	0	0	0	0
1	1	0	0	1
2	0	0	0	1
3	1	0	0	0
4	0	0	1	0
5	0	1	1	1
6	0	0	1	1
7	0	1	1	0
8	0	1	0	0
9	0	1	0	1

8-19 Analyze the counter in Figure 8-63 for a "lock-up" condition in which the counter cannot escape from an invalid state or states. An invalid state is one that is not in the counter's normal sequence.

Section 8–5

8–20 For each of the cascaded counter configurations in Figure 8–65, determine the frequency of the waveform at each point indicated by circled numbers, and determine the overall modulus.

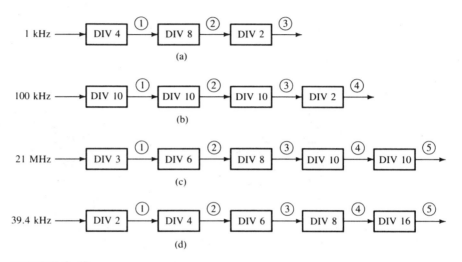

FIGURE 8–65

8–21 Expand the counter in Figure 8–37 to create a divide-by-10,000 counter and a divide-by-100,000 counter.

8–22 Design a modulus-1000 counter using 74LS160A decade counters.

8–23 Modify the design of the counter in Figure 8–40 to achieve a modulus of 30,000.

8–24 Repeat Problem 8–23 for a modulus of 50,000.

Section 8–6

8–25 Given a BCD decade counter with only the Q outputs available, show what decoding logic is required to decode each of the following states and how it should be connected to the counter. A HIGH output indication is required for each decoded state. The MSB is to the left.
(a) 0001 **(b)** 0011 **(c)** 0101 **(d)** 0111 **(e)** 1000

8–26 For the four-bit binary counter connected to the decoder in Figure 8–66, determine each of the decoder output waveforms in relation to the clock pulses.

8–27 If the counter in Figure 8–66 is asynchronous, determine where the decoding glitches occur on the decoder output waveforms.

8–28 Modify the circuit in Figure 8–66 to eliminate decoding glitches.

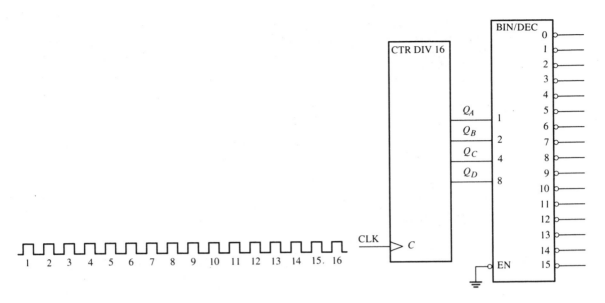

FIGURE 8-66

Section 8-7

8-29 Modify the digital clock in Figures 8-47, 8-48, and 8-49 so that it can be preset to any desired time.

8-30 Design an alarm circuit for the digital clock that can detect a predetermined time (hours and minutes only) and produce a signal to activate an audio alarm.

8-31 Modify the circuit in Figure 8-51 for a 1000-space parking garage and a 3000-space parking garage.

8-32 Implement the parallel-to-serial data conversion logic in Figure 8-52 with specific devices.

ANSWERS TO SECTION REVIEWS

Section 8-1
1. Each flip-flop after the first one is clocked by the output of the preceding flip-flop.
2. 14

Section 8-2
1. All flip-flops in a synchronous counter are clocked simultaneously.
2. The counter can be preset (initialized) to any given state.

3. Counter is enabled when ENP and ENT are both HIGH. RCO goes HIGH when final state in sequence is reached.

Section 8–3
1. 1001   2. 1111, 0000, 1111

Section 8–4
1. $J = 1$, $K = X$ (don't care)   2. $J = X$ (don't care), $K = 0$
3. (a) 1011_2
   (b) D (MSB): NC or SET. C: NC or RESET. B: NC or SET. A (LSB): SET or toggle.

Section 8–5
1. $n = 10$; $n = 14$
2. (a) Flip-flop and DIV 10.   (b) Flip-flop and DIV 16.
   (c) DIV 16 and DIV 10.   (d) DIV 16 and DIV 10 and flip-flop.

Section 8–6
1. (a) None, because there is a single bit change.
   (b) 0010_2, 0000_2.
   (c) None, because there is a single bit change.
   (d) 1110_2, 1100_2, 1000_2.

Section 8–7
1. Decodes count 12 to preset counter to 0001.
2. The decade counter advances from 0 to 9, and as it recycles from 9 back to 0, the flip-flop is toggled to the SET state. This produces a ten (10) on the display. In state 12, the decode NAND gate causes the decade counter to preset to 1. The flip-flop RESETS. This results in a one (01) on the display.

Section 8–8
1. C: Control, usually clock. M: mode. G: AND.
2. D

Shift registers are a type of sequential logic circuit closely related to counters. They are used basically for the storage and movement of digital data and, typically, do not possess a characteristic internal sequence as do counters.

In this chapter, the basic types of shift registers are studied and several applications are presented. Also, a new troubleshooting method is introduced.

Specific devices introduced in this chapter are as follows:

1. 7491A eight-bit serial shift register
2. 74164 eight-bit parallel out serial shift register
3. 74165 eight-bit parallel load shift register
4. 74195 four-bit parallel access shift register
5. 74194 four-bit bidirectional universal shift register
6. 74199 eight-bit bidirectional universal shift register

In this chapter, you will learn

☐ How a register stores data.
☐ The basic forms of data movement in shift registers.
☐ How data movement is controlled.
☐ How serial in–serial out, serial in–parallel out, parallel in–serial out, and parallel in–parallel out shift registers operate.
☐ How a bidirectional shift register operates.
☐ How a Johnson counter produces its characteristic sequence.
☐ How a ring counter operates.
☐ How a shift register can be used as a time-delay device.
☐ How to construct a ring counter from a shift register.
☐ How shift registers can be used to implement a serial-to-parallel data converter.
☐ The basic operation of a universal asynchronous receiver transmitter (UART).
☐ How a basic shift-register-controlled keyboard encoder operates.
☐ How to troubleshoot digital systems by "exercising" the system with a known test pattern.
☐ How shift registers can be represented by the ANSI/IEEE Std. 91–1984 symbols with dependency notation.

9

Shift Registers

9-1   SHIFT REGISTER FUNCTIONS

Shift registers are very important in applications involving the *storage* and *transfer* of data in a digital system. The basic difference between a register and a counter is that a register has no specified sequence of states except in certain very specialized applications. A register, in general, is used solely for *storing* and *shifting* data (1s and 0s) entered into it from an external source and possesses no characteristic internal sequence of states.

The storage capability of a register is one of its two basic functional characteristics and makes it an important type of *memory* device. Figure 9–1 illustrates the concept of "storing" a 1 or a 0 in a flip-flop. A 1 is applied to the input as shown, and a clock pulse is applied that stores the 1 by setting the flip-flop. When the 1 on the input is removed, the flip-flop remains in the SET state, thereby storing the 1. The same procedure applies to the storage of a 0, as also illustrated in Figure 9–1.

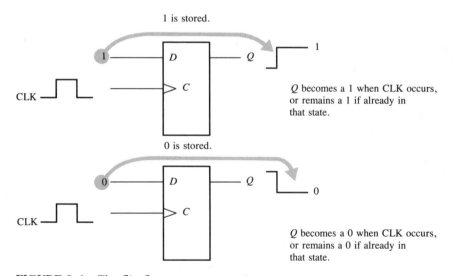

FIGURE 9-1   *The flip-flop as a storage element.*

The *storage capacity* of a register is the number of bits (1s and 0s) of digital data it can retain. Each stage of a shift register represents one bit of storage capacity, and therefore the number of stages in a register determines its total storage capacity.

Registers are commonly used for the *temporary* storage of data within a digital system. Registers are implemented with flip-flops or other storage devices to be introduced later. The *shift* capability of a register permits the movement of data from stage to stage within the register or into or out of the register upon application of clock pulses. Figure 9–2 shows symbolically the types of data movement in shift register operations. The block represents any arbitrary four-bit register, and the arrow indicates the direction and type of data movement.

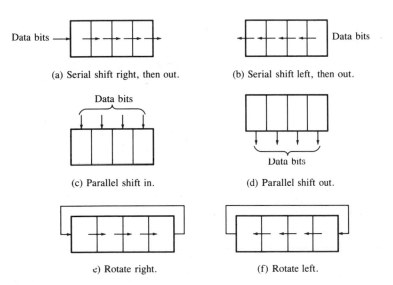

(a) Serial shift right, then out. (b) Serial shift left, then out.

(c) Parallel shift in. (d) Parallel shift out.

e) Rotate right. (f) Rotate left.

FIGURE 9–2 *Basic data movement in registers.*

SECTION REVIEW 9–1

1. Generally, what is the difference between a counter and a shift register?
2. What two principal functions are performed by a shift register?

9–2 SERIAL IN–SERIAL OUT SHIFT REGISTERS

This type of shift register accepts data serially—that is, one bit at a time on a single line. It produces the stored information on its output also in serial form. Let us first look at the serial entry of data into a typical shift register with the aid of Figure 9–3, which shows a four-bit device implemented with D flip-flops.

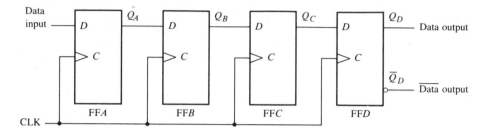

FIGURE 9–3   *Serial in–serial out shift register.*

With four stages, this register can store up to four bits of data; its *storage capacity* is four bits. We will illustrate the entry of the four-bit binary number 1010 into the register, beginning with the right-most bit. The 0 is put onto the data input line, making $D = 0$ for FFA. When the first clock is applied, FFA is RESET, thus "storing" the 0. Next the 1 is applied to the data input, making $D = 1$ for FFA and $D = 0$ for FFB because D of FFB is connected to the Q_A output.

When the second clock pulse occurs, the 1 on the data input is "shifted" into FFA because FFA SETS, and the 0 that was in FFA is "shifted" into FFB. The next 0 in the binary number is now put onto the data-input line, and a clock pulse is applied. The 0 is entered into FFA, the 1 stored in FFA is shifted into FFB, and the 0 stored in FFB is shifted into FFC. The last bit in the binary number, a 1, is now applied to the data input, and a clock pulse is applied. This time the 1 is entered into FFA, the 0 stored in FFA is shifted into FFB, the 1 stored in FFB is shifted into FFC, and the 0 stored in FFC is shifted into FFD. This completes the serial entry of the four-bit number into the shift register, where it can be stored for any length of time. Figure 9–4 illustrates each step in the shifting of the four bits into the register.

If we want to get the data out of the register, they must be shifted out serially and taken off the Q_D output. After CLK_4 in the data-entry operation described above, the right-most 0 in the number appears on the Q_D output. When clock pulse CLK_5 is applied, the second bit appears on the Q_D output. CLK_6 shifts the third bit to the output, and CLK_7 shifts the fourth bit to the output, as illustrated in Figure 9–5. Notice that while the original four bits are being shifted out, a new four-bit number can be shifted in. All 0s are shown being shifted in.

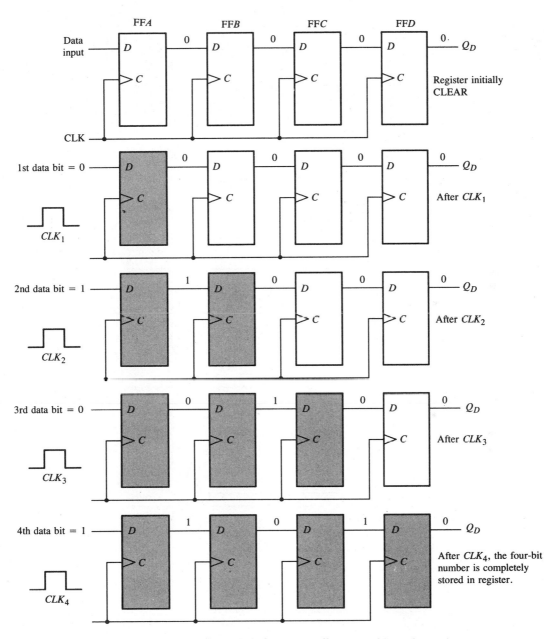

FIGURE 9–4 *Four bits (1010) being serially entered into the register.*

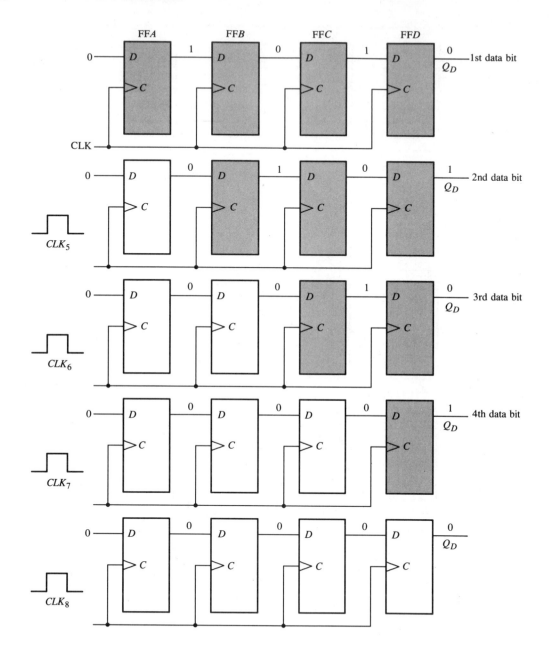

FIGURE 9–5 *Four bits (1010) being serially shifted out of the register.*

EXAMPLE 9–1   Show the states of the five-bit register in Figure 9–6(a) for the specified data input and clock waveforms. Assume that the register is initially cleared (all 0s).

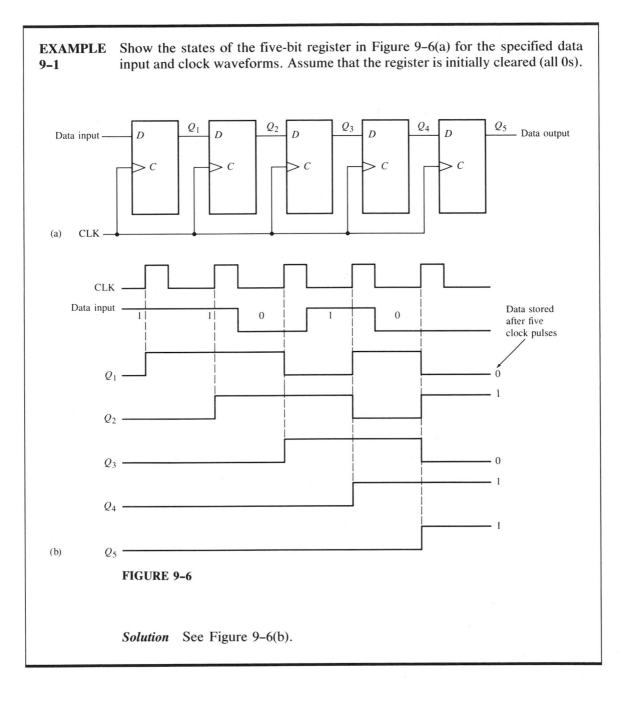

FIGURE 9–6

Solution   See Figure 9–6(b).

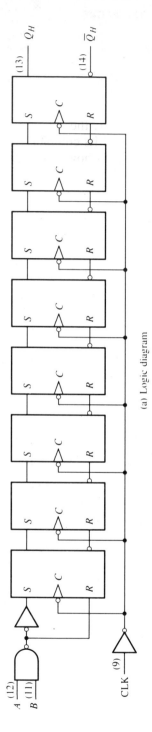

(a) Logic diagram

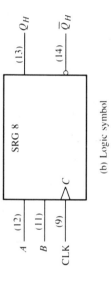

(b) Logic symbol

FIGURE 9-7 *The 7491A eight-bit serial in–serial out shift register.*

The 7491A Eight-Bit Shift Register

The 7491A is an example of an IC serial in–serial out shift register. The logic diagram is shown in Figure 9–7(a). As you can see, S-R flip-flops are used to implement this device. There are two gated data-input lines, A and B, for serial data entry. When data are entered on A, the B input must be HIGH, and vice versa. The serial data output is Q_H, and its complement is $\overline{Q_H}$.

A traditional logic block symbol is shown in Figure 9–7(b). The "SRG 8" designation means a shift register (SRG) with an eight-bit capacity.

SECTION REVIEW 9–2

1. Draw the logic diagram for the register in Figure 9–3 using J-K flip-flops to replace the D flip-flops.
2. How many clock pulses are required to serially enter a byte of data into an eight-bit shift register?

9–3   SERIAL IN–PARALLEL OUT SHIFT REGISTERS

Data bits are entered into this type of register in the same manner as discussed in the last section, that is, serially. The difference is the way in which the data bits are taken out of the register; in the parallel output register, the output of each stage is available. Once the data are stored, each bit appears on its respective output line and all bits are available simultaneously, rather than on a bit-by-bit

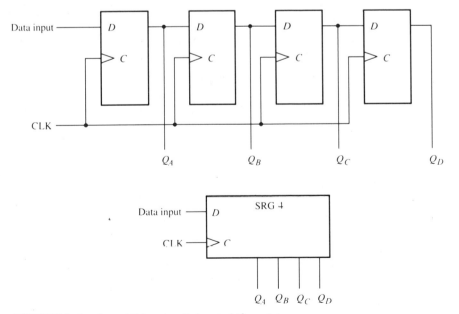

FIGURE 9–8 *A serial in–parallel out shift register.*

basis as with the serial output. Figure 9–8 shows a four-bit serial in–parallel out register and its logic block symbol.

EXAMPLE 9–2   Show the states of the four-bit register for the data input and clock waveforms in Figure 9–9(a). The register initially contains all 1s.

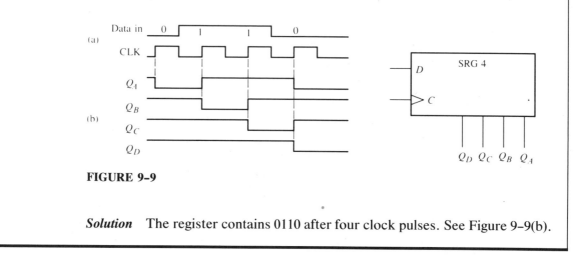

FIGURE 9–9

Solution   The register contains 0110 after four clock pulses. See Figure 9–9(b).

The 74164 Eight-Bit Serial In–Parallel Out Shift Register

The 74164 is an example of an IC shift register having serial in–parallel out operation. The logic diagram is shown in Figure 9–10(a), and a typical logic block symbol is shown in part (b). Notice that this device has two gated serial inputs, A and B, and a clear ($\overline{CLR}$) input that is active-LOW. The parallel outputs are Q_A through Q_H.

An example timing diagram for the 74164 is shown in Figure 9–11. Notice how the serial input data on A are shifted into and through the register after B goes HIGH.

SECTION REVIEW 9–3

1. The binary number 1101 is serially entered (right-most bit first) into a four-bit parallel out shift register that is initially clear. What are the Q outputs after two clock pulses?

2. How can a serial in–parallel out register be used as a serial in–serial out register?

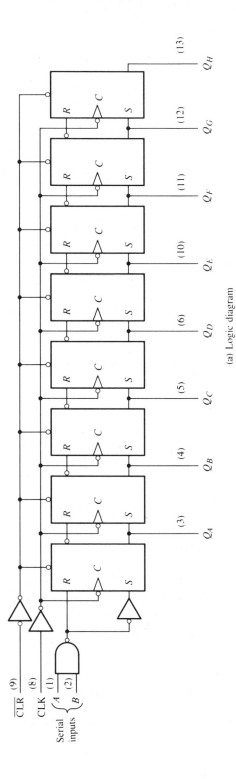

(a) Logic diagram

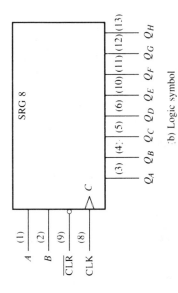

(b) Logic symbol

FIGURE 9–10 *The 74164 eight-bit serial in-parallel out shift register.*

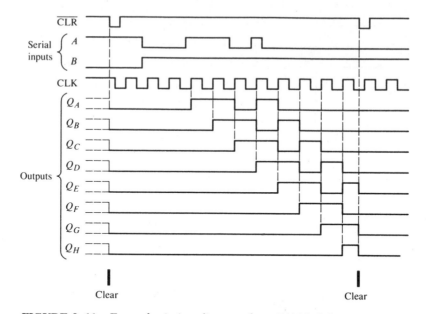

FIGURE 9–11   *Example timing diagram for a 74164 shift register.*

9–4   PARALLEL IN–SERIAL OUT SHIFT REGISTERS

For a register with parallel data inputs, the bits are entered simultaneously into their respective stages on parallel lines rather than on a bit-by-bit basis on one line as with serial data inputs. The serial output is executed as described in Section 9–2 once the data are completely stored in the register.

Figure 9–12 illustrates a four-bit parallel in–serial out register. Notice that there are four data-input lines, *A, B, C,* and *D,* and a SHIFT/$\overline{\text{LOAD}}$ input that allows <u>four bits</u> of data to be entered in parallel into the register. When SHIFT/$\overline{\text{LOAD}}$ is LOW, gates G_1 through G_3 are enabled, allowing each data bit to be applied to the *D* input of its respective flip-flop. When a clock pulse is applied, the flip-flops with $D = 1$ will SET and those with $D = 0$ will RESET, thereby storing all four bits simultaneously.

When SHIFT/$\overline{\text{LOAD}}$ is HIGH, gates G_1 through G_3 are disabled and gates G_4 through G_6 are enabled, allowing the data bits to shift right from one stage to the next. The OR gates allow either the normal shifting operation or the parallel data-entry operation, depending on which AND gates are enabled by the level on the SHIFT/$\overline{\text{LOAD}}$ input.

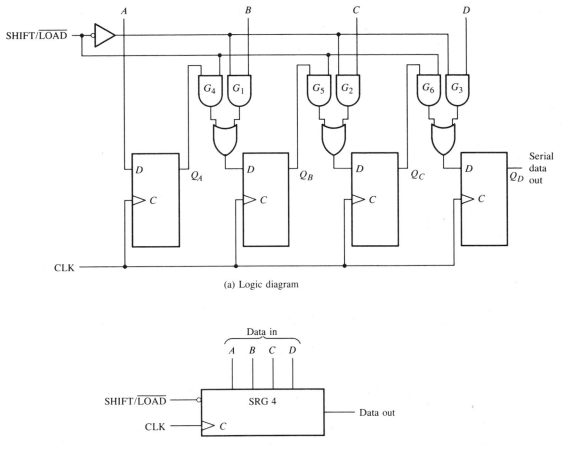

(a) Logic diagram

(b) Logic symbol

FIGURE 9–12 *A four-bit parallel in–serial out shift register.*

EXAMPLE 9–3

Show the data-output waveform for a four-bit register with the parallel input data and clock waveform given in Figure 9–13(a) on p. 424. Refer to Figure 9–12 for the logic diagram.

Solution On clock pulse 1, the parallel data (1010) are loaded into the register, making Q_D a 0. On clock pulse 2, the 1 from Q_C is shifted onto Q_D; on clock pulse 3, the 0 is shifted onto Q_D; on clock pulse 4, the next 1 is shifted onto Q_D; and on clock pulses 5 and 6, all data bits have been shifted out and only 0s remain in the register because no new data have been entered. See Figure 9–13(b).

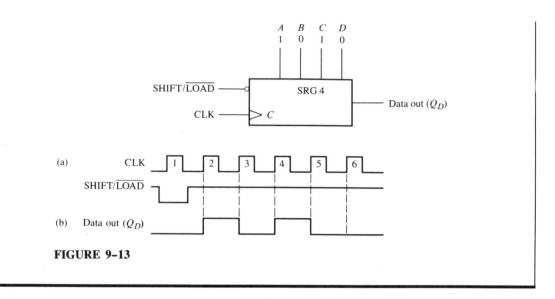

FIGURE 9–13

The 74165 Eight-Bit Parallel Load Shift Register

The 74165 is an example of an IC shift register that has a parallel in–serial out operation (it can also be operated as serial in–serial out). Figure 9–14 shows the internal logic diagram for this device and a typical logic block symbol.

A LOW on the SHIFT/LOAD (SH/LD) input enables all the NAND gates for parallel loading. When an input data bit is a 1, the flip-flop is asynchronously SET by a LOW out of the upper gate. When an input data bit is a 0, the flip-flop is asynchronously RESET by a LOW out of the lower gate. The clock is inhibited during parallel loading. A HIGH on the SH/LD input enables the clock, causing the data in the register to shift right.

Additionally, data can be entered serially on the SER input. Also, the clock can be inhibited anytime with a HIGH on the CLK INH input. The serial data outputs of the register are Q_H and its complement, $\overline{Q_H}$.

This implementation is different from the synchronous method of parallel loading previously discussed, demonstrating that there are usually several ways to accomplish the same function.

Figure 9–15 is a timing diagram showing an example of the operation of a 74165 shift register.

**SECTION
REVIEW
9–4**

1. Explain the function of the SHIFT/LOAD input.
2. Is the parallel load operation in a 74165 shift register synchronous or asynchronous? What does this mean?

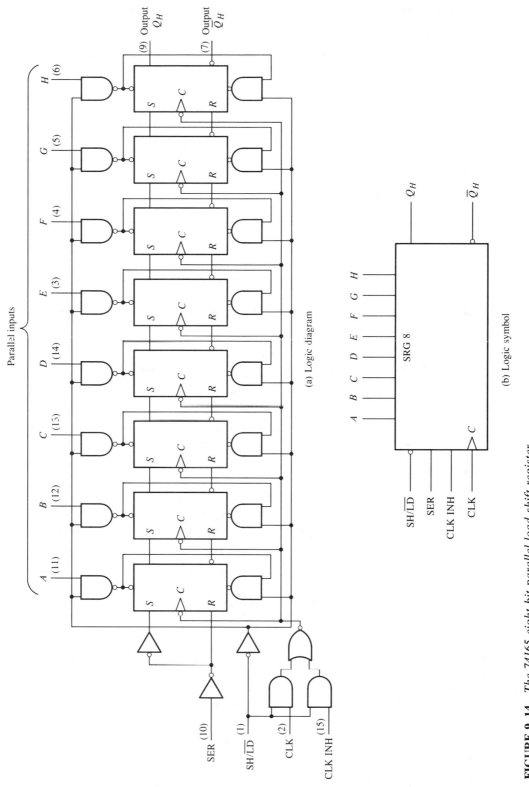

FIGURE 9-14 *The 74165 eight-bit parallel load shift register.*

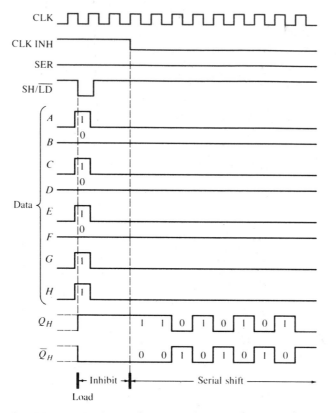

FIGURE 9–15 *Example timing diagram for a 74165 shift register.*

9–5   PARALLEL IN–PARALLEL OUT REGISTERS

Parallel entry of data was described in Section 9–4, and parallel output of data was also previously discussed. The parallel in–parallel out register employs both methods: Immediately following the simultaneous entry of all data bits, the bits appear on the parallel outputs. This type of register is shown in Figure 9–16.

The 74195 Four-Bit Parallel Access Shift Register

This device can be used for parallel in–parallel out operation. Since it also has a serial input, it can be used for serial in–serial out and serial in–parallel out operation. It can be used for parallel in–serial out operation by using Q_D as the output. A typical logic block symbol is shown in Figure 9–17.

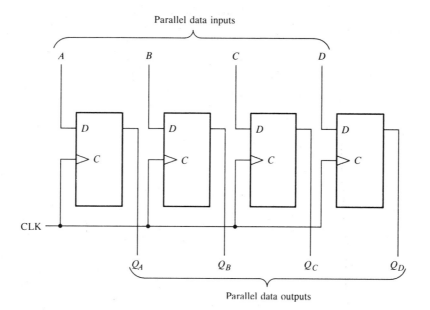

FIGURE 9-16 *A parallel in-parallel out register.*

FIGURE 9-17 *The 74195 four-bit parallel access shift register.*

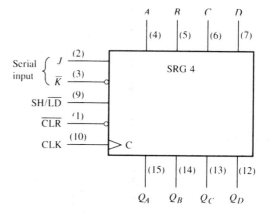

When the SHIFT/LOAD (SH/$\overline{\text{LD}}$) input is LOW, the data on the parallel inputs are entered synchronously on the positive transition of the clock. When the SH/$\overline{\text{LD}}$ input is HIGH, stored data will shift right (Q_A to Q_D) synchronously with the clock. J and $\overline{K}$ are the serial data inputs to the first stage of the register (Q_A); Q_D can be used for serial output data. The active-LOW clear is asynchronous.

The timing diagram in Figure 9–18 illustrates the operation of this register.

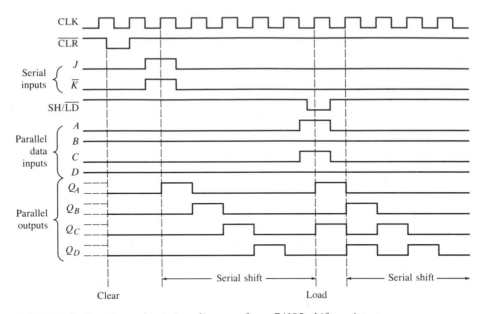

FIGURE 9–18 *Example timing diagram for a 74195 shift register.*

SECTION REVIEW 9–5

1. In Figure 9–16, $A = 1$, $B = 0$, $C = 0$, and $D = 1$. After three clock pulses, what are the data outputs?
2. For a 74195, $SH/\overline{LD} = 1$, $J = 1$, and $\overline{K} = 1$. What is Q_A after one clock pulse?

9–6   BIDIRECTIONAL SHIFT REGISTERS

A *bidirectional* shift register is one in which the data can be shifted either left or right. It can be implemented by using gating logic that enables the transfer of a data bit from one stage to the next stage to the right or to the left, depending on the level of a control line. A four-bit implementation is shown in Figure 9–19 for illustration. A HIGH on the RIGHT/$\overline{LEFT}$ control input allows data to be shifted to the right, and a LOW enables a left shift of data. An examination of the gating logic should make the operation apparent. When the RIGHT/$\overline{LEFT}$ control is HIGH, gates G_1 through G_4 are enabled, and the state of the Q output of each flip-flop is passed through to the D input of the *following* flip-flop. When a clock pulse occurs, the data are then effectively shifted one place to the *right*. When the RIGHT/$\overline{LEFT}$ control is LOW, gates G_5 through G_8 are enabled, and the Q output of each flip-flop is passed through to the D input of the *preceding* flip-flop. When a clock pulse occurs, the data are then effectively shifted one place to the *left*.

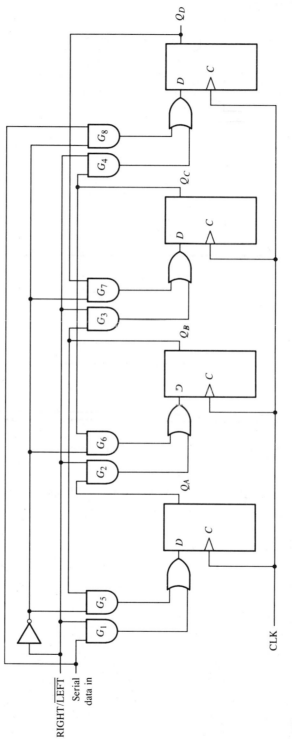

FIGURE 9-19 *Four-bit bidirectional shift register.*

429

EXAMPLE 9-4   Determine the state of the shift register of Figure 9–19 after each clock pulse for the given RIGHT/$\overline{\text{LEFT}}$ control input waveform in Figure 9–20(a). Assume that $Q_A = 1$, $Q_B = 1$, $Q_C = 0$, and $Q_D = 1$, and the *serial data-in* line is LOW.

Solution   See Figure 9–20(b).

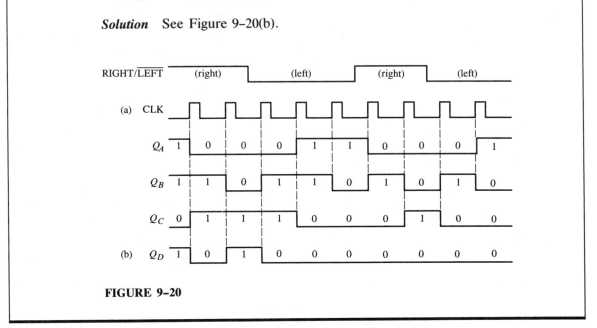

FIGURE 9-20

The 74194 Four-Bit Bidirectional Universal Shift Register

The 74194 is an example of a bidirectional shift register in integrated circuit form. A logic block symbol is shown in Figure 9–21.

FIGURE 9-21   *The 74194 four-bit bidirectional universal shift register.*

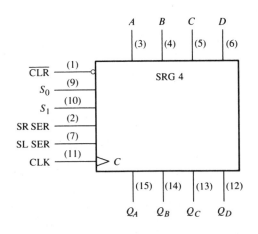

Parallel loading, which is synchronous with a positive transition of the clock, is accomplished by applying the four bits of data to the parallel inputs and a HIGH to the S_0 and S_1 inputs.

Shift right is accomplished synchronously with the positive edge of the clock when S_0 is HIGH and S_1 is LOW. Serial data in this mode are entered at the *shift-right serial input* (SR SER). When S_0 is LOW and S_1 is HIGH, data bits shift left synchronously with the clock, and new data are entered at the *shift-left serial input* (SL SER). An example timing diagram is shown in Figure 9–22. SR SER goes into the Q_A stage and SL SER goes into the Q_D stage.

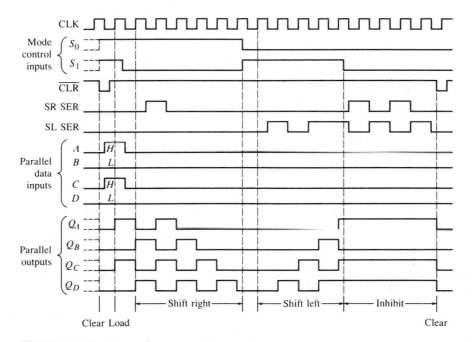

FIGURE 9-22 *Example timing diagram for a 74194 shift register.*

SECTION REVIEW 9-6

1. Assume that the bidirectional shift register in Figure 9–19 has the following contents: $Q_A = 1$, $Q_B = 1$, $Q_C = 0$, and $Q_D = 0$. There is a 1 on the serial data-in line. If RIGHT/$\overline{\text{LEFT}}$ is HIGH for three clock pulses and LOW for two more clock pulses, what are the contents after the fifth clock pulse?

9-7 SHIFT REGISTER COUNTERS

A shift register counter is basically a shift register with the serial output connected back to the serial input in order to produce special sequences. These devices are often classified as counters because they exhibit a specified sequence of states. Two of the most common types of shift register counters are introduced in this section: the Johnson counter and the ring counter.

The Johnson Counter

In a Johnson counter, the complement of the output of the last flip-flop is connected back to the D input of the first flip-flop (it can be implemented with S-R or J-K flip-flops as well). This *feedback* arrangement produces a unique sequence of states as shown in Table 9–1 for a four-bit device and in Table 9–2 for a five-bit device. Notice that the four-bit sequence has a total of *eight* states and that the five-bit sequence has a total of *ten* states. In general, an n-stage Johnson counter will produce a modulus of $2n$, where n is the number of stages in the counter.

The implementations for the four- and five-stage Johnson counters are shown in Figure 9–23. The implementation of a Johnson counter is very straightforward and is the same regardless of the number of stages. The Q output of each stage is connected to the D input of the next stage (assuming that D flip-flops are used). The single exception is that the $\overline{Q}$ output of the last stage is connected back to the D input of the first stage. As the sequences in Tables 9–1 and 9–2 show, the

TABLE 9–1 *Four-bit Johnson sequence.*

Clock Pulse	Q_A	Q_B	Q_C	Q_D
0	0	0	0	0
1	1	0	0	0
2	1	1	0	0
3	1	1	1	0
4	1	1	1	1
5	0	1	1	1
6	0	0	1	1
7	0	0	0	1

TABLE 9–2 *Five-bit Johnson sequence.*

Clock Pulse	Q_A	Q_B	Q_C	Q_D	Q_E
0	0	0	0	0	0
1	1	0	0	0	0
2	1	1	0	0	0
3	1	1	1	0	0
4	1	1	1	1	0
5	1	1	1	1	1
6	0	1	1	1	1
7	0	0	1	1	1
8	0	0	0	1	1
9	0	0	0	0	1

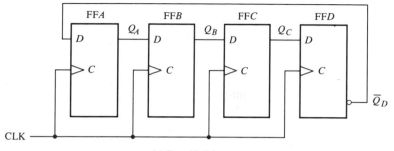

(a) Four-bit Johnson counter

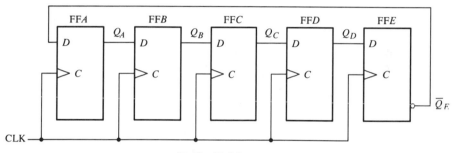

(b) Five-bit Johnson counter

FIGURE 9-23 *Johnson counters.*

counter will "fill up" with 1s from left to right, and then it will "fill up" with 0s again. One advantage of this type of sequence is that it is readily decoded with two-input AND gates.

Diagrams of the timing operations of both the four- and five-bit counters are shown in Figures 9-24 and 9-25, respectively.

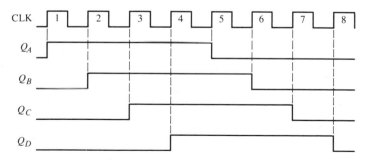

FIGURE 9-24 *Timing sequence for a four-bit Johnson counter.*

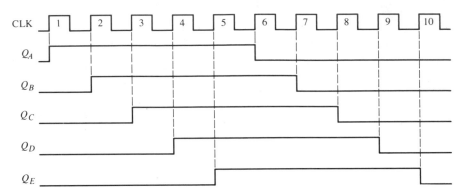

FIGURE 9–25   *Timing sequence for a five-bit Johnson counter.*

The Ring Counter

The ring counter utilizes one flip-flop for each state in its sequence, and it is therefore wasteful of flip-flops. It does have the advantage that decoding is not required for decimal conversion.

A logic diagram for a ten-bit ring counter is shown in Figure 9–26. The sequence for this ring counter is given in Table 9–3. Initially, a 1 is preset into the first flip-flop, and the rest of the flip-flops are cleared. Notice that the interstage connections are the same as those for a Johnson counter, except that Q rather than $\overline{Q}$ is fed back from the last stage. The ten outputs of the counter indicate directly the decimal count of the clock pulse. For instance, a 1 on Q_0 is a zero, a 1 on Q_1 is a one, a 1 on Q_2 is a two, a 1 on Q_3 is a three, and so on. You should verify for yourself that the 1 is always retained in the counter and simply shifted "around the ring," advancing one stage for each clock pulse.

Modified sequences can be achieved by having more than a single 1 in the counter, as illustrated in the following example.

TABLE 9–3   *Ring counter sequence (ten bits).*

Clock Pulse	Q_0	Q_1	Q_2	Q_3	Q_4	Q_5	Q_6	Q_7	Q_8	Q_9
0	1	0	0	0	0	0	0	0	0	0
1	0	1	0	0	0	0	0	0	0	0
2	0	0	1	0	0	0	0	0	0	0
3	0	0	0	1	0	0	0	0	0	0
4	0	0	0	0	1	0	0	0	0	0
5	0	0	0	0	0	1	0	0	0	0
6	0	0	0	0	0	0	1	0	0	0
7	0	0	0	0	0	0	0	1	0	0
8	0	0	0	0	0	0	0	0	1	0
9	0	0	0	0	0	0	0	0	0	1

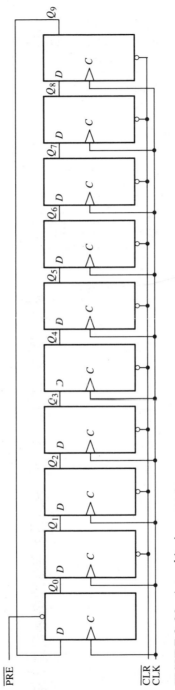

FIGURE 9-26 *A ten-bit ring counter.*

EXAMPLE 9–5   If a ten-bit ring counter similar to Figure 9–26 has the initial state 1010000000, determine the waveforms for each of the Q outputs.

Solution   See Figure 9–27.

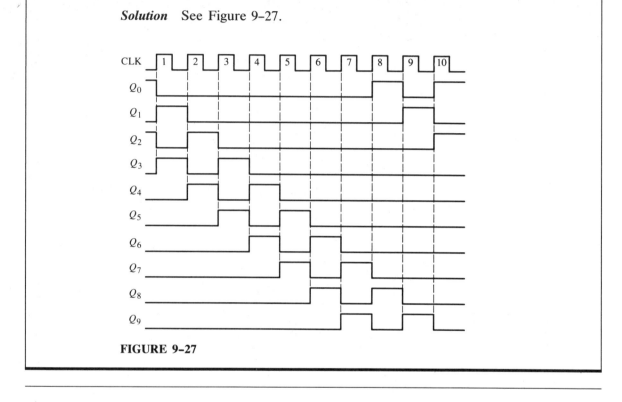

FIGURE 9–27

SECTION REVIEW 9–7

1. How many states are there in an eight-bit Johnson counter sequence?
2. Write the sequence of states for a three-bit Johnson counter starting with 000.

9–8   SHIFT REGISTER APPLICATIONS

Shift registers are found in an almost endless array of applications, a few of which are presented in this section.

Time Delay

The serial in–serial out shift register can be used to provide a time delay from input to output that is a function of both the number of stages (n) in the register and the clock frequency.

When a data pulse is applied to the serial input (*A* and *B* connected together), it enters the first stage on the triggering edge of the clock pulse. It is then shifted from stage to stage on each successive clock pulse until it appears on the serial output *n* clock periods later. This time-delay operation is illustrated in Figure 9–28 using a 7491A eight-bit shift register with a clock frequency of 1 MHz to achieve a time delay of 8 μs (8 × 1 μs). This time can be adjusted up or down by changing the clock frequency. The time delay can also be increased by cascading shift registers and decreased by taking the output from successively lower stages in the register, as illustrated in Example 9–6.

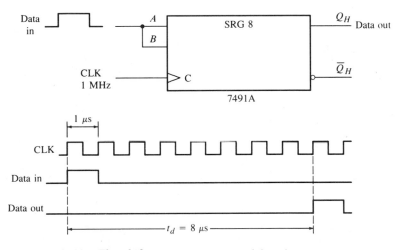

FIGURE 9–28 *The shift register as a time-delay device.*

EXAMPLE 9–6 Determine the amount of time delay between the serial input and each output in Figure 9–29. Show a timing diagram to illustrate.

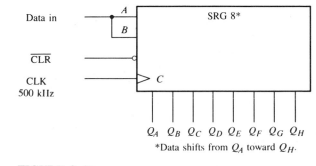

FIGURE 9–29

Solution The clock period is 2 μs. Thus, the time delay can be increased or decreased in 2-μs increments from a minimum of 2 μs to a maximum of 16 μs, as illustrated in Figure 9–30.

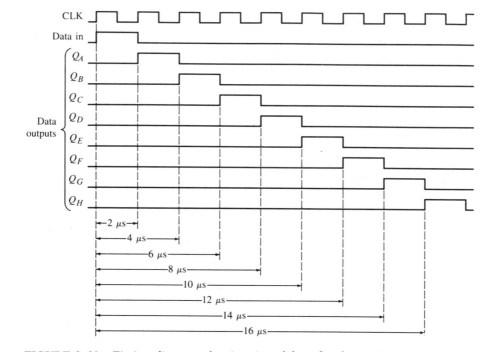

FIGURE 9–30 *Timing diagram showing time delays for the register in Figure 9–29.*

Ring Counter

If the output is connected back to the serial input, a shift register can be used as a ring counter. This application is illustrated in Figure 9–31 using a 74195 four-bit shift register.

Initially, a 1000_2 bit pattern (or any other pattern) can be synchronously preset into the counter by applying the bit pattern to the parallel data inputs and taking the SH/$\overline{\text{LD}}$ input LOW. After initialization, the 1 continues to circulate through the counter, as the timing diagram in Figure 9–32 shows.

Serial-to-Parallel Data Converter

Serial data transmission from one digital system to another is commonly used to reduce the number of wires in the transmission line. For example, eight bits can be sent serially over one wire, but it takes eight wires to send the same data in parallel.

FIGURE 9–31 *74195 connected as a ring counter.*

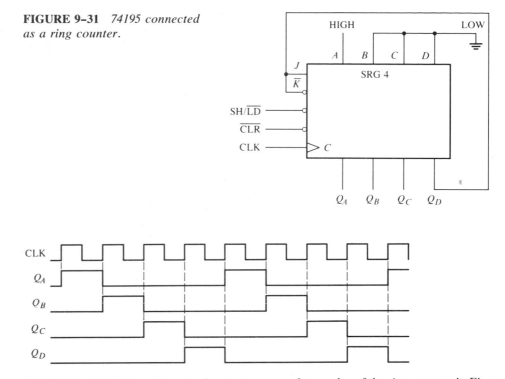

FIGURE 9–32 *Timing diagram showing two complete cycles of the ring counter in Figure 9–31 when it is initially preset to 1000_2.*

A computer or microprocessor-based system often requires incoming data to be in parallel format; thus the requirement for serial-to-parallel conversion.

A simplified serial-to-parallel data converter, in which two types of shift registers are used, is shown in Figure 9–33. In this case, use of specific devices is reserved as an end-of-chapter problem.

To illustrate the operation of this system, the serial data format shown in Figure 9–34 is used. It consists of eleven bits. The first bit (start bit) is always 0 and always begins with a HIGH-to-LOW transition. The next eight bits (D_0 through D_7) are the data bits (one of the bits can be parity), and the last two bits (stop bits) are always 1s. When no data are being sent, there is a continuous 1 on the serial data line.

The HIGH-to-LOW transition of the *start bit* SETS the *control flip-flop,* which enables the clock generator. After a fixed delay time, the clock generator begins producing a pulse waveform which is applied to the *data-input register* and to the *divide-by-8 counter.* The clock has a frequency precisely equal to that of the incoming serial data, and the first clock pulse after the start bit is coincident with the first data bit.

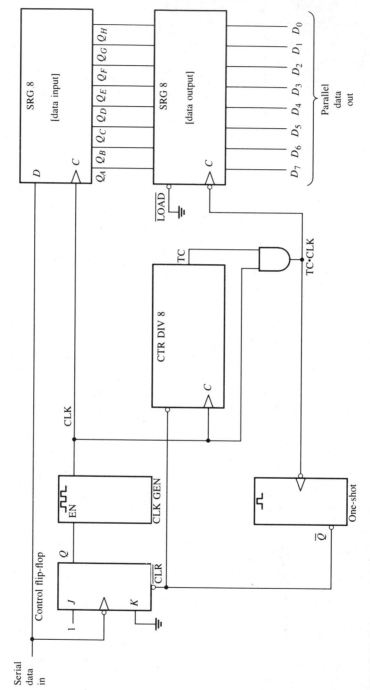

FIGURE 9–33 *Simplified logic diagram of a serial-to-parallel converter.*

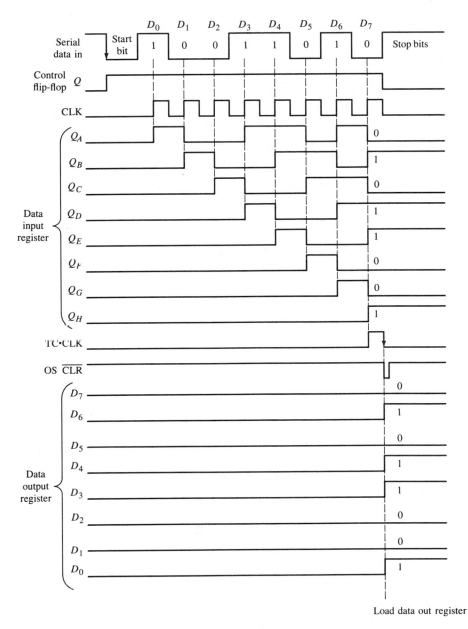

FIGURE 9–34 *Serial data format.*

FIGURE 9–35 *Timing diagram illustrating the operation of the serial-to-parallel data converter in Figure 9–33.*

The eight data bits (D_0 through D_7) are serially shifted into the data-input register. After the eighth clock pulse, a HIGH-to-LOW transition of the *terminal count* (TC) output of the counter ANDed with the clock loads the eight bits that are in the data-input register into the data-output register. This same transition also triggers the one-shot (OS), which produces a short-duration pulse to clear the counter and RESET the control flip-flop and thus disable the clock generator. The system is now ready for the next group of eleven bits, and it waits for the next HIGH-to-LOW transition at the beginning of the start bit. The timing diagram in Figure 9–35 (p. 441) illustrates the basic operation just discussed.

By basically reversing the process just covered, parallel-to-serial data conversion can be accomplished. However, since the serial data format must be produced, additional requirements must be taken into consideration. This is left as a challenging problem at the end of the chapter.

Universal Asynchronous Receiver Transmitter (UART)

As mentioned, computers and microprocessor-based systems often send and receive data in a parallel format. Frequently, these systems must communicate with external devices that send and/or receive *serial* data. An interfacing device used to accomplish these conversions is the UART. Figure 9–36 illustrates the UART in a general microprocessor-based system application.

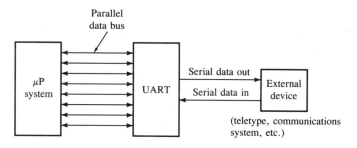

FIGURE 9–36   *UART interface.*

A UART includes a serial-to-parallel data converter such as we have discussed and a parallel-to-serial converter as shown in Figure 9–37. The *data bus* is basically a set of parallel conductors along which data move between the UART and the microprocessor system. The buffers interface the data registers with the data bus.

The UART receives data in serial format, converts the data to parallel format, and places them on the data bus. The UART also accepts parallel data from the data bus, converts the data to serial format, and transmits them to an external device.

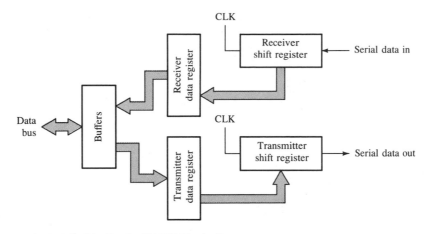

FIGURE 9–37 *Basic UART block diagram.*

Keyboard Encoder

The keyboard encoder is a good example of the application of a shift register used as a ring counter in conjunction with other devices. Recall that a simplified calculator keyboard encoder without data storage was presented in Chapter 6.

Figure 9–38 shows a simplified keyboard encoder for encoding a key closure in 64-key matrix organized in eight rows and eight columns. A 74199 eight-bit universal shift register is connected as a ring counter (Q_H to J, $\overline{K}$) with a fixed bit pattern of seven 1s and one 0 preset into it when the power is turned on. Two 74147 priority encoders (introduced in Chapter 6) are used as eight-line-to-three-line encoders (9 input HIGH, D output unused) to encode the ROW and COLUMN lines of the keyboard matrix. Another 74199 shift register is used as a parallel in–parallel out register into which the ROW/COLUMN code from the priority encoders is stored.

The basic operation of the keyboard encoder in Figure 9–38 is as follows: The *ring counter* "scans" the rows for a key closure as the clock signal shifts the 0 around the counter at a 5-kHz rate.

The 0 (LOW) is sequentially applied to each ROW line, while all other ROW lines are HIGH. All the ROW lines are connected to the *ROW encoder* inputs, so the three-bit output of the ROW encoder at any time is the binary representation of the ROW line that is LOW.

When there is a key closure, one COLUMN line is connected to one ROW line. When the ROW line is taken LOW by the ring counter, that particular COLUMN line is also pulled LOW. The *COLUMN encoder* produces a binary output corresponding to the COLUMN in which the key is closed. The three-bit ROW code plus the three-bit COLUMN code uniquely identifies the key that is closed. This six-bit code is applied to the inputs of the *key code register*. When a key is closed, the two one-shots produce a delayed clock pulse to parallel load the six-bit code into the key code register. This allows the contact bounce to die out. Also, a one-shot output inhibits the ring counter to prevent it from scanning while the data are being loaded into the key code register.

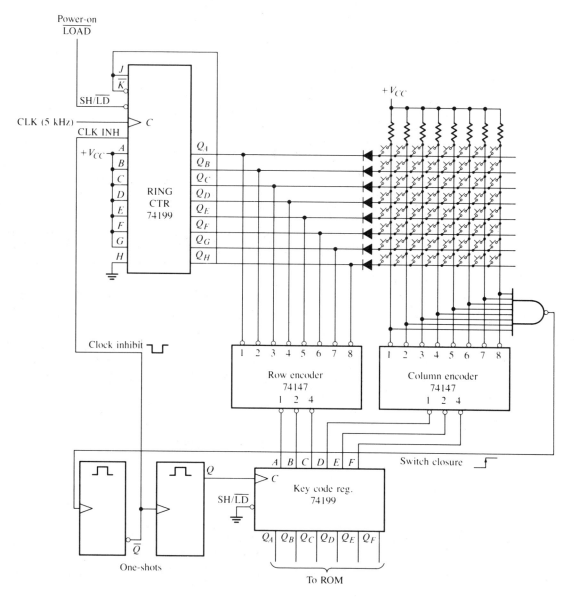

FIGURE 9–38 *Simplified keyboard encoding circuit.*

The six-bit code in the key code register is now applied to a ROM (read only memory) to be converted to ASCII or some other appropriate alphanumeric code that identifies the keyboard character. ROMs are studied in the next chapter.

SECTION REVIEW 9-8
1. In the keyboard encoder, how many times per second does the ring counter scan the keyboard?
2. What is the six-bit ROW/COLUMN code for the top row and the left-most column in the keyboard encoder?
3. What is the purpose of the diodes in the keyboard encoder? What is the purpose of the resistors?

9-9   TROUBLESHOOTING

Another basic method of troubleshooting sequential logic and other more complex digital systems uses a procedure of "exercising" the circuit under test with a known input waveform (stimulus) and then observing the output for the proper bit pattern.

The serial-to-parallel data converter in Figure 9–33 is used to illustrate this procedure. The main objective in "exercising" the circuit is to force all elements (flip-flops and gates) into all of their states to be certain that nothing is "stuck" in a given state. The *input test pattern,* in this case, must be designed to force each flip-flop in the registers into both states, clock the counter through all of its eight states, and take the control flip-flop, clock generator, one-shot, and gate through their paces.

The input test pattern that accomplishes this objective for the serial-to-parallel data converter is based on the serial data format in Figure 9–34. It consists of the pattern 10101010 in one group of data bits followed by 01010101 in the next group, as shown in Figure 9–39. These patterns are generated on a repetitive basis by a special test circuit. The basic test setup is shown in Figure 9–40.

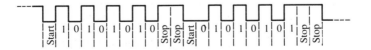

FIGURE 9-39   *Example test pattern.*

After both patterns have been run through the circuit under test, each flip-flop in the data-input register and in the data-output register has resided in both SET and RESET states; the counter has gone through its sequence (once for each bit pattern); and all the other devices have been exercised.

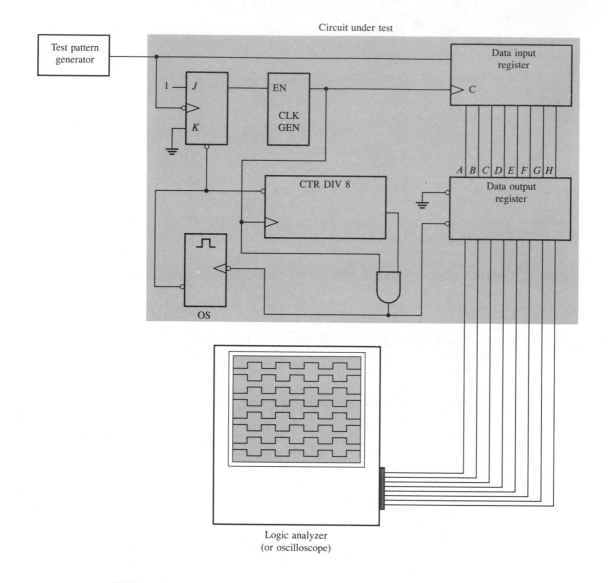

FIGURE 9–40 *Basic test setup for the serial-to-parallel data converter of Figure 9–33.*

To check for proper operation, each of the parallel data outputs is observed for an alternating pattern of 1s and 0s as the input test patterns are repetitively shifted into the data-input register and then loaded into the data-output register. The proper timing diagram is shown in Figure 9–41. Each output can be observed individually or in pairs with a dual-trace oscilloscope, or all eight outputs can be observed simultaneously with a logic analyzer configured for timing analysis.

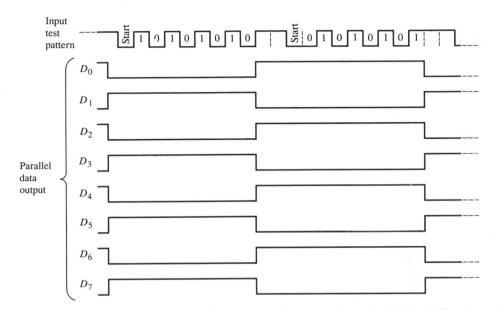

FIGURE 9–41 *Proper outputs for the circuit under test in Figure 9 40. Two full cycles of the input test pattern are shown.*

If one or more outputs of the data-output register are incorrect, then we must back up to the outputs of the data-input register. If these outputs are correct, then probably the data-output register is defective. If the data-input register outputs are also incorrect, the fault could be with the register itself or with any of the other logic, and additional investigation is necessary to isolate the problem.

Signature Analysis

Signature analysis is another effective troubleshooting method which is based on the comparison of measured bit patterns (called *signatures*) with documented signatures at various test points (*nodes*) in a system that is being tested.

Figure 9–42 shows a simplified model of a *signature analyzer* with a sixteen-bit internal signature register that includes four *feedback* tap positions. As illustrated, the signature analyzer *compresses* a node's waveform (bit pattern) into a four-digit signature display.

Table 9–4 shows a typical sixteen-character display set used for signature analysis. In this case, the standard hexadecimal characters are changed to eliminate possible confusion between the lower-case *b* and the digit 6, the capital *B* and the digit 8, and so on.

During troubleshooting, measured signatures are compared against documented ones that appear on the schematic diagram or on a test point list for the system under test. There is no diagnostic information in the signature itself; thus a signature display is useless without a known signature with which to compare it.

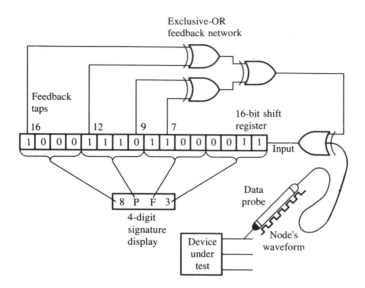

FIGURE 9–42 *Simplified signature analyzer.*

A typical signature analyzer is shown in Figure 9–43. The *measurement cycle* of the analyzer is controlled by an internal state called the *gate*. When the gate is open, the analyzer is taking a new signature measurement through the *data-probe input*. When the gate closes, the analyzer stops taking the new signature measurement and displays it, replacing the previous measurement. The

TABLE 9–4 *Signature display character set.*

Four-Bit Group	Displayed Character
0000	0
0001	1
0010	2
0011	3
0100	4
0101	5
0110	6
0111	7
1000	8
1001	9
1010	A
1011	C
1100	F
1101	H
1110	P
1111	U

FIGURE 9–43 *A typical signature analyzer. (Courtesy of Hewlett-Packard)*

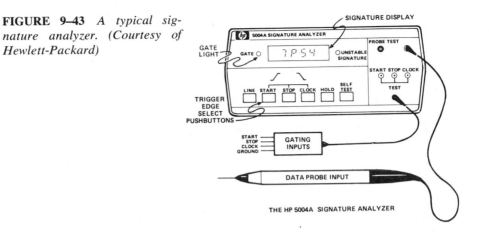

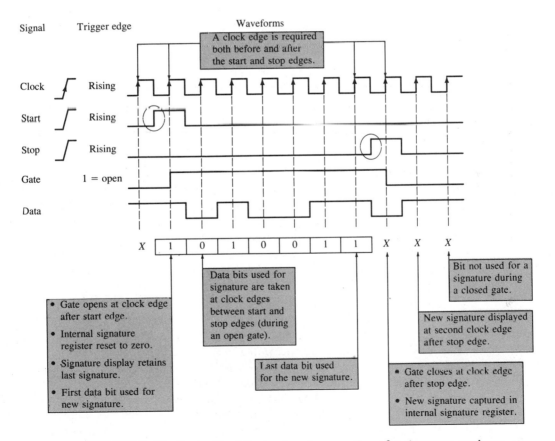

FIGURE 9–44 *Example of the basic gate operation of a signature analyzer.*

gate lamp on the front panel is *on* during a measurement cycle when the gate is open and is *off* when the gate is closed.

The analyzer gate is controlled by three inputs from the system or device under test: *start, stop,* and *clock.* Basic gate control for a signature measurement cycle is shown in Figure 9–44 (p. 449). The start and stop inputs are used to open and close the gate at predetermined triggering edges, which are selected by front panel controls. The clock synchronizes the analyzer to the system or device being tested. The logic level of the waveform (bit pattern) measured by the data-probe input is sampled only at the clock pulse edges. These logic levels are used as data bits that are compressed into the signature. The gate opens at the clock edge following the selected start edge and closes at the clock edge following the selected stop edge, as illustrated in Figure 9–44.

<table>
<tr><td>

**SECTION
REVIEW
9–9**

</td><td>

1. What is the purpose of providing a test input to a sequential logic circuit?
2. Generally, when an output waveform is found to be incorrect, what is the next step to be taken?
3. Define *signature analysis.*

</td></tr>
</table>

9–10   LOGIC SYMBOLS WITH DEPENDENCY NOTATION

Two examples of ANSI/IEEE Std. 91–1984 symbols with dependency notation for shift registers are presented in this section. This material can be treated as optional.

The logic symbol for a 74164 eight-bit parallel out serial shift register is shown in Figure 9–45. The common control inputs are shown on the notched block. The clear ($\overline{\text{CLR}}$) input is indicated by an R (for RESET) inside the block. Since there is no dependency prefix to link R with the clock ($C1$), the clear function is asynchronous. The right arrow symbol after $C1$ indicates data flow from Q_A to Q_H. The A and B inputs are ANDed, as indicated by the imbedded AND symbol, to provide the synchronous data input, $1D$ to the first stage (Q_A). Note the dependency of D on C, as indicated by the 1 suffix on C and the 1 prefix on D.

Figure 9–46 is the logic symbol for the 74194 four-bit bidirectional universal shift register. Starting at the top left side of the control block, note that the $\overline{\text{CLR}}$ input is active-LOW and is asynchronous (no prefix link with C). S_0 and S_1 are *mode* inputs that determine the *shift-right,* the *shift-left,* and the *parallel load* modes of operation as indicated by the $\frac{0}{3}$ dependency designation following the M. The $\frac{0}{3}$ represents the binary states of 0, 1, 2, and 3 on the S_0 and S_1 inputs. When one of these digits is used as a prefix for another input, a dependency is established. The "$1\rightarrow/2\leftarrow$" symbol on the clock input indicates the following: "$1\rightarrow$" indicates that a right shift (Q_A toward Q_D) occurs when the mode inputs (S_0, S_1) are in the binary 1 state ($S_0 = 1$, $S_1 = 0$). "$2\leftarrow$" indicates that a left shift (Q_D

toward Q_A) occurs when the mode inputs are in the binary 2 state ($S_0 = 0$, $S_1 = 1$). The shift-right serial input (SR SER) is both mode- and clock-dependent as indicated by $1,4D$. The parallel inputs (A, B, C, and D) are all mode-dependent (prefix 3 indicates parallel load mode) and clock-dependent as indicated by $3,4D$.

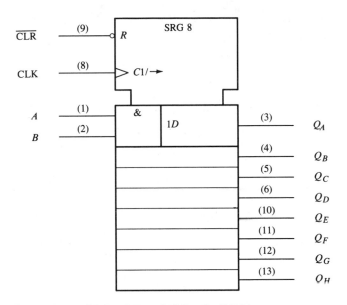

FIGURE 9-45 *Logic symbol for the 74164.*

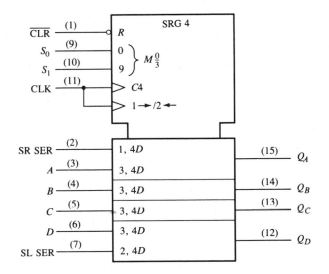

FIGURE 9-46 *Logic symbol for the 74194.*

The shift-left serial input (SL SER) is both mode- and clock-dependent as indicated by 2,4D.

The four modes for this device are summarized below:

Do nothing: $S_0 = 0, S_1 = 0$ (Mode 0)
Shift right: $S_0 = 1, S_1 = 0$ (Mode 1, as in 1,4D)
Shift left: $S_0 = 0, S_1 = 1$ (Mode 2, as in 2,4D)
Parallel load: $S_0 = 1, S_1 = 1$ (Mode 3, as in 3,4D)

**SECTION
REVIEW
9–10**

1. In Figure 9–46, are there any inputs that are dependent on the mode inputs being in the 0 state?
2. Is the parallel load synchronous with the clock?

SUMMARY

☐ Shift registers are basically storage devices in which data can be moved (shifted) in or out in several ways.
☐ The basic types of shift registers in terms of data movement are

1. Serial in–serial out
2. Serial in–parallel out
3. Parallel in–serial out
4. Parallel in–parallel out

☐ A bidirectional shift register is one in which data can be moved internally in either direction.
☐ A universal shift register is one that has both serial and parallel inputs and outputs.
☐ Shift register counters are shift registers with feedback that exhibit special sequences. Examples are the Johnson counter and the ring counter.
☐ The Johnson counter has 2n states in its sequence, where n is the number of stages.
☐ The ring counter has n states in its sequence.

SELF-TEST

1. Explain how a flip-flop can store a data bit.
2. How many clock pulses are required to shift a byte of data into and out of an eight-bit serial in–serial out shift register?
3. Repeat Problem 2 for an eight-bit serial in–parallel out shift register.
4. Draw a logic diagram for the register in Figure 9–3 using S-R flip-flops to replace the D flip-flops.
5. Show how the shift register in Figure 9–8 can be used as a three-bit serial in–serial out shift register.

6. The binary number 10110101 is serially shifted into an eight-bit parallel out shift register that has an initial content of 11100100. What are the Q outputs after two clock pulses? After four clock pulses? After eight clock pulses?

7. An eight-bit bidirectional shift register contains 10111011. Zeros are applied to the shift-right and shift-left serial data inputs. The shift register is in the shift-right mode for three clock pulses, then in the shift-left mode for six clock pulses, then in the shift-right mode for two clock pulses. What is the final state of the register?

8. How many flip-flops are required to implement a divide-by-10 Johnson counter?

9. What is the basic difference between a Johnson counter and a ring counter?

10. Write the sequence of states for a four-bit Johnson counter.

11. Write the sequence of states for an eight-bit ring counter with 1s in the first and fourth stages and 0s in the rest.

12. In Figure 9–28, what clock frequency is required for a 24-μs delay?

13. Draw the timing diagram for the ring counter in Figure 9–32 if 1s are preset in A and C and 0s in B and D.

14. What is the purpose of a UART?

15. Specify a type of one-shot, by device number, that can be used in the keyboard encoder of Figure 9–38.

16. Explain the purpose of the ring counter in Figure 9–38.

PROBLEMS

Section 9–1

9–1 Why are shift registers considered to be basic memory devices?

9–2 What is the storage capacity of a register that can retain two bytes of data?

Section 9–2

9–3 For the data input and clock in Figure 9–47, determine the states of each flip-flop in the shift register of Figure 9–3, and draw the Q waveforms. Assume that the register contains all 1s initially.

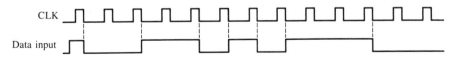

FIGURE 9–47

9–4 Repeat Problem 9–3 for the waveforms in Figure 9–48.

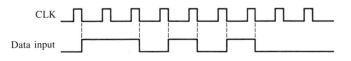

FIGURE 9–48

9–5   What is the state of the register in Figure 9–49 after each clock pulse if it starts in the 101001111000 state?

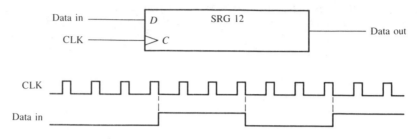

FIGURE 9–49

9–6   For the serial in–serial out shift register, determine the data-output waveform for the data-input and clock waveforms in Figure 9–50. Assume that the register is initially cleared.

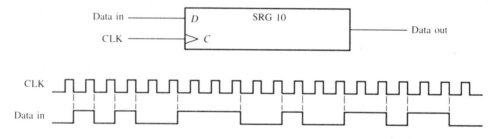

FIGURE 9–50

9–7   Repeat Problem 9–6 for the waveforms in Figure 9–51.

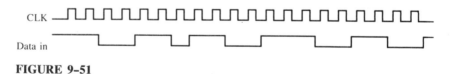

FIGURE 9–51

9–8   The data-output waveform in Figure 9–52 is related to the clock as indicated. What binary number is stored in the register if the first bit out is the LSB?

FIGURE 9–52
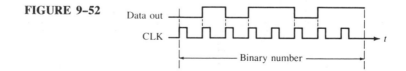

Section 9–3

9–9 Draw a complete timing diagram showing the parallel outputs for the shift register in Figure 9–8. Use the inputs in Figure 9–50 with the register initially clear.

9–10 Repeat Problem 9–9 for the input waveforms in Figure 9–51.

9–11 Sketch the Q_A through Q_H outputs for a 74164 shift register with the input waveforms shown in Figure 9–53.

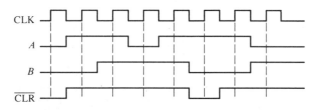

FIGURE 9–53

Section 9–4

9–12 The shift register in Figure 9–54(a) has SHIFT/$\overline{\text{LOAD}}$ and CLK inputs as shown in part (b). The serial data input (SER) is a 0. The parallel data inputs are $A = 1$, $B = 0$, $C = 1$, $D = 0$. Sketch the data-output waveform in relation to the inputs.

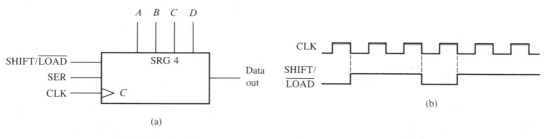

(a)

(b)

FIGURE 9–54

9–13 The waveforms in Figure 9–55 are applied to a 74165 shift register. The parallel inputs are all 0. Determine the Q_H waveform.

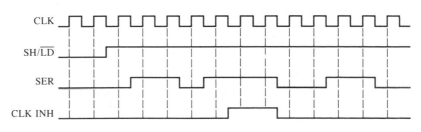

FIGURE 9–55

9–14   Repeat Problem 9–13 if the parallel inputs are all 1.

9–15   Repeat Problem 9–13 if the SER input is inverted.

Section 9–5

9–16   Determine all of the Q output waveforms for a 74195 four-bit shift register when the inputs are as shown in Figure 9–56.

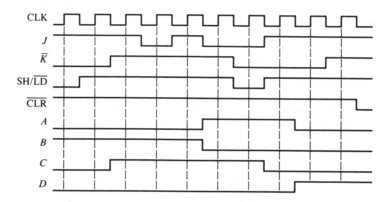

FIGURE 9–56

9–17   Repeat Problem 9–16 if the SH/$\overline{\text{LD}}$ input is inverted and the register is initially clear.

9–18   Use two 74195 shift registers to form an eight-bit shift register. Show the required connections.

Section 9–6

9–19   For the eight-bit bidirectional register in Figure 9–57, determine the state of the register after each clock pulse for the right/left control waveform given. A HIGH on this input enables a shift to the right, and a LOW enables a shift to the left. Assume that the register is initially storing a "76" in binary, with the right-most position being the LSB. "Data in" is LOW.

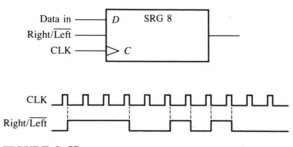

FIGURE 9–57

9–20 Repeat Problem 9–19 for the waveforms in Figure 9–58.

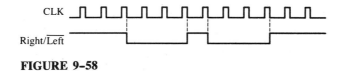

FIGURE 9–58

9–21 Use two 74194 four-bit bidirectional shift registers to create an eight-bit bidirectional shift register. Show the connections.

9–22 Determine the Q outputs of a 74194 with the inputs shown in Figure 9–59. A, B, C, and D are all HIGH.

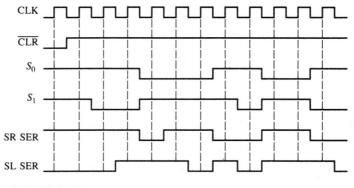

FIGURE 9–59

Section 9–7

9–23 How many flip-flops are required to implement each of the following in a Johnson counter configuration:

    (a) Divide-by-6   **(b)** Divide-by-10   **(c)** Divide-by-14

    (d) Divide-by-16   **(e)** Divide-by-20   **(f)** Divide-by-24

    (g) Divide-by-36

9–24 Draw the logic diagram for a divide-by-18 Johnson counter. Sketch the timing diagram and write the sequence in tabular form.

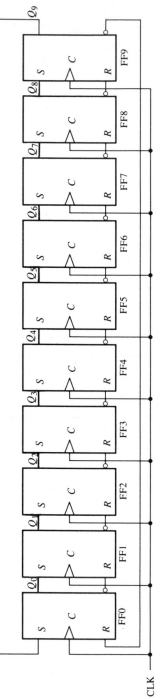

FIGURE 9–60

458

9–25 For the ring counter in Figure 9–60, draw the waveforms for each flip-flop output with respect to the clock. Assume that FF0 is initially SET and the rest RESET. Show at least ten clock pulses.

9–26 The waveform pattern in Figure 9–61 is required. Show a ring counter, and indicate how it can be preset to produce this waveform.

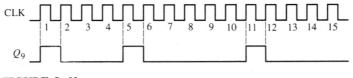

FIGURE 9–61

Section 9–8

9–27 Use 74195 four-bit shift registers to implement a sixteen-bit ring counter. Show the connections.

9–28 Specify the devices that can be used to implement the serial-to-parallel data converter in Figure 9–33. Draw the complete logic diagram showing any modifications necessary to accommodate the specific devices used.

9–29 Modify the serial-to-parallel converter in Figure 9–33 to provide sixteen-bit conversion.

9–30 Design an eight-bit parallel-to-serial data converter that produces the data format in Figure 9–34. Show a logic diagram and specify the devices.

9–31 What is the purpose of the power-on LOAD input in Figure 9–38?

9–32 Develop a power-on LOAD circuit for the keyboard encoder in Figure 9–38. This circuit must generate a short-duration LOW pulse when the power switch is turned on.

9–33 What happens when two keys are pressed simultaneously in Figure 9–38?

Section 9–9

9–34 Develop a test procedure for exercising the keyboard encoder in Figure 9–38. Specify the procedure on a step-by-step basis, indicating the output code from the key code register that should be observed at each step in the test.

9–35 Implement the test pattern generator used in Figure 9–40 to troubleshoot the serial-to-parallel converter.

9–36 What symptoms are observed for the following failures in the serial-to-parallel converter in Figure 9–40 (refer to Figure 9–33 for more detail):
  (a) AND gate output stuck in HIGH state
  (b) Clock generator output fails LOW
  (c) Third stage of data input register stuck in SET state
  (d) Terminal count output of counter stuck in HIGH state

ANSWERS TO SECTION REVIEWS

Section 9–1
1. A counter has a specified sequence of states, but a register does not.
2. Storage, data movement (shifting).

Section 9–2
1. FFA: Data input to J, $\overline{\text{Data input}}$ to K. FFB: Q_A to J, $\overline{Q}_A$ to K. FFC: Q_B to J, $\overline{Q}_B$ to K. FFD: Q_C to J, $\overline{Q}_C$ to K.
2. Eight.

Section 9–3
1. 0100   2. Take the serial output from the right-most flip-flop.

Section 9–4
1. When SHIFT/$\overline{\text{LOAD}}$ is HIGH, the data are shifted right one bit per clock pulse. When SHIFT/$\overline{\text{LOAD}}$ is LOW, the data on the parallel inputs are loaded into the register.
2. Asynchronous. The parallel load operation is not dependent on the clock.

Section 9–5
1. 1001   2. $Q_A = 1$

Section 9–6
1. 1111

Section 9–7
1. Sixteen   2. 000, 100, 110, 111, 011, 001, 000

Section 9–8
1. 625 scans/second   2. $ABCDEF = 001001$
3. The diodes provide unidirectional paths for pulling the ROWs LOW and preventing HIGHs on the row lines from being connected to the switch matrix. The resistors pull the column lines HIGH.

Section 9–9
1. To sequence the circuit through all of its states.
2. Check the input to that portion of the circuit. If the signal on that input is correct, the fault is isolated to the circuitry between the good input and the bad output.
3. Signature analysis is a troubleshooting technique whereby measured bit patterns are compared to documented bit patterns (signatures).

Section 9–10
1. No.   2. Yes, as indicated by the 4D label.

The last chapter covered shift registers, which are a type of storage device; in fact, a shift register is basically a memory. The memory devices covered in this chapter are generally used for longer-term storage of larger amounts of data than registers can provide.

Modern data-processing systems require the permanent or semipermanent storage of large amounts of data. Microprocessor-based systems rely on memories for their operation because memories are used to store programs and data for processing.

In this chapter, both semiconductor and magnetic memories and their applications are covered. Specific memory devices introduced are the following:

1. 74184 ROM BCD-to-binary converter
2. 74185 ROM binary-to-BCD converter
3. 7488 ROM
4. TMS47256 ROM
5. TMS2516 16K-bit EPROM
6. TMS4016 16K-bit byte-oriented static RAM
7. 74189 64-bit static RAM
8. 74PL689 programmable logic array
9. TMS4164 64K-bit dynamic RAM

In this chapter, you will learn

☐ The basic types of semiconductor and magnetic memories.
☐ What a ROM is and how it works.
☐ What a RAM is and how it works.
☐ How memories are used in microprocessor-based systems.
☐ The difference between a data *word* and a data *byte*.
☐ How ROMs are used to store fixed programs.
☐ How three-state interfacing is used to connect memories to other parts of a system.
☐ How data are stored and retrieved from a memory.
☐ How PROMs and EPROMs operate.
☐ How a RAM operates and how it differs from a ROM.
☐ The difference between static RAMs and dynamic RAMs.
☐ How dynamic RAMs are refreshed.
☐ How a magnetic bubble memory (MBM) works.
☐ Several methods of storing data on magnetic tape.
☐ The basic structure of floppy disks.
☐ The fundamentals of PLAs, FIFOs, and CCD memories.
☐ How to use a logic analyzer in a memory troubleshooting problem.

10

Memories

10-1   MEMORY CONCEPTS

Data are stored in a memory by a process called *writing* and are retrieved from the memory by a process called *reading*. Memories are made up of storage locations in which data can be stored. Each location is identified by an *address*. The number of storage locations can vary from a few in some memories to hundreds of thousands in others.

Each storage location can accommodate one or more bits. Generally, the total number of bits that a memory can store is its *capacity*. Sometimes the capacity is specified in terms of *bytes* (eight-bit groups).

Memories are made up of storage elements (flip-flops or capacitors in semiconductor memories and magnetic domains in magnetic storage), each of which stores one bit of data. A storage element is called a *cell*.

Figure 10-1 illustrates in a very simplified way the concepts of write, read, address, and storage capacity for a generalized memory.

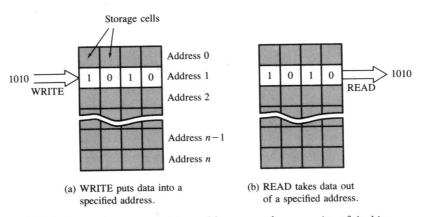

(a) WRITE puts data into a specified address.

(b) READ takes data out of a specified address.

FIGURE 10-1   *A memory with n addresses and a capacity of 4n bits.*

Types of Semiconductor Memories

The two basic types of semiconductor memories are the ROM and the RAM. The ROM is a *read only memory*. Data are permanently or semipermanently stored and can be read from the memory at any time. In a ROM in which the data are permanently stored, there is no write operation because specified data are either manufactured into the device or programmed into the device by the user and cannot be altered. In a ROM in which the data are semipermanently stored, the data can be altered by special methods, but there is no write operation. ROMs are *nonvolatile* memories; that is, the stored data are retained even when power is removed. Examples of ROM applications are look-up tables, conversions, and preprogrammed instructions.

The RAM is a *random access memory* that has both read and write capability. The term *random access* means that any storage location in the memory can be addressed (accessed) in any order to read or write data. Actually, most ROMs are also random access devices, so the terminology is somewhat misleading. More precisely, the RAM is a read/write random access memory, and the ROM is a read only random access memory. RAMs are *volatile* memories, so data are lost if the power is removed.

The ROM family   Semiconductor ROMs are manufactured with *bipolar technology* (such as TTL) or with *MOS* (metal oxide semiconductor) *technology.* Figure 10–2 shows how ROMs are categorized. The *mask* ROM is the type in which data are permanently stored in the memory during the manufacturing process. The PROM, or *programmable* ROM, is the type in which the data are electrically stored by the user with the aid of specialized equipment. Notice that both the mask ROM and the PROM can be of either technology. The EPROM, or *erasable* PROM, is strictly a MOS device.

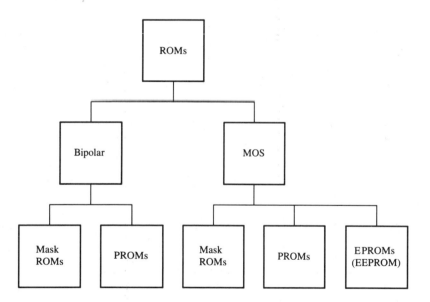

FIGURE 10–2   *The ROM family.*

The EPROM is electrically programmable by the user, but the stored data can be erased either by exposure to ultraviolet light or by electrical means. The latter type is called an *electrically erasable* PROM (EEPROM) or *electrically alterable* PROM (EAPROM).

The RAM family   RAMs are also manufactured with either bipolar or MOS technologies. Bipolar RAMs are all *static* RAMs; that is, the storage elements used in the memory are latches, so data can be stored for an indefinite period of time as long as the power is on. Some MOS RAMs are of the static type and some are *dynamic*. A dynamic memory is one in which data are stored on capacitors which require periodic recharging (*refreshing*) to retain the data. Figure 10–3 shows the categories of RAMs.

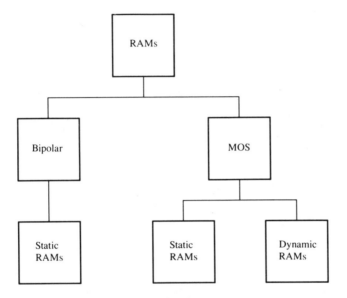

FIGURE 10–3   *The RAM family.*

How Memories Are Used in Microprocessor-Based Systems

Figure 10–4 shows a generalized basic block diagram of a typical microcomputer which forms the heart of microprocessor-based systems. As you can see, it consists of four functional blocks: *microprocessor, memory, input/output (I/O) interface*, and *I/O device*. The microprocessor unit within the microcomputer is interconnected to the memory and I/O interface with an *address bus*, a *data bus*, and a *control bus*.

   The function of the *memory* is to store binary data that are to be used or processed by the microprocessor. The function of the *I/O interface* is to get information into and out of the memory or microprocessor from the I/O device. The I/O device can be a keyboard, video terminal, card reader, printer, magnetic disk or tape unit, or other type of equipment.

   The *address bus* provides a path from microprocessor to memory and I/O interface. It allows the microprocessor to select the memory address from which to acquire data or in which data are to be stored. It also provides for communication with an I/O device for inputting or outputting data.

   The *data bus* provides a path over which data are transferred between the microprocessor, memory, and I/O interface.

FIGURE 10-4 *Basic micro-computer.*

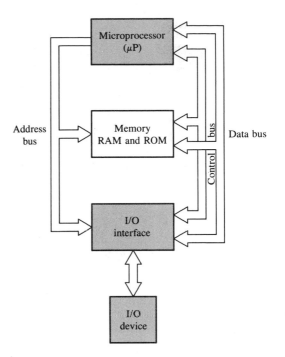

As mentioned, the microprocessor is connected to a memory with the address bus and the data bus. In addition, there are certain control signals that must be sent between the microprocessor and the memory, such as the *read* and the *write* controls. This is illustrated in Figure 10–5.

As indicated, the address bus is unidirectional; that is, the address bits go only one way—from the microprocessor to the memory. The data bus is bidirectional. This provides for data bits to be transferred from the memory to the microprocessor or from the microprocessor to the memory.

Words and bytes A *complete* unit of binary information or data is called a *word*. For instance, the number 200_{10} can be represented by eight bits as 11001000_2.

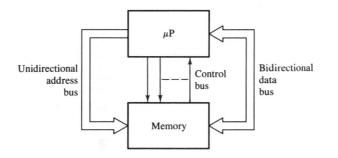

FIGURE 10-5 *A microprocessor with memory.*

This, then, is an eight-bit word because it completely represents the decimal number 200. Now consider the decimal number 32,768. This number cannot be represented with eight bits but requires 16 bits—1000000000000000_2. So a 16-bit word is required for representing decimal numbers from 0 to 65,535.

Most microprocessors handle bits in eight-bit groups called *bytes*. A byte may be a word or only part of a word. For instance, an *eight-bit* word mentioned above consists of *one byte*. However, a *16-bit word* consists of *two bytes*.

In the following example of the read and write operations, a memory capacity of 65,536 bits is used because it represents a common memory size used in many microprocessor-based systems. This memory capacity is traditionally designated as a 64K.

Read Operation

To transfer a byte of data from the memory to the microprocessor, a *read* operation must be performed, as illustrated in Figure 10–6. To begin, a *program counter* contains the 16-bit address of the byte to be read from the memory. This address is loaded into the *address buffer* and put onto the address bus.

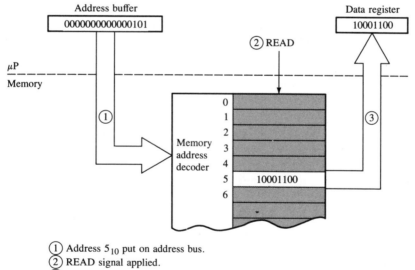

① Address 5_{10} put on address bus.
② READ signal applied.
③ Contents of address 5_{10} in memory put on data bus and stored by data register.

FIGURE 10–6 *Illustration of the r... d operation in a typical microcomputer.*

Once the address code is on the bus, the microprocessor control unit sends a *read* signal to the memory. At the memory, the address bits are decoded, and the desired memory location is selected. The *read* signal causes the *contents* of the selected address to be put on the data bus. The data byte is then loaded into the data register to be used by the microprocessor. This completes the read operation.

Note that each memory location contains one byte of data. When a byte is read from the memory, is it not destroyed but remains in the memory. This process of "copying" the contents of a memory location without destroying those contents is called *nondestructive readout*.

Write Operation

In order to transfer a byte of data from the microprocessor to the memory, a *write* operation is required, as illustrated in Figure 10–7.

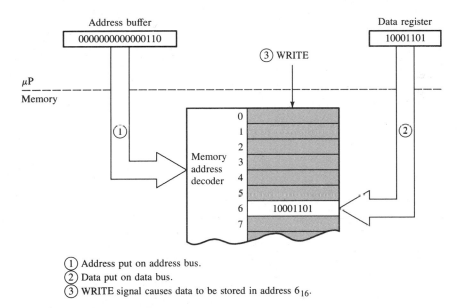

① Address put on address bus.
② Data put on data bus.
③ WRITE signal causes data to be stored in address 6_{16}.

FIGURE 10–7 *Illustration of the write operation.*

The memory is addressed in the same way as during a read operation. A data byte being held in the data register is put onto the data bus, and the microprocessor sends the memory a *write* signal. This causes the byte on the data bus to be stored at the selected location in the memory as specified by the 16-bit address code. The existing contents of that particular memory location are *replaced* by the new data byte. This completes the write operation.

Hexadecimal Representation of Address and Data

The only things a microprocessor recognizes are combinations of 1s and 0s. However, most literature on microprocessors uses the hexadecimal number system to simplify the representation of binary quantities.

For instance, the binary address 0000000000001111 can be written as 000F in hexadecimal. A 16-bit address can have a *minimum* hexadecimal value of 0000_{16} and a *maximum* value of $FFFF_{16}$. With this notation, a 64K memory can be shown in block form as in Figure 10–8. The lowest memory address is 0000_{16} and the highest address is $FFFF_{16}$.

FIGURE 10–8 *Representation of a 64K memory.*

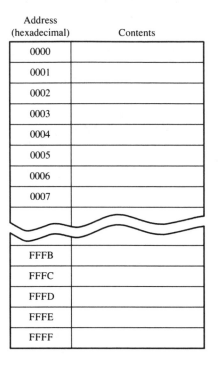

Address (hexadecimal)	Contents
0000	
0001	
0002	
0003	
0004	
0005	
0006	
0007	
FFFB	
FFFC	
FFFD	
FFFE	
FFFF	

A data byte can also be represented in hexadecimal. A data byte is eight bits and can represent decimal numbers from 0_{10} to 255_{10}, or it can represent up to 256_{10} instructions. For example, a microprocessor code that is 10001100 in binary is written as 8C in hexadecimal.

SECTION REVIEW 10–1

1. Explain the basic difference between ROM and RAM.
2. What does the term *nonvolatile* mean?
3. Explain the difference between static and dynamic memories.

10–2   READ ONLY MEMORIES (ROMs)

As mentioned before, a ROM contains permanently or semipermanently stored data which can be read from the memory, but which either cannot be changed at all or cannot be changed without specialized equipment. ROMs are used to store data that are used repeatedly in system applications, such as tables, conversions, or programmed instructions for system initialization and operation.

The Mask ROM

The mask ROM is referred to simply as a ROM. It is permanently programmed during the manufacturing process to provide widely used standard functions such as popular conversions or to provide user-specified functions. Once the memory is programmed, it cannot be changed. Most IC ROMs utilize the presence or absence of a transistor connection at a ROW/COLUMN junction to represent a 1 or a 0. ROMs can be either bipolar or MOS.

Figure 10–9(a) shows bipolar ROM cells. The presence of a connection from a ROW line to the *base* of a transistor represents a 1 at that location because when the ROW line is taken HIGH, all transistors with a base connection to that ROW line turn on and connect the HIGH (1) to the associated COLUMN lines. At ROW/COLUMN junctions where there are no base connections, the COLUMN lines remain LOW (0) when the ROW is addressed.

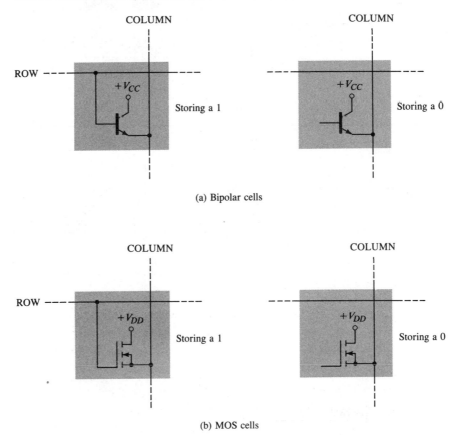

FIGURE 10–9 *ROM cells.*

Figure 10–9(b) illustrates MOS ROM cells. They are basically the same as the bipolar cells, except they are made with MOSFETs (metal oxide semiconductor field-effect transistors). The presence or absence of a *gate* connection at a junction permanently stores a 1 or a 0 as shown.

A Simple ROM

To illustrate the concept, Figure 10–10 shows a small, simplified ROM array. The light squares represent stored 1s, and the dark squares represent stored 0s. The basic read operation is as follows: When a binary address code is applied to

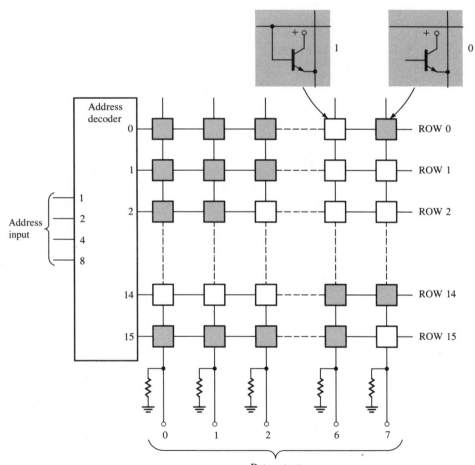

FIGURE 10–10 *A 16X8-bit ROM array.*

the *address input,* the corresponding ROW line goes HIGH. This HIGH is connected to the COLUMN lines through the transistors at each junction (cell) where a 1 is stored. At each cell where a 0 is stored, the COLUMN line stays LOW because of the terminating resistor. The COLUMN lines form the *data output.* The eight data bits stored in the selected ROW appear on the output lines.

As you can see, this example ROM is organized into 16 addresses, each of which stores 8 data bits. Thus, it is a 16X8 (16 by 8) ROM, and its total capacity is 128 bits.

EXAMPLE 10-1 Show a basic ROM, similar to that shown in Figure 10–10, programmed for four-bit binary-to-Gray conversion.

Solution Referring to Chapter 2 for the Gray code, the following table is developed for use in programming the ROM:

Binary				Gray			
B_3	B_2	B_1	B_0	G_3	G_2	G_1	G_0
0	0	0	0	0	0	0	0
0	0	0	1	0	0	0	1
0	0	1	0	0	0	1	1
0	0	1	1	0	0	1	0
0	1	0	0	0	1	1	0
0	1	0	1	0	1	1	1
0	1	1	0	0	1	0	1
0	1	1	1	0	1	0	0
1	0	0	0	1	1	0	0
1	0	0	1	1	1	0	1
1	0	1	0	1	1	1	1
1	0	1	1	1	1	1	0
1	1	0	0	1	0	1	0
1	1	0	1	1	0	1	1
1	1	1	0	1	0	0	1
1	1	1	1	1	0	0	0

The resulting ROM array is shown in Figure 10–11. You can see that a binary code on the address inputs produces the corresponding Gray code on the output lines (COLUMNs).

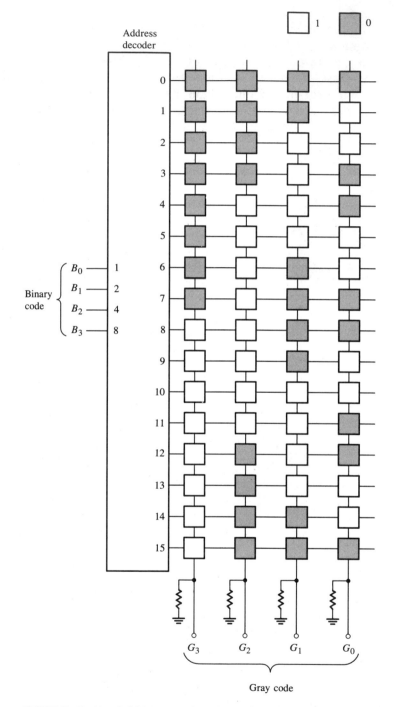

FIGURE 10–11 *ROM programmed as a binary-to-Gray code converter.*

Internal ROM Structure

Most IC ROMs have a somewhat more complex internal structure than that in the basic simplified example just covered. To illustrate how an IC ROM is structured, a 1024-bit device with a 256X4 organization is used. The logic symbol is shown in Figure 10–12. When any one of 256 binary codes (eight bits) is applied to the address inputs, four data bits appear on the outputs if the chip-select inputs are LOW.

FIGURE 10–12 *A 256X4 ROM logic symbol.*

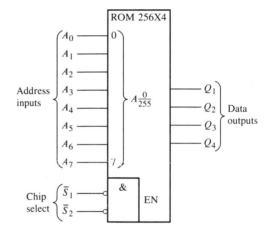

Although the 256X4 organization of this device implies that there are 256 rows and 4 columns in the memory array, this is not the case. The memory cell array is actually a 32X32 matrix (32 rows and 32 columns), as shown in the block diagram in Figure 10–13.

The ROM works as follows: Five of the eight address lines (A_0 through A_4) are decoded by the ROW decoder (often called the Y decoder) to select one of the 32 rows. Three of the eight address lines (A_5 through A_7) are decoded by the column decoder (often called the X decoder) to select four of the 32 columns. Actually, the X decoder consists of four 1-of-8 decoders (data selectors) as shown.

The result of this structure is that when an eight-bit address code (A_0 through A_7) is applied, a four-bit data word appears on the data outputs when the chip-select lines (S_1 and S_2) are LOW to enable the output buffers.

This type of internal structure (architecture) is typical of IC ROMs of various capacities. In fact, the 74187 ROM has this exact configuration.

A Small-Capacity ROM

Figure 10–14 shows the logic symbol for a 7488 as an example of a bipolar ROM. It is organized as a 32X8 memory; that is, it has 32 address locations, each of which has eight bits of storage. A five-bit address on inputs A_0 through A_4 selects one of the 32 locations (0 through 31). A LOW on the $\overline{S}$ input enables (EN) the device and

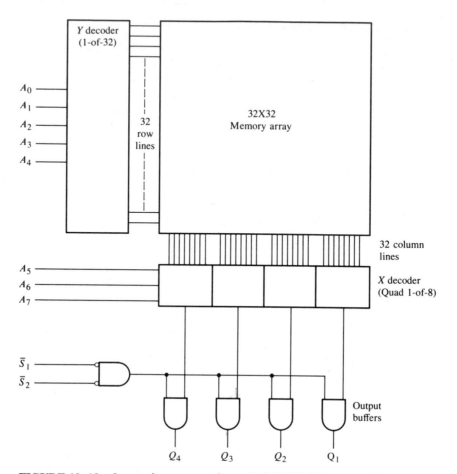

FIGURE 10–13 *Internal structure of a typical ROM. This particular example is a 1024-bit ROM with a 256X4 organization.*

FIGURE 10–14 *Logic symbol for the 7488 256-bit bipolar ROM.*

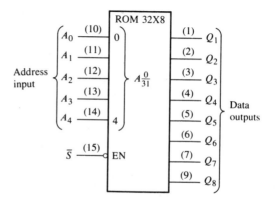

places the selected data byte on the outputs (Q_1 through Q_8). This enable input is called the *chip select*.

This ROM can be programmed to user specifications by the manufacturer.

A Large-Capacity ROM

As a comparison with the small bipolar ROM just discussed, the TMS47256 (TI) is an example of a large-capacity MOS ROM. It can store 262,144 bits with a 32,768X8 (32KX8) organization. That is, it has 32,768 address locations with eight bits at each location.

Figure 10–15 is a logic symbol for this device. Notice that it has 15 address lines. These are required to address the 32,768 locations ($2^{15} = 32,768$). There is an active-LOW chip select, $\overline{S}_1$, and on this particular device, a *chip enable/power down* input, $\overline{E}$. These inputs work as follows: $\overline{S}_1$ and $\overline{E}$ must both be LOW to enable the memory output. When $\overline{E}$ is HIGH, the device is put into a low power standby mode that reduces the current drain on the dc power supply.

FIGURE 10–15 *Logic symbol for the TMS47256 MOS static ROM.*

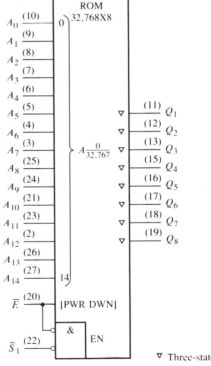

▽ Three-state output

Three-State Outputs and Buses

At this point, it is convenient to introduce the concept of the three-state output, which is commonly found on memory devices such as the ROM just presented.

The three-state outputs are indicated on logic symbols by a small inverted triangle (∇), as shown in Figure 10–15, and are used for compatibility with bus structures such as those found in microprocessor-based systems.

Physically, a *bus* is a set of conductive paths that serves to interconnect two or more functional components of a system or several diverse systems. Electrically, a bus is a collection of voltages, levels, and signals that allow the various devices connected to the bus to work properly together.

For example, a microprocessor is connected to memories and input/output devices by certain bus structures. This is illustrated by the block diagram in Figure 10–16. An *address bus* allows the microprocessor to address the memories, and the *data bus* provides for transfer of data between the microprocessor, the memories, and the input/output devices. The *control bus* allows the microprocessor to control data flow and timing for the various components.

FIGURE 10–16   *Block diagram of a basic microcomputer system with bus interconnections.*

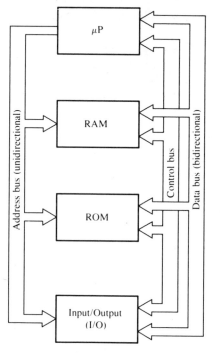

The physical bus is symbolically represented by wide lines with arrow heads indicating direction of data movement. Buses can also be used to interconnect various test instruments or other types of electronic systems.

Three-state interface to the bus   In a typical application, several devices are connected to one bus, for example, a microprocessor, a RAM, and a ROM. For this reason, *three-state* logic circuits are used to interface digital devices such as memories to a bus. Figure 10–17(a) shows the logic symbol for a noninverting three-state buffer with an active-HIGH *enable*. Part (b) of the figure shows one with an active-LOW *enable*.

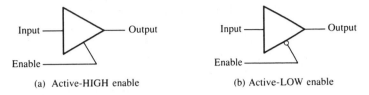

(a) Active-HIGH enable

(b) Active-LOW enable

FIGURE 10–17 *Three-state buffer symbols.*

The basic operation of a three-state buffer can be understood in terms of switching action as illustrated in Figure 10–18. When the enable input is active, the gate operates as a normal noninverting circuit. That is, the output is HIGH when the input is HIGH and LOW when the input is LOW, as shown in parts (a) and (b). The HIGH and LOW levels represent two of the states. The buffer operates in its *third* state *when the enable input is not active.* In this state, the circuit acts as an *open switch* and the output is completely *disconnected* from the input, as shown in part (c). This is sometimes called the high-impedance or high-Z state.

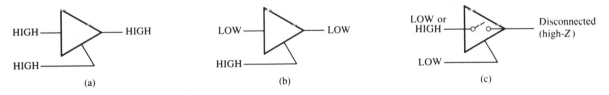

(a)

(b)

(c)

FIGURE 10–18 *Three-state buffer operation.*

Many microprocessors, memories, and other integrated circuit functions have three-state buffers that serve to interface with the buses. This is necessary when two or more devices are connected to a common bus. To prevent the devices from interfering with each other, the three-state buffers are used to disconnect all devices except the ones that are communicating at any given time.

ROM Access Time

A typical timing diagram that illustrates ROM access time is shown in Figure 10–19. The access time, t_a, of a ROM is the time from the application of a valid address code on the inputs until the appearance of valid output data. Access time can also be measured from the activation of *chip select* ($\overline{S}$) to the occurrence of valid output data when a valid address is already on the inputs.

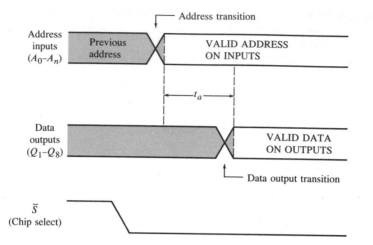

FIGURE 10–19 *ROM access time (t_a) from address change to data output with chip enable already active.*

A ROM Application

As mentioned before, some ROMs are programmed to perform widely used functions and are available "off the shelf." Examples are the 74184, a ROM device programmed as a *BCD-to-binary converter,* and the 74185, a ROM device programmed as a *binary-to-BCD converter.* Logic symbols for these devices used as six-bit converters are shown in Figure 10–20, and Figures 10–21 and 10–22 show how these devices can be expanded for conversions involving larger numbers of bits.

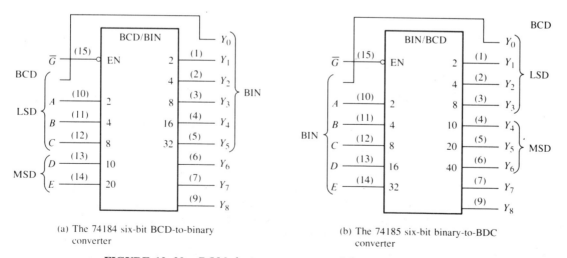

(a) The 74184 six-bit BCD-to-binary converter

(b) The 74185 six-bit binary-to-BDC converter

FIGURE 10–20 *ROM devices programmed for specific conversions.*

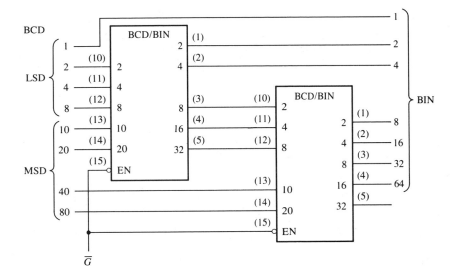

FIGURE 10–21   *74184s expanded for conversion of two BCD digits to binary.*

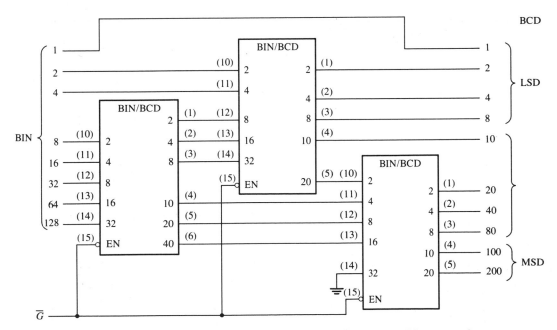

FIGURE 10–22   *74185s expanded for eight-bit binary-to-BCD conversion.*

**SECTION
REVIEW
10–2**

1. What is the bit storage capacity of a ROM with a 512X4 organization?
2. Explain the purpose of three-state outputs.
3. How many address bits are required for a 2048-bit memory?

10–3   PROGRAMMABLE ROMs (PROMs AND EPROMs)

PROMs are basically the same as mask ROMs, once they have been programmed. The difference is that PROMs come from the manufacturer unprogrammed and are custom programmed in the field to meet the user's needs. There are two general categories: PROMs and EPROMs.

PROMs

PROMs are available in both bipolar and MOS technologies and generally have four-bit or eight-bit output word formats with bit capacities ranging in excess of 250,000. PROMs employ some type of *fusing* process to store bits, whereby a memory link is fused open or left intact to represent a 0 or a 1. The fusing process is irreversible. Once a PROM is programmed, it cannot be changed.

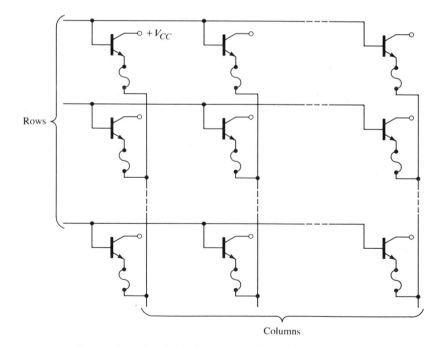

FIGURE 10–23 *Bipolar PROM array with fusible links. (All collectors are commonly connected to V_{CC}.)*

Figure 10–23 illustrates a bipolar PROM array with *fusible links*. The fuse links are manufactured into the PROM between the emitter of each cell's transistor and its column line. In the programming process, a sufficient current is injected through the fuse link to burn it open to create a stored 0. The link is left intact for a stored 1.

Three basic fuse technologies are used in PROMs: metal links, silicon links, and PN junctions. A brief description of each of these follows:

1. *Metal links* are made of a material such as nichrome. Each bit in the memory array is represented by a separate link. During programming, the link is either "blown" open or left intact. This is basically done by first addressing a given cell and then forcing a sufficient amount of current through the link to cause it to open.
2. *Silicon links* are formed by narrow, notched strips of polycrystalline silicon. Programming of these fuses requires melting of the links by passing a sufficient amount of current through them. This amount of current causes a high temperature at the fuse location that oxidizes the silicon and forms an insulation around the now open link.
3. *Shorted junction* or *avalanche-induced migration* technology consists basically of two PN junctions arranged back-to-back. During programming, one of the diode junctions is avalanched, and the resulting voltage and heat cause aluminum ions to migrate and short the junction. The remaining junction is then used as a forward-biased diode to represent a data bit.

PROM Programming

A PROM is normally programmed by plugging it into a special instrument called a *PROM programmer*. Basically, the programming is accomplished as shown by the simplified setup in Figure 10–24. An address is selected by the switch settings on

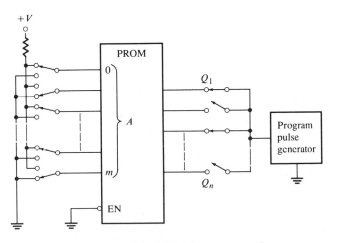

FIGURE 10–24 *Simplified PROM programming setup.*

the address lines, and then a pulse is applied to those output lines corresponding to bit locations where 0s are to be stored (the PROM starts out with all 1s). These pulses blow the fuse links, thus creating the desired bit pattern. The next address is then selected and the process is repeated. This sequence is done automatically by the PROM programmer.

EPROMs

An EPROM is an *erasable* PROM. Unlike an ordinary PROM, an EPROM can be reprogrammed by first erasing an existing program in the memory array. These devices use an NMOSFET array with an *isolated gate* structure.

The isolated transistor gate has no electrical connections and can store an electrical charge for indefinite periods of time. The data bits in this type of array are represented by the presence or absence of a stored gate charge. Erasure of a data bit is a process that removes the gate charge.

There are two basic types of erasable PROMs: the *ultraviolet light erasable* PROM (UV EPROM) and the *electrically erasable* PROM (EEPROM).

UV EPROMs   You can recognize this UV EPROM device by the transparent quartz lid on the package, as shown in Figure 10–25. The isolated gate in the FET of an ultraviolet EPROM is "floating" within an oxide insulating material. The programming process causes electrons to be removed from the floating gate. Erasure is done by exposure of the memory array chip to high-intensity ultraviolet radiation through the quartz window. This neutralizes the positive charge stored on the gate after several minutes to an hour of exposure time.

FIGURE 10–25   *Ultraviolet erasable PROM package. (Courtesy of Motorola Semiconductor Products)*

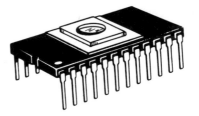

EEPROMs   *Electrically erasable* PROMs can be both erased and programmed with electrical pulses. This type of device is also known as an *electrically alterable* ROM (EAROM). Since it can be both electrically written into and electrically erased, the EEPROM can be programmed and erased in-circuit for reprogramming.

There are two types of EEPROMs: floating gate MOS and metal-nitride-oxide-silicon (MNOS). The application of a voltage on the control gate in the floating gate structure permits the storage and removal of charge from the floating gate.

A Specific PROM

The TMS2516 is an example of a MOS EPROM device. Its operation is representative of that of other typical EPROMs. As the logic symbol in Figure 10–26 shows, this device has 2048 addresses (2^{11} = 2048), each with eight bits. Notice that the eight outputs are three-state (∇).

FIGURE 10–26 *The logic symbol for a TMS2516 PROM.*

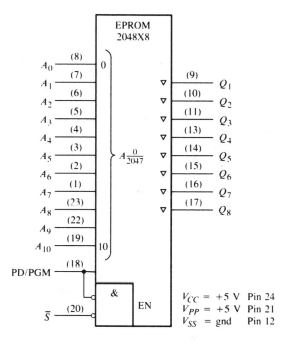

To *read* from the memory, the select input ($\overline{S}$) must be LOW and the power down/program (PD/PGM) input LOW. (The power-down mode was mentioned earlier in relation to the TMS47256 ROM.) To *erase* the stored data, the device is exposed to high-intensity ultraviolet light through the transparent lid. A typical 12 mW/cm² filterless UV lamp will erase the data in about 20 to 25 minutes. As in most EPROMs after erasure, all bits are 1s. Normal ambient light contains the correct wavelength of UV light for erasure. Therefore, the transparent lid must be kept covered.

To *program* the device, +25 V dc is applied to V_{PP} (which is normally +5 V), and $\overline{S}$ is HIGH. The eight data bits to be programmed into a given address are applied to the outputs (Q_1 through Q_8), and the address is selected on inputs A_0 through A_{10}. Next, a 10-ms to 55-ms wide HIGH level pulse is applied to the PD/PGM input. The addresses can be programmed in any order.

A timing diagram for the programming mode is shown in Figure 10–27. These signals are normally produced by an EPROM programmer.

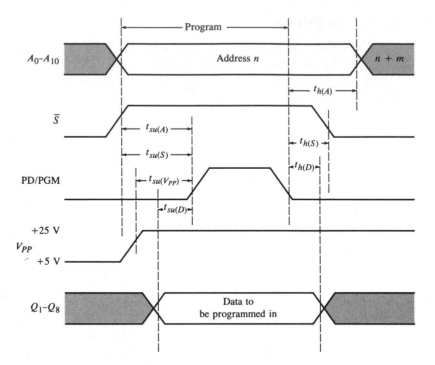

FIGURE 10–27 *Timing diagram for TMS2516 PROM programming cycle, with critical setup times (t_s) and hold times (t_h) indicated.*

SECTION REVIEW 10–3	1.	How do PROMs differ from ROMs?

1. How do PROMs differ from ROMs?
2. After erasure, all bits are (1s, 0s) in a typical EPROM.
3. What is the *normal* mode of operation for a PROM?

10–4   READ/WRITE RANDOM ACCESS MEMORIES (RAMs)

Data can be readily written into and read from a RAM at any selected address in any sequence. When data are *written* into a given address in the RAM, the data previously stored at that address are destroyed and replaced by the new data. When data are *read* from a given address in the RAM, the data at that address are not destroyed. This *nondestructive* read operation can be thought of as "copying" the contents of an address while leaving the contents intact.

The Static RAM Cell

As mentioned before, the storage cells in a static RAM are either bipolar or MOS latches. Once a data bit is stored in a cell, it remains indefinitely (unless power is lost or a new data bit is written in). Because data are lost if the power to the memory is lost, a static RAM is a type of *volatile* memory.

Figure 10–28 shows a functional logic diagram of a static RAM cell. The basic operation is as follows: The cell (or a group of cells) is selected by HIGHs on the ROW and COLUMN lines. When the READ/WRITE line is LOW (write), the

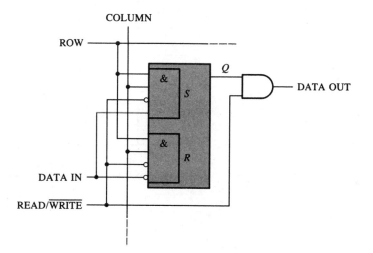

FIGURE 10–28 *A generalized logic diagram for a static RAM cell. A D-type latch can also be used to implement the static RAM cell.*

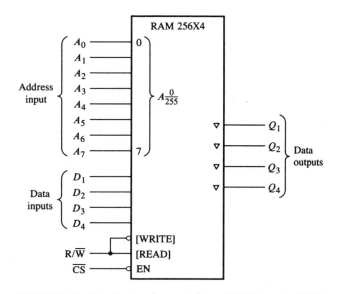

FIGURE 10–29 *Logic diagram for a 256X4 static RAM.*

input data bit is written into the cell by setting the latch for a 1 and resetting the latch for a 0. When the READ/$\overline{\text{WRITE}}$ line is HIGH (read), the latch is un-affected, but the stored data bit (Q) is gated to the *data-out* line.

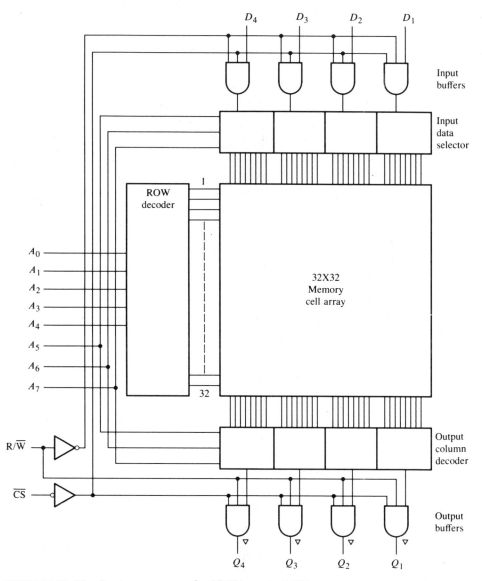

FIGURE 10–30 *Basic structure of a 256X4 static RAM.*

Basic Structure of a Static RAM

RAMs are addressed in basically the same way as ROMs. The main difference between the structures of RAMs and ROMs is that RAMs have data inputs and a read/write control.

To illustrate the generalized structure of a RAM, a 1024-bit device with a 256X4 organization is used. The logic symbol is shown in Figure 10–29 (p. 487).

In the read mode (R/$\overline{W}$ HIGH), four data bits from the selected address appear on the data outputs when the chip select ($\overline{CS}$) is LOW. In the write mode (R/$\overline{W}$ LOW), the four data bits that are applied to the data inputs are stored at the selected address.

Although the 256X4 organization of this device implies that there are 256 rows and 4 columns, the memory cell array actually is a 32X32 matrix (32 rows and 32 columns), as shown in Figure 10–30.

The static RAM works as follows: Five of the eight address lines (A_0 through A_4) are decoded by the ROW decoder to select one of the 32 rows. Three of the eight address lines (A_5 through A_7) are decoded by the *output column decoder*. In the read mode, the output buffers are enabled (the input buffers are disabled), and the four data bits from the selected address appear on the outputs. In the write mode, the input data buffers are enabled (the output buffers are disabled), and the four input data bits are routed through the *input data selector* by the address bits, A_5 through A_7, to the selected address for storage. The *chip select* must be LOW during read and write.

Variations of this basic structure and other RAM structures are possible, but this illustrates the basic static RAM operation and provides a comparison with the basic ROM structure in Figure 10–13.

A typical timing diagram for a read/write cycle is shown in Figure 10–31.

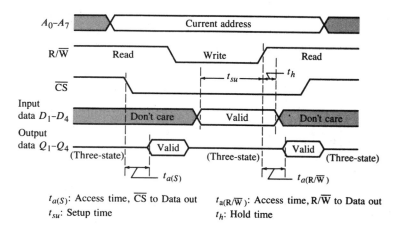

$t_{a(S)}$: Access time, $\overline{CS}$ to Data out $t_{a(R/\overline{W})}$: Access time, R/$\overline{W}$ to Data out
t_{su}: Setup time t_h: Hold time

FIGURE 10–31 *Basic read/write cycle timing for the static RAM in Figure 10–30.*

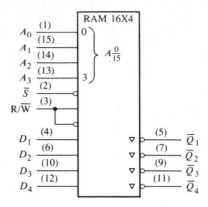

(a) The 74189 64-bit RAM (TTL)

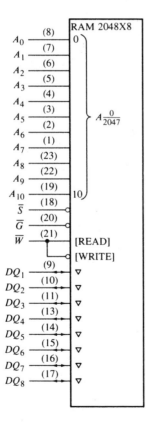

$\overline{W}$	$\overline{S}$	$\overline{G}$	DQ_1–DQ_8	Mode
0	0	X	Valid data	WRITE
1	0	0	Data output	READ
X	1	X	High-impedance*	Device disabled
1	0	1	High-impedance*	Output disabled

*Three-state output

(c) Truth table for the TMS4016

(b) The TMS4016 16,384-bit RAM (MOS)

FIGURE 10–32 *Specific static RAMs.*

Specific Static RAMs

As an example of a small-capacity static RAM, the 74189 TTL RAM is introduced. A logic symbol is shown in Figure 10–32(a).

This device has a 16X4 organization. The R/$\overline{\text{W}}$ input is HIGH for read and LOW for write. The four address inputs select one of the 16 four-bit locations. The chip select, $\overline{S}$, must be LOW for memory operation.

Because its capacity is small, the internal structure of this particular device is simpler than that discussed for the generalized RAM. For example, it does not have both ROW and COLUMN decoders. It has a ROW decoder (16 rows) only, because the array is actually a 16X4 matrix.

Another example of a static RAM is the TMS4016, which is a MOS device. As shown by the logic symbol in Figure 10–32(b), this memory has a 2048X8 organization; that is, there is storage capacity for 2048 eight-bit words. Since each memory location consists of eight bits, this is an example of a byte-organized static RAM (sometimes called *byte-wide*).

There are eleven address inputs to select the 2048 locations. The data-in/data-out (DQ_1 through DQ_8) are bidirectional terminals that provide for both data input during write and data output during read. The operation of this device is specified in the table in Figure 10–32(c).

The Dynamic RAM Cell

Dynamic memory cells store a data bit in a small *capacitor* rather than in a latch. The advantage of this type of cell is that it is very simple, thus allowing very large memory arrays to be constructed on a chip at a lower cost per bit than in static memories. The disadvantage is that the storage capacitor cannot hold its charge over an extended period of time and will lose the stored data bit unless its charge is *refreshed* periodically. This process of refreshing requires additional memory circuitry and complicates the operation of the dynamic RAM. Figure 10–33 shows a typical dynamic cell consisting of a single MOS transistor and a capacitor.

FIGURE 10–33 *A dynamic MOS RAM cell.*

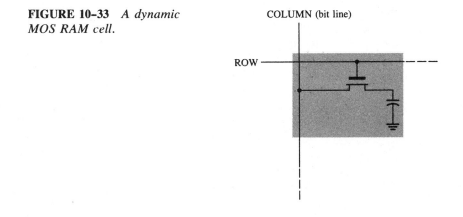

In this type of cell, the transistor acts as a switch. The basic simplified operation is illustrated in Figure 10–34 and is as follows: A LOW on the R/$\overline{W}$ line (write mode) enables the three-state *input buffer* and disables the *output buffer*. To write a 1 into the cell, the D_{IN} line is HIGH, and the transistor is turned on by a HIGH on the ROW line. The transistor acts as a closed switch connecting the capacitor to the BIT line. This allows the capacitor to charge to a positive voltage as shown in Figure 10–34(a). When a 0 is to be stored, a LOW is applied to the D_{IN} line. The capacitor remains uncharged, or if it is storing a 1, it discharges as indicated in Figure 10–34(b). When the ROW line is taken back LOW, the transistor turns off and disconnects the capacitor from the BIT line, thus "trapping" the charge (1 or 0) on the capacitor.

To read from the cell, the R/$\overline{W}$ line is HIGH, enabling the output buffer and disabling the input buffer. When the ROW line is taken HIGH, the transistor turns on and connects the capacitor to the BIT line and thus to the output buffer (sense amplifier), so the data bit appears on the *data-output line* (D_{OUT}). This is illustrated in Figure 10–34(c).

To refresh the memory cell, the R/$\overline{W}$ line is HIGH, the ROW line is HIGH, and the REFRESH line is HIGH. The transistor turns on, connecting the capacitor to the BIT line. The output buffer is enabled, and the stored data bit is applied to the input of the *refresh buffer* which is enabled by the HIGH on the REFRESH input. This produces a voltage on the BIT line corresponding to the stored bit, thus replenishing the capacitor as illustrated in Figure 10–34(d).

Basic Structure of a Dynamic RAM

As you have seen, the main difference between dynamic RAMs and static RAMs is the type of memory cell. Because the dynamic memory cell requires refreshing to retain data, additional circuitry is necessary. Several specific features common to most dynamic RAMs are now discussed. (Sometimes dynamic RAMs are called DRAMs.)

Address multiplexing   Most dynamic RAMs employ a technique called *address multiplexing* to reduce the number of address lines and thus the number of input/output pins on the package. Figure 10–35 shows a block diagram of a 16K (16,384) dynamic RAM which has been simplified to illustrate address multiplexing. This example memory has a 16K×1 organization (it stores 16,384 one-bit words).

The fourteen-bit address (2^{14} = 16,384) is applied seven bits at a time to the address inputs. First, the seven-bit ROW address is applied, and the RAS (row address strobe) latches the seven bits into the *row address latch*. Next, the seven-bit COLUMN address is applied to the address inputs, and the CAS (column address strobe) latches these seven bits into the *column address latch*. The seven-bit ROW address and the seven-bit COLUMN address are then decoded to select

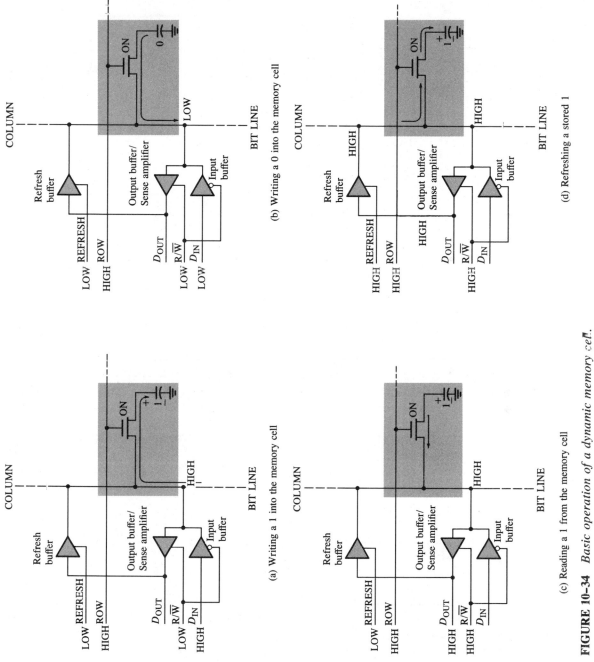

FIGURE 10-34 *Basic operation of a dynamic memory cell.*

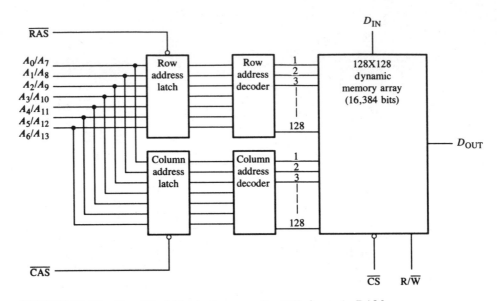

FIGURE 10–35 *Simplified block diagram of a 16K dynamic RAM.*

the appropriate memory cell for a read or write operation. The basic timing for this address multiplexing operation is shown in Figure 10–36.

Memory refresh   All dynamic RAMs require a refresh operation. The refresh logic can be either internal or external to the memory chip. Figure 10–37 shows a block diagram for the same basic generalized 16K memory just discussed, with the addition of refresh logic and somewhat greater detail.

In its simplest form, the memory refresh operation sequentially refreshes each row of cells, one row after the other, until all rows are refreshed. This is called *burst refresh* and must be repeated every 2 ms to 4 ms in a typical dynamic RAM. During refresh, data cannot be written into or read from the memory.

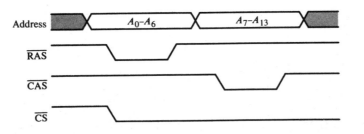

FIGURE 10–36 *Timing for address multiplexing in Figure 10–35.*

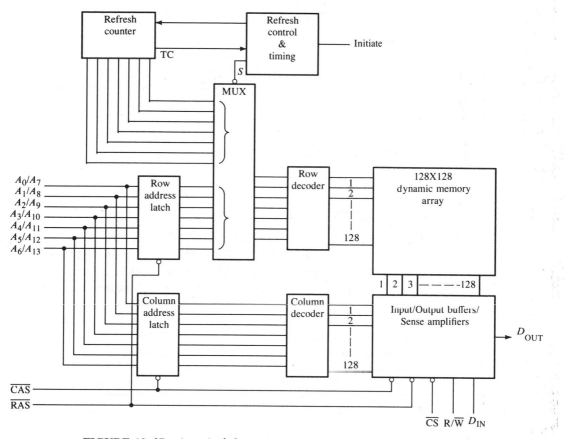

FIGURE 10–37 *A typical dynamic RAM with basic refresh circuitry.*

Another basic refresh method *distributes* the row refresh operations *between* the read and write operations, rather than doing them all at once in a burst. Still, each row must be refreshed within the specified refresh period.

Now let's see how refresh works in the memory of Figure 10–37. When a refresh operation is initiated, the seven output lines from the *refresh counter* go into the data selector (MUX) and are selected by the *refresh control* and applied to the *row decoder* in place of the row address lines.

The refresh counter is started and sequences through all of its states (0 through 127). Each row is sequentially selected, and the $\overline{RAS}$ signal causes each cell in each row to be refreshed, as previously discussed. Meanwhile, the $\overline{CAS}$ signal is HIGH, disabling the *output buffer* so that the changing data do not appear on the outputs as the rows are refreshed. When the refresh counter reaches its terminal count, the refresh operation terminates, and the memory is available for normal read/write cycling. A basic timing diagram illustrating the refresh operation is shown in Figure 10–38.

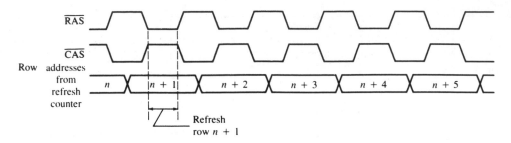

FIGURE 10–38 *Basic refresh timing diagram for a portion of the complete memory refresh operation.*

A Specific Dynamic RAM

The TMS4164 is a 65,536-bit (64K) dynamic RAM. It is organized as a 64KX1 memory. The logic symbol is shown in Figure 10–39. There are eight address lines (A_0 through A_7), a row address strobe ($\overline{RAS}$), a column address strobe ($\overline{CAS}$), and the read/write input ($\overline{W}$). The data input is D, and the data output is Q. This device uses a single +5-V supply.

When $\overline{RAS}$ goes LOW, the eight row address bits are latched into the RAM. Next the column bits are latched in when $\overline{CAS}$ goes LOW. A read cycle is initiated when $\overline{W}$ is HIGH, and a write cycle is initiated when $\overline{W}$ is LOW.

Larger-capacity dynamic RAMs with 256K bits, such as the TMS4256, are also commonly used. Also, 1M-bit dynamic RAMs are currently available.

FIGURE 10–39 *Logic symbol for a TMS4164 64KX1 dynamic RAM.*

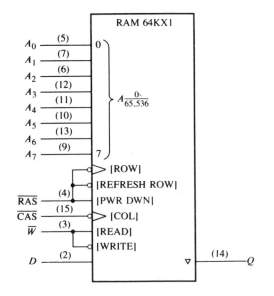

**SECTION
REVIEW
10–4**
1.   Explain the basic difference between static RAMs and dynamic RAMs.
2.   What is the reason for the refresh operation in dynamic RAMs?

10–5   MEMORY EXPANSION

Many RAMs are available in one-bit and four-bit word organizations as well as larger word sizes. In many applications, word sizes of eight bits or sixteen bits are required, and therefore the *expansion* of memory chips to increase the word size is necessary.

For example, a 16KX1 memory such as the one previously discussed can be expanded to a 16KX2 memory by connecting two memory chips as shown in Figure 10–40. Notice that the devices are essentially connected in parallel. The word capacity is still 16K, but the *word size* has been increased from one bit to two bits.

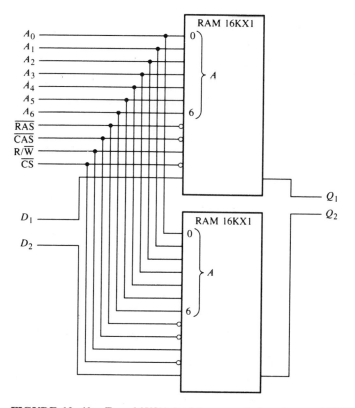

FIGURE 10–40   *Two 16KX1 RAMs expanded to form a 16KX2 RAM.*

EXAMPLE 10–2 Show how to expand the 16KX1 dynamic RAM to obtain a word length of four bits.

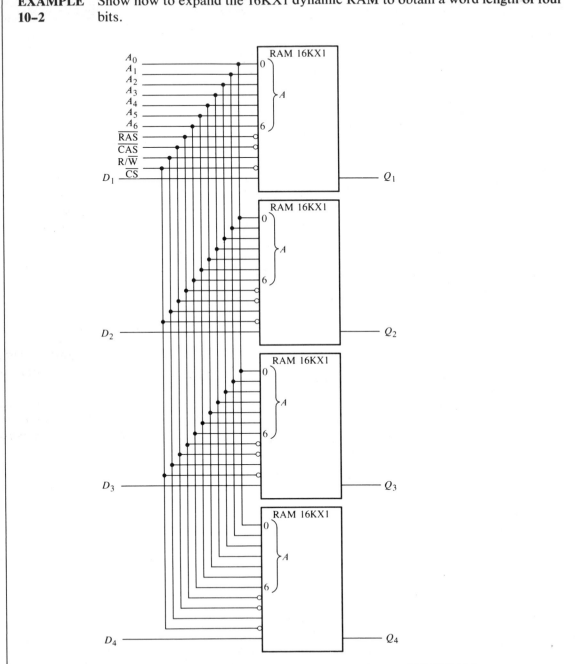

FIGURE 10–41 *Expansion of 16KX1 RAMs to form a 16KX4 RAM.*

Solution Four memories are connected as shown in Figure 10–41.

In addition to expansion to increase word size, memories can also be expanded to increase *word capacity*. For example, a 16KX4 memory has a 16K word capacity; that is, it can store 16,384 four-bit words. To expand to a word capacity of 32K (32,768) four-bit words, two 16KX4 memories are connected as shown in Figure 10–42. Notice that the seven multiplexed address lines (A_0

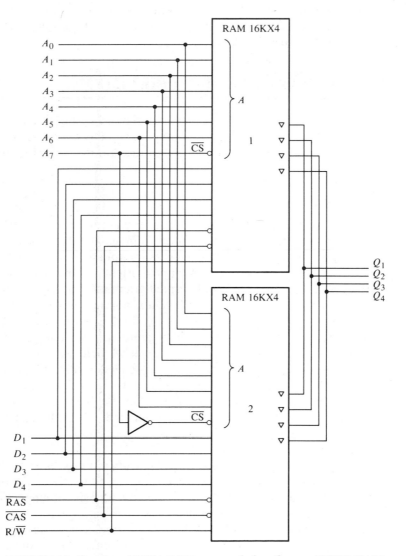

FIGURE 10–42 *Two 16KX4 RAMs expanded to form a 32KX4 RAM.*

through A_6) go to both memories and an eighth (MSB) address line, A_7, uses the chip select to enable memory 1 when it is 0 and memory 2 when it is 1. Therefore, memory addresses 0 through 16,383 are located in memory 1, and addresses 16,384 through 32,767 are located in memory 2. Because both memories have three-state outputs and one memory is disabled at all times, the corresponding output data lines can be connected together as shown.

SECTION REVIEW 10–5

1. How many 16KX1 RAMs are required to achieve a memory with a word capacity of 16K and a word length of eight bits?
2. To expand the 16KX8 memory in Problem 1 to a 32KX8 organization, how many more 16KX1 RAMs are required?

10–6 MAGNETIC BUBBLE MEMORIES (MBMs)

In an MBM, data bits are stored in the form of magnetic "bubbles" moving in thin films of magnetic material. The bubbles are actually cylindrical magnetic domains whose polarization is opposite to that of the thin magnetic film in which they are embedded.

The Magnetic Film

When a thin film of magnetic *garnet* is viewed by polarized light through a microscope, a pattern of "wavy" strips of magnetic domains can be seen. In one set of strips, the tiny internal magnets point up, and in the other areas, they point down. As a result, one set of strips appear bright and the other dark when exposed to the polarized light. This is illustrated graphically in Figure 10–43(a).

Now, if an *external magnetic field* is applied *perpendicular* to the film and slowly increased in strength, the wavy domain strips whose magnetization is opposite to that of the external field begin to narrow. This is illustrated in Figure 10–43(b). At a certain magnitude of external field strength, all these domains suddenly contract into small circular areas called "bubbles," as shown in part (c). These bubbles typically are only a few micrometers in diameter and act as tiny magnets floating in the external field. The bubbles can be easily moved and controlled within the film by rotating magnetic fields in the plane of the film or by current-carrying conductive elements.

Moving the bubbles   The bubbles can be made to move laterally in the film by the application of a magnetic field *parallel* to the film. To control the direction of movement, magnetic "paths" are created by deposits of magnetically conductive material on the surface of the thin film in a specific pattern.

For example, Figure 10–44 shows an asymmetrical chevron pattern. Various other patterns are also feasible. Imagine a magnetic field rotating counterclockwise, parallel with the thin garnet film. At successive points in its rotation, the

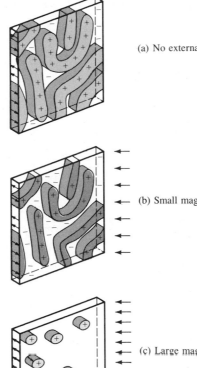

(a) No external magnetic field—wide magnetic domains.

(b) Small magnetic field—positive magnetic domains shrink.

(c) Large magnetic field—bubbles form.

FIGURE 10–43 *Creation of magnetic bubbles in a thin magnetic film by application of an external magnetic field.*

magnetic field is pointing right, down, left, and up as illustrated. As this happens, the chevrons are polarized in the direction of rotation of the magnetic field.

When a bubble is introduced at the right end of the chevron pattern, it moves step by step to the left as the field rotates. In one complete revolution of the field, a bubble moves from a position between two chevrons to a corresponding position between the next two chevrons, as illustrated in Figure 10–45. So, as the magnetic field continues to rotate, the bubbles continue to move along the chevron paths.

Physical Structure of an MBM

A typical MBM assembly is shown in Figure 10–46. The main components are the *thin film memory chip*, the *drive coils (orthogonal coils)*, the *permanent magnets*,

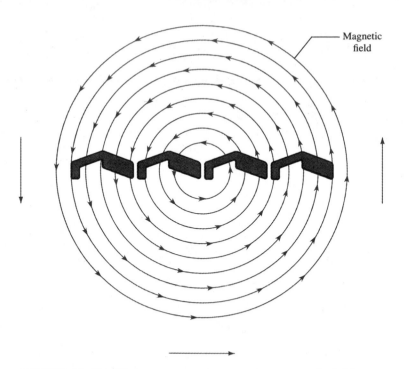

FIGURE 10–44 *Chevron pattern in a rotating magnetic field.*

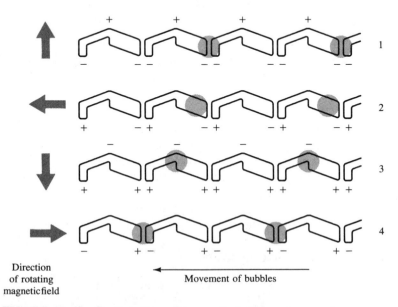

Direction
of rotating
magnetic field

Movement of bubbles

FIGURE 10–45 *Propagation of magnetic bubbles along a chevron pattern.*

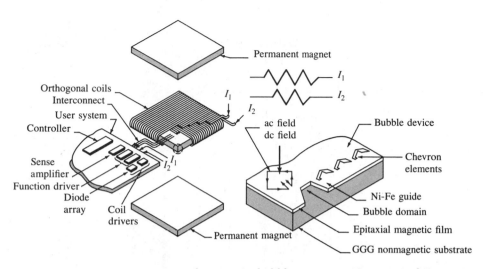

FIGURE 10–46 *Construction of magnetic bubble memory. (Courtesy of Texas Instruments, Inc.)*

and the *control electronics*. These components are assembled into a case which serves as a shield to protect the device from disruptive external magnetic fields.

Major/Minor Loop Architecture

Loops in an MBM are in the form of continuous, elongated paths of chevron patterns. The major/minor loop arrangement consists basically of one *major loop* and many *minor loops,* as shown in Figure 10–47.

In an MBM, 1s and 0s are represented by the presence or absence of bubbles. The minor loops are essentially the memory cell arrays that store the data bits. The major loop is primarily a path to get data from the minor loops, where it is stored to the output during *read,* and to get data from the input to the minor loops during *write.*

Five control functions are used in the read/write cycles of an MBM. These are *generation, transfer, replication, annihilation,* and *detection.*

The Read Cycle

Data are read from the MBM in blocks called *pages*. Basically, a page consists of a number of bits equal to the number of minor loops. For example, a typical MBM may have 150 minor loops, each of which is capable of storing 600 bits. In this case, the page size is 150 bits.

Each bit in a given page occupies the same relative location in each minor loop. During read, all of the bits in a page of data are shifted to the transfer gates and onto the major loop at the same time. Then they are serially shifted around the major loop to the *replicator/annihilator* where each bubble is "stretched" by the

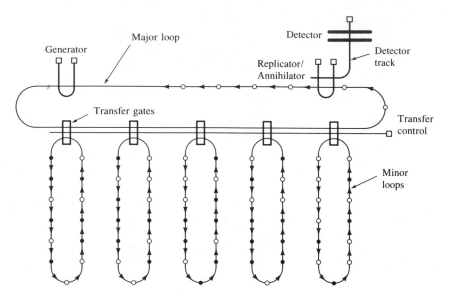

FIGURE 10–47 *Major/minor loop architecture (greatly simplified).*

replicator until it "splits" into two bubbles. One of the replicated bubbles (or no bubble, as the case may be) is transferred to the *detector* where the presence or absence of the bubble is sensed and translated to the appropriate logic level to represent a 1 or a 0. The other replicated bubble continues along the major loop and is transferred back onto the appropriate minor loop for storage. This process continues until each bit in the page is read. The replication/detection process results in a *nondestructive* readout.

The Write Cycle

Before a new page of data can be written into an address, the data currently stored at that address must be annihilated. This is done by a *destructive* readout in which the bubbles are not replicated. Now, the new page of data is produced by the *generator* one bit at a time. Each bit is injected onto the major loop until the entire page of data has been entered. It is then serially moved into position and transferred onto the minor loops for storage.

Applications

MBMs are capable of storing large amounts of data (up to 1 million bits). They are also *nonvolatile* so data are not lost if the power goes off. Because of these features, MBMs are competitive with semiconductor memories in many applications. The main disadvantage is that it takes much longer to get data into and out of an MBM compared to semiconductor memories because MBMs are serially accessed rather than randomly accessed. Typically, higher-capacity MBMs use an architecture called *block replicate* rather than the major/minor loop structure. This, however, is beyond our scope of coverage.

SECTION   **1.** Name the two types of loops in an MBM.
REVIEW   **2.** How are 1s and 0s represented?
10–6   **3.** What is a disadvantage of MBMs compared to semiconductor memories?

10–7   MAGNETIC SURFACE STORAGE DEVICES

In addition to the magnetic bubble memory, there are several other types of magnetic memories in use, including disks and tapes. These devices use a magnetic surface moving past a read/write head to store and retrieve data.

A simplified diagram of the magnetic surface read/write operation is shown in Figure 10–48. A data bit (1 or 0) is written on the magnetic surface by

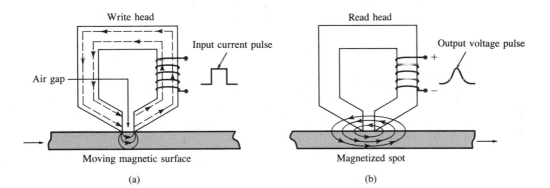

(a)   (b)

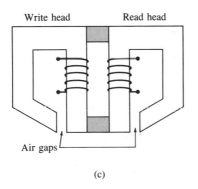

(c)

FIGURE 10–48 *Read/write function on a magnetic surface.*

magnetizing a small segment of the surface as it moves by the *write head.* The direction of the magnetic flux lines is controlled by the direction of the current pulse in the winding as shown in Figure 10–48(a). At the air gap in the write head, the magnetic flux takes a path through the surface of the storage device. This magnetizes a small spot on the surface in the direction of the field. A magnetized spot of one polarity represents a binary 1, and one of the opposite polarity represents a binary 0. Once a spot on the surface is magnetized, it remains until written over with an opposite magnetic field.

When the magnetic surface passes a *read head,* the magnetized spots produce magnetic fields in the read head which induce voltage pulses in the winding. The polarity of these pulses depends on the direction of the magnetized spot and indicates whether the stored bit is a 1 or a 0. This is illustrated in Figure 10–48(b). Very often the read and write heads are combined into a single unit as shown in part (c).

Magnetic Recording Formats

Several ways in which digital data can be represented for purposes of magnetic surface recording are *return-to-zero* (RZ), *non-return-to-zero* (NRZ), *biphase, Manchester,* and the *Kansas City standard.* These waveform representations are separated into *bit times,* the intervals during which the level or frequency of the waveform indicates a 1 or 0 bit. These bit times are definable by their relation to a basic system timing signal or *clock.*

Figure 10–49 shows an example of a *return-to-zero* (RZ) waveform. In this case, a fixed-width pulse occurring during a bit time represents a 1, and no pulse during a bit time is a 0. There is always a return to the 0 level after a 1 occurs. The period of the clock waveform determines the bit time interval.

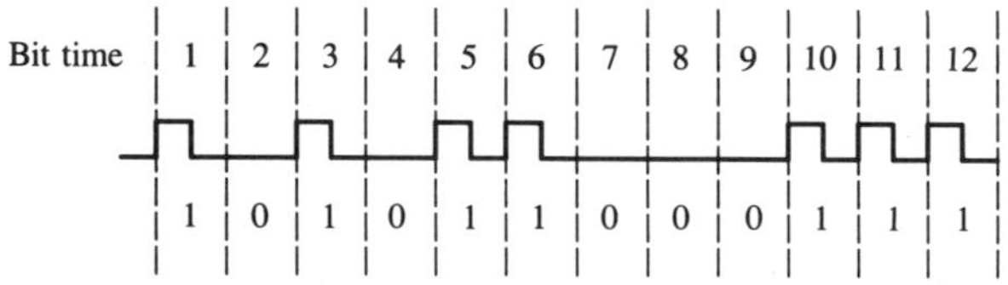

FIGURE 10–49   *An RZ waveform representing 101011000111.*

Figure 10–50 illustrates a *non-return-to-zero* (NRZ) waveform. In this case, a 1 or 0 level remains during the entire bit time. If two or more 1s occur in succession, the waveform does not return to the 0 level until a 0 occurs.

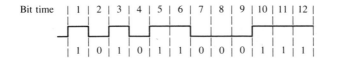

FIGURE 10–50   *An NRZ waveform representing 101011000111.*

Figure 10–51 is an illustration of a *biphase* waveform. In this type, a 1 is a HIGH level for the first half of a bit time and a LOW level for the second half, so a *HIGH-to-LOW transition occurring in the middle of a bit time is interpreted as a 1*. A 0 is represented by a LOW level during the first half of a bit time followed by a HIGH level during the second half, so a *LOW-to-HIGH transition in the middle of a bit time is interpreted as a 0.*

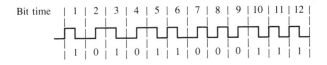

FIGURE 10–51 *A biphase waveform representing 101011000111.*

Manchester is another type of phase encoding in which a HIGH-to-LOW transition at the start of a bit time represents a 0 and no transition represents a 1. Figure 10–52 illustrates a Manchester waveform.

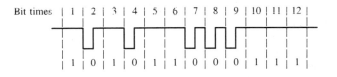
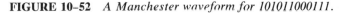

FIGURE 10–52 *A Manchester waveform for 101011000111.*

The *Kansas City* method uses two different frequencies to represent 1s and 0s. The standard 300 bits/second version uses eight cycles of 2400 Hz to represent a 1, and four cycles of 1200 Hz to represent a 0. This is illustrated in Figure 10–53.

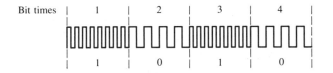

FIGURE 10–53 *A Kansas City standard code for 1010.*

We have all seen the tape cassettes used in small microcomputer systems to store data. Larger reel-type systems are used in many large computer systems. A typical tape format with nine tracks (some have seven) is shown in Figure 10–54(a). There is a separate read/write head for each track as indicated. The tape is divided in records with gaps between records for the starting and stopping of the tape. A record is organized into a format such as shown in Figure 10–54(b). The marker indicates the beginning of a record, the ID specifically identifies the record, and the stored data then appear followed by a checksum.

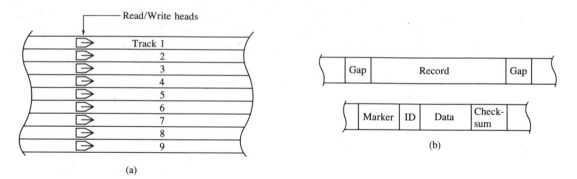

(a)

(b)

FIGURE 10–54 *Typical tape format.*

The *checksum* is the arithmetic sum of all the bytes in a record and is used by the computer to check for errors in the data as they are being taken from the tape. This is basically done by adding the data bytes as they are read from the tape and comparing this figure to the checksum.

The Floppy Disk

The floppy disk (diskette) is a small flexible, Mylar® disk with a magnetic surface. It is *permanently* housed in a square jacket for protective purposes as shown in Figure 10–55.

The surface of the disk is coated with a thin magnetic film in which binary data are stored in the form of minute magnetized regions. There are cutout areas

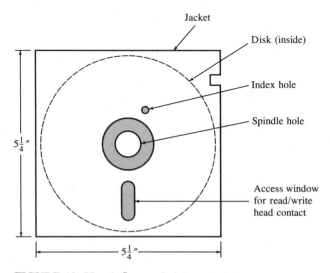

FIGURE 10–55 *A floppy disk in a jacket.*

in the jacket for the drive spindle, read/write head, and index position sensor. The index hole establishes a reference point for all the tracks on the disk. As the disk rotates at 360 rpm within the stationary jacket, the read/write head makes contact through the access window.

A typical $5\frac{1}{4}$-inch floppy disk is organized into 77 tracks, as shown in Figure 10–56(a). The disk is divided into 26 sectors, as shown in Figure 10–56(b), so that each of the 77 tracks is also divided into 26 equal-sized sectors. The longer outside track has the same number of sectors and the same sector length as does the shorter inside track. There is just more unused space between sectors in the longer track. Notice that the index hole appears between the first and the last sectors.

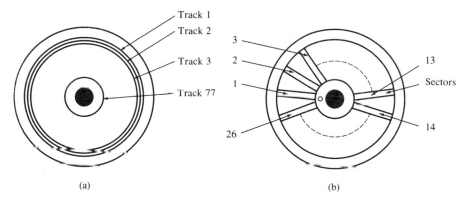

FIGURE 10–56 *Track and sector organization.*

Each sector can store 128 bytes of data. (Typically, 1 byte represents one character or numeral.) The total storage capacity of the disk is therefore

$$(128 \text{ bytes/sector})(26 \text{ sectors/track})(77 \text{ tracks}) \cong 256,000 \text{ bytes}$$

A typical sector format is shown in Figure 10–57, where each sector is divided into fields. The address mark passes the read/write head first and identifies the upcoming areas of the sector as the ID field. The ID field identifies the data field by sector and track number. The data mark indicates whether the upcoming data field contains a good record or a deleted record. The data field is the portion of the sector that contains the 128 data bytes. The average access time

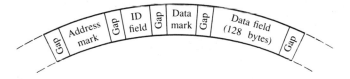

FIGURE 10–57 *A typical sector format for one track of a floppy disk.*

to a given sector is about 500 ms. This is much faster than for magnetic tape but much slower than the semiconductor memories.

Both the ID field and the data field typically contain two bytes for *cyclic redundancy check* (CRC). These bytes provide for error detection. The CRC is computed by the disk controller with the recorded data, using a special algorithm. Then it is compared to the CRC recorded with the data. If they are not the same, there is an error.

SECTION REVIEW 10-7

1. How are data bits stored on a magnetic surface?
2. How many tracks are there on a typical 5¼-inch floppy disk?
3. What does *CRC* mean?

10-8   SPECIAL MEMORIES AND APPLICATIONS

Programmable Logic Arrays (PLAs)

PLAs are similar in many ways to ROMs but are very different in terms of their internal structure. The PLA is either mask programmed (PLA) or field programmed by the user (FPLA) to produce specified logic functions.

Basically, a PLA consist of two arrays, one for AND logic and one for OR logic, and can be programmed to produce desired logic functions on the outputs. An example of a small PLA is shown in Figure 10-58. Each connection is mask programmable so that desired variables can be connected into each gate. This three-variable example is programmed to produce the indicated logic expressions by masking *open* certain connections. The masking process is done by the manufacturer.

A block diagram representing a specific PLA, the 74PL839, is shown in Figure 10-59. This device has fourteen inputs. The input buffer ($\triangleright$) provides the input variables and their complements on its outputs. The AND matrix ($\&$) consists of 32 product terms (AND), and the OR matrix ($\geq$) consists of six OR terms. The AND and OR matrices can be programmed for the desired sum-of-products output functions. The fused inputs ($\sim$) to the AND matrix can be programmed for the true inputs, or their complements, or both. The exclusive-OR array ($=1$) is for programming inversions of the output functions. This is accomplished by programming the fused input ($\sim$). A fused input left connected to ground (0) results in the input bit remaining uncomplemented ($1 \oplus 0 = 1, 0 \oplus 0 = 0$). A fused input blown open (1) results in inversion of the input bit ($1 \oplus 1 = 0, 0 \oplus 1 = 1$).

First In–First Out (FIFO) Memories

This type of memory is formed by an arrangement of shift registers. The term *FIFO* refers to the basic operation of this type of memory, in which the first data bit written into the memory is the first to be read out.

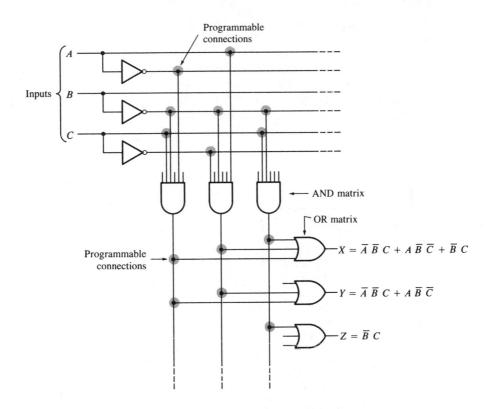

FIGURE 10–58 *Simple example of a PLA.*

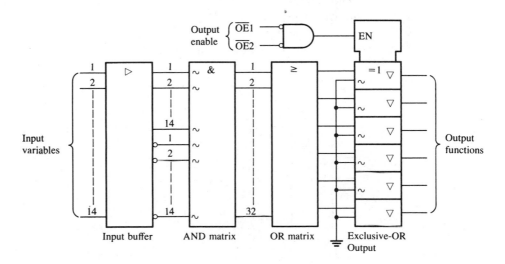

~ denotes programmable fused inputs.

FIGURE 10–59 *Block diagram of a 74PL839 PLA.*

There is one important difference between a conventional shift register and a FIFO memory register: In a conventional register, a data bit moves through the register only as new data bits are entered; but in a FIFO register, a data bit immediately goes through the register to the right-most bit location that is empty. This is illustrated in Figure 10–60.

Conventional shift register

Input	X	X	X	X	Output
0	0	X	X	X	→
1	1	0	X	X	→
1	1	1	0	X	→
0	0	1	1	0	→

X = unknown data bits.

In a conventional shift register, data stay to the left until "forced" through by additional data.

FIFO shift register

Input	—	—	—	—	Output
0	—	—	—	0	→
1	—	—	1	0	→
1	—	1	1	0	→
0	0	1	1	0	→

— = empty positions.

In a FIFO shift register, data "fall" through (go right).

FIGURE 10–60 *Comparison of conventional and FIFO register operation.*

Figure 10–61 is a block diagram of a typical FIFO serial memory. This particular memory has four serial 64-bit data registers and a 64-bit control register (marker register). When data are entered by a shift-in pulse, they move automatically under control of the marker register to the empty location closest to the output. Data cannot advance into occupied positions. However, when a data bit is shifted out by a shift-out pulse, the data bits remaining in the registers automatically move to the next position toward the output. In an *asynchronous* FIFO, data are shifted out independent of data entry with the use of two separate clocks.

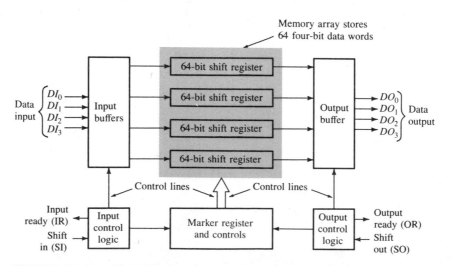

FIGURE 10–61 *Block diagram of a typical FIFO serial memory.*

Applications

One important application area for the FIFO is the case in which two systems of differing data rates must communicate. Data can be entered into a FIFO at one rate and be put out at another rate. Figure 10–62 illustrates how a FIFO might be used in these situations.

(a) Irregular telemetry data can be stored and retransmitted at a constant rate.

(b) Data input at a slow keyboard rate can be stored and then transferred at a higher rate for processing.

(c) Data input at a steady rate can be stored and then output in even bursts.

(d) Data in bursts can be stored and reformatted into a steady-rate output.

FIGURE 10–62 *The FIFO in data-rate buffering applications.*

CCD Memories

The CCD (charge-coupled device) memory stores data as charges on capacitors. Unlike the dynamic RAM, however, the storage cell does not include a transistor. High density is the main advantage of CCDs.

The CCD memory consists of long rows of semiconductor capacitors, called *channels*. Data are entered into a channel serially by depositing a small charge for a 0 and a large charge for a 1 on the capacitors. These charge *packets* are then shifted along the channel by clock signals as more data are entered.

As with the dynamic RAM, the charges must be refreshed periodically. This process is done by shifting the charge packets serially through a refresh circuit. Figure 10–63 shows the basic concept of a CCD channel.

Charge movement

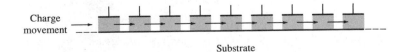

Substrate

FIGURE 10–63 *A CCD (charge-coupled device) channel.*

1. A PLA consists of arrays of _____.
2. What is a FIFO memory?
3. What does the term *CCD* stand for?

10–9   TROUBLESHOOTING

In this section, a specific memory problem is presented, and a specific troubleshooting approach using a logic analyzer is used to isolate the problem. The intent here is not to cover all details of the approach or of the use of a logic

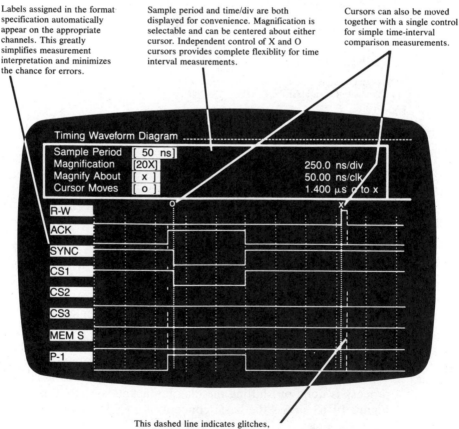

Labels assigned in the format specification automatically appear on the appropriate channels. This greatly simplifies measurement interpretation and minimizes the chance for errors.

Sample period and time/div are both displayed for convenience. Magnification is selectable and can be centered about either cursor. Independent control of X and O cursors provides complete flexiblity for time interval measurements.

Cursors can also be moved together with a single control for simple time-interval comparison measurements.

This dashed line indicates glitches, whether they appear on a static line or on an edge.

FIGURE 10–64   *An example of a timing analysis display on a logic analyzer.*

analyzer, but only to give you a general idea of how this instrument can be used in a typical troubleshooting situation.

Figure 10-64 shows an example timing format and the features of a typical logic analyzer display.

The Problem

Suppose that you are checking out a memory system and discover incorrect data out of a RAM. From your knowledge of this memory, you realize that if any address lines change during a write operation, erroneous data can result; so, you decide to check out this possibility.

Qualifying a Transition with a Pattern

You do not want to view just any address line transition. The only one of interest is the one (if any) that changes when the *write enable* is LOW. The *trace specification* on the logic analyzer can be set to qualify the transition with a pattern. For example, in this case the analyzer should trigger only if an address line transition occurs while the write enable ($\overline{\text{WE}}$) is LOW, as shown in Figure 10-65.

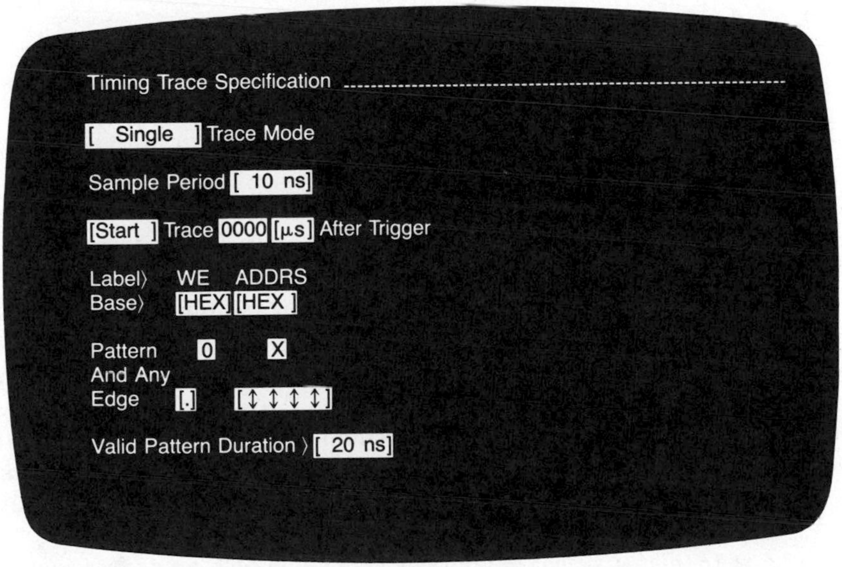

FIGURE 10-65 *This TRACE SPECIFICATION arms the logic analyzer to trigger when the WRITE ENABLE line is low and there is either a positive or negative transition on any address line.*

Figure 10–66 shows that the analyzer found the specified condition. Address line 0 makes a low-to-high transition when $\overline{WE}$ is LOW. The waveforms show that this does not occur each time $\overline{WE}$ is LOW, only intermittently.

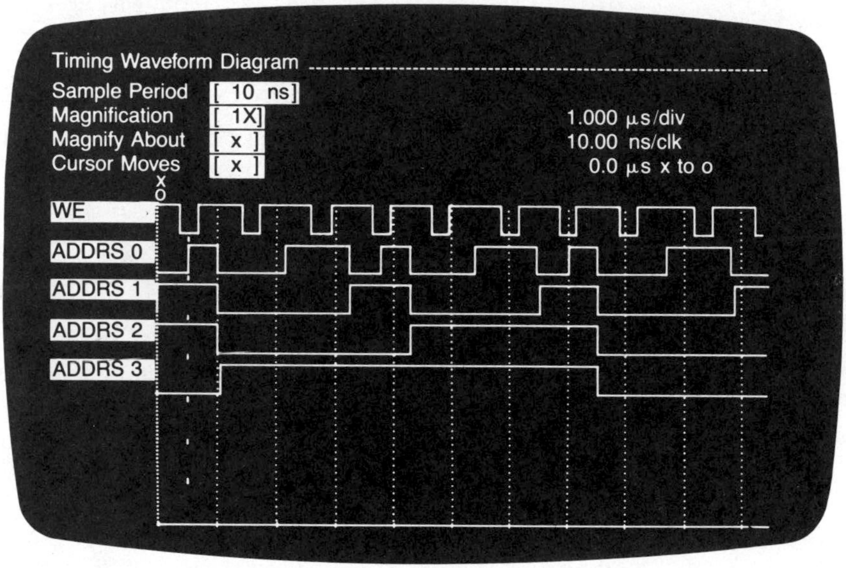

FIGURE 10–66   *ADDRESS line 0 goes from low to high during WRITE ENABLE. However, the display shows that it does not occur every cycle, indicating an intermittent problem.*

Measure the Time and Compare

By magnifying the trace and using the *dual cursors,* you can measure the time interval between the address line transition and the positive transition of the WE line. Since in this example the interval is 150 ns, as Figure 10–67 shows, the problem is probably not caused by timing errors because of the length of the time interval.

A second edge-to-edge transition interval can be checked by moving the cursors together. Figure 10–68 shows that the intervals are the same. Because the problem is intermittent but occurs at the same point in a write cycle, a glitch is suspected.

In this particular memory system under test, an *address counter* provides for the sequencing of the address lines. Therefore, the next step is to check out this address counter by looking at its *clock, enable,* and *reset* inputs.

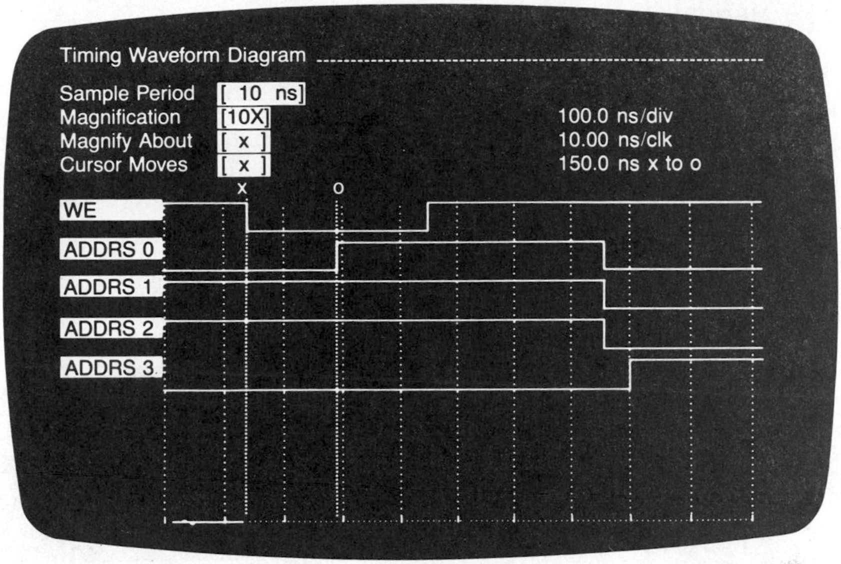

FIGURE 10–67 *Using movable cursors, the time interval between the ADDRESS line 0 transition and WRITE ENABLE going high is found to be 150 ns. This suggests something other than a timing margin problem.*

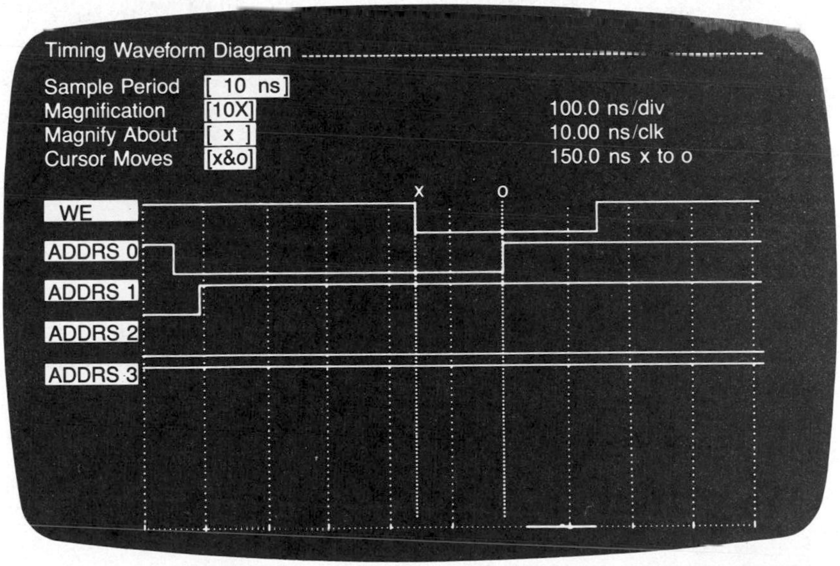

FIGURE 10–68 *Comparison of the time intervals in the first and third cycles, made by moving the two cursors simultaneously as a "time window," shows that the two intervals are the same.*

Trigger on a Glitch with Pattern Qualification

The analyzer can be armed to trigger on a glitch on any of the inputs to the address counter when the $\overline{WE}$ line is LOW. The resultant trace, shown in Figure 10–69, shows that there is indeed a glitch on the clock input to the address counter, and it occurs just prior to the unwanted transitions of address line 0. The problem is now basically isolated to the clock line of the counter. An investigation of the clock source must be made to identify the exact cause of the problem.

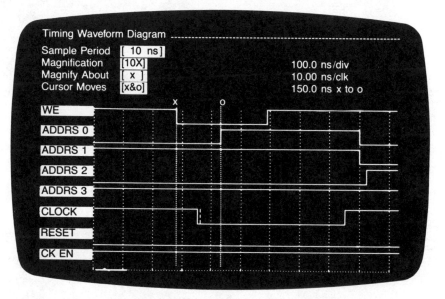

FIGURE 10–69 *Connecting to the address counter inputs reveals a glitch on the trailing edge of the CLOCK input. This occurs immediately preceding the transition on ADDRESS line 0 and is the source of the problem.*

SECTION REVIEW 10–9

1. How are glitches indicated on the logic analyzer display?

SUMMARY

☐ Data are stored in a memory by a process called *writing*.
☐ Data are retrieved from a memory by a process called *reading*.
☐ The *capacity* of a memory is the total number of bits that it can store.
☐ ROMs are read only memories in which data are permanently or semipermanently stored.

☐ RAMs (random access memories) are read/write memories.
☐ *Nonvolatility* means that a memory does not lose its data when power goes off. *Volatility* means that data are lost when power goes off.
☐ Each storage location in a memory is defined by a specific address.
☐ A *word* is a complete unit of binary data which can be any number of bits. A *byte* is eight bits of binary data.
☐ Semiconductor memories can be either bipolar or MOS.
☐ The three states of three-state logic are HIGH, LOW, and open (high impedance).
☐ Memory *access time* is the time required for valid data to appear on the outputs after the activation of the appropriate input.
☐ PROMs are user-programmable ROMs. EPROMs are erasable PROMs.
☐ RAMs can be either *static* or *dynamic*. A static RAM has latch storage cells. A dynamic RAM has capacitive storage cells.
☐ The organization of a memory is normally specified by the number of words (m) and the number of bits per word (n), as $m \times n$.
☐ In a magnetic bubble memory (MBM), data are stored as tiny magnetic domains called *bubbles*. Bubble memories are serially accessed.
☐ Two other types of magnetic memories are *tape* and *disk*.
☐ PLAs are programmable logic arrays in which AND-OR gate arrays can be mask programmed to produce specified logic functions.
☐ FIFOs are first in–first out memories in which the first bit written is the first bit read.
☐ CCD (charge-coupled device) memories use channels of semiconductor capacitors to store charges representing data bits.

SELF-TEST

1. What is the bit capacity of a memory that has 512 addresses and can store 8 bits at each address?
2. To retrieve data from a memory, a _____ operation must be performed.
3. (a) List three types of memories in the MOS ROM category.
   (b) List two types of memories in the bipolar ROM category.
4. List two types of memories in (a) the MOS RAM category and (b) the bipolar RAM category.
5. How many bytes does a 32-bit data word contain?
6. What is the purpose of the program counter in a microprocessor system?
7. What is the purpose of the address buffer in a microprocessor system?
8. How many bits can actually be stored in a 64K memory?
9. What is the total capacity of a ROM with 1024 rows and 4 columns?
10. How many address bits are required for a 512X4 memory?
11. What does the chip-select input of a memory do?
12. Explain the power down mode that is available on some memory devices.
13. For a certain ROM, the access time from address to data output is 450 ns, and the access time from chip select is 200 ns. The chip select is activated 100 ns before the valid address is applied. How long after chip select do valid output data appear?
14. A certain 256X4 RAM has an access time of 300 ns from address to data output. How long does it take to read data from all memory addresses if the address transition times are neglected?

15. Explain the purpose of the address latches in a dynamic RAM such as in Figure 10–35.
16. What is the function of the refresh counter in Figure 10–37?
17. How many 64KX4 RAMs are needed to form a 64KX8 memory? Is this an example of word-capacity expansion or word-length expansion?
18. What are the storage sections in an MBM called?
19. What is the purpose of checksum in a magnetic tape format?
20. How does a PLA differ from a ROM?

PROBLEMS

Section 10–1

10–1 Identify the ROM and the RAM in Figure 10–70.

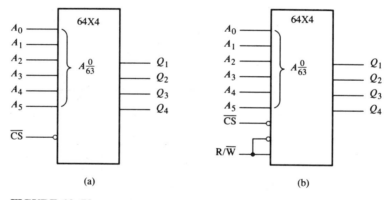

FIGURE 10–70

10–2 Explain why RAMs and ROMs are both random access memories.

10–3 Explain the purposes of the address bus, the data bus, and the control bus in a typical microprocessor-based system.

10–4 What memory address (0 through 256) is represented by each of the following hexadecimal numbers:
(a) $0A_{16}$ (b) $3F_{16}$ (c) CD_{16}

Section 10–2

10–5 For the ROM array in Figure 10–71, determine the outputs for all possible input combinations, and summarize them in tabular form (light cell is a 1, dark cell is a 0).

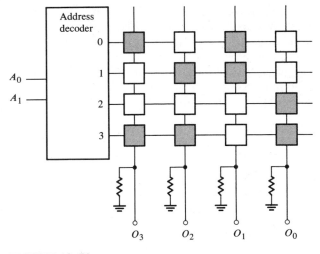

FIGURE 10–71

10–6   Determine the truth table for the ROM in Figure 10–72.

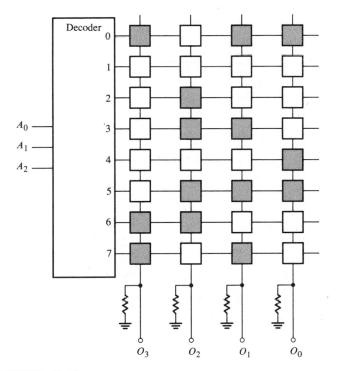

FIGURE 10–72

10–7 Using a procedure similar to that in Example 10–1, design a ROM for conversion of single-digit BCD to Excess-3 code.

10–8 What is the total bit capacity of a ROM that has fourteen address lines, eight data inputs, and eight data outputs?

10–9 Determine the output of the 74185 binary-to-BCD converter in Figure 10–73 for each of the following inputs:
(a) 001101_2 **(b)** 101110_2 **(c)** 110111_2

FIGURE 10–73

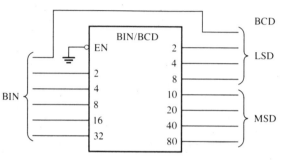

FIGURE 10–74

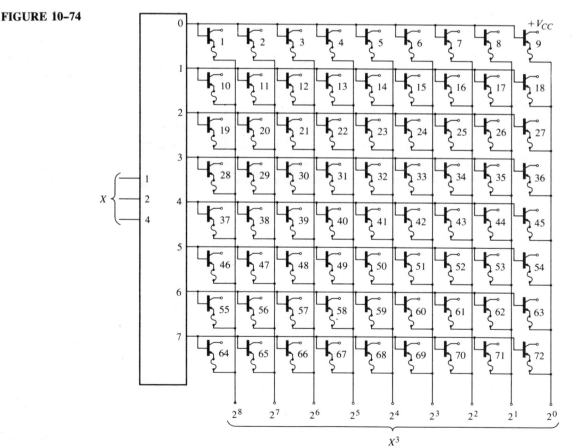

10–10 Expand the BCD-to-binary converter in Figure 10–22 to handle a nine-bit binary input.

Section 10–3

10–11 Assuming that the PROM matrix in Figure 10–74 is programmed by blowing a fuse link to create a 0, indicate the links to be blown to program an x^3 look-up table, where x is a number from 0 through 7.

10–12 Determine the addresses that are programmed and the contents of each address after the programming sequence in Figure 10–75 is applied to a TMS2516 PROM.

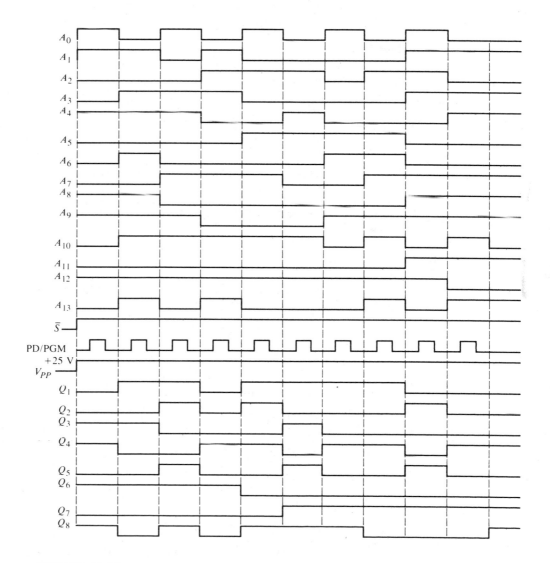

FIGURE 10–75

Section 10–4

10–13 A static memory cell such as the one in Figure 10–28 is storing a 0 (RESET). What is its state after each of the following conditions:
   (a) ROW = 1, COLUMN = 1, DATA IN = 1, READ/$\overline{\text{WRITE}}$ = 1
   (b) ROW = 0, COLUMN = 1, DATA IN = 1, READ/$\overline{\text{WRITE}}$ = 1
   (c) ROW = 1, COLUMN = 1, DATA IN = 1, READ/$\overline{\text{WRITE}}$ = 0

10–14 Draw a basic logic diagram for a 512X8-bit static RAM showing all the inputs and outputs.

10–15 Assuming that a 512X8 static RAM has a structure similar to that of the 256X4 RAM in Figure 10–30, determine the size of its memory cell array.

10–16 Redraw the block diagram in Figure 10–37 for a 32K memory.

Section 10–5

10–17 Use 74189 16X4 RAMs to build a 64X8 RAM. Show the logic diagram.

10–18 Use TMS4164 64KX1 dynamic RAMs to build a 256KX4 RAM.

10–19 What is the word size in the memory of Problem 10–17? Problem 10–18?

Section 10–6

10–20 A certain MBM has 250 minor loops, each of which stores 1000 bits. What is the total bit capacity? What is the page size?

10–21 What mechanism in an MBM permits nondestructive readout?

Section 10–7

10–22 Determine the sequence of bits represented by each of the waveforms in Figure 10–76.

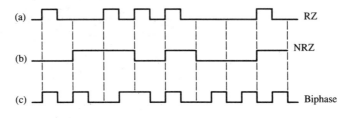

FIGURE 10–76

10–23 Repeat Problem 10–22 for the waveforms in Figure 10–77.

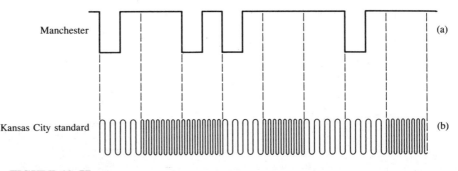

FIGURE 10–77

Section 10–8

10–24 In the simplified PLA in Figure 10–78, determine the points to be opened by mask programming in order to produce the following logic functions:

(a) $A\overline{B}C + \overline{A}\,\overline{B}\,\overline{C} + ABC$ (b) $\overline{A} + BC$ (c) $\overline{A} + \overline{B} + \overline{C}$

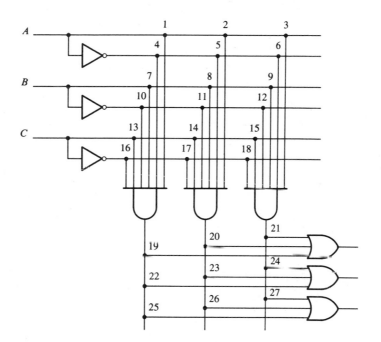

FIGURE 10–78

10–25 Complete the timing diagram in Figure 10–79 by showing the output waveforms for a FIFO serial memory like that shown in Figure 10–61.

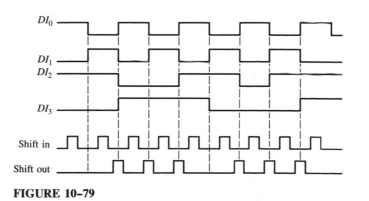

FIGURE 10–79

ANSWERS TO SECTION REVIEWS

Section 10–1

1. ROM has only a read operation; RAM has both read and write operations.
2. Data are not lost when power is removed.
3. Static: the storage cells are latches and can retain data indefinitely. Dynamic: the storage cells are capacitors and must be refreshed periodically.

Section 10–2

1. 2048 bits.   2. Three-state allows outputs to be disconnected from bus lines.
3. Eleven bits.

Section 10–3

1. PROMs are field-programmable; ROMs are not.   2. 1s.   3. Read.

Section 10–4

1. Static RAMs have latch or flip-flop storage cells that can retain data indefinitely. Dynamic RAMs have capacitive storage cells that must be periodically refreshed.
2. The refresh operation prevents data from being lost due to capacitive discharge. A stored bit is restored periodically by recharging the capacitor.

Section 10–5

1. Eight   2. Sixteen

Section 10–6

1. Major loop, minor loop.
2. 1 = presence of magnetic bubble; 0 = absence of magnetic bubble.
3. MBMs are slower than semiconductor memories (it takes longer to get data in and out).

Section 10–7

1. Magnetized spots with specified polarities.
2. 77   3. Cyclic redundancy check.

Section 10–8

1. AND gates and OR gates.
2. In FIFO, the first bit (or data word) in is the first one out.
3. Charge-coupled device.

Section 10–9

1. By short, vertical dashed lines.

Interfacing is the process of making two or more electronic devices or systems operationally compatible with each other so that they function together as required. Interfacing involves such considerations as input and output voltages and currents, loading, signal timing, and data and control formats.

Specific devices introduced in this chapter are the following:

1. SN75160A interface bus transceiver
2. TMS9914 bus controller
3. ADC0801 analog-to-digital converter
4. Bell 103 modem
5. TMS99532 modem

In this chapter, you will learn

☐ How to interface CMOS to TTL and TTL to CMOS.
☐ What a totem-pole output is.
☐ What an open-collector output is.
☐ How to drive lamps and LEDs.
☐ How to create a wired-AND function.
☐ How to interface digital devices to buses.
☐ The characteristics of the GPIB (IEEE–488).
☐ How to convert from digital to analog using the binary-weighted input method and the $R/2R$ ladder method.
☐ How to convert from analog to digital using the simultaneous method, the stairstep-ramp method, the tracking method, the single-slope method, the dual-slope method, and the successive-approximation method.
☐ How modems are used and how they function.
☐ How digital data are transmitted over telephone lines.
☐ Characteristics of the RS-232C, RS422A, and RS-423A interfaces.
☐ What a Schmitt trigger circuit is and how it can be used.

11

Interfacing and Data Transmission

11-1 INTERFACING LOGIC FAMILIES

In this section, the interfacing of TTL and CMOS logic devices is considered. Also, other practical situations in which logic devices are interfaced using methods such as *open-collector* logic are introduced.

When two different types of technologies are interfaced, the input and output voltages and currents of each are important parameters. In fact, the difference in these parameters creates the interfacing problem. Table 11–1 shows typical worst-case values of the input and output parameters for a CMOS family and several TTL families.

TABLE 11–1 *Worst-case values of interfacing parameters.*

Parameter	74H CMOS	74 TTL	74LS TTL	74AS TTL	74ALS TTL
$V_{IH(min)}$	3.5 V	2 V	2 V	2 V	2 V
$V_{IL(max)}$	1 V	0.8 V	0.8 V	0.8 V	0.8 V
$V_{OH(min)}$	4.9 V	2.4 V	2.7 V	2.7 V	2.7 V
$V_{OL(max)}$	0.1 V	0.4 V	0.4 V	0.4 V	0.4 V
$I_{IH(max)}$	1 μA	40 μA	20 μA	200 μA	20 μA
$I_{IL(max)}$	−1 μA	−1.6 mA	−400 μA	−2 mA	−100 μA
$I_{OH(max)}$	−4 mA	−400 μA	−400 μA	−2 mA	−400 μA
$I_{OL(max)}$	4 mA	16 mA	8 mA	20 mA	4 mA

CMOS-to-TTL Interfacing

This discussion is for the case in which CMOS is driving TTL. Table 11–1 shows that the minimum high output voltage $V_{OH(min)}$ for CMOS is 4.9 V. Since this value exceeds the minimum high input voltage $V_{IH(min)}$ of 2 V that is required by TTL, the CMOS is compatible with TTL in the high state.

CMOS has a maximum low output voltage $V_{OL(max)}$ of 0.1 V. Since this value is less than the maximum low input voltage $V_{IL(max)}$ of 0.8 V required by TTL, the CMOS is also compatible with TTL in the low state.

In terms of currents, CMOS can sink 4 mA, $I_{OL(max)}$, in the low output state with a guaranteed output voltage. When driving *standard* TTL, the CMOS gate must be able to sink 1.6 mA from each TTL input. This limits the fanout of the CMOS gate to *two* TTL inputs (2 × 1.6 mA = 3.2 mA), as shown in Figure 11–1(a).

When driving *low-power Schottky* TTL (74LS), the CMOS gate must be able to sink 400 μA from each TTL input. This limits the fanout of the CMOS gate to *ten* LS TTL inputs (10 × 400 μA = 4 mA), as shown in Figure 11–1(b).

When driving *advanced Schottky* TTL (74AS), the CMOS gate must be able to sink 2 mA from each TTL input. This limits the fanout of the CMOS gate to *two* AS TTL inputs (2 × 2 mA = 4 mA), as shown in Figure 11–1(c).

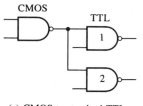

(a) CMOS-to-standard TTL.
Fanout = 2

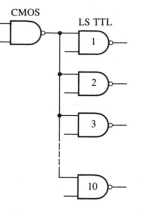

(b) CMOS-to-low power Schottky TTL.
Fanout = 10

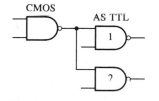

(c) CMOS-to-advanced Schottky TTL.
Fanout = 2

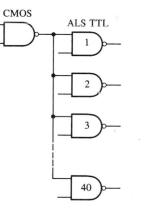

(d) CMOS-to-advanced low power
Schottky TTL.
Fanout = 40

FIGURE 11–1 *CMOS-to-TTL interfacing.*

Finally, when driving *advanced low-power Schottky* TTL (74ALS), the CMOS gate must be able to sink 100 μA from each TTL input. This limits the fanout of the CMOS gate to *forty* ALS TTL inputs (40 × 100 μA = 4 mA), as shown in Figure 11–1(d).

TTL-to-CMOS Interfacing

When TTL is driving CMOS, the interface is not as simple as CMOS-to-TTL. As you can see in Table 11–1, the TTL families have minimum high output voltages $V_{OH(\text{min})}$ of 2.4 V to 2.7 V. The minimum high input voltage required by CMOS is 3.5 V. Thus, the output voltage of TTL is not sufficient to drive CMOS in the high state. In the low state, the voltages are compatible, as you can verify in Table 11–1.

To properly interface TTL to CMOS, a *pull-up* resistor (R_p) to V_{CC} must be added as shown in Figure 11–2.

FIGURE 11–2   *TTL-to-CMOS interfacing.*

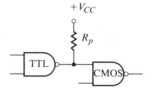

In the low state, the TTL driving gate must sink current from the resistor as well as from the CMOS inputs to which it is connected, as indicated in Figure 11–3. This sets the minimum value of R_p according to the following equation:

$$R_p = \frac{V_{CC} - V_{OL(\text{max})}}{I_{OL(\text{TTL})} + nI_{IL(\text{CMOS})}} \tag{11-1}$$

where n is the number of CMOS inputs being driven and $I_{OL(\text{TTL})} + nI_{IL(\text{CMOS})} = I_{Rp}$.

FIGURE 11–3   *Low-state current sinking.*

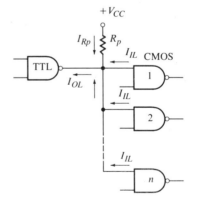

EXAMPLE 11–1   A 74LS TTL gate drives four 74HC CMOS gates. The minimum V_{CC} is 4.75 V. Determine the minimum value of pull-up resistor for interfacing these devices.

Solution   From Table 11–1, $V_{OL(\text{max})}$ = 0.4 V, I_{OL} = 8 mA, and I_{IL} = −1 µA. Thus,

$$R_p = \frac{V_{CC} - V_{OL(\text{max})}}{I_{OL} + 4I_{IL}}$$

$$= \frac{4.75 \text{ V} - 0.4 \text{ V}}{8 \text{ mA} - 4 \text{ } \mu\text{A}} = \frac{4.35 \text{ V}}{7.996 \text{ mA}}$$

$$= 544 \text{ } \Omega$$

TTL with Totem-Pole Outputs

Regular TTL output circuits have a "totem-pole" arrangement as shown in Figure 11–4(a). In the high output state, transistor Q_1 is on and transistor Q_2 is off. In the low output state, Q_1 is off and Q_2 is on.

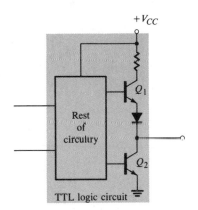

(a) TTL totem-pole output circuit

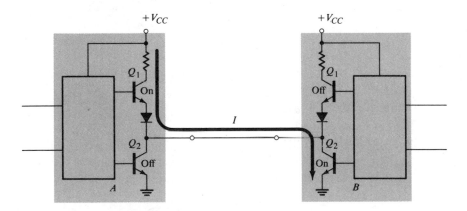

(b) Totem-pole outputs wired together may cause excessive current through Q_1 of device A and Q_2 of device B.

FIGURE 11–4 *TTL totem-pole outputs.*

Totem-pole outputs cannot be connected together because such a connection may produce excessive current and result in destruction of the devices. For example, in Figure 11–4(b), when Q_1 in device A and Q_2 in device B are both on, the output of device A is effectively *shorted* to ground through Q_2 of device B.

Also, TTL with the totem-pole output is limited in the amount of current that it can sink in the low state, $I_{OL(\text{max})}$, to 16 mA for standard TTL and 20 mA for AS TTL. In many applications, a gate must drive an external device that requires more current than this.

Lamp/LED Interfacing

Logic devices with *open-collector* outputs, rather than totem-pole outputs, are generally used for driving display lamps or LEDs because of their higher voltage and current-handling capability. However, totem-pole outputs can be used as long as the current required by the external device does not exceed the amount that the TTL driver can sink.

In an open-collector TTL gate, the collector of the output transistor is not connected internally but is available for connection to external loads. A simplified open-collector circuit is shown in Figure 11–5.

FIGURE 11–5 *TTL open-collector circuit.*

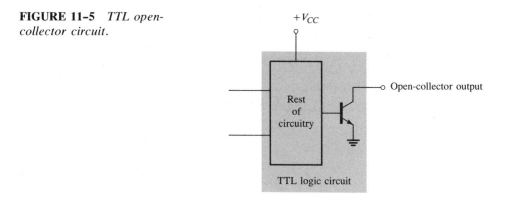

The open-collector output can be connected to either an LED or a lamp as shown in Figure 11–6. In part (a), the limiting resistor is used to keep the current

FIGURE 11–6 *Open-collector gates driving displays.*

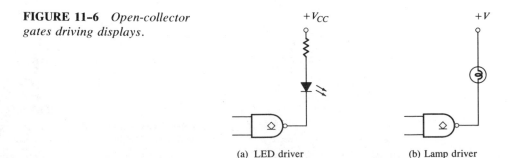

(a) LED driver   (b) Lamp driver

below the maximum LED current. When the output of the gate is LOW, the output transistor is sinking current and the LED is *on*. The LED is *off* when the gate output is HIGH. A typical open-collector buffer can sink up to about 40 mA.

In part (b) of Figure 11–6, the lamp requires no limiting resistor (the filament is resistive). Typically, up to +30 V can be used on the open collector, depending on the particular logic family. The standard designation used on logic symbols to denote an open-collector output is ◇ .

Wired-AND Interfacing

The open-collector outputs of several gates can be connected together to form what is known as a *wired-AND* function. For example, four open-collector AND gate outputs are wired together in Figure 11–7. When all input variables are HIGH, the output transistor of each gate is off and the output X is pulled HIGH. When one or more input variables are LOW, one or more gate outputs are LOW, forcing all the others LOW, so X is LOW. Thus, X is HIGH if and only if all input variables are HIGH. Therefore, the output X is the AND of all the inputs:

$$X = ABCDEFGH$$

FIGURE 11–7 *A wired-AND configuration.*

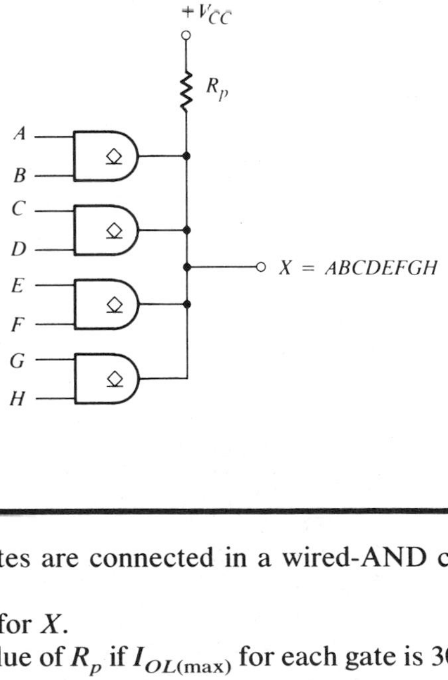

EXAMPLE 11–2
Three open-collector NAND gates are connected in a wired-AND configuration as shown in Figure 11–8.
(a) Write the logic expression for X.
(b) Determine the minimum value of R_p if $I_{OL(\text{max})}$ for each gate is 30 mA and $V_{OL(\text{max})}$ is 0.4 V.
Assume that the wired-AND circuit is driving four TTL inputs (-1.6 mA each).

FIGURE 11–8

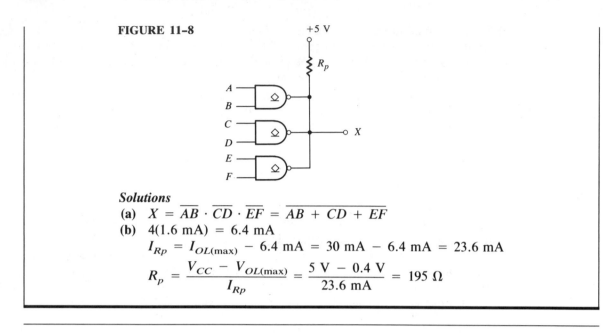

Solutions

(a) $X = \overline{AB} \cdot \overline{CD} \cdot \overline{EF} = \overline{AB + CD + EF}$

(b) $4(1.6 \text{ mA}) = 6.4 \text{ mA}$

$I_{Rp} = I_{OL(\text{max})} - 6.4 \text{ mA} = 30 \text{ mA} - 6.4 \text{ mA} = 23.6 \text{ mA}$

$R_p = \dfrac{V_{CC} - V_{OL(\text{max})}}{I_{Rp}} = \dfrac{5 \text{ V} - 0.4 \text{ V}}{23.6 \text{ mA}} = 195 \ \Omega$

SECTION REVIEW 11–1

1. How many unit loads (inputs of same family) can a 74H CMOS gate drive? Refer to Table 11–1.
2. Write the output logic expression for five open-collector inverters in a wired-AND configuration.

11–2   INTERFACING WITH BUSES

Buses and three-state buffers were introduced in the last chapter. As you have seen, many microprocessors, memories, and other logic functions have three-state outputs that are used for interfacing to bus structures. To prevent the devices connected to a common bus from interfering with each other, the three-state circuits are used to disconnect all devices except the ones that are communicating at any given time.

To illustrate the basic idea of bus interfacing, Figure 11–9 shows four devices connected to a four-bit *unidirectional* (one-way) bus with three-state interfacing. Data can flow only from devices *A* or *B* to devices *C* or *D*. When device *A* is sending data, device *B* is disconnected from the bus with the three-state buffers, and vice versa.

Figure 11–10 illustrates a *bidirectional* bus connecting several devices (only two are shown). Data can be transferred back and forth among the devices. When it is not sending or receiving data, a given device is disconnected from the bus by disabling its three-state buffers.

FIGURE 11–9 *Example of a unidirectional bus with three-state interfacing.*

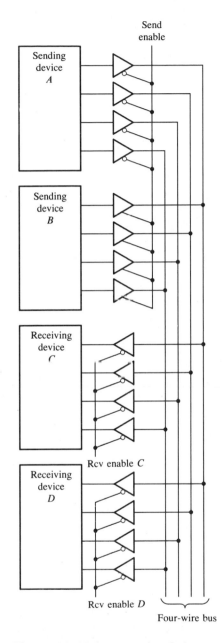

Notice that each device in Figure 11–10 has a pair of three-state circuits (transceivers) on each input/output (I/O) line. The reason for this is as follows: When a given device is sending data, the *output* three-state drivers are enabled and the input three-state receivers are disabled. When a given device is receiving data, the input three-state receivers are enabled and the output three-state drivers

FIGURE 11–10 *Bidirectional bus with three-state transceiver interfacing.*

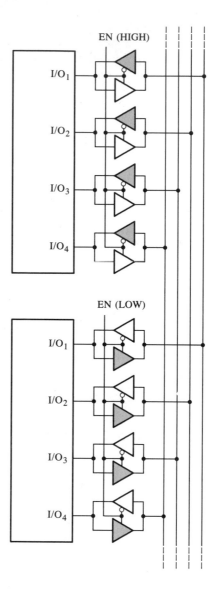

are disabled. This is illustrated in Figure 11–10, where the shaded three-state circuits are disabled.

A Specific Bidirectional Bus Transceiver

The SN75160A is an example of a three-state bus interface circuit. A logic diagram for this device is shown in Figure 11–11, where TE is *terminal enable*, PE is *peripheral enable*, D is driver, and R is receiver.

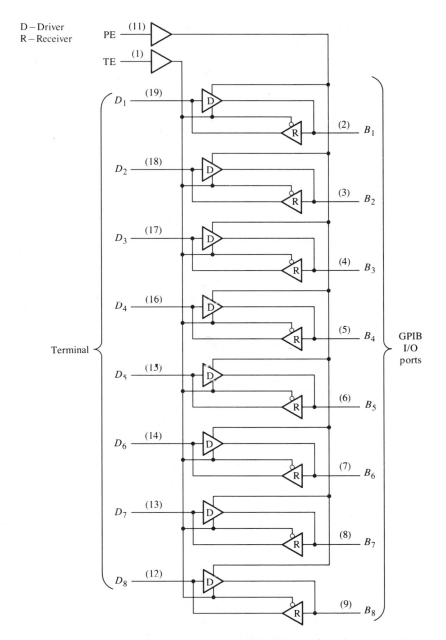

FIGURE 11-11 *Logic diagram of an SN75160A interface bus transceiver.*

The General-Purpose Interface Bus (GPIB)

As an example of a standard bus system, we will examine the GPIB. This bus is defined by IEEE Std. 488 and is therefore sometimes known as the IEEE–488 bus. This is a standard for interfacing with programmable instrumentation or control systems. It is an eight-bit system that operates to 1 MHz and can drive up to 20 meter lines.

The IEEE–488 standard defines three basic classes of devices that can be connected to the GPIB: *talker*, *listener*, and *controller*. There can also be devices that are combinations of these basic types.

An important application of the GPIB is in the interconnection of several test instruments to form an automated test system. Examples of *talkers* in this application are devices that produce information, such as digital multimeters and frequency counters. Examples of *listeners* are display devices and programmable instruments such as signal generators, multimeters, and power supplies. A *programmable multimeter,* for example, can be both a talker and a listener. It functions as a talker when outputting voltage measurements and as a listener when receiving program instructions. The *controller* is a device that determines when the other devices can use the bus.

The GPIB allows up to fifteen instruments within a localized area (within 20 meters) to communicate with each other at data rates up to 1 Mbyte/s. Each device connected to the bus is assigned a unique address to which it responds. Information sent on the bus consists of both device-dependent *data* and *commands*.

Physically, the GPIB consists of a 24-wire shielded cable. Eight lines carry data, eight lines carry control signals (commands), and eight are reserved for grounds. The eight commands are divided into three *data byte transfer controls* and five *interface management* lines. The standard connector diagram is shown in Figure 11–12, and the functions of the sixteen lines are summarized in Table 11–2.

Handshaking is a widely used interfacing term. It is basically a method or procedure by which two devices establish communication with each other and

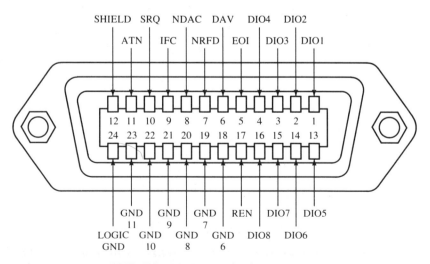

FIGURE 11–12   *IEEE–488 (GPIB) standard connector.*

TABLE 11–2 *GPIB control and data lines.*

DAV	Data Valid	Handshake line controlled by talker to show listener when valid data are present.
NDAC	Not Data Accepted	Handshake line. Listener sets this line HIGH when it has latched the data from the I/O lines.
NRFD	Not Ready for Data	Handshake line. Sent by listener to indicate readiness for next byte.
ATN	Attention	Sent by controller. When LOW, interface commands are being sent over DIO lines. When HIGH, these lines carry data.
REN	Remote Enable	Sent by controller to select control either from front panel or from bus.
IFC	Interface Clear	Sent by controller to set the interface system into a known state.
SRQ	Service Request	Set HIGH by a device to indicate a need for service.
EOI	End or Identify	If ATN is HIGH, this indicates the end of a message block. If ATN is LOW, the controller is requesting a parallel poll.
DIO8 . . . DIO1		DIO8 through DIO1 are the data input/output lines.

consists of a specified sequence of signals that are transferred between the two devices in a prescribed manner.

Three of the interface management lines of the GPIB operate as a three-line handshake between talker (or controller) and listeners. No new data are sent until each device addressed to listen has received the last byte and is ready for the next. This method ensures that the data rate is suited to the slowest active listener.

A typical system GPIB interface arrangement is shown in Figure 11–13 (p. 542). This particular system consists of five talker/listener devices which are interfaced to the common bus with TMS9914 controllers and bus transceivers.

SECTION REVIEW 11–2

1. Why must interfaces to common buses be made with three-state devices?
2. How many lines does the GPIB have (not including grounds)?

11–3 INTERFACING DIGITAL AND ANALOG SYSTEMS

Because most quantities in nature occur in *analog* form, conversions from digital to analog and from analog to digital are of great importance. In this section we will look at methods to convert from digital signals to analog signals, and vice versa.

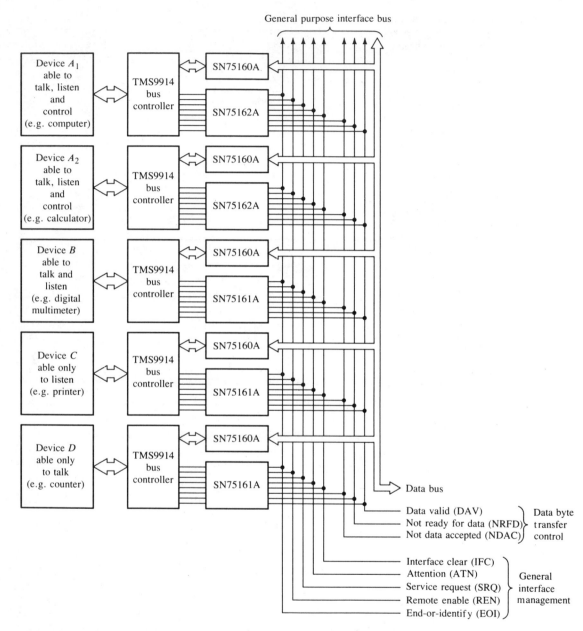

FIGURE 11–13 *A typical system GPIB interface arrangement.*

Digital and Analog Signals

You already know how a quantity can be expressed in digital form, but what is an *analog* quantity? An analog quantity is one that has a *continuous* set of values over a given range, as contrasted with *discrete* values for the digital case.

Almost any measurable quantity is analog in nature, such as temperature, pressure, speed, and time. To further illustrate the difference between an analog and a digital representation of a quantity, let us take the case of a voltage that varies over a range from 0 V to +15 V. The analog representation of this quantity takes in *all* values between 0 and +15 V.

In the case of a digital representation using a four-bit binary code, only 16 values can be defined. More values between 0 and +15 can be represented by using more bits in the digital code. So an analog quantity can be represented to some degree of accuracy with a digital code that specifies discrete values within the range. This concept is illustrated in Figure 11–14, where the analog function shown is a smoothly changing curve that takes on values between 0 V and +15 V. If a four-bit code is used to represent this curve, each binary number represents a discrete point on the curve.

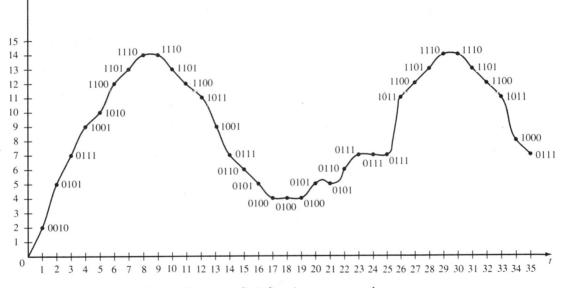

FIGURE 11–14 *Discrete (digital) points on an analog curve.*

In Figure 11–14, the voltage on the analog curve is measured or *sampled* at each of 35 equal intervals. The voltage at each of these intervals is represented by a four-bit code as indicated. At this point, we have a series of binary numbers representing various voltage values along the analog curve. This is the basic idea of *analog-to-digital* (A/D) conversion.

An approximation of the analog function in Figure 11–14 can be reconstructed from the sequence of digital numbers that has been generated. Obviously, there will be some error in the reconstruction because only certain values are represented (35 in this example) and not the continuous set of values. If the digital values at each of the 35 intervals are graphed as shown in Figure 11–15, we have a

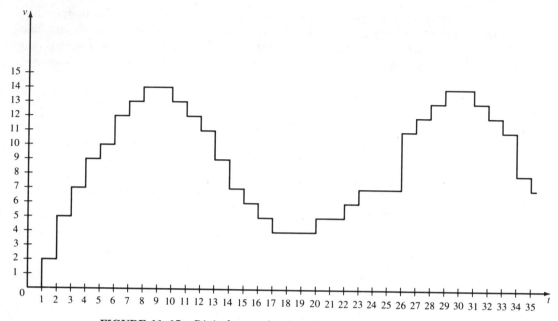

FIGURE 11–15 *Digital reproduction of an analog curve.*

reconstructed function. As you can see, the graph only approximates the original curve because values between the points are not known.

Digital-to-Analog (D/A) Conversion

Digital-to-analog conversion is an important interface process in many applications. An example is a voice signal that has been digitized for processing or transmission and must be changed back into an approximation of the original signal to ultimately drive a speaker.

Binary-weighted input D/A converter One method of D/A conversion uses a resistor network with values that represent the binary weights of the input bits of the digital code. Figure 11–16 shows a four-bit D/A converter of this type. The

FIGURE 11–16 *Four-bit binary-weighted input D/A converter.*

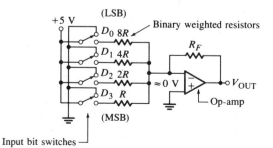

switch symbols represent transistor switches for inputting each of the four bits. The operational amplifier (op-amp) provides a very high impedance load to the resistor network, and its inverting input ($-$) looks like "virtual" ground so that the output is proportional to the current through the feedback resistor R_F (the sum of the input currents). Almost all the current is through R_F and into the low impedance output of the op-amp. The inverting input is approximately at 0 V.

The lowest-value resistor (R) corresponds to the highest binary-weighted input (2^3). Each of the other resistors is a multiple of R: $2R$, $4R$, and $8R$, corresponding to the binary weights 2^2, 2^1, and 2^0, respectively. One of the disadvantages of this type of D/A converter is the number of different resistor values. For example, an eight-bit converter requires eight resistors ranging from some value R to $128R$. This range of resistors requires tolerances of one part in 255 (less than 0.5%) to accurately convert the input, making this type of D/A converter very difficult to mass-produce.

EXAMPLE 11-3

Determine the output of the D/A converter in Figure 11–17(a) if the sequence of four-bit numbers in Figure 11–17(b) is applied to the inputs. D_0 is the LSB.

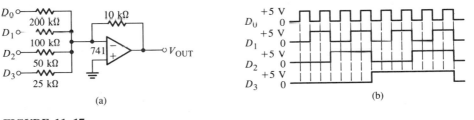

(a)　(b)

FIGURE 11–17

Solution First, we determine the output current I for each of the weighted inputs. Since the inverting input ($-$) of the op-amp is at 0 V (virtual ground) and a binary 1 corresponds to +5 V, the current through any of the input resistors is 5 V divided by the resistance value.

$$I_0 = 5 \text{ V}/200 \text{ k}\Omega = 0.025 \text{ mA}$$
$$I_1 = 5 \text{ V}/100 \text{ k}\Omega = 0.05 \text{ mA}$$
$$I_2 = 5 \text{ V}/50 \text{ k}\Omega = 0.1 \text{ mA}$$
$$I_3 = 5 \text{ V}/25 \text{ k}\Omega = 0.2 \text{ mA}$$

Almost none of this current goes into the inverting op-amp input because of its extremely high impedance. Therefore, all of the current goes through the feedback resistor R_F. Since one end of R_F is at 0 V (virtual ground), the drop across R_F equals the output voltage.

$$V_{OUT(D0)} = (10 \text{ k}\Omega)(-0.025 \text{ mA}) = -0.25 \text{ V}$$
$$V_{OUT(D1)} = (10 \text{ k}\Omega)(-0.05 \text{ mA}) = -0.5 \text{ V}$$
$$V_{OUT(D2)} = (10 \text{ k}\Omega)(-0.1 \text{ mA}) = -1 \text{ V}$$
$$V_{OUT(D3)} = (10 \text{ k}\Omega)(-0.2 \text{ mA}) = -2 \text{ V}$$

From Figure 11–17(b), the first input code is 0001_2 (binary 1). For this, the output voltage is -0.25 V. The next code is 0010_2 which produces an output voltage of -0.5 V. The next code is 0011_2 which produces an output voltage of -0.25 V $+ -0.5$ V $= -0.75$ V. Each successive binary code increases the output voltage by -0.25 V, so for this particular straight binary sequence on the inputs, the output is a *stairstep* waveform going from 0 V to -3.75 V in -0.25 V steps. This is shown in Figure 11–18.

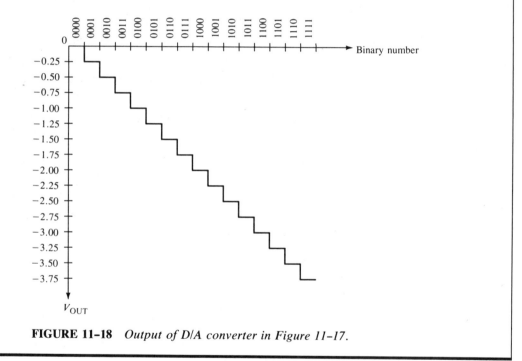

FIGURE 11–18   *Output of D/A converter in Figure 11–17.*

R/2R ladder D/A converter   Another method of D/A conversion is the *R/2R ladder,* as shown in Figure 11–19 for four bits. It overcomes one of the problems in the previous type in that it requires only two resistor values. Again, the switch symbols represent transistor switches.

Start by assuming that the D_3 switch is connected to $+5$ V and the others to ground. This represents 1000_2. A circuit analysis will show you that this reduces

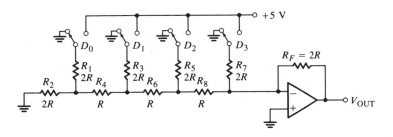

FIGURE 11–19 *An R/2R ladder D/A converter.*

to the equivalent form shown in Figure 11–20(a). There is essentially no current through the $2R$ *equivalent* resistance. Thus, all of the current ($I = 5$ V/$2R$) through R_7 also goes through R_F, and the output voltage is -5 V.

Part (b) of the figure shows the equivalent circuit when the D_2 switch is in the $+5$ V position and the others at ground. This represents 0100_2. If we thevenize looking from R_8, we get 2.5 V in series with R, as shown in part (b). This results in a current through R_F of $I = 2.5$ V/$2R$, which gives an output voltage of -2.5 V. Keep in mind that there is no current into the op-amp inverting input and that there is no current through the equivalent resistance to ground because it has 0 V across it due to the virtual ground.

Part (c) of the figure shows the equivalent circuit when the D_1 input is connected to $+5$ V and the others to ground. Again thevenizing looking from R_8, we get 1.25 V in series with R as shown. This results in a current through R_F of $I = 1.25$ V/$2R$, which gives an output voltage of -1.25 V.

In part (d) of the figure, the equivalent circuit representing the case where D_0 is connected to $+5$ V and the other inputs to ground is shown. Thevenizing from R_8 gives an equivalent of 0.625 V in series with R as shown. The resulting current through R_F is $I = 0.625$ V/$2R$, which gives an output voltage of -0.625 V.

Notice that each successively lower weighted input produces an output voltage that is halved, so that the output voltage is proportional to the binary weight of the input bits.

D/A performance characteristics   The performance characteristics of a D/A converter include resolution, accuracy, linear errors, monotonicity, and settling time, each of which is discussed below.

1. *Resolution.* The resolution of a D/A converter is the *reciprocal of the number of discrete steps* in the D/A output. This, of course, is dependent on the number of input bits. For example, a four-bit D/A converter has a resolution of one part in $2^4 - 1$ (one part in fifteen). Expressed as a percentage, this is $(1/15)100 = 6.67\%$. The total number of discrete steps equals $2^n - 1$, where n is the number of bits.

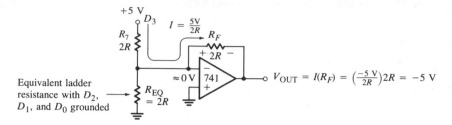

(a) Equivalent circuit for $D_3 = 1$, $D_2 = 0$, $D_1 = 0$, $D_0 = 0$

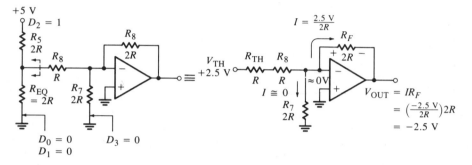

(b) Equivalent circuit for $D_3 = 0$, $D_2 = 1$, $D_1 = 0$, $D_0 = 0$

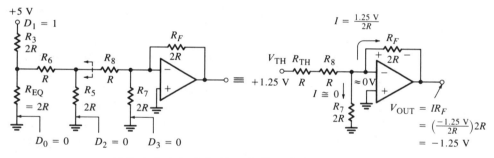

(c) Equivalent circuit for $D_3 = 0$, $D_2 = 0$, $D_1 = 1$, $D_0 = 0$

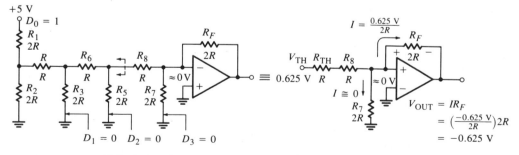

(d) Equivalent circuit for $D_3 = 0$, $D_2 = 0$, $D_1 = 0$, $D_0 = 1$

FIGURE 11-20 *Analysis of the R/2R ladder D/A converter. (The operational amplifier keeps the inverting ($-$) input near zero volts (≈ 0 V) due to negative feedback. Therefore, all current goes through R_F rather than into the inverting input.)*

EXAMPLE 11–4

Determine the resolution of **(a)** an eight-bit and **(b)** a twelve-bit D/A converter in terms of percentage.

Solutions
(a) For the eight-bit converter,

$$\frac{1}{2^8 - 1} \times 100 = \frac{1}{255} \times 100 = 0.392\%$$

(b) For the 12-bit converter,

$$\frac{1}{2^{12} - 1} \times 100 = \frac{1}{4095} \times 100 = 0.0244\%$$

2. *Accuracy.* Accuracy is a comparison of the *actual* output of a D/A converter with the *expected* output. It is expressed as a percentage of a full-scale or maximum output voltage. For example, if a converter has a full-scale output of 10 V and the accuracy is $\pm 0.1\%$, then the maximum error for any output voltage is $(10 \text{ V})(0.001) = 10$ mV. Ideally, the accuracy should be at most $\pm \frac{1}{2}$ of an LSB. For an eight-bit converter, 1 LSB is $1/256 = 0.0039$ (0.39% of full scale). The accuracy should be approximately $\pm 0.2\%$.

3. *Linear errors.* A linear error is the deviation from the ideal straight-line output of a D/A converter. A special case is an *offset error*, which is the amount of output voltage when the input bits are all zeros.

4. *Monotonicity.* A D/A converter is *monotonic* if it does not take any reverse steps when it is sequenced over its entire range of input bits.

5. *Settling time.* This is normally defined as the time it takes a D/A converter to settle within $\pm \frac{1}{2}$ LSB of its final value when a change occurs in the input code.

Analog-to-Digital (A/D) Conversion

Analog-to-digital conversion is the process by which an analog quantity is converted to digital form. A/D conversion is necessary when measured quantities must be in digital form for processing in a computer or for display or storage. Several types of A/D conversion methods are now examined.

Simultaneous A/D converter   This method (also called *flash* conversion) utilizes parallel differential comparators that compare reference voltages with the analog input voltage. When the analog voltage exceeds the reference voltage for a given comparator, a HIGH is generated. Figure 11–21 shows a three-bit converter which uses seven comparator circuits; a comparator is not needed for the all-zero condition. A four-bit converter of this type requires fifteen comparators. In general, $2^n - 1$ comparators are required for conversion to an n-bit binary code.

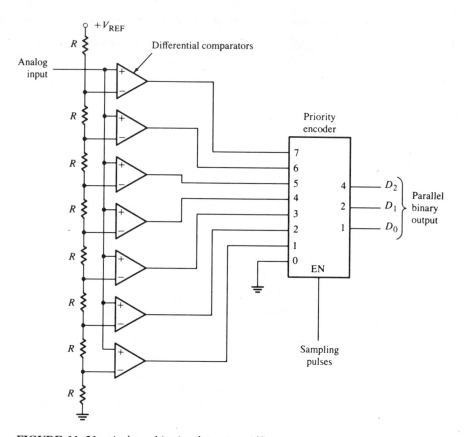

FIGURE 11–21 *A three-bit simultaneous A/D converter.*

The large number of comparators necessary for a reasonably sized binary number is one of the disadvantages of the simultaneous A/D converter. Its chief advantage is that it provides a *fast conversion time.*

The reference voltage for each comparator is set by the resistive voltage-divider network. The output of each comparator is connected to an input of the *priority encoder.* The encoder is sampled by a pulse on the enable input, and a three-bit binary code proportional to the value of the analog input appears on the encoder's outputs. The binary code is determined by the highest-order input having a HIGH level.

The *sampling rate* determines the accuracy with which the sequence of digital codes represents the analog input of the A/D converter. The more samples taken in a given unit of time, the more accurately the analog signal is represented in digital form.

The following example illustrates the basic operation of the simultaneous A/D converter in Figure 11–21.

EXAMPLE 11-5

Determine the binary code output of the three-bit simultaneous A/D converter for the analog input signal in Figure 11–22 and the sampling pulses (encoder enable) shown. $V_{REF} = +8$ V.

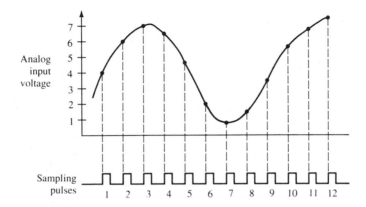

FIGURE 11-22 *Sampling of values on an analog waveform for conversion to digital.*

Solution The resulting A/D output sequence is listed as follows and shown in the waveform diagram of Figure 11–23 in relation to the sampling pulses:

100, 101, 110, 110, 100, 010, 000, 001, 011, 101, 110, 111

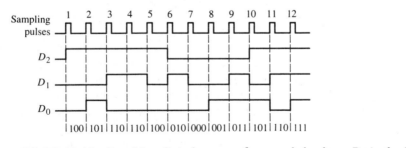

FIGURE 11-23 *Resulting digital outputs for sampled values. D_0 is the LSB.*

Stairstep-ramp A/D converter This method of A/D conversion is also known as the *digital-ramp* or the *counter* method. It employs a D/A converter and a binary counter to generate the digital value of an analog input. Figure 11–24 shows a diagram of this type of converter.

Assume that the counter begins RESET and the output of the D/A converter is zero. Now assume that an analog voltage is applied to the input. When it

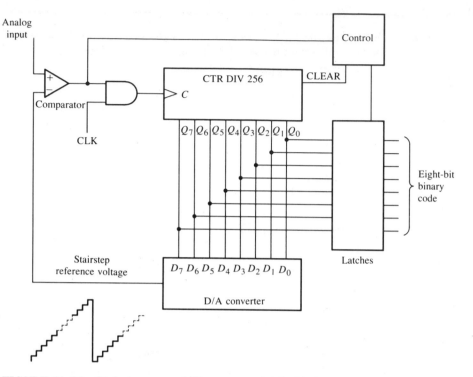

FIGURE 11–24   *Stairstep-ramp A/D converter (eight bits).*

exceeds the reference voltage (output of D/A), the comparator switches to a HIGH output state and enables the AND gate. The clock pulses begin advancing the counter through its binary states, producing a stairstep reference voltage from the D/A converter. The counter continues to advance from one binary state to the next, producing successively higher steps in the reference voltage. When the stairstep reference voltage reaches the analog input voltage, the comparator output will go LOW and disable the AND gate, thus cutting off the clock pulses to stop the counter. The state of the counter at this point equals the number of steps in the reference voltage at which the comparison occurs. This binary number, of course, represents the value of the analog input. The control logic loads the binary count into the latches and resets the counter, thus beginning another count sequence to sample the input value.

This method is slower than the simultaneous method because, in the worst case of maximum input, the counter must sequence through its maximum number of states before a conversion occurs. For an eight-bit conversion, this means a maximum of 256 counter states. Figure 11–25 illustrates a conversion sequence for a four-bit conversion. Notice that for each sample, the counter must count from *zero* up to the point at which the stairstep reference voltage reaches the analog input voltage. The conversion time varies, depending upon the analog voltage.

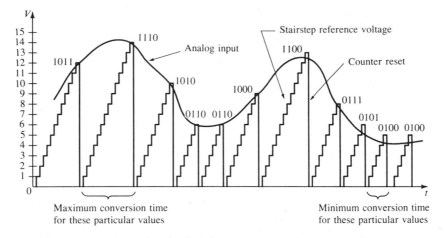

FIGURE 11–25 *Example of a four-bit conversion showing an analog input and the stairstep reference voltage.*

Tracking A/D converter   This method uses an *up/down counter* and is faster than the stairstep-ramp method because the counter is not reset after each sample, but rather tends to *track* the analog input. Figure 11–26 shows a typical eight-bit tracking A/D converter.

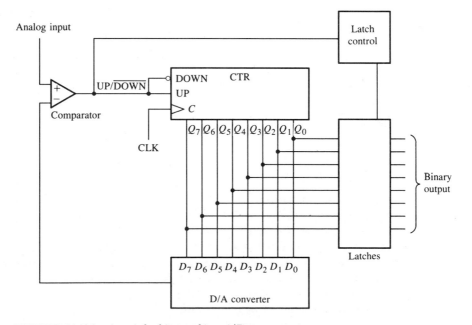

FIGURE 11–26 *An eight-bit tracking A/D converter.*

As long as the D/A output reference voltage is less than the analog input, the comparator output is HIGH, putting the counter in the UP mode, causing it to produce an up sequence of binary counts. This causes an *increasing* stairstep reference voltage out of the D/A converter, which continues until the stairstep reaches the value of the input voltage.

When the reference voltage equals the analog input, the comparator's output switches LOW and puts the counter in the DOWN mode, causing it to back up one count. If the analog input is decreasing, the counter will continue to back down in its sequence and effectively *track* the input. If the input is increasing, the counter will back down one count after the comparison occurs and then will begin counting up again. When the input is constant, the counter backs down one count when a comparison occurs. The reference output is now less than the analog input, and the comparator output goes HIGH, causing the counter to count up. As soon as the counter increases one state, the reference voltage becomes greater than the input, switching the comparator to its LOW state. This causes the counter to back down one count. This back-and-forth action continues as long as the analog input is a constant value, thus causing an *oscillation* between two binary states in the A/D output. This is a disadvantage of this type of converter.

Figure 11–27 illustrates the tracking action of this type of A/D converter for a four-bit conversion.

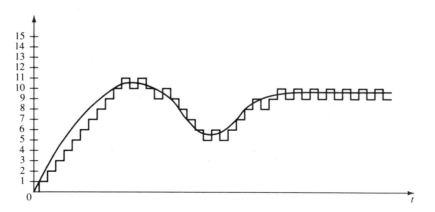

FIGURE 11–27 *An A/D converter tracking action.*

Single-slope A/D converter Unlike the previous two methods, this type of converter does not require a D/A converter. It uses a linear ramp generator to produce a constant-slope reference voltage. A diagram is shown in Figure 11–28.

At the beginning of a conversion cycle, the counter is RESET and the ramp generator output is 0 V. The analog input is greater than the reference voltage at this point and therefore produces a HIGH output from the comparator. This HIGH enables the clock to the counter and starts the ramp generator.

Assume that the slope of the ramp is 1 V/ms. It will increase until it equals the analog input; at this point the ramp is RESET and the binary or BCD count is

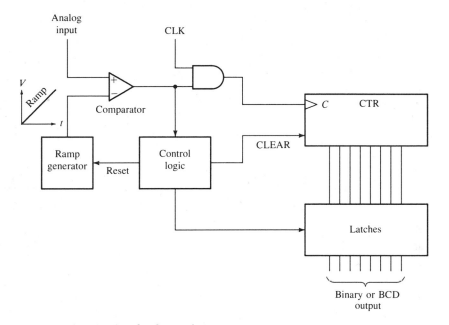

FIGURE 11-28 *Single-slope A/D converter.*

stored in the latches by the control logic. Let us assume that the analog input is 2 V at the point of comparison. This means that the ramp is also 2 V and has been running for 2 ms. Since the comparator output has been HIGH for 2 ms, 200 clock pulses have been allowed to pass through the gate to the counter (assuming a clock frequency of 100 kHz). At the point of comparison, the counter is in the binary state representing decimal 200. With proper scaling and decoding, this binary number can be displayed as 2.00 V. This basic concept is used in some digital voltmeters.

Dual-slope A/D converter   The operation of this type of A/D converter is similar to that of the single-slope type except that a variable-slope ramp and a fixed-slope ramp are both used. This type of converter is common in digital voltmeters and other types of measurement instruments.

A ramp generator (integrator), A_1, is used to produce the dual-slope characteristic, the purpose of which we will now discuss. A block diagram of a dual-slope A/D converter is shown in Figure 11–29 for reference.

We will start by assuming that the counter is RESET and the output of the integrator is zero. Now assume that a positive input voltage is applied to the input through the switch (S_1) as selected by the control logic. Since the inverting input of A_1 is at virtual ground, and assuming that V_{in} is constant for a period of time, there will be constant current through the input resistor R and therefore through the capacitor C. C will charge linearly because the current is constant, and as a

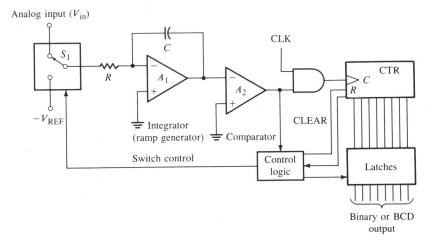

Analog input (V_{in})

CLK

CTR

$-V_{REF}$

Integrator
(ramp generator)   Comparator

CLEAR

Switch control

Control
logic

Latches

Binary or BCD
output

FIGURE 11–29   *Dual-slope A/D converter.*

result, there will be a negative-going linear voltage ramp on the output of A_1, as illustrated in Figure 11–30(a).

When the counter reaches a specified count, it will be reset, and the control logic will switch the negative reference voltage ($-V_{REF}$) to input A_1 as shown in Figure 11–30(b). At this point, the capacitor is charged to a negative voltage ($-V$) *proportional* to the input analog voltage.

Now the capacitor discharges linearly due to the constant current from the $-V_{REF}$ as shown in Figure 11–30(c). This produces a positive-going ramp on the A_1 output, starting at $-V$ and with a *constant slope* that is independent of the charge voltage.

As the capacitor discharges, the counter advances from its reset state. The time it takes the capacitor to discharge to zero depends on the initial voltage $-V$ (proportional to V_{in}) because the discharge rate (slope) is constant. When the integrator (A_1) output voltage reaches zero, the comparator (A_2) switches to the LOW state and disables the clock to the counter. The binary count is latched, thus completing one conversion cycle. The binary count is proportional to V_{in} because the time it takes the capacitor to discharge depends only on $-V$, and the counter records this interval of time.

Successive-approximation A/D converter   This is perhaps the most widely used method of A/D conversion. It has a much shorter conversion time than the other methods with the exception of the simultaneous method. It also has a fixed conversion time that is the same for any value of the analog input.

Figure 11–31 shows a basic block diagram of a four-bit successive-approximation A/D converter. It consists of a D/A converter, successive-approximation register (SAR), and comparator. The basic operation is as follows: The bits of the

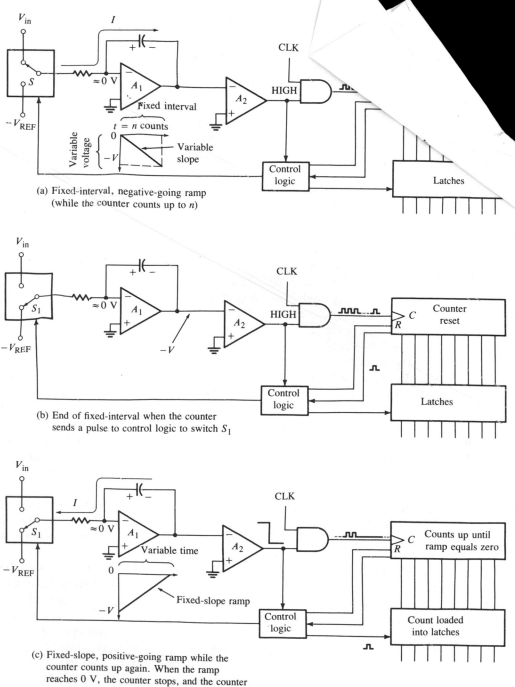

(a) Fixed-interval, negative-going ramp
(while the counter counts up to n)

(b) End of fixed-interval when the counter
sends a pulse to control logic to switch S_1

(c) Fixed-slope, positive-going ramp while the
counter counts up again. When the ramp
reaches 0 V, the counter stops, and the counter
output is loaded into latches.

FIGURE 11–30 *Dual-slope conversion.*

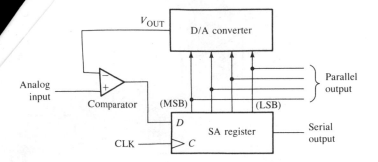

FIGURE 11–31 *Successive-approximation A/D converter.*

D/A converter are enabled one at a time, starting with the most significant bit (MSB). As each bit is enabled, the comparator produces an output that indicates whether the analog input voltage is greater or less than the output of the D/A. If the D/A output is greater than the analog input, the comparator's output is LOW, causing the bit in the register to RESET. If the D/A output is less than the analog input, the bit is retained in the register.

The system does this with the MSB first, then the next most significant bit, then the next, and so on. After all the bits of the D/A have been tried, the conversion cycle is complete.

In order to better understand the operation of this type of A/D converter, we will take a specific example of a four-bit conversion. Figure 11–32 illustrates the step-by-step conversion of a given analog input voltage (5 V in this case). We will assume that the D/A converter has the following output characteristic: $V_{OUT} = 8$ V for the 2^3 bit (MSB), $V_{OUT} = 4$ V for the 2^2 bit, $V_{OUT} = 2$ V for the 2^1 bit, and $V_{OUT} = 1$ V for the 2^0 bit (LSB).

Figure 11–32(a) shows the first step in the conversion cycle with the MSB = 1. The output of the D/A is 8 V. Since this is *greater* than the analog input of 5 V, the output of the comparator is LOW, causing the MSB in the SA register to be RESET to a 0.

Figure 11–32(b) shows the second step in the conversion cycle with the 2^2 bit equal to a 1. The output of the D/A is 4 V. Since this is *less* than the analog input of 5 V, the output of the comparator switches to a HIGH, causing this bit to be retained in the SA register.

Figure 11–32(c) shows the third step in the conversion cycle with the 2^1 bit equal to a 1. The output of the D/A is 6 V because there is a 1 on the 2^2 bit input and on the 2^1 bit input, so 4 V + 2 V = 6 V. Since this is *greater* than the analog input of 5 V, the output of the comparator switches to a LOW, causing this bit to be RESET to a 0.

Figure 11–32(d) shows the fourth and final step in the conversion cycle with the 2^0 bit equal to a 1. The output of the D/A is 5 V because there is a 1 on the 2^2 bit input and on the 2^0 bit input, so 4 V + 1 V = 5 V.

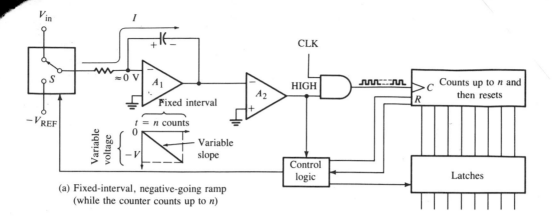

(a) Fixed-interval, negative-going ramp
(while the counter counts up to n)

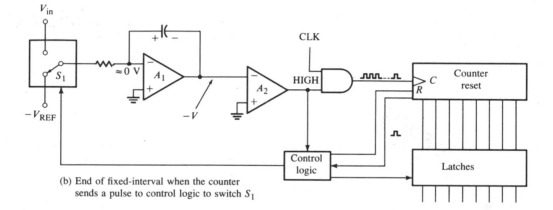

(b) End of fixed-interval when the counter
sends a pulse to control logic to switch S_1

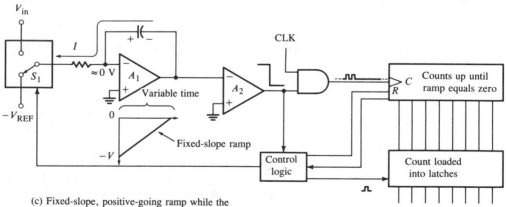

(c) Fixed-slope, positive-going ramp while the
counter counts up again. When the ramp
reaches 0 V, the counter stops, and the counter
output is loaded into latches.

FIGURE 11–30 *Dual-slope conversion.*

557

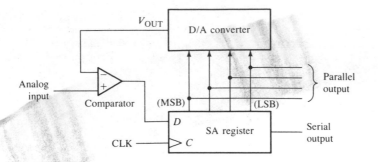

FIGURE 11–31 *Successive-approximation A/D converter.*

D/A converter are enabled one at a time, starting with the most significant bit (MSB). As each bit is enabled, the comparator produces an output that indicates whether the analog input voltage is greater or less than the output of the D/A. If the D/A output is greater than the analog input, the comparator's output is LOW, causing the bit in the register to RESET. If the D/A output is less than the analog input, the bit is retained in the register.

The system does this with the MSB first, then the next most significant bit, then the next, and so on. After all the bits of the D/A have been tried, the conversion cycle is complete.

In order to better understand the operation of this type of A/D converter, we will take a specific example of a four-bit conversion. Figure 11–32 illustrates the step-by-step conversion of a given analog input voltage (5 V in this case). We will assume that the D/A converter has the following output characteristic: $V_{OUT} = 8$ V for the 2^3 bit (MSB), $V_{OUT} = 4$ V for the 2^2 bit, $V_{OUT} = 2$ V for the 2^1 bit, and $V_{OUT} = 1$ V for the 2^0 bit (LSB).

Figure 11–32(a) shows the first step in the conversion cycle with the MSB = 1. The output of the D/A is 8 V. Since this is *greater* than the analog input of 5 V, the output of the comparator is LOW, causing the MSB in the SA register to be RESET to a 0.

Figure 11–32(b) shows the second step in the conversion cycle with the 2^2 bit equal to a 1. The output of the D/A is 4 V. Since this is *less* than the analog input of 5 V, the output of the comparator switches to a HIGH, causing this bit to be retained in the SA register.

Figure 11–32(c) shows the third step in the conversion cycle with the 2^1 bit equal to a 1. The output of the D/A is 6 V because there is a 1 on the 2^2 bit input and on the 2^1 bit input, so 4 V + 2 V = 6 V. Since this is *greater* than the analog input of 5 V, the output of the comparator switches to a LOW, causing this bit to be RESET to a 0.

Figure 11–32(d) shows the fourth and final step in the conversion cycle with the 2^0 bit equal to a 1. The output of the D/A is 5 V because there is a 1 on the 2^2 bit input and on the 2^0 bit input, so 4 V + 1 V = 5 V.

The four bits have all been tried, thus completing the conversion cycle. At this point the binary code in the register is 0101_2, which is the binary value of the analog input of 5 V. Another conversion cycle now begins and the basic process is repeated. The SAR is cleared at the beginning of each cycle.

FIGURE 11–32 *Successive-approximation conversion process.*

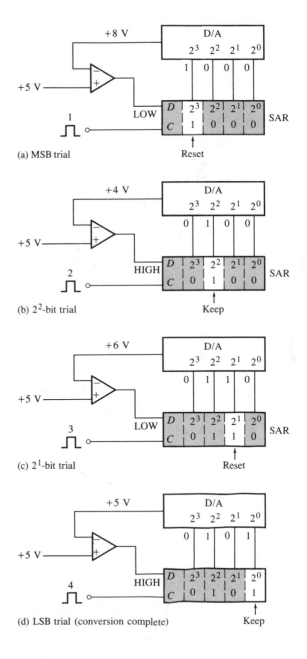

(a) MSB trial

(b) 2^2-bit trial

(c) 2^1-bit trial

(d) LSB trial (conversion complete)

A specific A/D converter   The ADC0801 is an example of a successive-approximation analog-to-digital converter. A block diagram is shown in Figure 11–33. This device operates from a +5 V supply and has a resolution of eight bits with a conversion time of 100 μs. Also, it has guaranteed monotonicity and an on-chip clock generator. The data outputs are three-stated so that it can be interfaced with a microprocessor bus system.

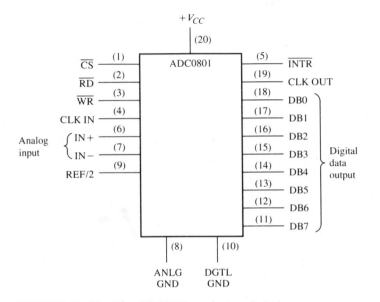

FIGURE 11–33   *The ADC0801 analog-to-digital converter.*

A detailed logic diagram of the ADC0801 is shown in Figure 11–34, and a description of its basic operation is as follows: The ADC0801 contains the equivalent of a 256-resistor D/A converter network. The successive-approximation logic sequences the network to match the analog differential input voltage $(V_{in}+ - V_{in}-)$ with an output from the resistive network. The MSB is tested first. After eight comparisons (64 clock periods), an eight-bit binary code is transferred to an output latch, and the interrupt ($\overline{INTR}$) output goes LOW. The device can be operated in a free-running mode by connecting the $\overline{INTR}$ output to the write ($\overline{WR}$) input and holding the conversion start ($\overline{CS}$) LOW. To ensure start-up under all conditions, a LOW $\overline{WR}$ input is required during the power-up cycle. Taking $\overline{CS}$ low anytime after that will interrupt a conversion in process.

When the $\overline{WR}$ input goes LOW, the internal successive-approximation register (SAR) and the eight-bit shift register are RESET. As long as both $\overline{CS}$ and $\overline{WR}$ remain LOW, the analog-to-digital converter remains in a RESET state. One to eight clock periods after $\overline{CS}$ or $\overline{WR}$ makes a LOW-to-HIGH transition, conversion starts.

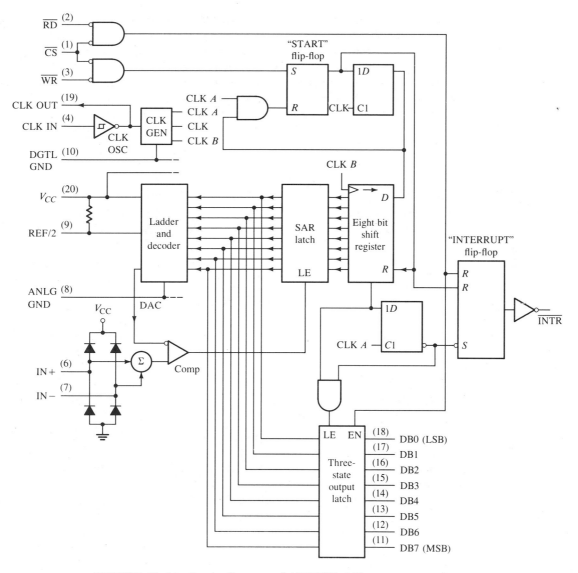

FIGURE 11–34 *Logic diagram of ADC0801 A/D converter.*

When the $\overline{CS}$ and $\overline{WR}$ inputs are LOW, the start flip-flop is SET and the interrupt flip-flop and eight-bit register are RESET. The HIGH is ANDed with the next clock pulse, which puts a HIGH on the reset input of the start flip-flop. If either $\overline{CS}$ or $\overline{WR}$ has gone HIGH, the set signal to the start flip-flop is removed, causing it to be RESET. A HIGH is placed on the D input of the eight-bit shift register and the conversion process is started. If the $\overline{CS}$ and $\overline{WR}$ inputs are still LOW, the start flip-flop, the eight-bit shift register, and the SAR remain RESET. This action allows for wide $\overline{CS}$ and $\overline{WR}$ inputs, with conversion starting from one to eight clock periods after one of the inputs goes HIGH.

When the HIGH input has been clocked through the eight-bit shift register, completing the SAR search, it is applied to an AND gate controlling the output latches and to the D input of a flip-flop. On the next clock pulse, the digital word is transferred to the three-state output latches and the interrupt flip-flop is SET. The output of the interrupt flip-flop is inverted to provide an $\overline{INTR}$ output that is HIGH during conversion and LOW when conversion is complete.

When a LOW is at both the $\overline{CS}$ and $\overline{RD}$ inputs, the three-state output latch is enabled, the output code is applied to the DB0 through DB7 lines, and the interrupt flip-flop is RESET. When either the $\overline{CS}$ or the $\overline{RD}$ input returns to a HIGH, the DB0 through DB7 outputs are disabled. The interrupt flip-flop remains RESET.

SECTION REVIEW 11–3

1. What is the fastest method of analog-to-digital conversion?
2. What is the resolution, expressed in percentage, of (a) a six-bit converter and (b) an eight-bit converter?

11–4 MODEMS AND INTERFACES

When digital data are sent from one computer (or other digital system) to another that is a great distance away, the telephone network is often used to transmit the data. The digital systems that originate and receive the data in these situations are referred to as *data terminal equipment* (DTE).

Since the telephone network is designed for carrying only analog signals in the audio (voice) range, the digital data must be converted to a form that is compatible with the telephone system. Thus, the digital data stream from the DTE is converted to *tones* (sine waves) having frequencies within the 300-Hz to 3000-Hz bandwidth required by the telephone company. The *modem* functions as an interface device between the DTE and the telephone network to perform the required conversion from digital data to tones, and vice versa.

At the transmitting end, digital data from the DTE are converted to tones and transmitted over the telephone channel by the modem. At the receiving end, the tones are converted back to digital data by the modem and passed to the receiving DTE. The modem does not change the content of the data; it changes only the form for transmission. This concept is shown in Figure 11–35.

The term *modem* is a contraction of *modulator-demodulator*. Modems are classified as *data communication equipment* (DCE).

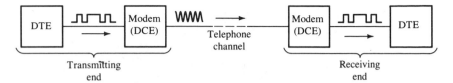

FIGURE 11-35 *Digital data transmission using modems.*

Low-Speed Modem Operation

Low-speed modems are typically used with personal computers on directly dialed telephone lines. They operate at a data rate of 300 bits/second (bps). For transmission, the voice band is divided into two separate bands so that data can be transmitted in both directions simultaneously. This is called *full-duplex* transmission. One modem uses one band, and the other modem uses the second band.

In each band, one frequency represents a 0 (called a *space*), and another frequency represents a 1 (called a *mark*). As Figure 11-36 shows, the *originate modem* transmits 0s at 1070 Hz and 1s at 1270 Hz. The *answer modem* transmits 0s at 2025 Hz and 1s at 2225 Hz.

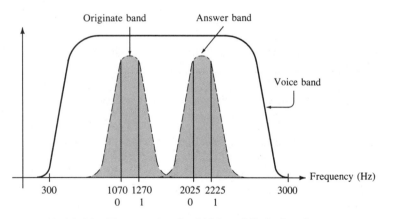

FIGURE 11-36 *Frequencies for 300 bps full-duplex data transmission.*

As a modem transmits data, the frequencies are shifted from one of the frequencies to the other to represent the sequence of 0s and 1s in the digital data stream. This type of modulation is called *frequency shift keying* (FSK). A simplified example of FSK operation is shown in Figure 11-37.

Modem Interfaces

A modem must interface with both the telephone network and the DTE. The interface to the telephone network consists basically of only two wires, the "tip"

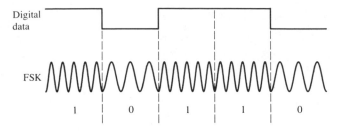

FIGURE 11–37   *Example of frequency shift keying (FSK).*

and the "ring." The modem-to-telephone interface must, of course, conform to the voltage, current, power, and frequency requirements established by the phone company.

The interface between the modem and the DTE is specified by an Electronic Industries Association (EIA) standard such as RS-232C, which is discussed later.

Certain procedures (protocol) must be adhered to in establishing communications between the two ends of the data system. First, the DTE at the transmitting end indicates to the modem that it wants to transmit. The transmitting modem then signals the receiving modem to determine if it is ready to receive. After the transmitting modem knows that the receiving modem and DTE are ready, it notifies the transmitting DTE. This process is called *handshaking*. The transmitting DTE then begins passing data to the modem, which FSK-modulates the data and transmits them to the receiving modem, which demodulates them and passes them to the receiving DTE.

Asynchronous and Synchronous Operation

In *asynchronous* 300 bps operation, no timing information (clock) is sent. There is a *start* bit that is used as a synchronizing signal for each character transmitted. Two *stop* bits and a parity bit follow each character, as the format in Figure 11–38 indicates. As the data come into the receiver, an internal oscillator determines the input timing independently of the data. The timing oscillator frequencies of the transmitter and receiver must stay within certain limits, of course.

In *synchronous* operation, a special code is transmitted at the beginning of each block of data to keep the receiver oscillator in step with the transmitted data.

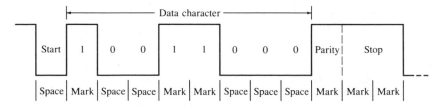

FIGURE 11–38   *Typical asynchronous data format.*

The Bell 103–Type Modem

The Bell 103 modem has, by widespread usage, become somewhat of a U.S. standard for low-speed (300 bps), asynchronous, full-duplex modems. (*Full-duplex* means that data can be transmitted in both directions simultaneously.) The frequency assignments for this type of modem are shown in Figure 11–39(a), and a basic block diagram is shown in part (b). Notice that the two-wire telephone line is terminated in a line-matching transformer.

As you can see in Figure 11–39, the *receive bandpass filter* and the *transmit bandpass filter* have center frequencies of 1170 Hz and 2125 Hz, respectively, in the *answer* modem and 2125 Hz and 1170 Hz, respectively, in the *originate* modem, thus permitting full-duplex operation. In the receiver section, the *limiter* eliminates variations in the amplitude of the filter output. The *delay detector*

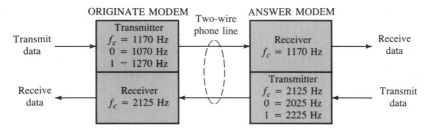

(a) Frequency assignments ($f_c \equiv$ center frequency)

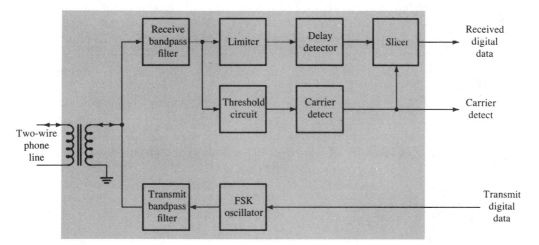

(b) Bell 103–type modem block diagram

FIGURE 11–39 *Low-speed, asynchronous, full-duplex modem.*

determines whether the frequency represents a 1 or a 0 and then produces a proportional output. The *slicer* clips the top and bottom of the detected signal and produces a digital output with proper logic voltage levels.

In the transmitter section, the *FSK oscillator* changes the digital data stream into the appropriate audio tones, which are then passed through the bandpass filter and onto the telephone lines.

Rather than transformer coupling directly to the phone lines, some modems use *acoustical coupling*. The acoustical coupler has a microphone and speaker which interface with the telephone handset and couple the transmitted and received signals by sound waves.

An IC Modem

The TMS99532 is an example of an integrated circuit modem that is Bell 103-compatible. Some of its features are as follows:

1. All filtering, modulation, and demodulation are on the chip.
2. Originate and answer modes.
3. Data rates are from zero to 300 bps.
4. The crystal-controlled oscillator is on the chip.
5. It has TTL-compatible interfacing.

Figure 11–40 illustrates the TMS99532 modem in an acoustical interface application.

The RS-232C Interface

When a modem is connected to a DTE in order to interface with the telephone lines, the modem and the DTE must also be properly interfaced with each other. The most common U.S. standard, as previously mentioned, has been recommended by the EIA (Electronic Industries Association) for interfacing DTEs and DCEs. The standard, called the RS-232C, specifies the basic signals for sending and receiving data, status information, and control. Also, standard designations, connector pin assignments, and voltage levels are specified.

The RS-232C voltage levels are $+3$ V to $+25$ V for a space (0), and -3 V to -25 V for a mark (1). The recommended maximum cable length is 50 feet, which is determined by the maximum specified cable capacitance of 2500 pF.

Table 11–3 lists all of the RS-232C signals with pin numbers for a 25-pin connector, EIA designations, and abbreviations. The ten most often used signals are indicated with **boldface** type. Not all of the RS-232C signals are used in every application, but the minimum DTE-to-DCE interface connections are *transmitted data* (TD), *received data* (RD), *signal ground* (SG), and *chassis ground* (GND).

RS-232C limitations   There are several drawbacks to the RS-232C standard, some of which are as follows:

1. The 50-foot length limitation is a serious restriction in some applications.

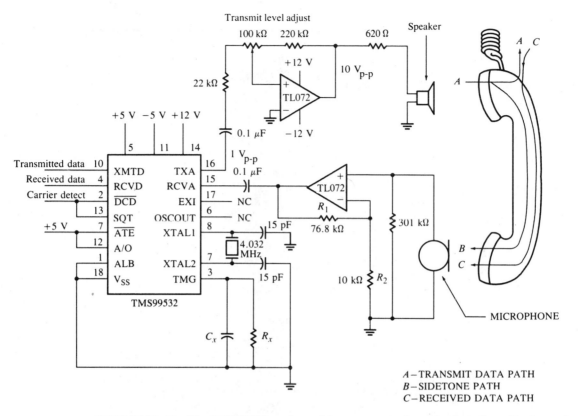

FIGURE 11–40 *The TMS99532 modem with an acoustical coupler interface. (Redrawn by permission from Texas Instruments, Inc.)*

2. The voltage levels are not compatible with logic levels in the computers or other DTEs with which the modem must interface, and additional interface devices must be used to overcome this.

3. All of the data and control signals are referenced to the same signal ground (pin 7). Although this method of *unbalanced transmission* works fine in many applications, there is sometimes a difference in ground potential between the two ends of a long cable, which can affect signal quality and result in errors.

Other Interfaces

The RS-232 was introduced in 1962 and is still widely used throughout industry. It was developed for transmission over short distances at relatively slow data rates. Many current higher-performance systems require transmission at faster data rates over longer lines and are making the RS-232C inadequate for many applications.

TABLE 11-3 *RS-232C signals.*

Pin Number	EIA Ckt.	Signal Description	Common Abbreviation	From DCE	To DCE
1	AA	**Protective (Chassis) Ground**	**GND**	X	X
2	BA	**Transmitted Data**	**TD**		X
3	BB	**Received Data**	**RD**	X	
4	CA	**Request to Send**	**RTS**		X
5	CB	**Clear to Send**	**CTS**	X	
6	CC	**Data Set Ready**	**DSR**	X	
7	AB	**Signal Ground/Common Return**	**SG**	X	X
8	CF	**Received Line Signal Detector**	**DCD**	X	
9		Reserved			
10		Reserved			
11		Unassigned			
12	SCF	Secondary Received Line Signal Detector		X	
13	SCB	Secondary Clear to Send		X	
14	SBA	Secondary Transmitted Data			X
15	DB	Transmitter Signal Element Timing (DCE)		X	
16	SBB	Secondary Received Data		X	
17	DD	Receiver Signal Element Timing		X	
18		Unassigned			
19	SCA	Secondary Request to Send			X
20	CD	**Data Terminal Ready**	**DTR**		X
21	CG	Signal Quality Detector	SQ	X	
22	CE	**Ring Indicator**	**RI**	X	
23	CH	Data Signal Rate Selector (DTE)			X
23	CI	Data Signal Rate Selector (DCE)		X	
24	DA	Transmitter Signal Element Timing (DTE)			X
25		Unassigned			

Current loop The signal interface in teleprinters is called the *current loop*. This method uses the presence or absence of current to represent 1s and 0s. A current loop can be either 20 mA or 60 mA, depending on the type of equipment. There are no standard pin numbers for current-loop interfacing, but data rates up to 9600 bps and cable lengths up to 1500 feet can be handled.

RS-422A This standard eliminates some of the problems with the RS-232C. It uses two wires for each signal, thus producing *balanced transmission* for greater data rates (up to 10 Mbps) and eliminating the ground problems.

RS-423A This standard defines an adaptor for use between RS-232C and RS-422A, thus permitting RS-232C equipment to be interfaced with RS-422A equipment.

RS-366 This standard provides for automatic dialing under computer control.

Table 11–4 summarizes the features of RS-232C, RS-422A, and RS-423A standards.

TABLE 11–4

Specification		RS-232C	RS-423A	RS-422A
Mode of operation		Single-ended	Single-ended	Differential
Number of drivers and receivers allowed on one line		1 driver, 1 receiver	1 driver, 10 receivers	1 driver, 10 receivers
Maximum cable length		50 feet	4000 feet	4000 feet
Maximum data rate		20 kb/s	100 kb/s	10 Mb/s
Maximum voltage applied to driver output		± 25 V	± 6 V	-0.25 V to 6 V
Driver output signal	Loaded	± 5 V	± 3.6 V	± 2 V
	Unloaded	± 15 V	± 6 V	± 5 V
Driver load		3 kΩ to 7 kΩ	450 Ω min	100 Ω
Maximum driver output current (High-impedance state)	Power on	—	—	—
	Power off	$V_{max}/300$ Ω	± 100 μA	± 100 μA
Receiver input voltage range		± 15 V	± 12 V	-7 V to 7 V
Receiver input sensitivity		± 3 V	± 200 mV	± 200 mV
Receiver input resistance		3 kΩ to 7 kΩ	4 kΩ min	4 kΩ min

SOURCE: *Interface Circuits Data Book,* Texas Instruments, Inc.

SECTION REVIEW 11–4

1. What is a modem?
2. What function do modems perform?
3. What type of signal does a modem output to the telephone network?

11–5   THE SCHMITT TRIGGER AS AN INTERFACE CIRCUIT

The Schmitt trigger operates with two *threshold* or trigger points. When an increasing input signal reaches the *upper threshold point* (UTP), the circuit switches and the output goes to its HIGH level. When the input signal decreases to the *lower threshold point* (LTP), the circuit switches back and the output goes to its LOW level. This operation is illustrated in Figure 11–41(a).

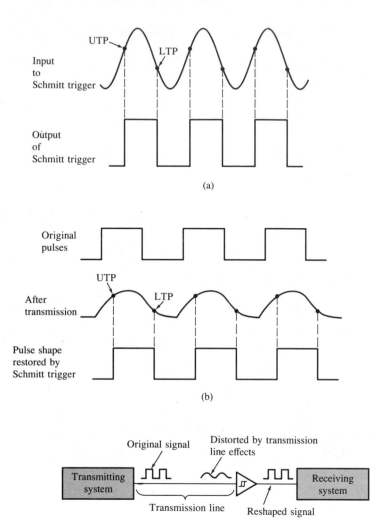

FIGURE 11–41 *Schmitt trigger operation.*

As you can see, the Schmitt trigger circuit can be used to convert a sine wave into a pulse waveform. More importantly, however, in interfacing applications it is useful in reshaping pulses that have been distorted in transmission as depicted in Figure 11–41(b). Part (c) of the figure shows a simplified diagram of a data transmission system to illustrate this application.

Many standard gates are available in Schmitt trigger versions. The small *square loop* symbol on the inverter or gate symbol indicates a Schmitt trigger circuit as shown in Figure 11–42.

FIGURE 11–42 *Typical logic symbols for Schmitt trigger inverter and gate.*

SECTION REVIEW 11–5

1. How does a Schmitt trigger logic gate differ from a regular logic gate?

SUMMARY

- ☐ No extra circuitry is required for CMOS-to-TTL interfacing.
- ☐ A pull-up resistor is required for TTL-to-CMOS interfacing.
- ☐ TTL totem-pole outputs cannot be safely connected together.
- ☐ Open-collector outputs can be wire-ANDed.
- ☐ Three-state bus transceivers are used for bidirectional bus interfacing.
- ☐ The *general-purpose interface bus* (GPIB) is defined by the IEEE–488 standard.
- ☐ Two methods of digital-to-analog (D/A) conversion are the *binary-weighted input* method and the *R/2R resistive ladder* method.
- ☐ Resolution of a D/A converter is the reciprocal of the number of steps: $1/(2^n - 1)$.
- ☐ The fastest method of analog-to-digital (A/D) conversion is the simultaneous method, sometimes called the *flash* method.
- ☐ Other methods of A/D conversion are the *ramp, tracking, single-slope, dual-slope*, and *successive-approximation*.
- ☐ A modem is a modulator-demodulator used to interface data terminal equipment (DTE) to communications channels.
- ☐ The RS-232C is an EIA standard for interfacing DTEs to DCEs.

SELF-TEST

1. Define the term *interfacing*.
2. What is a major interfacing consideration when CMOS is driving TTL?
3. Which interface requires additional circuitry, CMOS-to-TTL or TTL-to-CMOS? What is required and why?
4. What type of TTL circuit must be used in a wired-AND configuration?
5. Show the difference between totem-pole outputs and open-collector outputs by sketching the basic circuits.
6. What type of TTL circuit should normally be used to drive an indicator lamp?
7. What is a *bidirectional bus*?
8. Name the three basic types of devices defined to operate on the IEEE–488 bus.
9. Under the IEEE–488 standard, how many devices can be connected to the GPIB?
10. Distinguish between analog and digital signals.

11. What is the percent resolution of a five-bit D/A converter?
12. What does the term *monotonicity* mean in relation to D/A conversion?
13. What is the advantage of the tracking-type A/D converter over the stairstep-ramp converter?
14. What is the bandwidth of the telephone network?
15. Explain why digital signals cannot be directly transmitted over the telephone lines.
16. What does *FSK* mean?
17. What is an acoustical interface?
18. What is the purpose of the RS-232C interface standard?

PROBLEMS

Section 11–1

11–1   A 74H CMOS gate drives one standard TTL input and five low-power Schottky inputs. Is the fanout of the CMOS gate exceeded?

11–2   How much current can a 74H CMOS gate sink in the low output state?

11–3   Determine the minimum value of R_p required for interfacing a standard TTL with ten 74H CMOS gates. Show the circuit.

11–4   How many 74H CMOS gates can a standard TTL gate drive when $R_p = 3.3 \text{ k}\Omega$?

11–5   Find the value of limiting resistor for a 20-mA LED driven by an LS TTL gate with open collector. Neglect the resistance of the LED, and use $V_{CC} = +5$ V.

11–6   For the wired-AND configuration in Figure 11–43, determine the maximum current that each gate must sink.

FIGURE 11–43

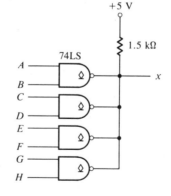

Section 11–2

11–7   Determine the signals on the bus line in Figure 11–44 for the data-input and enable waveforms shown.

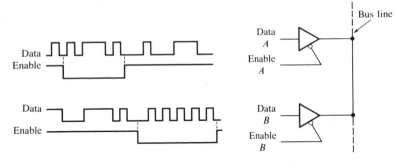

FIGURE 11–44

11–8 List the handshake lines in the GPIB.

Section 11–3

11–9 Determine the output of the D/A converter in Figure 11–45(a) if the sequence of four-bit numbers in part (b) is applied to the inputs.

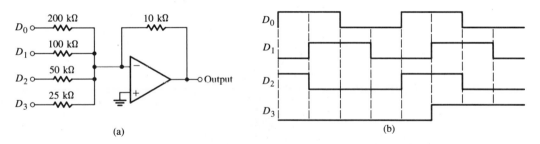

FIGURE 11–45

11–10 Repeat Problem 11–9 for the inputs in Figure 11–46.

FIGURE 11–46

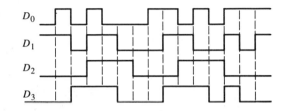

11–11 Determine the resolution for each of the following D/A converters:
(a) five-bit **(b)** ten-bit **(c)** eighteen-bit

11–12 Determine the binary output code of a three-bit simultaneous A/D converter for the analog input signal in Figure 11–47. The sampling rate is 100 kHz.

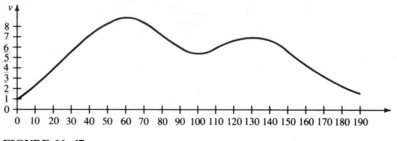

FIGURE 11–47

11–13 Repeat Problem 11–12 for the analog waveform in Figure 11–48.

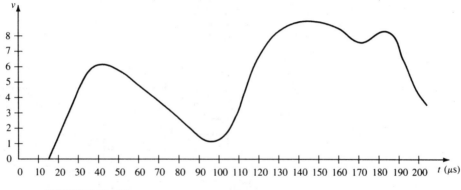

FIGURE 11–48

11–14 For a three-bit stairstep-ramp A/D converter, the reference voltage advances one step every microsecond. Determine the encoded binary sequence for the analog signal in Figure 11–49.

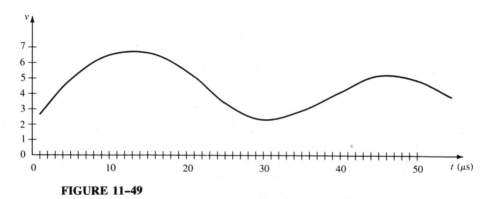

FIGURE 11–49

11-15 For a four-bit stairstep-ramp A/D converter, assume that the clock period is 1 microsecond. Determine the binary sequence on the output for the input signal in Figure 11–50.

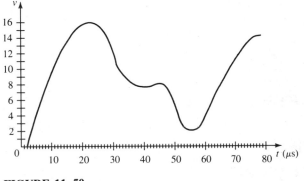

FIGURE 11–50

11-16 Repeat Problem 11–14 for a tracking A/D converter.

11-17 For a certain four-bit successive-approximation A/D converter, the maximum ladder output is $+8$ V. If a constant $+6$ V is applied to the analog input, determine the sequence of binary states for the SA register.

Section 11–4

11-18 Determine the digital data stream corresponding to the simplified FSK pattern represented in Figure 11–51.

FIGURE 11–51 *FSK pattern.*

11-19 Modems are necessary to connect computers to telephone lines because
   (a) The telephone bandwidth is too high for digital signals.
   (b) The telephone company rules require them.
   (c) Asynchronous transmission is not permitted over the phone lines.
   (d) None of the above.
   (e) If you chose (d), state the correct reason.

Section 11-5

11-20 For a Schmitt trigger circuit with UTP = 1 V and LTP = 0.5 V, sketch the output for the input signal in Figure 11-52.

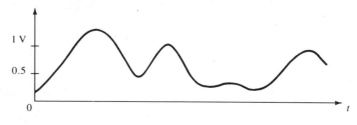

FIGURE 11-52

ANSWERS TO SECTION REVIEWS

Section 11-1
1. 4000   **2.** $Y = \overline{ABCDE}$

Section 11-2
1. So that all devices, except those communicating at any given time, can be disconnected from the bus.
2. Sixteen.

Section 11-3
1. Simultaneous (flash).   **2.** **(a)** 1.59%   **(b)** 0.39%

Section 11-4
1. Modulator/demodulator for transmitting and receiving digital data.
2. Interfaces between data terminal equipment (DTE) and telephone network.
3. Tones in the 300-Hz to 3000-Hz bandwidth.

Section 11-5
1. It has threshold trigger points.

In Chapter 2, binary addition, subtraction, multiplication, and division were covered, and the 1's and 2's complement methods were introduced. This chapter expands that coverage to include *signed* binary numbers. Signed BCD addition is also covered.

A typical *arithmetic logic unit* (ALU) is introduced with applications in a simple processor. Two microprocessors, the 6800 and the 8085A, are discussed to introduce the concept of the microprocessor. However, details of microprocessor operation are left for future courses dealing specifically with these important devices.

In this chapter, you will learn

☐ How to handle signed binary arithmetic.
☐ How a specific ALU (the 74S381) operates.
☐ How the ALU, in conjunction with registers and control logic, performs arithmetic and logic operations.
☐ How to handle signed BCD addition.
☐ How to implement BCD addition with logic elements.
☐ What a microprocessor is.
☐ The basic organization of the 6800 and the 8085A microprocessors.

12

Arithmetic Processes with an Introduction to Microprocessors

12–1   BINARY REPRESENTATION OF SIGNED NUMBERS

The capability of handling both positive and negative numbers is a requirement of any *arithmetic logic unit* (ALU). A signed number consists of both *sign* and *magnitude* information. The sign indicates whether a number is positive or negative, and the magnitude is the value of the number.

In binary systems, the sign is represented by including an additional bit with the magnitude bits. Conventionally, a 0 represents a *positive* sign and a 1 represents a *negative* sign. For example, in an eight-bit number the left-most bit is the sign bit and the remaining seven bits are magnitude bits. This binary sign and magnitude representation is as follows for decimal numbers $+107$ and -20:

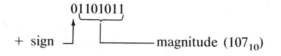

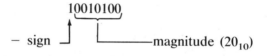

A *byte,* that is, eight bits of information, is very common in many microprocessor systems. With an eight-bit sign and magnitude number, a range of values from $01111111_2 = +127_{10}$ to $11111111_2 = -127_{10}$ can be represented. Two or more bytes can be used to represent larger numbers. For example, 16 bits include one sign bit and 15 magnitude bits. With this format, a range of numbers from $0111111111111111_2 = +32,767_{10}$ to $1111111111111111_2 = -32,767_{10}$ can be represented.

1's Complement Form of Negative Numbers

Negative numbers are represented in 1's complement form by *inverting all the magnitude bits and leaving the sign bit as is.* For example,

$$-53_{10} = 10110101 \longrightarrow 11001010$$
$$\text{true form} \qquad \text{1's complement form}$$

EXAMPLE 12–1   Convert each negative binary number to 1's complement form:
(a) 11000100   (b) 10001001   (c) 11110110   (d) 11000011

Solutions
(a) 11000100 $\longrightarrow$ 10111011   (b) 10001001 $\longrightarrow$ 11110110
(c) 11110110 $\longrightarrow$ 10001001   (d) 11000011 $\longrightarrow$ 10111100

2's Complement Form of Negative Numbers

Most digital systems use 2's complement to represent negative numbers for arithmetic operations. One method of obtaining the 2's complement of a negative number is to add 1 to the 1's complement, as discussed in Chapter 2. Only the magnitude bits are complemented; the sign bit is left a 1. For example,

$$11101010 \longrightarrow \begin{array}{l} 10010101 \quad \text{1's complement} \\ \underline{+\ 1} \\ 10010110 \quad \text{2's complement} \end{array}$$

A second method of obtaining the 2's complement was also discussed in Chapter 2. Again, only the magnitude bits are affected. The procedure is as follows: Starting with the right-most bit, the bits remain uncomplemented up to and including the first 1. The remaining magnitude bits are inverted.

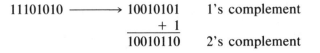

```
          ┌─complemented
          │  ┌─uncomplemented
10110110      true form
11001010      2's complement
```

EXAMPLE 12-2 Convert each of the following negative signed binary numbers to 2's complement form using both of the methods just discussed:
(a) 10010010 **(b)** 10110000 **(c)** 11110111 **(d)** 10000000

Solutions
(a) Taking the 1's complement and adding 1 yields

$$10010010 \longrightarrow \begin{array}{l} 11101101 \quad \text{1's complement} \\ \underline{+\ 1} \\ 11101110 \quad \text{2's complement} \end{array}$$

Using the second method, we have

$$\underbrace{10010}_{\text{complemented}}\underbrace{010}_{\text{uncomplemented}} \longrightarrow 11101110$$

(b) Take the 1's complement and add 1.

$$10110000 \longrightarrow \begin{array}{l} 11001111 \quad \text{1's complement} \\ \underline{+\ 1} \\ 11010000 \quad \text{2's complement} \end{array}$$

Use the second method:

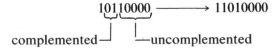

$$\underbrace{1011}_{\text{complemented}}\underbrace{0000}_{\text{uncomplemented}} \longrightarrow 11010000$$

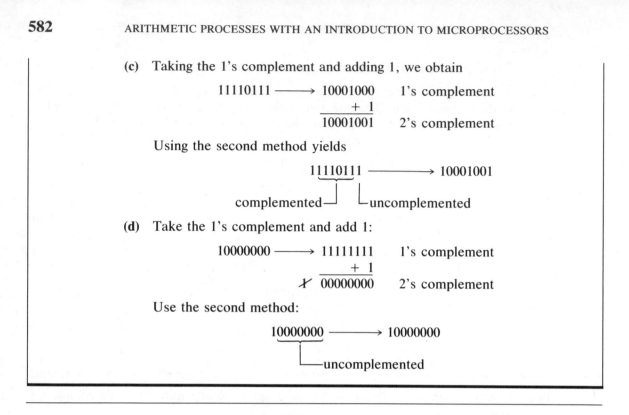

(c) Taking the 1's complement and adding 1, we obtain

$$11110111 \longrightarrow \begin{array}{r} 10001000 \\ +\ 1 \\ \hline 10001001 \end{array} \quad \begin{array}{l} \text{1's complement} \\ \\ \text{2's complement} \end{array}$$

Using the second method yields

$$11110111 \longrightarrow 10001001$$

complemented⌐   ⌐uncomplemented

(d) Take the 1's complement and add 1:

$$10000000 \longrightarrow \begin{array}{r} 11111111 \\ +\ 1 \\ \hline \chi\ 00000000 \end{array} \quad \begin{array}{l} \text{1's complement} \\ \\ \text{2's complement} \end{array}$$

Use the second method:

$$10000000 \longrightarrow 10000000$$

└─uncomplemented

SECTION REVIEW 12–1

1. Determine the decimal equivalent of each of the signed binary numbers (all are in true form).
   (a) 0011011   **(b)** 11011101   **(c)** 01010001
2. Express each decimal number in 2's complement.
   (a) -38 **(b)** -59 **(c)** -273

12–2 SIGNED BINARY ARITHMETIC

Binary Addition

In this section you will see how signed binary numbers are added. The 2's complement will be used to represent negative numbers.

There are four cases that must be considered when adding two numbers:

1. Both numbers positive
2. Positive number and smaller negative number
3. Positive number and larger negative number
4. Both numbers negative

We will take one case at a time. Eight bits are used to represent each number.

Both numbers positive   In this case, both sign bits are zero and a 2's complement is not required. To illustrate, we will add $+7$ and $+4$:

$$
\begin{array}{r@{\qquad}l}
7 & 00000111 \\
+\ \ 4 & 00000100 \\
\hline
11 & 00001011
\end{array}
$$

Positive number and smaller negative number   In this case, the true binary form of the positive number is added to the 2's complement of the negative number. The sign bits are included in the addition, and the result will be positive. To illustrate, we will add $+15$ and -6:

$$
\begin{array}{r@{\qquad}l}
15 & 00001111 \\
+\ \ -6 & 11111010 \\
\hline
9 & \cancel{}\ 00001001
\end{array}
$$

└─discard carry

Notice that the sign of the sum is positive (0) as it should be.

Positive number and larger negative number   Again, the true binary form of the positive number is added to the 2's complement of the negative number. The sign bits are included in the addition, and the result will be negative. To illustrate, we will add $+16$ and -24:

$$
\begin{array}{r@{\qquad}ll}
16 & 00010000 \\
+\ \ -24 & 11101000 \\
\hline
-8 & 11111000 & \text{2's complement of } -8
\end{array}
$$

Notice that the result automatically comes out in 2's complement because it is a negative number.

Both numbers negative   In this case, the 2's complements of both numbers are added and, of course, the sum is a negative number in 2's complement form. To illustrate, we will add -5 and -9:

$$
\begin{array}{r@{\qquad}ll}
-5 & 11111011 \\
+\ \ -9 & 11110111 \\
\hline
-14 & \cancel{}\ 11110010 & \text{2's complement of } -14
\end{array}
$$

└─discard carry

Overflow

When the number of bits in the sum exceeds the number of bits in each of the numbers added, *overflow* results, as illustrated by the following example.

	Decimal	*Binary*
EXAMPLE	+9	01001
12–3	+ +8	+ 01000
	+17	10001

sign incorrect ———┘ └——— magnitude incorrect

The overflow condition can occur *only* when both numbers are positive or both numbers are negative. *It is indicated by an incorrect sign bit.*

Summary of Signed Addition

The following is a summary of the four cases of signed binary addition.

Both numbers positive:

1. Add both numbers in *true* (uncomplemented) form, including the sign bit.
2. The sign bit of the sum will be 0 (+).
3. Overflow into the sign bit is possible. The sign bit and the magnitude of the sum will be incorrect if an overflow occurs.

Both numbers negative:

1. Take the 1's or 2's complement of the magnitude of both numbers. Leave the sign bits as they are.
2. Add the numbers in their complement form, including sign bits.
3. Add the end-around carry in the case of the 1's complement method; drop the carry in the case of the 2's complement method. The sign bit of the sum will be 1 (−), and the magnitude of the sum will be in complement form.
4. Overflow into the sign bit is possible. The sign bit and magnitude of sum will be incorrect if an overflow occurs.

Larger number positive, smaller number negative:

1. Take the 1's or 2's complement of the magnitude of the negative number. Leave the sign bit as is. A positive number remains in true form.
2. Add the numbers, including the sign bits.
3. Add the end-around carry for the 1's complement method; drop the carry for the 2's complement method. The sign bit of the sum will be a 0 (+), and the magnitude will be in true form.
4. No overflow is possible.

Larger number negative, smaller number positive:

1. Take the 1's or 2's complement of the magnitude of the negative number. Leave the sign bit as is. A positive number remains in true form.
2. Add the numbers, including the sign bits.
3. No carries will occur. The sum will have the proper sign bit, and the magnitude will be in complement form.
4. No overflow is possible.

Now that we have examined the basic arithmetic processes by which *two* numbers can be added, let us look at the addition of a string of numbers added two at a time. This can be accomplished by adding the first two numbers, then adding the third number to the sum of the first two, then adding the fourth number to this result, and so on. The addition of several numbers taken two at a time is illustrated in the following example, which is the basic way adders operate in microprocessor systems.

EXAMPLE 12–4   Add the numbers 2, 4, 6, 5, 8, and 9.

Solution

$$
\begin{array}{r}
2 \\
+\ 4 \\
\hline
6 \leftarrow \text{first sum} \\
+\ 6 \\
\hline
12 \leftarrow \text{second sum} \\
+\ 5 \\
\hline
17 \leftarrow \text{third sum} \\
+\ 8 \\
\hline
25 \leftarrow \text{fourth sum} \\
+\ 9 \\
\hline
34 \leftarrow \text{final sum}
\end{array}
$$

Binary Subtraction

Subtraction is a special case of addition. For example, subtracting +6 (subtrahend) from +9 (minuend) is equivalent to *adding* −6 to +9.

Basically the subtraction operation *changes the sign of the subtrahend and adds it to the minuend*. The 2's complement method can be used in subtraction so that all operations require only addition. The four cases that were discussed in relation to the addition of signed numbers apply to the subtraction process because subtraction can be essentially reduced to an addition process. The following example will illustrate.

EXAMPLE 12–5

Perform each subtraction in binary for each pair of decimal numbers using the 2's complement method.

(a) $8 - 3$ (b) $12 - (-9)$ (c) $-25 - 8$
(d) $-58 - (-32)$ (e) $17 - 28$

Solutions

(a)
$$\begin{array}{r} 8 \\ -\quad 3 \\ \hline +5 \end{array} \equiv \begin{array}{r} 8 \\ +\ -3 \\ \hline +5 \end{array}$$

00001000	
+11111101	2's complement of -3
$\cancel{1}$ 00000101	$+5$

(b)
$$\begin{array}{r} 12 \\ -\ -9 \\ \hline +21 \end{array} \equiv \begin{array}{r} 12 \\ +\ 9 \\ \hline +21 \end{array}$$

00001100	
+00001001	
00010101	$+21$

(c)
$$\begin{array}{r} -25 \\ -\quad 8 \\ \hline -33 \end{array} \equiv \begin{array}{r} -25 \\ +\ -8 \\ \hline -33 \end{array}$$

11100111	2's complement of -25
+11111000	2's complement of $-\ 8$
11011111	2's complement of -33

(d)
$$\begin{array}{r} -58 \\ -\ -32 \\ \hline -26 \end{array} \equiv \begin{array}{r} -58 \\ +\ 32 \\ \hline -26 \end{array}$$

11000110	2's complement of -58
+00100000	
11100110	2's complement of -26

(e)
$$\begin{array}{r} 17 \\ -\quad 28 \\ \hline -11 \end{array} \equiv \begin{array}{r} 17 \\ +\ -28 \\ \hline -11 \end{array}$$

00010001	
+11100100	2's complement of -28
11110101	2's complement of -11

Binary Multiplication

In binary multiplication the two numbers are the *multiplicand* and the *multiplier,* and the result is the *product.* The magnitudes of the numbers must be in true form, and the sign bits are not used during the multiplication process.

The sign of the product depends on the signs of the two numbers that are being multiplied. If the two numbers have the same sign, either both positive or both negative, the product is positive. This is illustrated with decimal numbers as follows:

$$(+3)(+5) = +15 \qquad (-4)(-5) = +20$$
$$(-10)(+3) = -30 \qquad (+6)(-8) = -48$$

In the multiplication process, the sign bits of the two numbers are checked and the resulting sign of the product is stored prior to the actual multiplication. Once the product sign is determined, the multiplicand and multiplier sign bits are discarded, leaving only the magnitudes to be multiplied.

The multiplication process   Multiplication of binary numbers in arithmetic units is often accomplished by a series of additions and shifts similar to the way you multiply in longhand. So, as in subtraction, only an adder is required. The following example illustrates binary multiplication of two unsigned seven-bit numbers.

EXAMPLE 12-6

Multiply 1010011 (83_{10}) and 0111011 (59_{10}). The seven bits represent the magnitudes of the numbers and are given in true form. The signed bits are not included here.

Solution

$$
\begin{array}{r}
1010011 \\
\times\ 0111011 \\
\end{array}
\quad\text{multiplicand} \\
\quad\text{multiplier}
$$

$$
\left.
\begin{array}{r}
1010011 \\
1010011 \\
0000000 \\
1010011 \\
1010011 \\
1010011 \\
+\ 0000000 \\
\end{array}
\right\}\ \text{partial products}
$$

$$
\overline{1001100100001}\quad\text{final product } (4897_{10})
$$

Notice that the number of partial products equals the number of multiplier bits. Also, each partial product is shifted one bit to the left with respect to the previous partial product. The final product, which is the sum of the shifted partial products, can have up to twice the number of bits as the original numbers.

Most arithmetic logic units can add only two numbers at a time. Because of this, each partial product is added to the sum of the previous partial products, and the sum of the partial products is accumulated. For example, the first partial product and the second partial product are added; then the third partial product is added to this sum; and so on.

Now we will go through the previous multiplication example as a typical arithmetic unit would handle it.

EXAMPLE 12-7

Repeat the multiplication in Example 12–6, except this time keep a running total of the partial product sum so that only two numbers will have to be added at a time.

Solution

```
        1010011      multiplicand
      × 0111011      multiplier

        1010011      1st partial product
      + 1010011      2nd partial product

       11111001      sum of 1st and 2nd
      + 0000000      3rd partial product

      011111001      sum
      + 1010011      4th partial product

     1110010001      sum
      + 1010011      5th partial product

    100011000001     sum
      + 1010011      6th partial product

    1001100100001    sum
      + 0000000      7th partial product

    1001100100001    final product
```

The multiplier and multiplicand must be in true form when multiplied. In a computer using the 2's complement system, any negative number is already in 2's complement form and must be converted to its true form before multiplying. A negative number is converted back to its true form by taking the 2's complement of the 2's complement. For example, let us convert a signed binary number from its true form to 2's complement and then back to the true form:

$$10101110 \longrightarrow 11010010 \longrightarrow 10101110$$
$$\text{true} \qquad\qquad \text{2's comp.} \qquad\qquad \text{true}$$

To summarize the multiplication process in a typical 2's complement arithmetic unit, the following steps are listed:

1. Obtain the multiplicand and multiplier.
2. If either number or both are negative, convert them to true form.
3. Examine both sign bits and store the sign of the product. (If both signs are the same, the product is positive; if signs are different, the product is negative.)
4. Examine each multiplier magnitude bit beginning with the right-most bit. When the multiplier bit is a 1, shift the partial product sum one place to the right and add the multiplicand magnitude bits. This is, in effect, the same as shifting each new partial product one place to the left as in the previous example.

If the multiplier bit is a 0, the partial product sum is shifted one place to the right and no addition is performed. This is effectively the same as adding an all-zeros partial product.

5. Attach the predetermined sign bit to the final product. If the sign is negative, put the product into 2's complement form.

The following example will illustrate the steps in this process.

EXAMPLE 12-8 Multiply the following two signed binary numbers: $+107_{10}$ and -51_{10}. The negative number is in 2's complement form.

$$01101011 \times 11001101$$

Solution

Step 1. Put the negative number in true form:

$$11001101 \longrightarrow 10110011$$
$$\text{2's complement} \qquad \text{true number}$$

Step 2. Determine the sign of the product. Because the signs are different, the product will have a negative sign bit (1).

Step 3. Multiply the magnitudes:

	1101011	multiplicand magnitude (107_{10})
	0110011	multiplier magnitude (51_{10})
1st multiplier bit = 1	1101011	1st partial product
	1101011	shift right
2nd multiplier bit = 1	+ 1101011	2nd partial product
	101000001	sum
	101000001	shift right
3rd multiplier bit = 0	101000001	shift right (3rd part. prod. = 0)
4th multiplier bit = 0	101000001	shift right (4th part. prod. = 0)
5th multiplier bit = 1	+ 1101011	5th partial product
	11111110001	sum
	11111110001	shift right
6th multiplier bit = 1	+ 1101011	6th partial product
	1010101010001	final product
		(7th part. prod. = 0)

Step 4. Attach the sign bit and take the 2's complement of the final product because it is negative:

$$1101010101111_2 = 5457_{10}$$

Multiplying by repeated addition Another way to multiply two numbers is to simply add the multiplicand a number of times equal to the value of the multiplier. For example, to multiply 4 and 8, we can add 8 four times as follows:

$$
\begin{array}{r}
8 \\
+\ 8 \\
\hline
16 \\
+\ 8 \\
\hline
24 \\
+\ 8 \\
\hline
32 \quad \text{final product}
\end{array}
$$

This can be done with binary numbers in digital systems as follows:

1. First, both magnitudes are put into true form, and the sign of the product is stored as previously described.
2. Bring down the multiplicand and decrement the multiplier. (Subtract 1 from the multiplier by adding the 2's complement of 1.)
3. Check the multiplier for zero value. If it is zero, then the multiplication is complete. If the multiplier is not zero, add the multiplicand to the previous multiplicand sum.
4. This process continues until the multiplier equals zero. At that point, the multiplicand has been added a number of times equal to the value of the multiplicand. The final product is available and the sign is attached.

EXAMPLE 12–9 Multiply 1011010 (90_{10}) and 0000011 (3_{10}). Both numbers are true magnitudes. The smaller number is the multiplier.

Solution

Step 1. Bring down the multiplicand and decrement the multiplier:

multiplicand = 1011010

$$
\begin{array}{rl}
0000011 & = 3 \\
+\,1111111 & = \text{2's complement of 1} \\
\hline
\cancel{1}\,0000010 & = 2
\end{array}
$$

Step 2. Check the multiplier for zero value:

$$\text{multiplier} = 0000010 = 2$$

Step 3. Add the multiplicand to itself:

$$
\begin{array}{r}
1011010 \\
+\,1011010 \\
\hline
10110100
\end{array}
$$

Step 4. Decrement the multiplier:

$$0000010 = 2$$
$$+ 1111111 = \text{2's complement of 1}$$
$$\cancel{1}\ 0000001 = 1$$

Step 5. Check the multiplier for zero value:

$$\text{multiplier} = 0000001 = 1$$

Step 6. Add the multiplicand to the previous multiplicand sum:

$$10110100$$
$$+ 1011010$$
$$100001110$$

Step 7. Decrement the multiplier:

$$0000001 = 1$$
$$+ 1111111 = \text{2's complement of 1}$$
$$\cancel{1}\ 0000000 = 0$$

Step 8. Check the multiplier for zero value:

$$\text{multiplier} = 0000000 = 0$$

Step 9. The multiplication is complete and the final product is the last multiplicand sum:

$$100001110$$

This product is 270_{10}, which is correct because the multiplicand is 90_{10} and the multiplier is 3_{10}.

Binary Division

In division, the two numbers are called the *dividend* and the *divisor*. The result is the *quotient*. The sign of the quotient is determined from the signs of the dividend and divisor. If both signs are the same, the quotient is positive. If the signs differ, the quotient is negative.

The division process Division in arithmetic units can be accomplished by repeated subtraction. Since subtraction can be done with 2's complement addition, the division process requires only an adder. Keep in mind that the quotient is the number of times that the divisor goes into the dividend. The process is as follows:

1. Obtain the divisor and dividend. Initialize the quotient to zero.
2. Subtract the divisor from the dividend to get the partial remainder. If this remainder is positive, increment the quotient. (Add 1 to the quo-

tient.) A positive partial remainder indicates that the divisor goes into the dividend (or partial remainder). If the partial remainder is negative, the divisor does not go into the dividend (or partial remainder), and therefore the division is complete.

3. If the partial remainder is positive, repeat step 2 by subtracting the divisor from the partial remainder.

The following example will illustrate this process.

EXAMPLE 12–10

Divide 01100100 by 00110010. The decimal equivalent of this division is $+100_{10}$ divided by $+50_{10}$.

Solution

Step 1. Initialize the quotient to zero. Subtract the divisor from the dividend using 2's complement addition.

$$
\begin{array}{ll}
01100100 & \text{dividend} \\
+\ 11001110 & \text{2's complement of divisor} \\
\hline
00110010 & \text{positive 1st partial remainder} \\
\end{array}
$$

$+$ sign ⟶

Step 2. A positive partial remainder indicates that the divisor goes into the dividend and therefore the quotient is incremented by one. The new quotient is 00000001.

Step 3. Subtract the divisor from the 1st partial remainder:

$$
\begin{array}{ll}
00110010 & \text{1st partial remainder} \\
+\ 11001110 & \text{2's complement of divisor} \\
\hline
00000000 & \text{positive 2nd partial remainder} \\
\end{array}
$$

$+$ sign ⟶

Step 4. Increment the quotient to 00000010.

Step 5. Subtract the divisor from the 2nd partial remainder:

$$
\begin{array}{ll}
00000000 & \text{2nd partial remainder} \\
+\ 11001110 & \text{2's complement of divisor} \\
\hline
11001110 & \text{negative 3rd partial remainder} \\
\end{array}
$$

$-$ sign ⟶

The negative remainder indicates that the divisor will not go into the 2nd partial remainder. Therefore, the final quotient is $00000010 = 2_{10}$ and the division is complete.

In this particular example, the zero 2nd partial remainder indicated that the division was complete. However, a zero remainder does not always occur, and therefore the check for a *negative partial remainder* is a more general test.

1. Add the following signed binary numbers:
   (a) $01101101 + 00001110$ **(b)** $10011101 + 00111010$
2. Perform the following signed binary subtractions:
   (a) $01110110 - 10111001$ **(b)** $11011011 - 10111101$

12-3 IMPLEMENTATION OF SIGNED BINARY ARITHMETIC

First, as an example of an *arithmetic logic unit* (ALU), a specific device is examined.

The 74S381 Arithmetic Logic Unit/Function Generator

The 74S381 performs eight different operations, including binary arithmetic and logic functions on two 4-bit words as well as clear and preset operations. This device is represented in Figure 12-1 by the ANSI/IEEE Std. 91-1984 logic symbol.

The eight operations or modes (M) are selected by the three select inputs S_0, S_1, and S_2. The $\frac{0}{7}$ label after the M indicates that there are eight modes (0 through 7). The $\overline{CP}$ (carry propagate) and $\overline{CG}$ (carry generate) outputs provide look-ahead-carry functions for cascading with look-ahead-carry generators (74182) to expand the number of bits handled. In two of the eight modes, all the F (function)

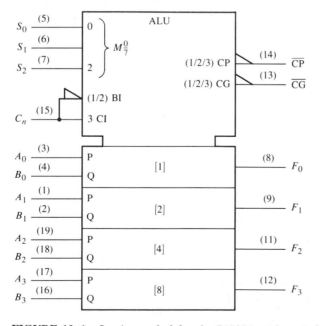

FIGURE 12-1 *Logic symbol for the 74S381 arithmetic logic unit/function generator.*

outputs can be made LOW (clear) or HIGH (preset) with the select inputs. The C_n input acts as a carry input (CI) for an addition and as a borrow input (BI) for a subtraction. The (1/2)BI label indicates that BI is active for modes 1 and 2 (*B* minus *A* and *A* minus *B*). The 3CI label indicates that CI is active for mode 3 (*A* plus *B*).

Table 12–1 lists the functions that this device can perform. Since both multiplication and division can be performed using addition or subtraction, a complete arithmetic processor can be constructed using appropriate external logic circuits in addition to the ALU.

A Basic Processor Unit

A simplified processor using generalized registers and control logic in conjunction with an ALU is used to illustrate, on a simplified basis, the essential steps in arithmetic operations with eight-bit binary numbers (*operands*). The simplified processor is shown in Figure 12–2. This system illustrates how the typical microprocessor performs addition and subtraction as well as logic operations.

Most microprocessors do not have direct (hardware) multiplication and division. Instead, they are *programmed* to do multiplication, for example, with a

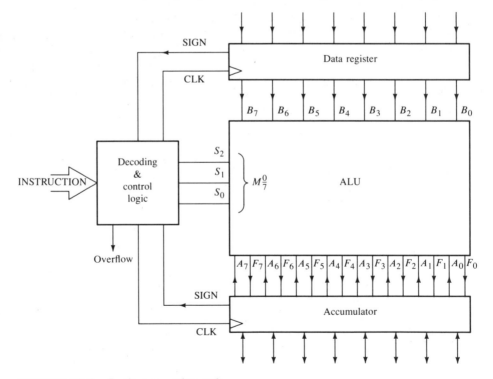

FIGURE 12–2 *Basic processing unit.*

TABLE 12–1 *Functions of the 74S381 ALU.*

Mode	S_2	S_1	S_0	Operation	Comments
0	0	0	0	Clear	F outputs LOW
1	0	0	1	$B - A$	B minus A
2	0	1	0	$A - B$	A minus B
3	0	1	1	$A + B$	A plus B
4	1	0	0	$A \oplus B$	A exclusive-OR B
5	1	0	1	$A + B$	A OR B
6	1	1	0	AB	A AND B
7	1	1	1	Preset	F outputs HIGH

series of addition and shifts and to do division with a series of subtractions and comparisons, as was discussed in the previous section.

Addition   In an addition operation, negative numbers are normally handled internally in the ALU using the 2's complement method. The two 8-bit numbers to be added (A_7 through A_0 and B_7 through B_0) are loaded into the *accumulator* and the *data register,* respectively. The sign logic checks the signs bits (A_7 and B_7) of both numbers to determine the potential for overflow (both positive or both negative). The control logic produces the appropriate select inputs to the ALU for an add operation ($S_2 = 0$, $S_1 = 1$, and $S_0 = 1$) in response to an external instruction.

The ALU performs the addition and produces the *sum* on its function (F) outputs. These outputs are then stored back in the accumulator, replacing the previous operand. The sign of the sum is checked for overflow. When several numbers are to be added, the next number is loaded in register B and the process is repeated. Figure 12–3 illustrates the process with two specific numbers as examples.

Subtraction   In a subtraction operation, the two 8-bit numbers are loaded into the registers. Again, the necessary complementation is handled internally in the ALU. The function-select inputs to the control logic produce the appropriate inputs to the ALU for a subtract operation ($S_2 = 0$, $S_1 = 1$, and $S_0 = 0$) in response to an external instruction. The remainder of the process is the same as for addition, with the result (difference) placed in the accumulator, replacing the previous operand. Figure 12–4 illustrates with specific numbers.

SECTION REVIEW 12–3

1.   What are the two basic functions of the accumulator?
2.   Why do most microprocessors not have a multiplication or division instruction?

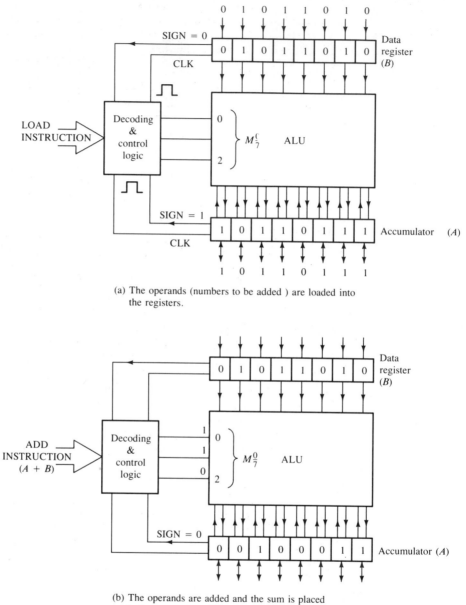

(a) The operands (numbers to be added) are loaded into
the registers.

(b) The operands are added and the sum is placed
in the accumulator [90 + (−55) = 35].

FIGURE 12–3 *A basic addition operation.*

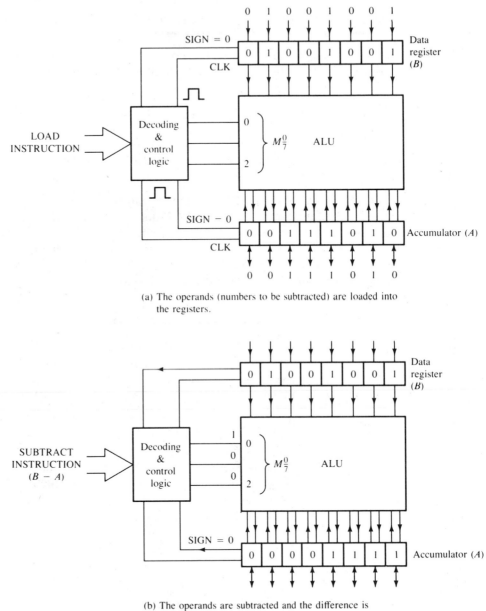

(a) The operands (numbers to be subtracted) are loaded into the registers.

(b) The operands are subtracted and the difference is placed in the accumulator ($73 - 58 = 15$).

FIGURE 12–4 *A basic subtraction operation.*

12-4 ADDITION OF SIGNED BCD NUMBERS

BCD is a numerical code, and many applications require arithmetic operations with BCD. Two BCD (8421) numbers are added as follows:

1. Add the two numbers using binary addition.
2. If the four-bit sum is equal to or less than 9, it is *valid*.
3. If the four-bit sum is greater than 9 or if a carry is generated from the four-bit sum, the sum is *invalid*.
4. To adjust the invalid sum, add 0110_2 to the four-bit sum. If a carry results from this addition, add it to the next higher-order BCD digit.

Addition of signed BCD numbers can be performed by using 9's or 10's complement methods. A negative BCD number can be expressed by taking the 9's or 10's complement. Recall that the 10's complement is the 9's complement plus 1.

Addition of signed BCD numbers basically follows the rules listed except for the complement operation for negative numbers. The following example shows addition of both 8421 BCD and Excess-3 signed numbers using the 10's complement method.

EXAMPLE 12-11

(a)

Decimal	BCD (8421)	
5	*10's comp. of −3:*	
+(−3)	10011	true BCD
2	10110	9's comp.
	+1	
	10111	10's comp.
	00101	+5
	+10111	−3
	11100	invalid
	+0110	add 6
	✗00010	+2
	└────drop carry	

(b)

Decimal	BCD (8421)	
−8	*10's comp. of −8:*	
+ 6	11000	true BCD
−2	10001	9's comp.
	+1	
	10010	10's comp.
	10010	−8
	+00110	+6
	11000	−2 (10's comp.)

(c)

Decimal	BCD (8421)	
24	*10's comp. of −17:*	
+(−17)		
+7	100010111	true BCD
	100000010	9's comp.
	+1	
	110000011	10's comp.
	000100100	+24
	+110000011	−17
	110100111	MSD invalid
	+0110	add 6
	000000 111	+7

A BCD Adder

Figure 12–5 shows a basic BCD adder. Functionally, the system operates as follows: Two 4-bit BCD digits are entered into the adder along with any input carry. The resulting sum is compared to 9_{10} (1001_2), and if it is equal to or less than 9 and no output carry occurs, it is gated through the sum selection logic to the final outputs. If the sum is greater than 9 and if an output carry occurs, the sum is invalid. The correct sum adder is enabled, adding 6 (0110_2) to the invalid sum; the corrected sum is then gated through the sum selection logic to the final outputs. The carry outputs of both adders are ORed and go to the next higher stage (if any).

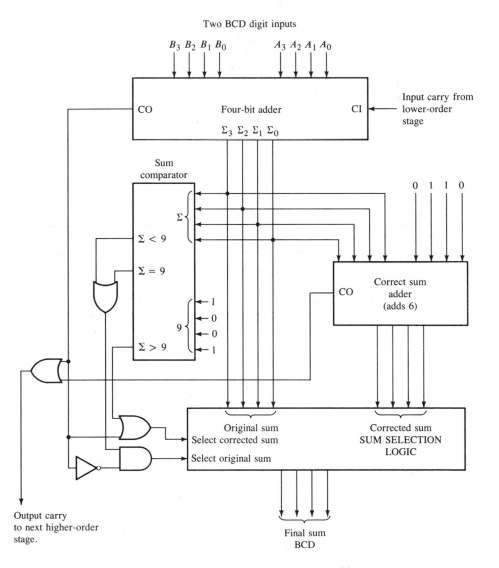

FIGURE 12–5 *Basic block diagram for single-digit BCD adder.*

1. What two conditions indicate that a BCD sum is invalid?
2. Which of the following BCD sums are invalid:
   (a) 0101   (b) 1010   (c) 1001

12–5   MICROPROCESSORS

Processing units similar to the one introduced in Section 12–3 form the heart of all microprocessors. Most microprocessors also have several additional registers and controls for programming and interfacing with memories and peripheral devices.

The basic architecture (internal organization) of two popular families of microprocessors are used as examples: the Motorola 6800 and the Intel 8085A. The purpose of this coverage is to provide only a brief introduction to microprocessors. A detailed coverage is beyond the scope of this book.

The 6800

A generalized block diagram of the 6800 microprocessor is shown in Figure 12–6. Notice that all the elements are interconnected by an eight-bit common internal bus structure. This microprocessor can handle data in groups of eight bits from the data bus, and it has a sixteen-bit address bus. Thus, up to 65,536 memory locations can be addressed. All memory (other than registers) is external to this particular microprocessor, thus requiring additional chips to form a complete microcomputer system.

The registers in the 6800 are as follows: two 8-bit accumulators, one 16-bit index register, one 16-bit program counter, one 16-bit stack pointer, one 8-bit condition code register, and one 8-bit instruction register. The 16-bit registers are actually treated as a combination of two 8-bit registers. The H label refers to the high-order eight bits, the L label to the low-order eight bits. A brief description of the function of each element in the microprocessor follows.

Accumulator   As you learned in Section 12–3, one function of the accumulator is to store an operand prior to an operation by the ALU, and the other function is to temporarily store the result of the operation.

Program counter   This sixteen-bit counter produces the sequence of memory addresses in which the program instructions are stored. The content of the program counter is always the memory address from which the next instruction is to be taken.

Instruction register   When the contents of the program counter go out onto the address bus to the memory, an eight-bit instruction is read from that memory location and is transferred on the data bus, to be temporarily stored in the

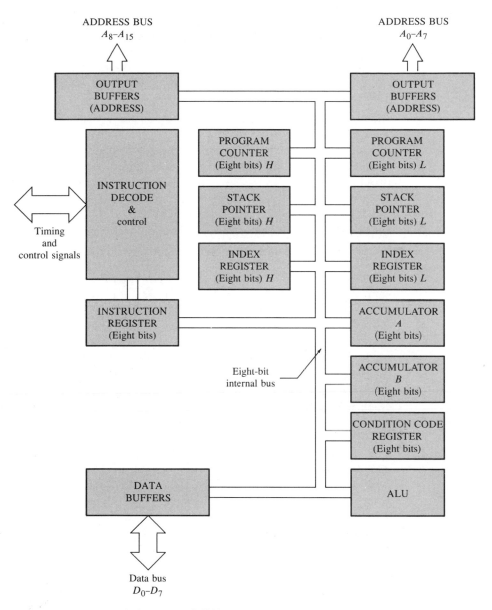

FIGURE 12–6 *Block diagram of 6800 microprocessor.*

instruction register until it is decoded and executed by the instruction decode and control element.

Instruction decode and control An *instruction* is a binary code that tells the microprocessor what it is to do. A sequence of many different instructions designed to accomplish a specific result is a *program*. In other words, a program is a step-by-step procedure used by the microprocessor to carry out a specified task.

The instruction decode and control element decodes an instruction code that has been transferred on the data bus from the memory. The instruction code is often called an *op code*. When the op code is decoded, the instruction decoder provides the control unit with this information so that it can produce the proper signals and timing sequence to *execute* or carry out the instruction.

Condition code register   The basic purpose of this element is to indicate the status of the contents of the accumulator or certain other conditions within the microprocessor. For example, it can indicate a zero result, a negative result, the occurrence of a carry, and the occurrence of an overflow from the accumulator.

Index register   The register is used in conjunction with one of the several addressing modes in the 6800 called *indexed addressing*.

Stack pointer   This register is used in conjunction with a portion of RAM that is set aside for use as a memory *stack*. The stack is used mainly to temporarily store the contents of all the registers when a subroutine is called or an interrupt requires the microprocessor to temporarily cease its current operation to respond to a request from an external device that needs to send data or requires service from the microprocessor. When an interrupt request has been serviced, the contents of the stack are transferred back into the appropriate registers and the microprocessor continues where it left off. The stack pointer is used to address the stack portion of RAM for storing and retrieving data in such a situation.

A logic symbol of the 6800 showing input and outputs with pin numbers appears in Figure 12–7.

The 6800 was the first of the Motorola microprocessor family and is still a widely used device. It requires three other chips to form a complete microcomputer system: an external clock generator, a RAM, and a ROM. Some additional devices in the family that have evolved from the 6800 are now briefly described.

The 6802 microprocessor is basically the same as the 6800 except that it has an on-chip clock generator and 32 bytes of RAM. The 6808 is the same as the 6802 without the on-chip memory. The 6801 is basically an eight-bit single-chip microcomputer having on-chip clock generation, two kbytes of ROM, and 128 bytes of RAM. The 6805 is also an eight-bit single-chip microcomputer with additional enhancements. Finally, the 68000 is a third-generation microprocessor that can handle sixteen bits of data at a time rather than eight.

The 8085A

A generalized block diagram of the Intel 8085A microprocessor is shown in Figure 12–8. The 8085A has an eight-bit data bus which is shared with a sixteen-bit

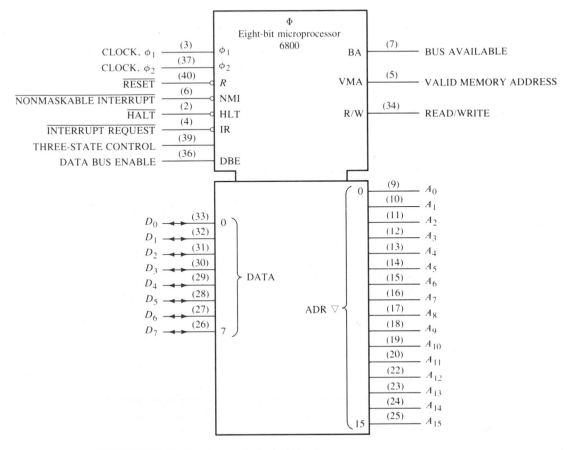

FIGURE 12–7 *Logic symbol of 6800 microprocessor.*

address bus. Notice that like the 6800, the 8085A also has a sixteen-bit program counter, a sixteen-bit stack pointer, and an eight-bit instruction register. The flag register is basically equivalent to the condition code register in the 6800.

The 8085A has a single 8-bit accumulator and six 8-bit general-purpose registers (*B, C, D, E, H,* and *L*). Two general-purpose registers in the register array can be combined to form 16-bit registers.

The *incrementer/decrementer* is associated with the register array and allows each register to be incremented (increased by 1) or decremented (decreased by 1) independent of the ALU and accumulator. The single *temporary register* associated with the accumulator holds one of the operands for an ALU operation.

The 8085A operates on a single 5-V supply. The required clock signals are created internally by connecting an external crystal between *X*1 and *X*2. Also

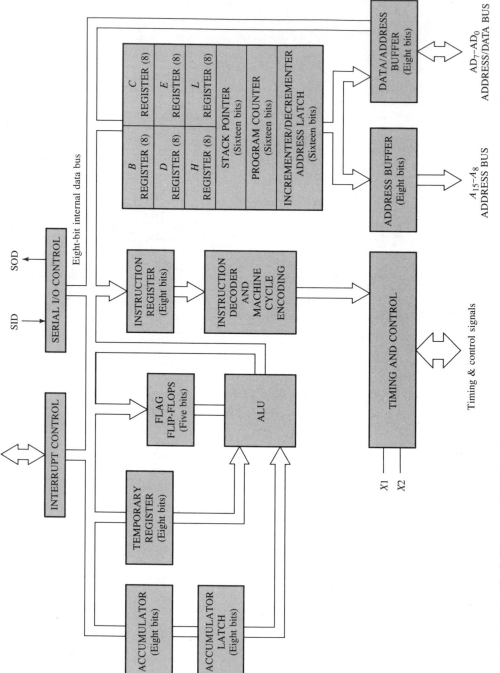

FIGURE 12-8 *Block diagram of 8085A microprocessor.*

notice that the 8085A has serial input data (SID) and serial output data (SOD) capability for use in simple forms of serial communications, such as a teletype.

A logic symbol showing inputs and outputs with pin numbers appears in Figure 12–9.

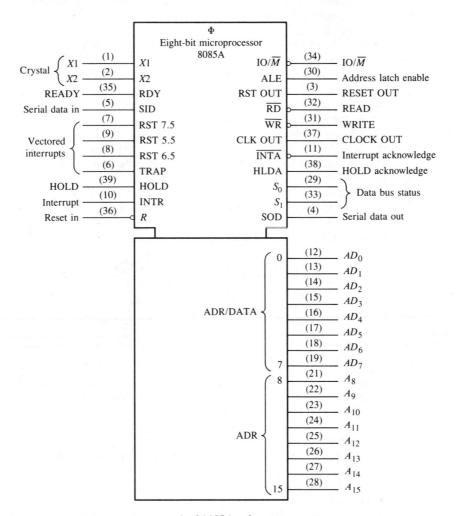

FIGURE 12–9 *Logic symbol of 8085A microprocessor.*

SECTION REVIEW 12–5

1. When someone says that a certain microprocessor is an eight-bit device, what does this mean?
2. How does a microprocessor address a memory?

SUMMARY

- ☐ A positive binary number is represented by a 0 sign bit.
- ☐ A negative binary number is represented by a 1 sign bit.
- ☐ For arithmetic operations, negative binary numbers are represented in 1's complement or 2's complement form.
- ☐ In an addition operation, an overflow is possible when both numbers are positive or when both numbers are negative. An incorrect sign bit in the sum indicates the occurrence of an overflow.
- ☐ An ALU is an *arithmetic logic unit* that performs addition, subtraction, and logic operations.
- ☐ An *accumulator* is a register that stores an operand prior to an operation and then stores the result of the operation.
- ☐ The Motorola 6800 is a microprocessor with one 8-bit data bus, one 16-bit address bus, one 16-bit program counter, one 16-bit stack pointer, one 16-bit index register, two 8-bit accumulators, one 8-bit condition code register, and one 8-bit instruction register.
- ☐ The Intel 8085A is a microprocessor with one 8-bit data bus that is shared with a 16-bit address bus, one 8-bit temporary register, six 8-bit general-purpose registers, one 16-bit stack pointer, one 16-bit program counter, one 8-bit instruction register, and one 8-bit accumulator.

SELF-TEST

1. How are signed binary numbers represented?
2. In an addition operation, how are negative binary numbers handled?
3. Explain *overflow* and list the conditions for which an overflow is possible.
4. How can overflow be detected?
5. List the operations performed by the 74S381 ALU.
6. In what mode is the 74S381 when $S_2 = 1$, $S_1 = 1$, and $S_0 = 0$?
7. The *result* of an arithmetic or logic operation in an ALU is placed in a register called the _____.
8. The two numbers on which an arithmetic or logic operation is to be performed are called _____.
9. How is an invalid BCD sum adjusted?
10. Explain the purposes of the data bus and the address bus in microprocessor-based systems.
11. Describe briefly the purpose of each of the following microprocessor elements:
    (a) Program counter (b) Stack pointer (c) Index register
    (d) Condition code register (e) Instruction register
12. Why does a microprocessor generally have more address bus lines than data bus lines?

PROBLEMS

Section 12-1

12-1 Convert each negatively signed true binary number to 1's complement form.
(a) 10010001   (b) 11100111   (c) 10000101   (d) 11110111

12-2 Convert each negatively signed true binary number to 2's complement form.
(a) 10000110   (b) 11111000   (c) 10101010   (d) 10001111

Section 12-2

12-3 Convert each decimal number to signed binary, and perform each addition using the 2's complement method when appropriate. Leave negative results in 2's complement form.
(a) 33 + 15   (b) 56 + (-27)   (c) 99 + (-75)
(d) -46 + 25   (e) -110 + (-84)

12-4 Perform the additions using the 2's complement method. Each signed binary number is shown in true (uncomplemented) form. Leave negative results in 2's complement form.
(a) 01001 + 10110   (b) 00110 + 10011   (c) 10100 + 10010
(d) 11101 + 10001   (e) 11000 + 00111   (f) 10101 + 01010

12-5 Repeat Problem 12-4 using the 1's complement method.

12-6 Perform each subtraction by converting to signed binary and using the 2's complement method. Leave negative results in 2's complement form.
(a) 43 - 27   (b) 20 - (-14)   (c) 30 - (-45)

12-7 Multiply the signed binary numbers using repeated addition.

Multiplicand: 011110110
Multiplier: 100000100

12-8 Divide 010000100 by 001100011.

Section 12-3

12-9 Determine the output $(F_3 F_2 F_1 F_0)$ of the 74S381 ALU for each of the following input conditions:
(a) $A_3 A_2 A_1 A_0 = 0110$, $B_3 B_2 B_1 B_0 = 1010$, and $S_2 S_1 S_0 = 001$
(b) $A_3 A_2 A_1 A_0 = 1110$, $B_3 B_2 B_1 B_0 = 0110$, and $S_2 S_1 S_0 = 110$
(c) $A_3 A_2 A_1 A_0 = 1000$, $B_3 B_2 B_1 B_0 = 0100$, and $S_2 S_1 S_0 = 101$

12-10 What are the contents of the accumulator in Figure 12-2 after each of the following operations:
(a) Operand A: 10001011; operand B: 01101010; Mode: 100 $(S_2 S_1 S_0)$
(b) Operand A: 11011111; operand B: 01100011; Mode: 010
Assume that the mode selection is the same as for the 74S381 ALU. For operands, the right-most bit is the LSB and the left-most bit is the sign.

Section 12-4

12-11 Add the following BCD numbers:
(a) 0110 + 0010   (b) 1000 + 0001   (c) 0101 + 0111
(d) 0111 + 0110

12–12 Convert the following decimal numbers to BCD and perform the indicated addition:

(a) 7 + 2   (b) 12 + 9   (c) 49 + 36   (d) 192 + 148

Section 12–5

12–13 Which of the following statements is (are) true:

(a) The 6800 is a complete microcomputer.

(b) The 6800 has an eight-bit address bus.

(c) The 6800 has two accumulators.

(d) The 6800 has a sixteen-bit data bus.

(e) The 6800 and the 8085A are both eight-bit microprocessors.

12–14 If the program counter in a microprocessor contains $00FA_{16}$, what is the address (in hexadecimal) of the previously addressed byte in memory?

12–15 The hexadecimal representations of two numbers are $0A_{16}$ and $1C_{16}$. If these numbers are added in the ALU, what is the hexadecimal representation of the contents of the accumulator after the operation?

12–16 Draw the basic block diagram of a microcomputer.

ANSWERS TO SECTION REVIEWS

Section 12–1

1. (a) +27   (b) −93   (c) +81
2. (a) 1011010   (b) 1000101   (c) 1011101111

Section 12–2

1. (a) 01111011   (b) 00011101
2. (a) 00101111   (b) 1001100

Section 12–3

1. Stores one operand before operation; stores result after operation.
2. Because multiplication and division can be performed using addition or subtraction by properly programming the microprocessor.

Section 12–4

1. Greater than 9; a carry is generated.
2. (b) 1010

Section 12–5

1. It has an eight-bit data bus.
2. By sending an address code out on the address bus to the memory.

A
Integrated Circuit Technologies

A-1 INTEGRATED CIRCUIT INVERTERS

TTL Inverter

Transistor-transistor logic (TTL or T²L) is one of the most widely used integrated circuit technologies. The basic element in TTL circuits is the *bipolar transistor*. The *logic function* of an inverter or any type of gate is always the same regardless of the type of circuit technology that is used. TTL is only *one* way to build an inverter or any other logic function. Figure A–1 shows a TTL circuit for an inverter; Q_1 is the input coupling transistor, and D_1 is the input clamp diode. Transistor Q_2 is called a *phase splitter,* and the combination of Q_3 and Q_4 forms the output circuit often referred to as a *totem-pole* arrangement. The operation of each of these components in the circuit, including diode D_2, will now be discussed.

When the input is a HIGH, the base-emitter junction of Q_1 is reverse biased, and the base-collector junction is forward biased. This permits current through R_1 and the base-collector junction of Q_1 into the base of Q_2, thus driving Q_2 into saturation. As a result, Q_3 is turned on by Q_2, and its collector voltage, which is

609

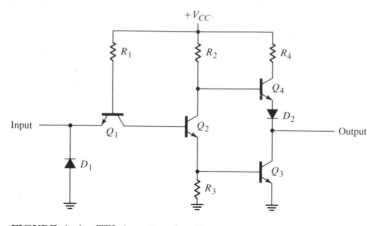

FIGURE A-1   *TTL inverter circuit.*

the output, is near ground potential. We therefore have a LOW output for a HIGH input. At the same time, the collector of Q_2 is a sufficiently low voltage level to keep Q_4 off.

When the input is LOW, the base-emitter junction of Q_1 is forward biased, and the base-collector junction is reverse biased. There is current through R_1 and the base-emitter junction of Q_1 out to the LOW input. A LOW provides a path to ground for the current. There is no current into the base of Q_2, so it is off. The collector of Q_2 is HIGH, thus turning Q_4 on. A saturated Q_4 provides a low impedance path from V_{CC} to the output; we therefore have a HIGH on the output for a LOW on the input. At the same time, the emitter of Q_2 is at ground potential, keeping Q_3 off.

The purpose of diode D_1 in the TTL circuit is to prevent negative spikes of voltage on the input from damaging Q_1. Diode D_2 insures that Q_4 will turn off when Q_2 is on (HIGH input). In this condition the collector voltage of Q_2 is equal to the V_{BE} of Q_3 plus the V_{CE} of Q_2. Diode D_2 provides an additional V_{BE} equivalent drop in series with the base-emitter junction of Q_4 to insure its turn-off when Q_2 is on.

The operation of the TTL inverter for the two input states is illustrated in Figure A–2. In the upper circuit, the base of Q_1 is 2.1 V above ground so that Q_2 and Q_3 are on. In the lower circuit, the base of Q_1 is about 0.7 V above ground—not enough to turn Q_2 and Q_3 on.

CMOS Inverter

Another common integrated circuit technology uses the MOSFET as its basic element and is called *complementary* MOS logic. CMOS uses both p- and n-channel enhancement MOSFETs, as shown in the inverter circuit in Figure A–3.

When a HIGH is applied to the input, the p-channel MOSFET Q_1 is off, and the n-channel MOSFET Q_2 is on. This connects the output to ground through the *on* resistance of Q_2, resulting in a LOW output. When a LOW is applied to the

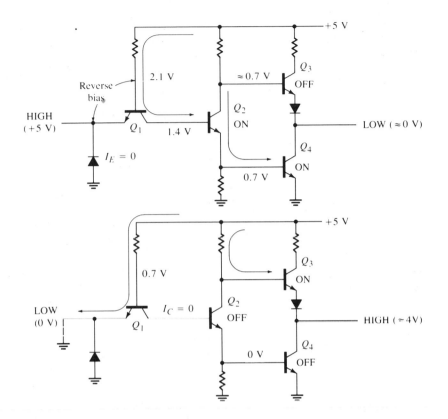

FIGURE A-2 *TTL inverter operation.*

FIGURE A-3 *CMOS inverter circuit.*

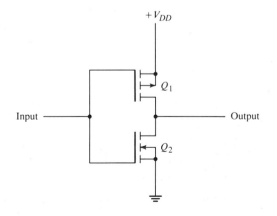

input, Q_1 is on and Q_2 is off. This connects the output to $+V_{DD}$ through the *on* resistance of Q_1, resulting in a HIGH output. The operation is illustrated in Figure A–4.

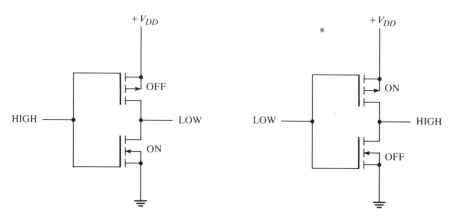

FIGURE A–4   *CMOS inverter operation.*

A–2   INTEGRATED CIRCUIT NAND GATES

TTL NAND Gate

A two-input TTL NAND gate is shown in Figure A–5. Basically, it is the same as the inverter circuit except for the additional input emitter of Q_1. TTL technology utilizes *multiple-emitter transistors* for the input devices. These can be compared to the diode arrangement, as shown in Figure A–6.

Perhaps you can understand the operation of this circuit better by visualizing Q_1 in Figure A–5 replaced by the diode arrangement. A LOW on either input A or

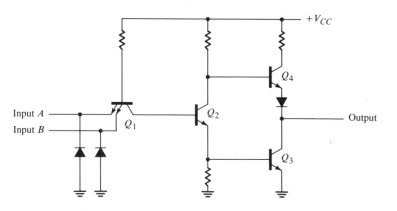

FIGURE A–5   *TTL NAND gate circuit.*

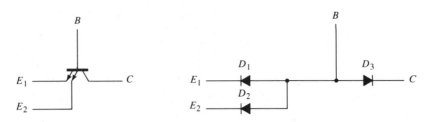

FIGURE A-6 *Diode equivalent of a TTL multiple-emitter transistor.*

input B forward-biases the respective diode and reverse-biases D_3 (Q_1 base-collector junction). This action keeps Q_2 off and results in a HIGH output in the same way as described for the TTL inverter. Of course, a LOW on both inputs will do the same thing.

A HIGH on *both* inputs reverse-biases both input diodes and forward-biases D_3 (Q_1 base-collector junction). This action turns Q_2 on and results in a LOW output in the same way as described for the TTL inverter.

You should recognize this operation as that of the NAND function: The output is LOW if and only if all inputs are HIGH.

CMOS NAND Gate

Figure A-7 shows a CMOS NAND gate with two inputs. Notice the arrangement of the complementary pairs (*n*- and *p*-channel MOSFETs).

The operation is as follows: When both inputs are LOW, Q_1 and Q_2 are *on*, and Q_3 and Q_4 are *off*. The output is pulled HIGH through the *on* resistance of Q_1 and Q_2 in parallel. When input A is LOW and input B is HIGH, Q_1 and Q_4 are

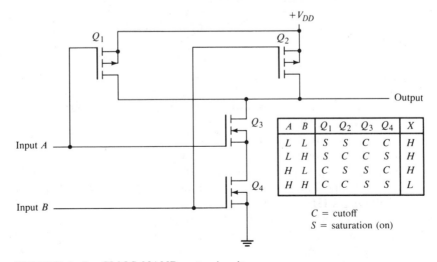

A	B	Q_1	Q_2	Q_3	Q_4	X
L	L	S	S	C	C	H
L	H	S	C	C	S	H
H	L	C	S	S	C	H
H	H	C	C	S	S	L

C = cutoff
S = saturation (on)

FIGURE A-7 *CMOS NAND gate circuit.*

on, and Q_2 and Q_3 are *off*. The output is pulled HIGH through the low *on* resistance of Q_1.

When input *A* is HIGH and input *B* is LOW, Q_1 and Q_4 are *off*, and Q_2 and Q_3 are *on*. The output is pulled HIGH through the low *on* resistance of Q_2.

Finally, when both inputs are HIGH, Q_1 and Q_2 are *off*, and Q_3 and Q_4 are *on*. In this case, the output is pulled LOW through the *on* resistance of Q_3 and Q_4 in series to ground.

A–3   INTEGRATED CIRCUIT NOR GATES

TTL NOR Gate

A two-input TTL NOR gate is shown in Figure A–8. Compare this to the NAND gate circuit and notice the additional transistors.

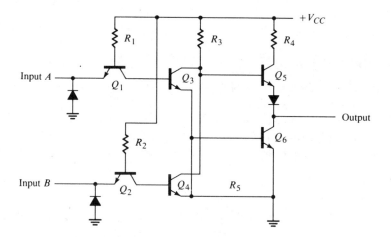

FIGURE A–8 *TTL NOR gate circuit.*

Both Q_1 and Q_2 are input transistors. Q_3 and Q_4 are in parallel and act as a phase splitter. Of course you should recognize Q_5 and Q_6 as the ordinary TTL totem-pole output circuit.

The operation is as follows: When both inputs are LOW, the forward-biased base-emitter junctions of the input transistors pull current away from the phase-splitter transistors Q_3 and Q_4, keeping them *off*. As a result, Q_5 is *on* and Q_6 is *off*, producing a HIGH output.

When input *A* is LOW and input *B* is HIGH, Q_3 is *off* and Q_4 is *on*. Q_4 turns on Q_6 and turns off Q_5, producing a LOW output.

When input *A* is HIGH and input *B* is LOW, Q_3 is *on* and Q_4 is *off*. Q_3 turns on Q_6 and turns off Q_5, producing a LOW output.

When both inputs are HIGH, both Q_3 and Q_4 are *on*. This has the same effect as either one being *on*, turning Q_6 *on* and Q_5 *off*. The result is still a LOW output. You should recognize this operation as that of the NOR function. (The output is LOW when any of the inputs are HIGH.)

CMOS NOR Gate

Figure A–9 shows a CMOS NOR gate with two inputs. Notice the arrangement of the complementary pairs.

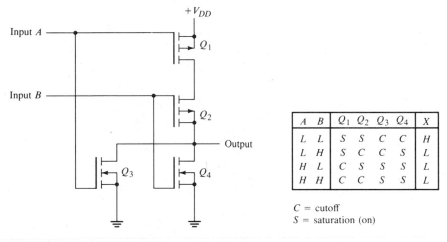

A	B	Q_1	Q_2	Q_3	Q_4	X
L	L	S	S	C	C	H
L	H	S	C	C	S	L
H	L	C	S	S	S	L
H	H	C	C	S	S	L

C = cutoff
S = saturation (on)

FIGURE A–9 *CMOS NOR gate circuit.*

The operation is as follows: When both inputs are LOW, Q_1 and Q_2 are *on*, and Q_3 and Q_4 are *off*. As a result, the output is pulled HIGH through the *on* resistance of Q_1 and Q_2 in series.

When input A is LOW and input B is HIGH, Q_1 and Q_3 are *on*, and Q_2 and Q_4 are *off*. The output is pulled LOW through the low *on* resistance of Q_3 to ground.

When input A is HIGH and input B is LOW, Q_1 and Q_3 are *off*, and Q_2 and Q_4 are *on*. The output is pulled LOW through the *on* resistance of Q_4 to ground.

When both inputs are HIGH, Q_1 and Q_2 are *off*, and Q_3 and Q_4 are *on*. The output is pulled LOW through the *on* resistance of Q_3 and Q_4 in parallel to ground.

A-4   TTL AND GATES and OR GATES

Figure A–10 shows a TTL two-input AND gate and a two-input OR gate. Compare these to the NAND and NOR gate circuits and notice the additional circuitry that is shaded. In each case, this transistor arrangement provides an inversion so the NAND becomes AND and the NOR becomes OR.

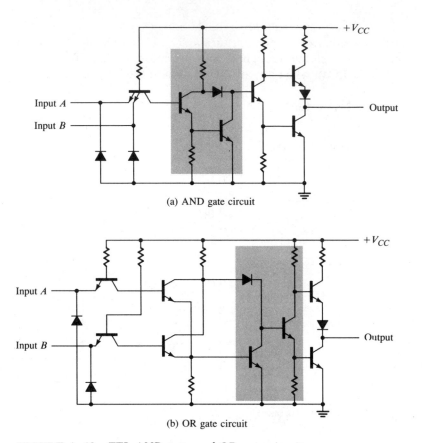

(a) AND gate circuit

(b) OR gate circuit

FIGURE A–10   *TTL AND gate and OR gate circuits.*

A–5   OPEN-COLLECTOR GATES

The TTL gates described in the previous sections all had the totem-pole output circuit. Another type of output available in TTL integrated circuits is the open-collector output. A TTL inverter with an open collector is shown in Figure A–11(a). The other gates are also available with open collector outputs. When input A is HIGH and input B is LOW, Q_1 and Q_3 are *off*, and Q_2 and Q_4 are *on*. The output is pulled LOW through the *on* resistance of Q_4 to ground.

When both inputs are HIGH, Q_1 and Q_2 are *off*, and Q_3 and Q_4 are *on*.

Notice that the output is the collector of transistor Q_3 with nothing connected to it, hence the name *open collector*. In order to get the proper HIGH and LOW logic levels out of the circuit, an *external pull-up resistor* must be connected to V_{CC} from the collector of Q_3, as shown in Figure A–11(b). When Q_3 is *off*, the output is pulled up to V_{CC} through the external resistor. When Q_3 is *on*, the output is connected to near-ground through the saturated transistor.

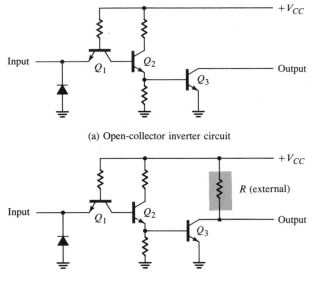

(a) Open-collector inverter circuit

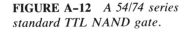

(b) With external pull-up resistor

FIGURE A–11 *TTL inverter with open-collector output.*

A-6   MORE TTL CIRCUITS

Figure A–12 shows the basic TTL gate circuit that was discussed earlier. It is a *current-sinking* type of logic that draws current from the load when in the LOW output state and sources negligible current to the load when in the HIGH output state. This is a *standard* 54/74 TTL two-input NAND gate and is used for comparison with the other types of TTL circuits.

FIGURE A–12 *A 54/74 series standard TTL NAND gate.*

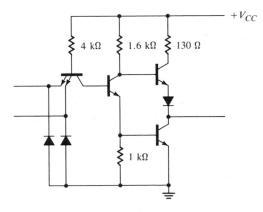

Low-Power TTL (54L/74L)

The 54L/74L series of TTL circuits is designed for low power consumption. A typical gate circuit is shown in Figure A–13. Notice that the circuit's resistor values are considerably higher than those of the standard gate in Figure A–12. This, of course, results in less current and therefore less power but increases the switching time of the gate. The typical power dissipation of a standard 54/74 gate is 10 mW, and that of a 54L/74L gate is 1 mW. The savings in power, however, is paid for in loss of speed. A typical standard 54/74 gate has a propagation delay time of 10 ns, compared to 33 ns for a 54L/74L gate.

FIGURE A–13 *A 54L/74L series TTL NAND gate.*

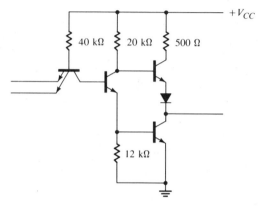

High-Speed TTL (54H/74H)

The 54H/74H series of TTL circuits is designed to provide less propagation delay than the standard gates, but the cost is higher power consumption. Notice that the 54H/74H circuit in Figure A–14 has lower resistor values than standard TTL. Typical propagation delay for a 54H/74H gate is 6 ns, but the power consumption is 22 mW.

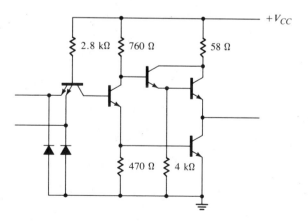

FIGURE A–14 *A 54H/74H series TTL NAND gte.*

Schottky TTL (54S/74S)

The 54S/74S series of TTL circuits provides a faster switching time than the 54H/74H series but requires much less power. These circuits incorporate Schottky diodes to prevent the transistors from going into saturation, thereby decreasing the time for a transistor to turn on or off.

Typical propagation delay for a 54S/74S gate is 3 ns, and the power dissipation is 19 mW. Figure A–15 shows a Schottky gate circuit.

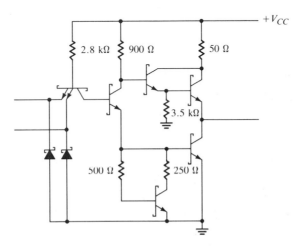

FIGURE A–15   *54S/74S series TTL NAND gate.*

Low-Power Schottky TTL (54LS/74LS)

This TTL series compromises speed to obtain a lower power dissipation. A NAND gate circuit is shown in Figure A–16. Notice that the input uses Schottky

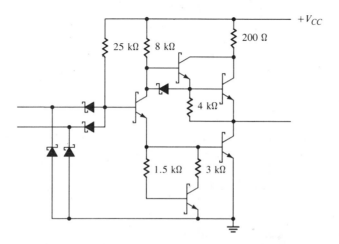

FIGURE A–16   *A 54LS/74LS series TTL NAND gate.*

diodes rather than the conventional transistor input. The typical power dissipation for a gate is 2 mW, and the propagation delay is 9.5 ns.

Advanced Schottky and Advanced Low-Power Schottky (AS/ALS)

These technologies are advanced versions of the S and LS series. The typical static power dissipation is 8.5 mW for the AS series and 1 mW for the ALS series. The typical propagation delay time is 4 ns for the ALS and 1.5 ns for the AS.

A-7   ECL CIRCUITS

ECL stands for *emitter-coupled logic* which, like TTL, is a bipolar technology. The typical ECL circuit, shown in Figure A–17(a), consists of a differential amplifier input circuit, a bias circuit, and emitter-follower outputs.

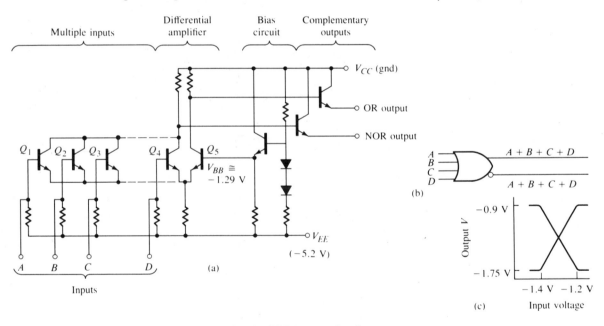

FIGURE A–17 *An ECL OR/NOR gate circuit.*

The emitter-follower outputs provide the OR logic function and its NOR complement, as indicated by Figure A–17(b).

Because of the low output impedance of the emitter-follower and the high input impedance of the differential amplifier input, high fan-out operation is possible. In this type of circuit, saturation is not possible. This results in higher power consumption and limited voltage swing (less than 1 V). The lack of saturation in the ECL circuits permits high-frequency switching.

The V_{CC} pin is normally connected to ground, and -5.2 V from the power supply is connected to V_{EE} for best operation. Notice that in Figure A–17(c) the output varies from a LOW level of -1.75 V to a HIGH level of -0.9 V with respect to ground. In positive logic, a 1 is the HIGH level (less negative), and a 0 is the LOW level (more negative).

ECL Circuit Operation

Beginning with LOWs on all inputs (typically -1.75 V), assume that Q_1 through Q_4 are off because the base-emitter junctions are reverse biased, and that transistor Q_5 is conducting (but not saturated). The bias circuit holds the base of Q_5 at -1.29 V, and therefore its emitter is approximately 0.8 V below the base at

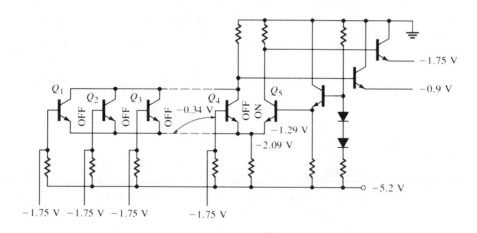

(a)

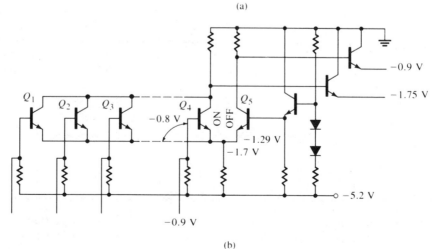

(b)

FIGURE A–18 *Basic ECL gate operation.*

-2.09 V. The voltage differential from base to emitter of the input transistors Q_1 through Q_4 is -2.09 V $- (-1.75$ V$) = -0.34$ V. This is less than the forward-bias voltage of these transistors, and they are therefore off. This condition is shown in Figure A–18(a) (p. 621).

When any one or all of the inputs are raised to the HIGH level (-0.9 V), that transistor (or transistors) will conduct. When this happens, the voltage at the emitters of Q_1 through Q_5 increases from -2.09 V to -1.7 V (one base-emitter drops below the -0.9 V base). Since the base of Q_5 is held at a constant -1.29 V by the bias circuit, Q_5 turns off. The resulting collector voltages are coupled through the emitter-followers to the output terminals. Because of the differential action of Q_1 through Q_4 with Q_5, Q_5 is off when one or more of the Q_1 through Q_4 transistors conduct, thus providing simultaneous complementary outputs. This is shown in Figure A–18(b).

When all of the inputs are returned to the LOW state, Q_1 through Q_4 are again cut off and Q_5 conducts.

Noise Margin

As you know, the noise margin of a gate is the measure of its immunity to undesired voltage fluctuations (noise). Typical ECL circuits have noise margins from about 0.2 V to 0.25 V. This is less than that for TTL and makes ECL unreliable in high-noise environments.

Comparison of ECL with Schottky TTL

Table A–1 shows a comparison of typical values of key parameters for Schottky TTL and ECL.

TABLE A–1

	Propagation Delay Time	Power Dissipation per Gate	Voltage Swing	dc Supply Voltage	Flip-Flop Clock Freq.
54S/74S	3 ns	19 mW	3.0 V	+5 V	125 MHz
ECL	1 ns	60 mW	0.85 V	−5.2 V	500 MHz

A-8   I²L CIRCUITS

Integrated injection logic (I²L) is a bipolar technology that allows extremely high component densities on a chip (up to ten times that of TTL). I²L is being used for complex LSI functions such as microprocessors and is simpler to fabricate than either TTL or MOS. It also has a low power requirement and reasonably good switching speeds that are improving all the time.

The basic I²L gate is extremely simple, as indicated in Figure A–19. Transistor Q_1 acts as a current source and active pull-up, and the multiple-collector transistor Q_2 operates as an inverter.

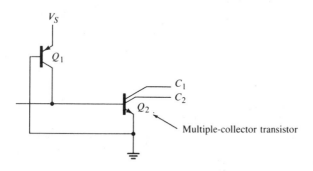

FIGURE A–19 *I²L gate.*

Because the base of Q_1 and the emitter of Q_2 are common and the collector of Q_1 and the base of Q_2 are common, the entire I²L gate (constructed on a silicon chip) takes only the space of a single TTL multiple-emitter transistor.

Transistor Q_1 is called a *current-injector transistor* because when its emitter is connected to an external power source, it can supply current into the base of Q_2. Switching action of Q_2 is accomplished by steering the injector current as follows: A LOW on the base of Q_2 will pull the injector current away from the base of Q_2 and through the low impedance path(s) provided by the driving gate(s), thus turning Q_2 OFF. When the output transistor of the driving gate is OFF (open), it corresponds to a HIGH input; this causes the injector current to be steered into the base of Q_2, which turns it ON. This action is illustrated in Figure A–20. Figure A–21 shows an example of an I²L implementation of a logic function.

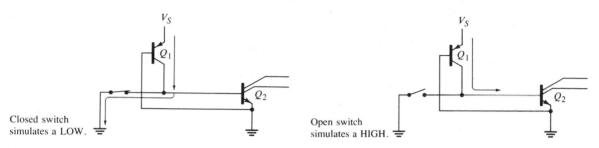

FIGURE A–20 *Basic I²L operation.*

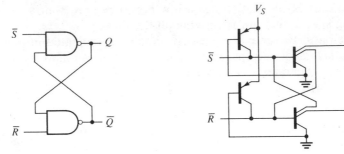

FIGURE A–21   *An I²L latch.*

A-9   CMOS CIRCUITS

CMOS stands for *complementary metal-oxide semiconductor*. In Chapter 3, CMOS gates were introduced and compared to the TTL circuits. CMOS is a popular type of MOS logic and is available in some LSI and MSI functions as well as SSI. Many of these devices are pin-compatible with equivalent TTL devices. In this section we will look further at MOS technology.

Figure A–22 shows a CMOS two-input NOR gate and a two-input NAND gate. As you recall, CMOS consists of both *n*-channel and *p*-channel MOS transistors arranged in a *complementary* connection.

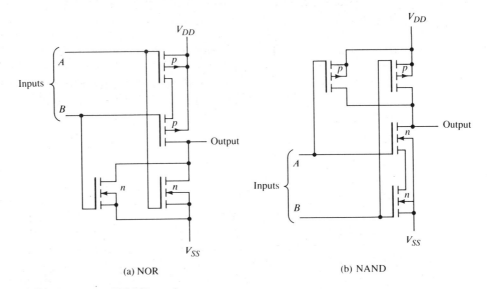

(a) NOR   (b) NAND

FIGURE A–22   *CMOS two-input gates.*

CMOS speed is comparable to that of the 54L/74L series of TTL and is several times slower than Schottky TTL speed. CMOS has very good noise immunity, much better than that of TTL, and its quiescent power dissipation is nearly zero.

Power Dissipation

With no input signals, the p-channel and the n-channel transistors are not conducting simultaneously; thus there is leakage current only from the positive (V_{DD}) to the negative (V_{SS}) power supply connection. Typically, this leakage current is 0.5 nA per gate, producing a power of 2.5 nW for a 5-V supply.

When signals are applied to the inputs, additional power is consumed to charge and discharge on-chip parasitic capacitances as well as load capacitance. Also, there is a short time during transitions when both the n- and p-type transistors are partially conducting. Dynamic power consumption occurs when these conditions exist and is proportional to the frequency at which the circuit is switching, the load capacitance, and the square of the supply voltage.

It has been found that the power consumption of a CMOS gate exceeds that of a Schottky low-power gate somewhere between 500 kHz and 2 MHz.

Supply Voltage

Unlike TTL, which is intolerant of large ranges in dc supply voltage, CMOS can be operated over a supply voltage range of 3 V to 15 V. Low power consumption and wide supply voltage range make CMOS ideal for battery-operated equipment.

Propagation Delay

CMOS devices are slow compared to TTL and very sensitive to capacitive loading. TTL with an output impedance of about 100 Ω is not affected much by an increase in capacitive loading. CMOS, however, has an output impedance of about 1 kΩ and is ten times more sensitive to capacitive loading. Also, propagation delay increases with an increase in supply voltage.

Noise Immunity

The noise immunity of a CMOS gate is approximately 45% of the supply voltage. For example, the noise margin is 2.25 V for a 5-V supply, 4.5 V for a 10-V supply, and so forth. In comparison, TTL exhibits a noise immunity that is typically 1 V and which can be as low as 0.4 V.

Also, the inherent capacitance of CMOS circuits produces delays and as a result acts as a noise filter. Ten-nanosecond voltage spikes tend to disappear in a chain of CMOS gates but are amplified by the faster-switching TTL gates.

Because of these features, CMOS is widely used in industrial equipment where there is a great deal of electrically and electromagnetically produced noise.

A-10   PMOS AND NMOS

MOS circuits are used largely in LSI (large-scale integration) functions such as long shift registers, large memories, and microprocessor products. Such usage is a result of the low power consumption and very small chip area required for MOS transistors. Of all the MOS technologies, CMOS is the only one that provides SSI functions at the gate and flip-flop level in a variety comparable to that of TTL.

PMOS

PMOS was one of the first high-density MOS circuit technologies to be produced. It utilizes enhancement-mode *p*-channel MOS transistors to form the basic gate building blocks. Figure A–23 shows a basic PMOS gate that produces the NOR function in positive logic.

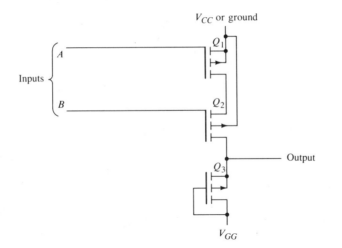

FIGURE A–23   *Basic PMOS gate.*

The operation of the PMOS gate is as follows: The supply voltage V_{GG} is a negative voltage, and V_{CC} is a positive voltage or ground (0 V). Transistor Q_3 is permanently biased to create a constant drain-to-source resistance. Its sole purpose is to function as a current-limiting resistor. If a HIGH (V_{CC}) is applied to input A or B, then Q_1 or Q_2 is off and the output is pulled down to a voltage near V_{GG}, which represents a LOW. When a LOW voltage (V_{GG}) is applied to both input A and input B, both transistors Q_1 and Q_2 are turned on. This causes the output to go to a HIGH level (near V_{CC}). Since a LOW output occurs when either or both inputs are HIGH, and a HIGH output occurs only when all inputs are LOW, we have a NOR gate.

NMOS

NMOS devices were developed as processing technology improved, and now most memories and microprocessors use NMOS. The n-channel MOS transistor is used in NMOS circuits, as shown in Figure A-24 for a NAND gate and a NOR gate.

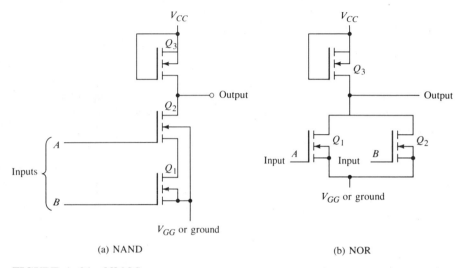

(a) NAND   (b) NOR

FIGURE A-24 *NMOS gates.*

In Figure A-24(a), Q_3 acts as a resistor to limit current. When a LOW (V_{GG} or ground) is applied to one or both inputs, then at least one of the transistors (Q_1 or Q_2) is off, and the output is pulled up to a HIGH level near V_{CC}. When HIGHs (V_{CC}) are applied to both inputs A and B, both Q_1 and Q_2 conduct and the output is LOW. This action, of course, identifies this circuit as a NAND gate.

In Figure A-24(b), Q_3 again acts as a resistor. A HIGH on either input turns Q_1 or Q_2 on, pulling the output LOW. When both inputs are LOW, both transistors are off and the output is pulled up to a HIGH level.

PROBLEMS

A-1   Describe one major difference between a bipolar integrated circuit and an MOS integrated circuit.

A-2   Explain why an open TTL input acts as a HIGH.

A-3   List five series of TTL circuits.

A–4   The fan-out of a standard TTL gate is 10 unit loads; this, of course, means that it can drive 10 other *standard* TTL gates. How many 54LS/74LS gates can the standard gate drive?

A–5   If two unused inputs of a TTL gate are connected to an input being driven by another TTL gate, how many other inputs can be driven by this gate? Assume that the fan-out is 10 unit loads.

A–6   How much current does a 54/74 gate sink if it is driving seven 54/74 loads?

A–7   What is the chief advantage of ECL over other IC technologies? What is a second advantage?

A–8   In what type of application should ECL not be used? Why?

A–9   If the frequency of operation of a CMOS device is increased, what happens to the dynamic power consumption?

A–10 Does CMOS or TTL perform better in a high-noise environment? Why?

A–11 Why are MOS devices shipped in conductive foam?

A–12 What type of MOS technology is predominant in LSI devices such as memories and microprocessors?

Data
Sheets

D2684, DECEMBER 1982 – REVISED MARCH 1984

- **Package Options Include Both Plastic and Ceramic Chip Carriers in Addition to Plastic and Ceramic DIPs**

- **Dependable Texas Instruments Quality and Reliability**

description

These devices contain four independent 2-input NAND gates. They perform the Boolean functions $Y = \overline{A \cdot B}$ or $Y = \overline{A} + \overline{B}$ in positive logic.

The SN54HC00 is characterized for operation over the full military temperature range of $-55\,°C$ to $125\,°C$. The SN74HC00 is characterized for operation from $-40\,°C$ to $85\,°C$.

FUNCTION TABLE (each gate)

INPUTS		OUTPUT
A	**B**	**Y**
H	H	L
L	X	H
X	L	H

logic symbol

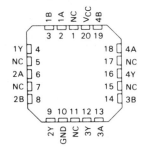

SN54HC00 . . . J PACKAGE
SN74HC00 . . . J OR N PACKAGE
(TOP VIEW)

```
1A  [ 1    14 ]  VCC
1B  [ 2    13 ]  4B
1Y  [ 3    12 ]  4A
2A  [ 4    11 ]  4Y
2B  [ 5    10 ]  3B
2Y  [ 6     9 ]  3A
GND [ 7     8 ]  3Y
```

SN54HC00 . . . FH OR FK PACKAGE
SN74HC00 . . . FH OR FN PACKAGE
(TOP VIEW)

```
         1B  1A  NC  VCC 4B
          3   2   1  20  19

1Y  [ 4                18 ]  4A
NC  [ 5                17 ]  NC
2A  [ 6                16 ]  4Y
NC  [ 7                15 ]  NC
2B  [ 8                14 ]  3B

          9  10  11  12  13
         2Y GND NC  3Y  3A
```

NC – No internal connection

Pin numbers shown are for J and N packages.

maximum ratings, recommended operating conditions, and electrical characteristics

See Table I, page 2-4.

(Courtesy of Texas Instruments Inc.)

switching characteristics over recommended operating free-air temperature range (unless otherwise noted), C_L = 50 pF (see Note 1)

PARAMETER	FROM (INPUT)	TO (OUTPUT)	V_{CC}	T_A = 25°C			SN54HC00		SN74HC00		UNIT
				MIN	TYP	MAX	MIN	MAX	MIN	MAX	
t_{pd}	A or B	Y	2 V		45	90		135		115	ns
			4.5 V		9	18		27		23	
			6 V		8	15		23		20	
t_t		Y	2 V		38	75		110		95	ns
			4.5 V		8	15		22		19	
			6 V		6	13		19		16	

C_{pd}	Power dissipation capacitance per gate	No load, T_A = 25°C	20 pF typ

NOTE 1: For load circuit and voltage waveforms, see page 1-14.

(Courtesy of Texas Instruments Inc.)

631

D2661, APRIL 1982—REVISED DECEMBER 1983

- **Package Options Include Both Plastic and Ceramic Chip Carriers in Addition to Plastic and Ceramic DIPs**

- **Dependable Texas Instruments Quality and Reliability**

description

These devices contain four independent 2-input NAND gates. They perform the Boolean functions $Y = \overline{A \cdot B}$ or $Y = \overline{A} + \overline{B}$ in positive logic.

The SN54ALS00A and SN54AS00 are characterized for operation over the full military temperature range of $-55\,°C$ to $125\,°C$. The SN74ALS00A and SN74AS00 are characterized for operation from $0\,°C$ to $70\,°C$.

FUNCTION TABLE (each gate)

INPUTS		OUTPUT
A	B	Y
H	H	L
L	X	H
X	L	H

logic symbol

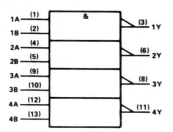

Pin numbers shown are for J and N packages.

SN54ALS00A, SN54AS00 . . . J PACKAGE
SN74ALS00A, SN74AS00 . . . N PACKAGE
(TOP VIEW)

```
        ___  ___
1A  [ 1      14 ]  V_CC
1B  [ 2      13 ]  4B
1Y  [ 3      12 ]  4A
2A  [ 4      11 ]  4Y
2B  [ 5      10 ]  3B
2Y  [ 6       9 ]  3A
GND [ 7       8 ]  3Y
```

SN54ALS00A, SN54AS00 . . . FH PACKAGE
SN74ALS00A, SN74AS00 . . . FN PACKAGE
(TOP VIEW)

```
       1B 1A NC VCC 4B
        3  2  1  20 19
1Y [ 4              18 ] 4A
NC [ 5              17 ] NC
2A [ 6              16 ] 4Y
NC [ 7              15 ] NC
2B [ 8              14 ] 3B
        9 10 11 12 13
       2Y GND NC 3Y 3A
```

NC—No internal connection

(Courtesy of Texas Instruments Inc.)

632

absolute maximum ratings over operating free-air temperature range (unless otherwise noted)

Supply voltage, V_{CC} . . . . 7 V
Input voltage . . . . 7 V
Operating free-air temperature range: SN54ALS00A . . . . $-55\,°C$ to $125\,°C$
SN74ALS00A . . . . $0\,°C$ to $70\,°C$
Storage temperature range . . . . $-65\,°C$ to $150\,°C$

recommended operating conditions

		SN54ALS00A			SN74ALS00A			UNIT
		MIN	NOM	MAX	MIN	NOM	MAX	
V_{CC}	Supply voltage	4.5	5	5.5	4.5	5	5.5	V
V_{IH}	High-level input voltage	2			2			V
V_{IL}	Low-level input voltage			0.8			0.8	V
I_{OH}	High-level output current			-0.4			-0.4	mA
I_{OL}	Low-level output current			4			8	mA
T_A	Operating free-air temperature	-55		125	0		70	°C

electrical characteristics over recommended operating free-air temperature range (unless otherwise noted)

PARAMETER	TEST CONDITIONS		SN54ALS00A			SN74ALS00A			UNIT
			MIN	TYP†	MAX	MIN	TYP†	MAX	
V_{IK}	$V_{CC} = 4.5$ V,	$I_I = -18$ mA			-1.5			-1.5	V
V_{OH}	$V_{CC} = 4.5$ V to 5.5 V,	$I_{OH} = -0.4$ mA	$V_{CC}-2$			$V_{CC}-2$			V
V_{OL}	$V_{CC} = 4.5$ V,	$I_{OL} = 4$ mA		0.25	0.4		0.25	0.4	V
	$V_{CC} = 4.5$ V,	$I_{OL} = 8$ mA					0.35	0.5	
I_I	$V_{CC} = 5.5$ V,	$V_I = 7$ V			0.1			0.1	mA
I_{IH}	$V_{CC} = 5.5$ V,	$V_I = 2.7$ V			20			20	µA
I_{IL}	$V_{CC} = 5.5$ V,	$V_I = 0.4$ V			-0.1			-0.1	mA
I_O‡	$V_{CC} = 5.5$ V,	$V_O = 2.25$ V	-15		-70	-15		-70	mA
I_{CCH}	$V_{CC} = 5.5$ V,	$V_I = 0$ V		0.5	0.85		0.5	0.85	mA
I_{CCL}	$V_{CC} = 5.5$ V,	$V_I = 4.5$ V		1.5	3		1.5	3	mA

†All typical values are at $V_{CC} = 5$ V, $T_A = 25\,°C$.
‡The output conditions have been chosen to produce a current that closely approximates one half of the true short-circuit output current, I_{OS}.

switching characteristics (see Note 1)

PARAMETER	FROM (INPUT)	TO (OUTPUT)	$V_{CC} = 4.5$ V to 5.5 V, $C_L = 50$ pF, $R_L = 500\ \Omega$, $T_A =$ MIN to MAX				UNIT
			SN54ALS00A		SN74ALS00A		
			MIN	MAX	MIN	MAX	
t_{PLH}	A or B	Y	3	14	3	11	ns
t_{PHL}	A or B	Y	2	10	2	8	ns

NOTE 1: For load circuit and voltage waveforms, see page 1-12.

(Courtesy of Texas Instruments Inc.)

- Package Options Include Both Plastic and Ceramic Chip Carriers in Addition to Plastic and Ceramic DIPs

- Dependable Texas Instruments Quality and Reliability

TYPE	TYPICAL MAXIMUM CLOCK FREQUENCY (C_L = 50 pF)	TYPICAL POWER DISSIPATION PER FLIP-FLOP
'ALS74A	50 MHz	6 mW
'AS74	134 MHz	26 mW

description

These devices contain two independent D-type positive-edge-triggered flip-flops. A low level at the Preset or Clear inputs sets or resets the outputs regardless of the levels of the other inputs. When Preset and Clear are inactive (high), data at the D input meeting the setup time requirements are transferred to the outputs on the positive-going edge of the clock pulse. Clock triggering occurs at a voltage level and is not directly related to the rise time of the clock pulse. Following the hold time interval, data at the D input may be changed without affecting the levels at the outputs.

The SN54ALS74A and SN54AS74 are characterized for operation over the full military temperature range of −55°C to 125°C. The SN74ALS74A and SN74AS74 are characterized for operation from 0°C to 70°C.

SN54ALS74A, SN54AS74 . . . J PACKAGE
SN74ALS74A, SN74AS74 . . . N PACKAGE
(TOP VIEW)

```
1CLR [1    14] VCC
1D   [2    13] 2CLR
1CLK [3    12] 2D
1PRE [4    11] 2CLK
1Q   [5    10] 2PRE
1Q̄   [6     9] 2Q
GND  [7     8] 2Q̄
```

SN54ALS74A, SN54AS74 . . . FH PACKAGE
SN74ALS74A, SN74AS74 . . . FN PACKAGE
(TOP VIEW)

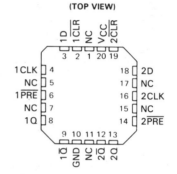

NC—No internal connection

FUNCTION TABLE

INPUTS				OUTPUTS	
PRESET	CLEAR	CLOCK	D	Q	Q̄
L	H	X	X	H	L
H	L	X	X	L	H
L	L	X	X	H*	H*
H	H	↑	H	H	L
H	H	↑	L	L	H
H	H	L	X	Q_0	$\bar{Q}_0$

* The output levels in this configuration are not guaranteed to meet the minimum levels for V_{OH} if the lows at Preset and Clear are near V_{IL} maximum. Furthermore, this configuration is nonstable; that is, it will not persist when either Preset or Clear returns to its inactive (high) level.

logic symbol

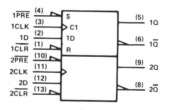

Pin numbers shown are for J and N packages.

absolute maximum ratings over operating free-air temperature range (unless otherwise noted)

Supply voltage, V_{CC} . . . . 7 V
Input voltage . . . . 7 V
Operating free-air temperature range: SN54ALS74A, SN54AS74 . . . . −55°C to 125°C
SN74ALS74A, SN74AS74 . . . . 0°C to 70°C
Storage temperature range . . . . −65°C to 150°C

(Courtesy of Texas Instruments Inc.)

recommended operating conditions

			SN54AS74			SN74AS74			UNIT
			MIN	NOM	MAX	MIN	NOM	MAX	
V_{CC}	Supply voltage		4.5	5	5.5	4.5	5	5.5	V
V_{IH}	High-level input voltage		2			2			V
V_{IL}	Low-level input voltage				0.8			0.8	V
I_{OH}	High-level output current				−2			−2	mA
I_{OL}	Low-level output current				20			20	mA
f_{clock}	Clock frequency		0		90	0		105	MHz
t_w	Pulse duration	$\overline{PRE}$ or $\overline{CLR}$ low	4			4			ns
		CLK high	4			4			
		CLK low	5.5			5.5			
t_{su}	Setup time before CLK↑	Data	4.5			4.5			ns
		$\overline{PRE}$ or $\overline{CLR}$ inactive	2			2			
t_h	Hold time, data after CLK↑		0			0			ns
T_A	Operating free-air temperature		−55		125	0		70	°C

electrical characteristics over recommended operating free-air temperature range (unless otherwise noted)

PARAMETER		TEST CONDITIONS		SN54AS74			SN74AS74			UNIT
				MIN	TYP[†]	MAX	MIN	TYP[†]	MAX	
V_{IK}		V_{CC} = 4.5 V,	I_I = −18 mA			−1.2			−1.2	V
V_{OH}		V_{CC} = 4.5 V to 5.5 V,	I_{OH} = −2 mA	V_{CC}−2			V_{CC}−2			V
V_{OL}		V_{CC} = 4.5 V,	I_{OL} = 20 mA		0.25	0.5		0.25	0.5	V
I_I		V_{CC} = 5.5 V,	V_I = 7 V			0.1			0.1	mA
I_{IH}	CLK or D	V_{CC} = 5.5 V,	V_I = 2.7 V			20			20	μA
	$\overline{PRE}$ or $\overline{CLR}$					40			40	
I_{IL}	CLK or D	V_{CC} = 5.5 V,	V_I = 0.4 V			−0.5			−0.5	mA
	$\overline{PRE}$ or $\overline{CLR}$					−1.8			−1.8	
I_{IO}[‡]		V_{CC} = 5.5 V,	V_O = 2.25 V	−30		−112	−30		−112	mA
I_{CC}		V_{CC} = 5.5 V	See Note 1		10.5	16		10.5	16	mA

[†]All typical values are at V_{CC} = 5 V, T_A = 25°C.
[‡]The output conditions have been chosen to produce a current that closely approximates one half of the true short-current output current, I_{OS}.
NOTE 1: I_{CC} is measured with D, CLK, and $\overline{PRE}$ grounded, then with D, CLK, and $\overline{CLR}$ grounded.

switching characteristics (see Note 2)

PARAMETER	FROM (INPUT)	TO (OUTPUT)	V_{CC} = 4.5 V to 5.5 V, C_L = 50 pF, R_L = 500 Ω, T_A = MIN to MAX				UNIT
			SN54AS74		SN74AS74		
			MIN	MAX	MIN	MAX	
f_{max}			90		105		MHz
t_{PLH}	$\overline{PRE}$ or $\overline{CLR}$	Q or $\overline{Q}$	3	8.5	3	7.5	ns
t_{PHL}			3.5	11.5	3.5	10.5	
t_{PLH}	CLK	Q or $\overline{Q}$	3.5	9	3.5	8	ns
t_{PHL}			4.5	10.5	4.5	9	

NOTE 2: For load circuit and voltage wavforms, see page 1-12 of the TTL Data Book, Volume 3.

(Courtesy of Texas Instruments Inc.)

D2661, APRIL 1982—REVISED FEBRUARY 1984

- 8-Line to 1-Line Multiplexers
  Can Perform As:
    Boolean Function Generators
    Parallel-to-Serial Converters
    Data Source Selectors
- Input Clamping Diodes Simplify System Design
- Fully Compatible With Most TTL Circuits
- Package Options Include Both Plastic and Ceramic
  Chip Carriers in Addition to Plastic and Ceramic DIPs
- Dependable Texas Instruments Quality and Reliability

description

These monolithic data selectors/multiplexers provide full binary decoding to select one of eight data sources. The strobe input ($\overline{G}$) must be at a low logic level to enable the inputs. A high level at the strobe terminal forces the W output high and the Y output low.

The SN54ALS151 and SN54AS151 are characterized for operation over the full military temperature range of −55 °C to 125 °C. The SN74ALS151 and SN74AS151 are characterized for operation from 0 °C to 70 °C.

SN54ALS151, SN54AS151 . . . J PACKAGE
SN74ALS151, SN74AS151 . . . N PACKAGE
(TOP VIEW)

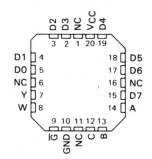

SN54ALS151, SN54AS151 . . . FH PACKAGE
SN74ALS151, SN74AS151 . . . FN PACKAGE
(TOP VIEW)

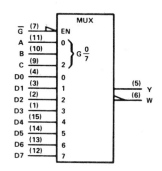

NC — No internal connection

FUNCTION TABLE

INPUTS				OUTPUTS	
SELECT			STROBE		
C	B	A	$\overline{G}$	Y	W
X	X	X	H	L	H
L	L	L	L	D0	$\overline{D0}$
L	L	H	L	D1	$\overline{D1}$
L	H	L	L	D2	$\overline{D2}$
L	H	H	L	D3	$\overline{D3}$
H	L	L	L	D4	$\overline{D4}$
H	L	H	L	D5	$\overline{D5}$
H	H	L	L.	D6	$\overline{D6}$
H	H	H	L	D7	$\overline{D7}$

H = high level, L = low level, X = irrelevant
D0, D1 . . . D7 = the level of the D respective input

logic symbol

Pin numbers shown are for J and N packages.

(Courtesy of Texas Instruments Inc.)

logic diagram (positive logic)

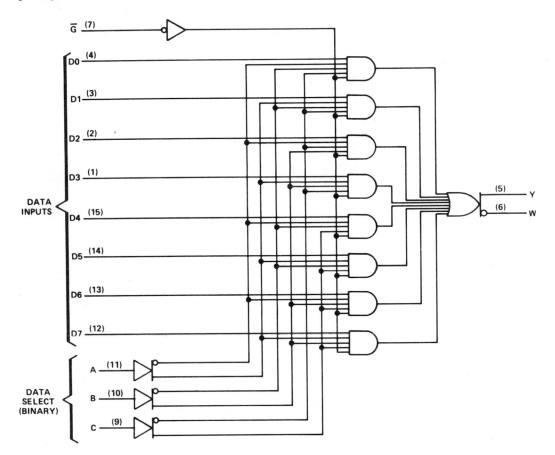

Pin numbers shown are for J and N packages.

absolute maximum ratings over operating free-air temperature range (unless otherwise noted)

Supply voltage, V_{CC} . . . . 7 V
Input voltage . . . . 7 V
Operating free-air temperature range: SN54ALS151, SN54AS151 . . . . −55 °C to 125 °C
SN74ALS151, SN74AS151 . . . . 0 °C to 70 °C
Storage temperature range . . . . −65 °C to 150 °C

(Courtesy of Texas Instruments Inc.)

637

recommended operating conditions

		SN54ALS151			SN74ALS151			UNIT
		MIN	NOM	MAX	MIN	NOM	MAX	
V_{CC}	Supply voltage	4.5	5	5.5	4.5	5	5.5	V
V_{IH}	High-level input voltage	2			2			V
V_{IL}	Low-level input voltage			0.8			0.8	V
I_{OH}	High-level output current			-1			-2.6	mA
I_{OL}	Low-level output current			12			24	mA
T_A	Operating free-air temperature	-55		125	0		70	°C

electrical characteristics over recommended operating free-air temperature range (unless otherwise noted)

PARAMETER	TEST CONDITIONS		SN54ALS151			SN74ALS151			UNIT
			MIN	TYP[†]	MAX	MIN	TYP[†]	MAX	
V_{IK}	$V_{CC} = 4.5$ V,	$I_I = -18$ mA			-1.5			-1.5	V
V_{OH}	$V_{CC} = 4.5$ V to 5.5 V,	$I_{OH} = -0.4$ mA	$V_{CC}-2$			$V_{CC}-2$			V
	$V_{CC} = 4.5$ V,	$I_{OH} = -1$ mA	2.4	3.3					
	$V_{CC} = 4.5$ V,	$I_{OH} = -2.6$ mA				2.4	3.2		
V_{OL}	$V_{CC} = 4.5$ V,	$I_{OL} = 12$ mA		0.25	0.4		0.25	0.4	V
	$V_{CC} = 4.5$ V,	$I_{OL} = 24$ mA					0.35	0.5	
I_I	$V_{CC} = 5.5$ V,	$V_I = 7$ V			0.1			0.1	mA
I_{IH}	$V_{CC} = 5.5$ V,	$V_I = 2.7$ V			20			20	μA
I_{IL}	$V_{CC} = 5.5$ V,	$V_I = 0.4$ V			-0.1			-0.1	mA
I_O‡	$V_{CC} = 5.5$ V,	$V_O = 2.25$ V	-30		-112	-30		-112	mA
I_{CC}	$V_{CC} = 5.5$ V,	Inputs at 4.5 V		7.5	12		7.5	12	mA

[†]All typical values are at $V_{CC} = 5$ V, $T_A = 25$°C.

‡The output conditions have been chosen to produce a current that closely approximates one half of the true short-circuit output current, I_{OS}.

switching characteristics (see Note 1)

PARAMETER	FROM (INPUT)	TO (OUTPUT)	$V_{CC} = 4.5$ V to 5.5 V, $C_L = 50$ pF, $R_L = 500\ \Omega$, $T_A = $ MIN to MAX				UNIT
			SN54ALS151		SN74ALS151		
			MIN	MAX	MIN	MAX	
t_{PLH}	A, B, or C	Y	4	21	4	18	ns
t_{PHL}			8	28	8	24	
t_{PLH}	A, B, or C	W	7	28	7	24	ns
t_{PHL}			7	26	7	23	
t_{PLH}	Any D	Y	3	12	3	10	ns
t_{PHL}			5	18	5	15	
t_{PLH}	Any D	W	3	18	3	15	ns
t_{PHL}			4	18	4	15	
t_{PLH}	$\overline{G}$	Y	4	21	4	18	ns
t_{PHL}			4	23	4	19	
t_{PLH}	$\overline{G}$	W	5	23	5	19	ns
t_{PHL}			5	26	5	23	

NOTE 1: For load circuit and voltage waveforms, see page 1-12.

(Courtesy of Texas Instruments Inc.)

D2661, APRIL 1982 REVISED APRIL 1985

- **Designed Specifically for High-Speed Memory Decoders and Data Transmission Systems**

- **Incorporates 3 Enable Inputs to Simplify Cascading and/or Data Reception**

- **Package Options Include Both Plastic and Ceramic Chip Carriers in Addition to Plastic and Ceramic DIPs**

- **Dependable Texas Instruments Quality and Reliability**

description

The 'ALS138 and 'AS138 circuits are designed to be used in high-performance memory-decoding or data-routing applications requiring very short propagation delay times. In high-performance memory systems this decoder can be used to minimize the effects of system decoding. When employed with high-speed memories utilizing a fast enable circuit, the delay times of this decoder and the enable time of the memory are usually less than the typical access time of the memory. This means that the effective system delay introduced by the Schottky-clamped system decoder is negligible.

The conditions at the binary select inputs and the three enable inputs select one of eight input lines. Two active-low and one active-high enable inputs reduce the need for external gates or inverters when expanding. A 24-line decoder can be implemented without external inverters and a 32-line decoder requires only one inverter. An enable input can be used as a data input for demultiplexing applications.

The SN54ALS138 and SN54AS138 are characterized for operation over the full military temperature range of −55 °C to 125 °C. The SN74ALS138 and SN74AS138 are characterized for operation from 0 °C to 70 °C.

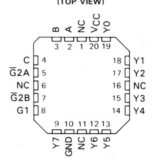

SN54ALS138, SN54AS138 . . . J PACKAGE
SN74ALS138, SN74AS138 . . . N PACKAGE
(TOP VIEW)

SN54ALS138, SN54AS138 . . . FK PACKAGE
SN74ALS138, SN74AS138 . . . FN PACKAGE
(TOP VIEW)

NC — No internal connection

(Courtesy of Texas Instruments Inc.)

639

SN54ALS138, SN54AS138, SN74ALS138, SN74AS138
3-LINE TO 8-LINE DECODERS/DEMULTIPLEXERS

logic symbols (alternatives)

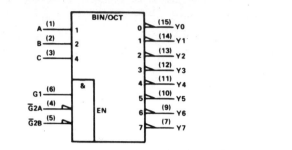 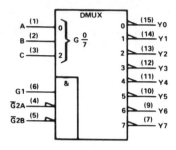

logic diagram (positive logic)

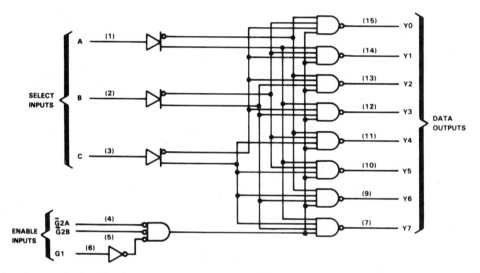

Pin numbers shown are for J and N packages.

(Courtesy of Texas Instruments Inc.)

640

FUNCTION TABLE

ENABLE INPUTS			SELECT INPUTS			OUTPUTS							
G1	$\overline{G2A}$	$\overline{G2B}$	C	B	A	Y0	Y1	Y2	Y3	Y4	Y5	Y6	Y7
X	H	X	X	X	X	H	H	H	H	H	H	H	H
X	X	H	X	X	X	H	H	H	H	H	H	H	H
L	X	X	X	X	X	H	H	H	H	H	H	H	H
H	L	L	L	L	L	L	H	H	H	H	H	H	H
H	L	L	L	L	H	H	L	H	H	H	H	H	H
H	L	L	L	H	L	H	H	L	H	H	H	H	H
H	L	L	L	H	H	H	H	H	L	H	H	H	H
H	L	L	H	L	L	H	H	H	H	L	H	H	H
H	L	L	H	L	H	H	H	H	H	H	L	H	H
H	L	L	H	H	L	H	H	H	H	H	H	L	H
H	L	L	H	H	H	H	H	H	H	H	H	H	L

absolute maximum ratings over operating free-air temperature range (unless otherwise noted)

Supply voltage, V_{CC} . . . . 7 V
Input voltage . . . . 7 V
Operating free-air temperature range; SN54ALS138, SN54AS138 . . . . −55°C to 125°C
SN74ALS138, SN74AS138 . . . . 0°C to 70°C
Storage temperature range . . . . −65°C to 150°C

recommended operating conditions

		SN54ALS138			SN74ALS138			UNIT
		MIN	NOM	MAX	MIN	NOM	MAX	
V_{CC}	Supply voltage	4.5	5	5.5	4.5	5	5.5	V
V_{IH}	High-level input voltage	2			2			V
V_{IL}	Low-level input voltage			0.8			0.8	V
I_{OH}	High-level output current			−0.4			−0.4	mA
I_{OL}	Low-level output current			4			8	mA
T_A	Operating free-air temperature	−55		125	0		70	°C

electrical characteristics over recommended operating free-air temperature range (unless otherwise noted)

PARAMETER	TEST CONDITIONS		SN54ALS138			SN74ALS138			UNIT
			MIN	TYP[†]	MAX	MIN	TYP[†]	MAX	
V_{IK}	$V_{CC} = 4.5$ V,	$I_I = -18$ mA			−1.5			−1.5	V
V_{OH}	$V_{CC} = 4.5$ V to 5.5 V,	$I_{OH} = -0.4$ mA	$V_{CC} - 2$			$V_{CC} - 2$			V
V_{OL}	$V_{CC} = 4.5$ V,	$I_{OL} = 4$ mA		0.25	0.4		0.25	0.4	V
	$V_{CC} = 4.5$ V,	$I_{OL} = 8$ mA					0.35	0.5	
I_I	$V_{CC} = 5.5$ V,	$V_I = 7$ V			0.1			0.1	mA
I_{IH}	$V_{CC} = 5.5$ V,	$V_I = 2.7$ V			20			20	μA
I_{IL}	$V_{CC} = 5.5$ V,	$V_I = 0.4$ V			−0.1			−0.1	mA
I_O[‡]	$V_{CC} = 5.5$ V,	$V_O = 2.25$ V	−30		−112	−30		−112	mA
I_{CC}	$V_{CC} = 5.5$ V			5	10		5	10	mA

[†]All typical values are at $V_{CC} = 5$ V, $T_A = 25$°C.
[‡]The output conditions have been chosen to produce a current that closely approximates one half of the true short-circuit output current, I_{OS}.

(Courtesy of Texas Instruments Inc.)

TYPES SN54ALS190, SN54ALS191, SN74ALS190, SN74ALS191
SYNCHRONOUS 4-BIT UP/DOWN DECADE AND BINARY COUNTERS

D2661, DECEMBER 1982—REVISED DECEMBER 1983

- Single Down/Up Count Control Line
- Look-Ahead Circuitry Enhances Speed of Cascaded Counters
- Fully Synchronous in Count Modes
- Asynchronously Presettable with Load Control
- Package Options Include Both Plastic and Ceramic Chip Carriers in Addition to Plastic and Ceramic DIPS
- Dependable Texas Instruments Quality and Reliability

SN54ALS190, SN54ALS191 . . . J PACKAGE
SN74ALS190, SN74ALS191 . . . N PACKAGE
(TOP VIEW)

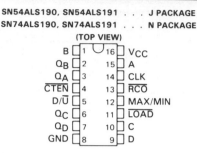

SN54ALS190, SN54ALS191 . . . FH PACKAGE
SN74ALS190, SN74ALS191 . . . FN PACKAGE
(TOP VIEW)

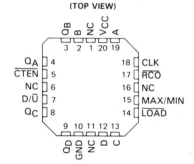

NC — no internal connection.

descriptions

The 'ALS190 and 'ALS191 are synchronous, reversible up/down counters. The 'ALS190 is a 4-bit decade counter and the 'ALS191 is a 4-bit binary counter. Synchronous counting operation is provided by having all flip-flops clocked simultaneously so that the outputs change coincident with each other when so instructed by the steering logic. This mode of operation eliminates the output counting spikes normally associated with asynchronous (ripple clock) counters.

The outputs of the four flip-flops are triggered on a low-to-high-level transition of the clock input if the enable input ($\overline{\text{CTEN}}$) is low. A high at $\overline{\text{CTEN}}$ inhibits counting. The direction of the count is determined by the level of the down/up (D/$\overline{\text{U}}$) input. When D/$\overline{\text{U}}$ is low, the counter counts up and when D/$\overline{\text{U}}$ is high, it counts down.

These counters feature a fully independent clock circuit. Changes at the control inputs ($\overline{\text{CTEN}}$ and D/$\overline{\text{U}}$) that will modify the operating mode have no effect on the contents of the counter until clocking occurs. The function of the counter will be dictated solely by the condition meeting the stable setup and hold times.

These counters are fully programmable; that is, the outputs may each be preset to either level by placing a low on the load input and entering the desired data at the data inputs. The output will change to agree with the data inputs independently of the level of the clock input. This feature allows the counters to be used as modulo-N dividers by simply modifying the count length with the preset inputs.

The CLK, D/$\overline{\text{U}}$, and LOAD inputs are buffered to lower the drive requirement, which significantly reduces the loading on, or current required by, clock drivers, etc., for long parallel words.

Two outputs have been made available to perform the cascading function: ripple clock and maximum/minimum count. The latter output produces a high-level output pulse with a duration approximately equal to one complete cycle of the clock while the count is zero (all outputs low) counting down or maximum (9 or 15) counting up. The ripple clock output produces a low-level output pulse under those same conditions but only while the clock input is low. The counters can be easily cascaded by feeding the ripple clock output to the enable input of the succeeding counter if parallel clocking is used, or to the clock input if parallel enabling is used. The maximum/minimum count output can be used to accomplish look-ahead for high-speed operation.

The SN54ALS190 and SN54ALS191 are characterized for operation over the full military temperature range of −55 °C to 125 °C. The SN74ALS190 and SN74ALS191 are characterized for operation from 0 °C to 70 °C.

(Courtesy of Texas Instruments Inc.)

'ALS191 logic symbol

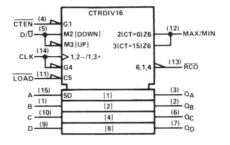

'ALS191 logic diagram (positive logic)

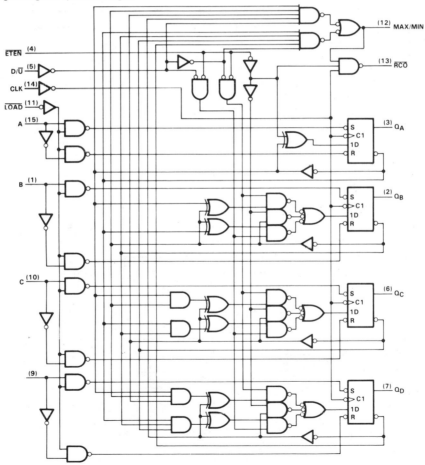

Pin numbers shown are for J and N packages.

(Courtesy of Texas Instruments Inc.)

typical load, count, and inhibit sequences

'ALS190

Illustrated below is the following sequence:

1. Load (preset) to BCD seven.
2. Count up to eight, nine (maximum), zero, one, and two.
3. Inhibit.
4. Count down to one, zero (minimum), nine, eight, and seven.

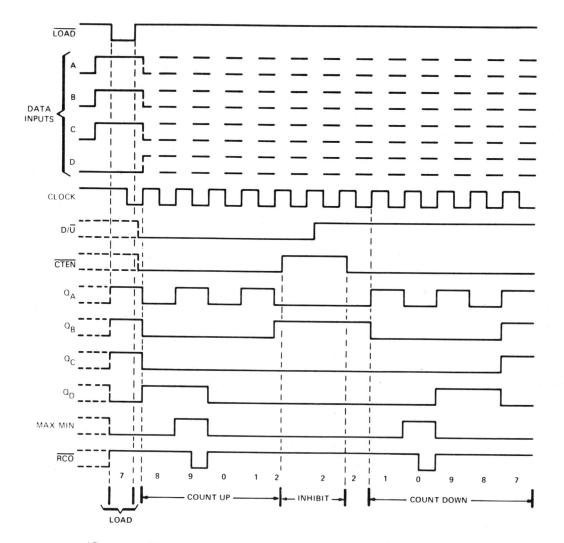

(Courtesy of Texas Instruments Inc.)

typical load, count, and inhibit sequences

'ALS191

Illustrated below is the following sequence:

1. Load (preset) to binary thirteen.
2. Count up to fourteen, fifteen (maximum), zero, one, and two.
3. Inhibit.
4. Count down to one, zero (minimum), fifteen, fourteen, and thirteen.

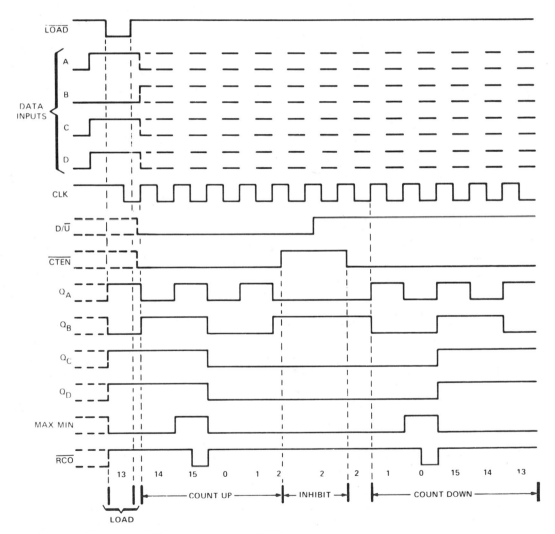

(Courtesy of Texas Instruments Inc.)

absolute maximum ratings over operating free-air temperature range (unless otherwise noted)

Supply voltage, V_{CC} . . . . 7 V
Input voltage . . . . 7 V
Operating free-air temperature range: SN54ALS190, SN54ALS191 . . . . −55°C to 125°C
                              SN74ALS190, SN74ALS191 . . . . 0°C to 70°C
Storage temperature range . . . . −65°C to 150°C

recommended operating conditions

				SN54ALS190 SN54ALS191			SN74ALS190 SN74ALS191			UNIT
				MIN	NOM	MAX	MIN	NOM	MAX	
V_{CC}	Supply voltage			4.5	5	5.5	4.5	5	5.5	V
V_{IH}	High-level input voltage			2			2			V
V_{IL}	Low-level input voltage					0.8			0.8	V
I_{OH}	High-level output current					−0.4			−0.4	mA
I_{OL}	Low-level output current					4			8	mA
f_{clock}	Clock frequency	'ALS190		0		20	0		25	MHz
		'ALS191		0		25	0		30	
t_w	Pulse duration	CLK high or low	'ALS190	25			20			ns
			'ALS191	20			16.5			
		$\overline{LOAD}$ low		25			20			
t_{su}	Setup time	Data before $\overline{LOAD}\uparrow$		25			20			ns
		$\overline{CTEN}$ before CLK↑		25			20			
		D/$\overline{U}$ before CLK↑		20			20			
		$\overline{LOAD}$ inactive before CLK↑		20			20			
t_h	Hold time	Data after $\overline{LOAD}\uparrow$		5			5			ns
		$\overline{CTEN}$ after CLK↑		0			0			
		D/$\overline{U}$ after CLK↑		0			0			
T_A	Operating free-air temperature			−55		125	0		70	°C

electrical characteristics over recommended operating free-air temperature range (unless otherwise noted)

PARAMETER		TEST CONDITIONS		SN54ALS190 SN54ALS191			SN74ALS190 SN74ALS191			UNIT
				MIN	TYP[†]	MAX	MIN	TYP[†]	MAX	
V_{IK}		$V_{CC} = 4.5$ V,	$I_I = −18$ mA			−1.5			−1.5	V
V_{OH}		$V_{CC} = 4.5$ V to 5.5 V,	$I_{OH} = −0.4$ mA	$V_{CC}−2$			$V_{CC}−2$			V
V_{OL}		$V_{CC} = 4.5$ V,	$I_{OL} = 4$ mA		0.25	0.4		0.25	0.4	V
		$V_{CC} = 4.5$ V	$I_{OL} = 8$ mA					0.35	0.5	
I_I		$V_{CC} = 5.5$ V,	$V_I = 7$ V			0.1			0.1	mA
I_{IH}		$V_{CC} = 5.5$ V,	$V_I = 2.7$ V			20			20	μA
I_{IL}	$\overline{CTEN}$ or CLK	$V_{CC} = 5.5$ V,	$V_I = 0.4$ V			−0.2			−0.2	mA
	All others					−0.1			−0.1	
I_O[‡]		$V_{CC} = 5.5$ V,	$V_O = 2.25$ V	−30		−112	−30		−112	mA
I_{CC}		$V_{CC} = 5.5$ V,	All inputs at 0 V		12	22		12	22	mA

[†]All typical values are at $V_{CC} = 5$ V, $T_A = 25$°C.

[‡]The output conditions have been chosen to produce a current that closely approximates one half of the true short-circuit output current, I_{OS}.

(Courtesy of Texas Instruments Inc.)

switching characteristics (see Note 1)

PARAMETER	FROM (INPUT)	TO (OUTPUT)	V_{CC} = 4.5 V to 5.5 V, C_L = 50 pF, R_L = 500 Ω, T_A = MIN to MAX				UNIT
			SN54ALS190 SN54ALS191		SN74ALS190 SN74ALS191		
			MIN	MAX	MIN	MAX	
f_{max}	'ALS190		20		25		MHz
	'ALS191		25		30		
t_{PLH}	$\overline{LOAD}$	Any Q	8	34	8	30	ns
t_{PHL}			8	34	8	30	
t_{PLH}	A, B, C, D	Any Q	4	25	4	21	ns
t_{PHL}			4	25	4	21	
t_{PLH}	CLK	$\overline{RCO}$	5	24	5	20	ns
t_{PHL}			5	24	5	20	
t_{PLH}	CLK	Any Q	3	22	3	18	ns
t_{PHL}			3	22	3	18	
t_{PLH}	CLK	MAX/MIN	8	34	8	31	ns
t_{PHL}			8	34	8	31	
t_{PLH}	$D/\overline{U}$	$\overline{RCO}$	15	42	15	37	ns
t_{PHL}			10	33	10	28	
t_{PLH}	$D/\overline{U}$	MAX/MIN	8	30	8	25	ns
t_{PHL}			8	30	8	25	
t_{PLH}	$\overline{CTEN}$	$\overline{RCO}$	4	21	4	18	ns
t_{PHL}			4	21	4	18	

NOTE 1: For load circuit and voltage waveforms, see page 1-12.

(Courtesy of Texas Instruments Inc.)

647

- 2K X 8 Organization, Common I/O

- Single + 5-V Supply

- Fully Static Operation (No Clocks, No Refresh)

- JEDEC Standard Pinout

- 24-Pin 600 Mil (15.2 mm) Package Configuration

- Plug-in Compatible with 16K 5 V EPROMs

- 8-Bit Output for Use in Microprocessor-Based Systems

- 3-State Outputs with $\overline{S}$ for OR-ties

- $\overline{G}$ Eliminates Need for External Bus Buffers

- All Inputs and Outputs Fully TTL Compatible

- Fanout to Series 74, Series 74S or Series 74LS TTL Loads

- N-Channel Silicon-Gate Technology

- Power Dissipation Under 385 mW Max

- Guaranteed dc Noise Immunity of 400 mV with Standard TTL Loads

- 4 Performance Ranges:

	ACCESS TIME (MAX)
TMS4016-12	120 ns
TMS4016-15	150 ns
TMS4016-20	200 ns
TMS4016-25	250 ns

TMS4016 . . . NL PACKAGE
(TOP VIEW)

```
        ┌────∪────┐
  A7 ┃ 1      24 ┃ VCC
  A6 ┃ 2      23 ┃ A8
  A5 ┃ 3      22 ┃ A9
  A4 ┃ 4      21 ┃ W̄
  A3 ┃ 5      20 ┃ Ḡ
  A2 ┃ 6      19 ┃ A10
  A1 ┃ 7      18 ┃ S̄
  A0 ┃ 8      17 ┃ DQ8
 DQ1 ┃ 9      16 ┃ DQ7
 DQ2 ┃ 10     15 ┃ DQ6
 DQ3 ┃ 11     14 ┃ DQ5
 VSS ┃ 12     13 ┃ DQ4
        └─────────┘
```

PIN NOMENCLATURE	
A0 – A10	Addresses
DQ1 – DQ8	Data In/Data Out
$\overline{G}$	Output Enable
$\overline{S}$	Chip Select
VCC	+ 5-V Supply
VSS	Ground
$\overline{W}$	Write Enable

description

The TMS4016 static random-access memory is organized as 2048 words of 8 bits each. Fabricated using proven N-channel, silicon-gate MOS technology, the TMS4016 operates at high speeds and draws less power per bit than 4K static RAMs. It is fully compatible with Series 74, 74S, or 74LS TTL. Its static design means that no refresh clocking circuitry is needed and timing requirements are simplified. Access time is equal to cycle time. A chip select control is provided for controlling the flow of data-in and data-out and an output enable function is included in order to eliminate the need for external bus buffers.

Of special importance is that the TMS4016 static RAM has the same standardized pinout as TI's compatible EPROM family. This, along with other compatible features, makes the TMS4016 plug-in compatible with the TMS2516 (or other 16K 5 V EPROMs). Minimal, if any modifications are needed. This allows the microprocessor system designer complete flexibility in partitioning his memory board between read/write and non-volatile storage.

The TMS4016 is offered in the plastic (NL suffix) 24-pin dual-in-line package designed for insertion in mounting hole rows on 600-mil (15.2 mm) centers. It is guaranteed for operation from 0 °C to 70 °C.

(Courtesy of Texas Instruments Inc.)

TMS4016
2048-WORD BY 8-BIT STATIC RAM

operation

addresses (A0 — A10)

The eleven address inputs select one of the 2048 8-bit words in the RAM. The address-inputs must be stable for the duration of a write cycle. The address inputs can be driven directly from standard Series 54/74 TTL with no external pull-up resistors.

output enable ($\overline{G}$)

The output enable terminal, which can be driven directly from standard TTL circuits, affects only the data-out terminals. When output enable is at a logic high level, the output terminals are disabled to the high-impedance state. Output enable provides greater output control flexibility, simplifying data bus design.

chip select ($\overline{S}$)

The chip-select terminal, which can be driven directly from standard TTL circuits, affects the data-in/data-out terminals. When chip select and output enable are at a logic low level, the D/Q terminals are enabled. When chip select is high, the D/Q terminals are in the floating or high-impedance state and the input is inhibited.

write enable ($\overline{W}$)

The read or write mode is selected through the write enable terminal. A logic high selects the read mode; a logic low selects the write mode. $\overline{W}$ must be high when changing addresses to prevent erroneously writing data into a memory location. The $\overline{W}$ input can be driven directly from standard TTL circuits.

data-in/data-out (DQ1 — DQ8)

Data can be written into a selected device when the write enable input is low. The D/Q terminal can be driven directly from standard TTL circuits. The three-state output buffer provides direct TTL compatibility with a fan-out of one Series 74 TTL gate, one Series 74S TTL gate, or five Series 74LS TTL gates. The D/Q terminals are in the high impedance state when chip select ($\overline{S}$) is high, output enable ($\overline{G}$) is high, or whenever a write operation is being performed. Data-out is the same polarity as data-in.

(Courtesy of Texas Instruments Inc.)

logic symbol[†]

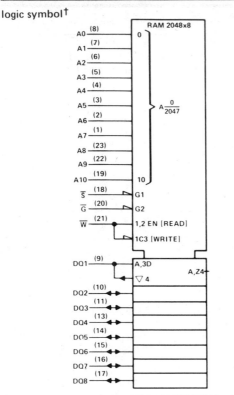

FUNCTION TABLE

$\overline{W}$	$\overline{S}$	$\overline{G}$	DQ1-DQ8	MODE
L	L	X	VALID DATA	WRITE
H	L	L	DATA OUTPUT	READ
X	H	X	HI-Z	DEVICE DISABLED
H	L	H	HI-Z	OUTPUT DISABLED

[†] This symbol is in accordance with IEEE Std 91/ANSI Y32.14 and recent decisions by IEEE and IEC. See explanation on page 10-1.

absolute maximum ratings over operating free-air temperature range (unless otherwise noted) [†]

Supply voltage, V_{CC} (see Note 1) . . . . −0.5 V to 7 V
Input voltage (any input) (see Note 1) . . . . −1 V to 7 V
Continuous power dissipation . . . . 1 W
Operating free-air temperature range . . . . 0°C to 70°C
Storage temperature range . . . . −55°C to 150°C

[†] Stresses beyond those listed under "Absolute Maximum Ratings" may cause permanent damage to the device. This is a stress rating only and functional operation of the device at these or any other conditions beyond those indicated in the "Recommended Operating Conditions" section of this specification is not implied. Exposure to absolute-maximum-rated conditions for extended periods may affect device reliability.

NOTE 1: Voltage values are with respect to the V_{SS} terminal.

recommended operating conditions

PARAMETER	MIN	NOM	MAX	UNIT
Supply voltage, V_{CC}	4.5	5	5.5	V
Supply voltage, V_{SS}		0		V
High-level input voltage, V_{IH}	2		5.5	V
Low-level input voltage, V_{IL} (see Note 2)	−1		0.8	V
Operating free-air temperature, T_A	0		70	°C

NOTE 2: The algebraic convention, where the more negative (less positive) limit is designated as minimum, is used in this data sheet for logic voltage levels only.

(Courtesy of Texas Instruments Inc.)

electrical characteristics over recommended operating free-air temperature range (unless otherwise noted)

PARAMETER		TEST CONDITIONS		MIN	TYP†	MAX	UNIT
V_{OH}	High level voltage	$I_{OH} = -1$ mA,	$V_{CC} = 4.5$ V	2.4			V
V_{OL}	Low level voltage	$I_{OL} = 2.1$ mA,	$V_{CC} = 4.5$ V			0.4	V
I_I	Input current	$V_I = 0$ V to 5.5 V				10	μA
I_{OZ}	Off-state output current	$\overline{S}$ or $\overline{G}$ at 2 V or $\overline{W}$ at 0.8 V, $V_O = 0$ V to 5.5 V				10	μA
I_{CC}	Supply current from V_{CC}	$I_O = 0$ mA, $T_A = 0$°C (worst case)	$V_{CC} = 5.5$ V,		40	70	mA
C_i	Input capacitance	$V_I = 0$ V,	f = 1 MHz			8	pF
C_O	Output capacitance	$V_O = 0$ V,	f = 1 MHz			12	pF

†All typical values are at $V_{CC} = 5$ V, $T_A = 25$°C.

timing requirements over recommended supply voltage range and operating free-air temperature range

PARAMETER		TMS4016-12		TMS4016-15		TMS4016-20		TMS4016-25		UNIT
		MIN	MAX	MIN	MAX	MIN	MAX	MIN	MAX	
$t_{c(rd)}$	Read cycle time	120		150		200		250		ns
$t_{c(wr)}$	Write cycle time	120		150		200		250		ns
$t_{w(W)}$	Write pulse width	60		80		100		120		ns
$t_{su(A)}$	Address setup time	20		20		20		20		ns
$t_{su(S)}$	Chip select setup time	60		80		100		120		ns
$t_{su(D)}$	Data setup time	50		60		80		100		ns
$t_{h(A)}$	Address hold time	0		0		0		0		ns
$t_{h(D)}$	Data hold time	5		10		10		10		ns

switching characteristics over recommended voltage range, $T_A = 0$°C to 70°C with output loading of Figure 1 (see notes 3 and 4)

PARAMETER		TMS4016-12		TMS4016-15		TMS4016-20		TMS4016-25		UNIT
		MIN	MAX	MIN	MAX	MIN	MAX	MIN	MAX	
$t_{a(A)}$	Access time from address		120		150		200		250	ns
$t_{a(S)}$	Access time from chip select low		60		75		100		120	ns
$t_{a(G)}$	Access time from output enable low		50		60		80		100	ns
$t_{v(A)}$	Output data valid after address change	10		15		15		15		ns
$t_{dis(S)}$	Output disable time after chip select high		40		50		60		80	ns
$t_{dis(G)}$	Output disable time after output enable high		40		50		60		80	ns
$t_{dis(W)}$	Output disable time after write enable low		50		60		60		80	ns
$t_{en(S)}$	Output enable time after chip select low	5		5		10		10		ns
$t_{en(G)}$	Output enable time after output enable low	5		5		10		10		ns
$t_{en(W)}$	Output enable time after write enable high	5		5		10		10		ns

NOTES: 3. $C_L = 100$ pF for all measurements except $t_{dis(W)}$ and $t_{en(W)}$.
$C_L = 5$ pF for $t_{dis(W)}$ and $t_{en(W)}$.
4. t_{dis} and t_{en} parameters are sampled and not 100% tested.

(Courtesy of Texas Instruments Inc.)

Error Detection and Correction

C-1 ERROR-DETECTION CODES

The *2-out-of-5* code is sometimes used in communications work. It utilizes five bits to represent the ten decimal digits, so it is a form of BCD code. Each code word has exactly two 1s, which facilitates decoding and provides for better error detection than the single parity bit method. If other than two 1s appear, an error is indicated.

The *63210 BCD* code is also characterized by having exactly two 1s in each of the five-bit groups. Like the 2-out-of-5 code, it provides reliable error detection and is used in some applications.

The *biquinary* code (two-five) is used in certain counters and is composed of a two-bit group and a five-bit group, each with a single 1. Its weights are 50 43210. The two-bit group, having weights 50, indicates whether the number represented is less than, equal to, or greater than 5. The five-bit group indicates the count above or below 5.

The *ring counter* code has ten bits, one for each decimal digit, and a single 1 makes error detection possible. It is easy to decode but wastes bits and requires

TABLE C-1 *Some codes with error-detection properties.*

Decimal	2-out-of-5	63210	5043210	9876543210
0	00011	00110	01 00001	0000000001
1	00101	00011	01 00010	0000000010
2	00110	00101	01 00100	0000000100
3	01001	01001	01 01000	0000001000
4	01010	01010	01 10000	0000010000
5	01100	01100	10 00001	0000100000
6	10001	10001	10 00010	0001000000
7	10010	10010	10 00100	0010000000
8	10100	10100	10 01000	0100000000
9	11000	11000	10 10000	1000000000

more circuitry to implement than the four- or five-bit codes. The name is derived from the fact that the code is generated by a certain type of shift register, or "ring counter." Its weights are 9876543210.

Each of these codes is listed in Table C-1. You should realize that this is not an exhaustive coverage of all codes but simply an introduction to some of them.

C-2   HAMMING ERROR-CORRECTION CODE

This section discusses a method, generally known as the *Hamming code,* which not only provides for the detection of a bit error, but also identifies which bit is in error so that it can be corrected. The code uses a number of parity bits (dependent on the number of information bits) located at certain positions in the code group.

The Hamming code is constructed as follows for *single-error* correction.

Number of Parity Bits

If the number of information bits is designated m, then the number of parity bits, p, is determined by the following relationship:

$$2^p \geq m + p + 1 \tag{C-1}$$

For example, if we have four information bits, then p is found by trial and error using Equation (C-1). Let $p = 2$. Then

$$2^p = 2^2 = 4$$

and

$$m + p + 1 = 4 + 2 + 1 = 7$$

Since 2^p must be equal to or greater than $m + p + 1$, the relationship in Equation (C–1) is *not* satisfied. We have to try again. Let $p = 3$. Then

$$2^p = 2^3 = 8$$

and

$$m + p + 1 = 4 + 3 + 1 = 8$$

This value of p satisfies the relationship of Equation (C–1), and therefore three parity bits are required to provide single-error correction for four information bits. It should be noted here that error detection and correction are provided for *all* bits, both parity and information, in a code group.

Placement of the Parity Bits in the Code

Now that we have found the number of parity bits required in our particular example, we must arrange the bits properly in the code. At this point you should realize that in this example, the code is composed of the four information bits and the three parity bits. The left-most bit is designated *bit 1*, the next bit is *bit 2*, and so on, as shown below:

bit 1, bit 2, bit 3, bit 4, bit 5, bit 6, bit 7

The parity bits are located in the positions that are numbered corresponding to ascending powers of two (1, 2, 4, 8, . . .) as indicated:

$$P_1, \quad P_2, \quad M_1, \quad P_3, \quad M_2, \quad M_3, \quad M_4$$

The symbol P_n designates a particular parity bit, and M_n designates a particular information bit, where n is the position number.

Assignment of Parity Bit Values

Finally, we must properly assign a 1 or 0 value to each parity bit. Since each parity bit provides a check on certain other bits in the total code, we must know the value of these others in order to assign the parity bit value. To do this, first number each bit position in *binary*; that is, write the binary number for each decimal position number (as shown in the second two rows of Table C–2). Next, indicate the parity and information bit locations, as shown in the first row of Table

TABLE C–2 *Bit position table for a seven-bit error-correcting code.*

	P_1	P_2	M_1	P_3	M_2	M_3	M_4
Bit designation	P_1	P_2	M_1	P_3	M_2	M_3	M_4
Bit position	1	2	3	4	5	6	7
Binary position number	001	010	011	100	101	110	111
Information bits (M_n)							
Parity bits (P_n)							

C–2. Notice that the binary position number of parity bit P_1 has a 1 for its right-most digit. This parity bit checks all bit positions, including itself, that have *1s in the same location in the binary position numbers*. Therefore, parity bit P_1 checks bit positions 1, 3, 5, and 7.

The binary position number for parity bit P_2 has a 1 for its middle bit. It checks all bit positions, including itself, that have 1s in this same position. Therefore, parity bit P_2 checks bit positions 2, 3, 6, and 7.

The binary position number for parity bit P_3 has a 1 for its left-most bit. It checks all bit positions, including itself, that have 1s in this same position. Therefore, parity bit P_3 checks bit positions 4, 5, 6, and 7.

In each case, *the parity bit is assigned a value to make the quantity of 1s in the set of bits that it checks odd or even,* depending on which is specified. The following examples should make this procedure clear.

EXAMPLE C–1

Determine the single error-correcting code for the BCD number 1001 (information bits) using even parity.

Solution

Step 1. Find the number of parity bits required. Let $p = 3$. Then

$$2^p = 2^3 = 8$$
$$m + p + 1 = 4 + 3 + 1 = 8$$

Three parity bits are sufficient.

$$\text{Total code bits} = 4 + 3 = 7$$

Step 2. Construct a bit position table:

Bit designation Bit position Binary position number	P_1 1 001	P_2 2 010	M_1 3 011	P_3 4 100	M_2 5 101	M_3 6 110	M_4 7 111
Information bits			1		0	0	1
Parity bits	0	0		1			

Parity bits are determined in the following steps.

Step 3. Determine the parity bits as follows:
P_1 checks bit positions 1, 3, 5, and 7 and must be a 0 in order to have an even number of 1s (2) in this group.
P_2 checks bit positions 2, 3, 6, and 7 and must be a 0 in order to have an even number of 1s (2) in this group.

P_3 checks bit positions 4, 5, 6, and 7 and must be a 1 in order to have an even number of 1s (2) in this group.

Step 4. These parity bits are entered into the table, and the resulting combined code is 0011001.

EXAMPLE C-2   Determine the single error-correcting code for the information code 10110 for odd parity.

Solution

Step 1. Determine the number of parity bits required. In this case the number of information bits, *m*, is five.

From the previous example we know that $p = 3$ will not work. Try $p = 4$.

$$2^p = 2^4 = 16$$

$$m + p + 1 = 5 + 4 + 1 = 10$$

Four parity bits are sufficient.

$$\text{Total code bits} = 5 + 4 = 9$$

Step 2. Construct a bit position table:

Bit designation	P_1	P_2	M_1	P_3	M_2	M_3	M_4	P_4	M_5
Bit position	1	2	3	4	5	6	7	8	9
Binary position number	0001	0010	0011	0100	0101	0110	0111	1000	1001
Information bits			1		0	1	1		0
Parity bits	1	0		1				1	

Parity bits are determined in the following steps.

Step 3. Determine the parity bits as follows:

P_1 checks bit positions 1, 3, 5, 7, and 9 and must be a 1 to have an odd number of 1s (3) in this group.

P_2 checks bit positions 2, 3, 6, and 7 and must be a 0 to have an odd number of 1s (3) in this group.

P_3 checks bit positions 4, 5, 6, and 7 and must be a 1 to have an odd number of 1s (3) in this group.

P_4 checks bit positions 8 and 9 and must be a 1 to have an odd number of 1s (1) in this group.

Step 4. These parity bits are entered into the table, and the resulting combined code is 101101110.

C-3 DETECTING AND CORRECTING AN ERROR

Now that a method for constructing an error-correcting code has been covered, how do we use it to locate and correct an error? Each parity bit, along with its corresponding group of bits, must be checked for the proper parity. If there are three parity bits in a code word, then three parity checks are made. If there are four parity bits, four checks must be made, and so on. Each parity check will yield a good or a bad result. The total result of all the parity checks indicates the bit, if any, that is in error, as follows:

Step 1. Start with the group checked by P_1.
Step 2. Check the group for proper parity. A 0 represents a good parity check, and 1 represents a bad check.
Step 3. Repeat step 2 for each parity group.
Step 4. The binary number formed by the results of each parity check designates the position of the code bit that is in error. This is the *error position code*. The first parity check generates the least significant bit (LSB). If all checks are good, there is no error.

EXAMPLE C-3

Assume that the code word in Example C-1 (0011001) is transmitted and that 0010001 is received. The receiver does not "know" what was transmitted and must look for proper parities to determine if the code is correct. Designate any error that has occurred in transmission if *even parity* is used.

Solution First, make a bit position table:

Bit designation	P_1	P_2	M_1	P_3	M_2	M_3	M_4
Bit position	1	2	3	4	5	6	7
Binary position number	001	010	011	100	101	110	111
Received code	0	0	1	0	0	0	1

First parity check:
   P_1 checks positions 1, 3, 5, and 7.
   There are two 1s in this group.
   Parity check is good. ────────────────────────────→ 0 (LSB)
Second parity check:
   P_2 checks positions 2, 3, 6, and 7.
   There are two 1s in this group.
   Parity check is good. ────────────────────────────→ 0
Third parity check:
   P_3 checks positions 4, 5, 6, and 7.

There is one 1 in this group.

Parity check is bad. ────────────────────────────→ 1 (MSB)

Result:

The error position code is 100 (binary 4). This says that the bit in the number 4 position is in error. It is a 0 and should be a 1. The corrected code is 0011001, which agrees with the transmitted code.

EXAMPLE C-4

The code 101101010 is received. Correct any errors. There are four parity bits and odd parity is used.

Solution First, make a bit position table:

Bit designation	P_1	P_2	M_1	P_3	M_2	M_3	M_4	P_4	M_5
Bit position	1	2	3	4	5	6	7	8	9
Binary position number	0001	0010	0011	0100	0101	0110	0111	1000	1001
Received code	1	0	1	1	0	1	0	1	0

First parity check:

P_1 checks positions 1, 3, 5, 7, and 9.

There are two 1s in this group.

Parity check is bad. ────────────────────────────→ 1 (LSB)

Second parity check:

P_2 checks positions 2, 3, 6, and 7.

There are two 1s in this group.

Parity check is bad. ────────────────────────────→ 1

Third parity check:

P_3 checks positions 4, 5, 6, and 7.

There are two 1s in this group.

Parity check is bad. ────────────────────────────→ 1

Fourth parity check:

P_4 checks positions 8 and 9.

There is one 1 in this group.

Parity check is good. ────────────────────────────→ 0 (MSB)

Result:

The error position code is 0111 (binary 7). This says that the bit in the number 7 position is in error. The corrected code is therefore 101101110.

Conversions

Decimal	BCD(8421)	Octal	Binary	Decimal	BCD(8421)	Octal	Binary
0	0000	0	0	25	00100101	31	11001
1	0001	1	1	26	00100110	32	11010
2	0010	2	10	27	00100111	33	11011
3	0011	3	11	28	00101000	34	11100
4	0100	4	100	29	00101001	35	11101
5	0101	5	101	30	00110000	36	11110
6	0110	6	110	31	00110001	37	11111
7	0111	7	111	32	00110010	40	100000
8	1000	10	1000	33	00110011	41	100001
9	1001	11	1001	34	00110100	42	100010
10	00010000	12	1010	35	00110101	43	100011
11	00010001	13	1011	36	00110110	44	100100
12	00010010	14	1100	37	00110111	45	100101
13	00010011	15	1101	38	00111000	46	100110
14	00010100	16	1110	39	00111001	47	100111
15	00010101	17	1111	40	01000000	50	101000
16	00010110	20	10000	41	01000001	51	101001
17	00010111	21	10001	42	01000010	52	101010
18	00011000	22	10010	43	01000011	53	101011
19	00011001	23	10011	44	01000100	54	101100
20	00100000	24	10100	45	01000101	55	101101
21	00100001	25	10101	46	01000110	56	101110
22	00100010	26	10110	47	01000111	57	101111
23	00100011	27	10111	48	01001000	60	110000
24	00100100	30	11000	49	01001001	61	110001

Decimal	BCD(8421)	Octal	Binary	Decimal	BCD(8421)	Octal	Binary
50	01010000	62	110010	75	01110101	113	1001011
51	01010001	63	110011	76	01110110	114	1001100
52	01010010	64	110100	77	01110111	115	1001101
53	01010011	65	110101	78	01111000	116	1001110
54	01010100	66	110110	79	01111001	117	1001111
55	01010101	67	110111	80	10000000	120	1010000
56	01010110	70	111000	81	10000001	121	1010001
57	01010111	71	111001	82	10000010	122	1010010
58	01011000	72	111010	83	10000011	123	1010011
59	01011001	73	111011	84	10000100	124	1010100
60	01100000	74	111100	85	10000101	125	1010101
61	01100001	75	111101	86	10000110	126	1010110
62	01100010	76	111110	87	10000111	127	1010111
63	01100011	77	111111	88	10001000	130	1011000
64	01100100	100	1000000	89	10001001	131	1011001
65	01100101	101	1000001	90	10010000	132	1011010
66	01100110	102	1000010	91	10010001	133	1011011
67	01100111	103	1000011	92	10010010	134	1011100
68	01101000	104	1000100	93	10010011	135	1011101
69	01101001	105	1000101	94	10010100	136	1011110
70	01110000	106	1000110	95	10010101	137	1011111
71	01110001	107	1000111	96	10010110	140	1100000
72	01110010	110	1001000	97	1000111	141	1100001
73	01110011	111	1001001	98	10011000	142	1100010
74	01110100	112	1001010	99	10011001	143	1100011

Powers of Two

2^n	n	2^{-n}
1	0	1.0
2	1	0.5
4	2	0.25
8	3	0.125
16	4	0.062 5
32	5	0.031 25
64	6	0.015 625
128	7	0.007 812 5
256	8	0.003 906 25
512	9	0.001 953 125
1 024	10	0.000 976 562 5
2 048	11	0.000 488 281 25
4 096	12	0.000 244 140 625
8 192	13	0.000 122 070 312 5
16 384	14	0.000 061 035 156 25
32 768	15	0.000 030 517 578 125
65 536	16	0.000 015 258 789 062 5
131 072	17	0.000 007 629 394 531 25
262 144	18	0.000 003 814 697 265 625
524 288	19	0.000 001 907 348 632 812 5
1 048 576	20	0.000 000 953 674 316 406 25
2 097 152	21	0.000 000 476 837 158 203 125
4 194 304	22	0.000 000 238 418 579 101 562 5
8 388 608	23	0.000 000 119 209 289 550 781 25
16 777 216	24	0.000 000 059 604 644 775 390 625
33 554 432	25	0.000 000 029 802 322 387 695 312 5
67 108 864	26	0.000 000 014 901 161 193 847 656 25
134 217 728	27	0.000 000 007 450 580 596 923 828 125
268 435 456	28	0.000 000 003 725 290 298 461 914 062 5
536 870 912	29	0.000 000 001 862 645 149 230 957 031 25
1 073 741 824	30	0.000 000 000 931 322 574 615 478 515 625
2 147 483 648	31	0.000 000 000 465 661 287 307 739 257 812 5
4 294 967 296	32	0.000 000 000 232 830 643 653 869 628 906 25
8 589 934 592	33	0.000 000 000 116 415 321 826 934 814 453 125
17 179 869 184	34	0.000 000 000 058 207 660 913 467 407 226 562 5
34 359 738 368	35	0.000 000 000 029 103 830 456 733 703 613 281 25
68 719 476 736	36	0.000 000 000 014 551 915 228 366 851 806 640 625
137 438 953 472	37	0.000 000 000 007 275 957 614 183 425 903 320 312 5
274 877 906 944	38	0.000 000 000 003 637 978 807 091 712 951 660 156 25
549 755 813 888	39	0.000 000 000 001 818 989 403 545 856 475 830 078 125
1 099 511 627 776	40	0.000 000 000 000 909 494 701 772 928 237 915 039 062 5
2 199 023 255 552	41	0.000 000 000 000 454 747 350 886 464 118 957 519 531 25
4 398 046 511 104	42	0.000 000 000 000 227 373 675 443 232 059 478 759 765 625
8 796 093 022 208	43	0.000 000 000 000 113 686 837 721 616 029 739 379 882 812 5
17 592 186 044 416	44	0.000 000 000 000 056 843 418 860 808 014 869 689 941 406 25
35 184 372 088 832	45	0.000 000 000 000 028 421 709 430 404 007 434 844 970 703 125
70 368 744 177 664	46	0.000 000 000 000 014 210 854 715 202 003 717 422 485 351 562 5
140 737 488 355 328	47	0.000 000 000 000 007 105 427 357 601 001 858 711 242 675 781 25
281 474 976 710 656	48	0.000 000 000 000 003 552 713 678 800 500 929 355 621 337 890 625
562 949 953 421 312	49	0.000 000 000 000 001 776 356 839 400 250 464 677 810 668 945 312 5
1 125 899 906 842 624	50	0.000 000 000 000 000 888 178 419 700 125 232 338 905 334 472 656 25
2 251 799 813 685 248	51	0.000 000 000 000 000 444 089 209 850 062 616 169 452 667 236 328 125
4 503 599 627 370 496	52	0.000 000 000 000 000 222 044 604 925 031 308 084 726 333 618 164 062 5
9 007 199 254 740 992	53	0.000 000 000 000 000 111 022 302 462 515 654 042 363 166 809 082 031 25
18 014 398 509 481 984	54	0.000 000 000 000 000 055 511 151 231 257 827 021 181 583 404 541 015 625
36 028 797 018 963 968	55	0.000 000 000 000 000 027 755 575 615 628 913 510 590 791 702 270 507 812 5
72 057 594 037 927 936	56	0.000 000 000 000 000 013 877 787 807 814 456 755 295 395 851 135 253 906 25
144 115 188 075 855 872	57	0.000 000 000 000 000 006 938 893 903 907 228 377 647 697 925 567 626 953 125
288 230 376 151 711 744	58	0.000 000 000 000 000 003 469 446 951 953 614 188 823 848 962 783 813 476 562 5
576 460 752 303 423 488	59	0.000 000 000 000 000 001 734 723 475 976 807 094 411 924 481 391 906 738 281 25
1 152 921 504 606 846 976	60	0.000 000 000 000 000 000 867 361 737 988 403 547 205 962 240 695 953 369 140 625
2 305 843 009 213 693 952	61	0.000 000 000 000 000 000 433 680 868 994 201 773 602 981 120 347 976 684 570 312 5
4 611 686 018 427 387 904	62	0.000 000 000 000 000 000 216 840 434 497 100 886 801 490 560 173 988 342 285 156 25
9 223 372 036 854 775 808	63	0.000 000 000 000 000 000 108 420 217 248 550 443 400 745 280 086 994 171 142 578 125
18 446 744 073 709 551 616	64	0.000 000 000 000 000 000 054 210 108 624 275 221 700 372 640 043 497 085 571 289 062 5
36 893 488 147 419 103 232	65	0.000 000 000 000 000 000 027 105 054 312 137 610 850 186 320 021 748 542 785 644 531 25
73 786 976 294 838 206 464	66	0.000 000 000 000 000 000 013 552 527 156 088 805 425 093 160 010 874 271 392 822 265 625
147 573 952 589 676 412 928	67	0.000 000 000 000 000 000 006 776 263 578 034 402 712 546 580 005 437 135 696 411 132 812 5
295 147 905 179 352 825 856	68	0.000 000 000 000 000 000 003 388 131 789 017 201 356 273 290 002 718 567 848 205 566 406 25
590 295 810 358 705 651 712	69	0.000 000 000 000 000 000 001 694 065 894 508 600 678 136 645 001 359 283 924 102 783 203 125
1 180 591 620 717 411 303 424	70	0.000 000 000 000 000 000 000 847 032 947 254 300 339 068 322 500 679 641 962 051 391 601 562 5
2 361 183 241 434 822 606 848	71	0.000 000 000 000 000 000 000 423 516 473 627 150 169 534 161 250 339 820 981 025 695 800 781 25
4 722 366 482 869 645 213 696	72	0.000 000 000 000 000 000 000 211 758 236 813 575 084 767 080 625 169 910 490 512 847 900 390 625

Solutions to Self-Tests

CHAPTER 1

1. 1 or 0
2. HIGH = 0 (negative logic)
3. Bit-binary digit
4. Negative-going
5. Negative-going
6. A *periodic* waveform repeats itself at fixed time intervals. A *nonperiodic* waveform is nonrepetitive.
7. *Period* is the time interval for a periodic waveform to repeat, or the time for one cycle.
8. *Frequency* is the rate at which a periodic waveform repeats itself. The unit of frequency is the *hertz,* Hz.
9. Duty cycle increases when pulse width increases with respect to the period.
10. Boolean algebra
11. Inverter
12. AND: the output is HIGH only when all inputs are HIGH. OR: the output is HIGH when one or more inputs are HIGH.
13. A flip-flop can retain (store) a binary state, but a gate cannot.
14. Comparison: Determines $>$, $=$, $<$ relationships of two numbers.
    Arithmetic: Add, subtract, multiply, and divide.
    Encoding: Converts information into coded form.
    Decoding: Converts coded data to a familiar form.
    Counting: Counts events, divides frequency.

Register: Stores digital data.
Multiplexing: Puts data from several sources onto a single line in a time sequence.
Demultiplexing: Reverse of multiplexing.
15. Dual-in-line package, a type of integrated circuit package (DIP).
16. The microprocessor, memory, and I/O interface.
17. *Architecture* is the internal structure and arrangement of a system.
18. Input/output
19. *Troubleshooting* is the systematic approach to isolating and identifying a fault in a circuit or system.
20. Oscilloscope, logic analyzer, signature analyzer, logic probe, pulser.

CHAPTER 2

1. The base of the decimal number system is 10 (ten); 10 digits
2. (a) $28_{10} = 2 \times 10 + 8 \times 1$
   (b) $389_{10} = 3 \times 100 + 8 \times 10 + 9 \times 1$
   (c) $1473_{10} = 1 \times 1000 + 4 \times 100 + 7 \times 10 + 3 \times 1$
   (d) $10,956_{10} = 1 \times 10,000 + 9 \times 100 + 5 \times 10 + 6 \times 1$
3. (a) $\begin{array}{r} 99 \\ -57 \\ \hline 42 \end{array}$ (b) $\begin{array}{r} 999 \\ -892 \\ \hline 107 \end{array}$ (c) $\begin{array}{r} 9999 \\ -3347 \\ \hline 6652 \end{array}$
4. (a) $\begin{array}{r} 9 \\ -8 \\ \hline 1 \\ +1 \\ \hline 2 \end{array}$ (b) $\begin{array}{r} 99 \\ -57 \\ \hline 42 \\ +1 \\ \hline 43 \end{array}$ (c) $\begin{array}{r} 9999 \\ -1964 \\ \hline 8035 \\ +1 \\ \hline 8036 \end{array}$ (d) $\begin{array}{r} 9999 \\ -8840 \\ \hline 1159 \\ +1 \\ \hline 1160 \end{array}$
5. 100000, 100001, 100010, 100011, 100100, 100101, 100110, 100111, 101000, 101001, 101010.
6. (a) $1101_2 = 1 \times 2^3 + 1 \times 2^2 + 1 \times 2^0 = 8 + 4 + 1 = 13_{10}$
   (b) $100101_2 = 1 \times 2^5 + 1 \times 2^2 + 1 \times 2^0 = 32 + 4 + 1 = 37_{10}$
   (c) $11011101_2 = 1 \times 2^7 + 1 \times 2^6 + 1 \times 2^4 + 1 \times 2^3 + 1 \times 2^2 + 1 \times 2^0$
       $= 128 + 64 + 16 + 8 + 4 + 1 = 221_{10}$
   (d) $1011.011_2 = 1 \times 2^3 + 1 \times 2^1 + 1 \times 2^0 + 1 \times 2^{-2} + 1 \times 2^{-3}$
       $= 8 + 2 + 1 + 0.25 + 0.125 = 11.375_{10}$
   (e) $1110.1011_2 = 1 \times 2^3 + 1 \times 2^2 + 1 \times 2^1 + 1 \times 2^{-1} + 1 \times 2^{-3} + 1 \times 2^{-4}$
       $= 8 + 4 + 2 + 0.5 + 0.125 + 0.0625 = 14.6875_{10}$

7. **(a)**
$$\begin{array}{r} 110_2 \\ +011_2 \\ \hline 1001_2 \end{array}$$

(b)
$$\begin{array}{r} 11010_2 \\ +01111_2 \\ \hline 101001_2 \end{array}$$

(c)
$$\begin{array}{r} 110_2 \\ -010_2 \\ \hline 100_2 \end{array}$$

(d)
$$\begin{array}{r} 11011_2 \\ -10110_2 \\ \hline 00101_2 \end{array}$$

(e)
$$\begin{array}{r} 111_2 \\ \times 101_2 \\ \hline 111 \\ +1110 \\ \hline 100011_2 \end{array}$$

(f)
$$\begin{array}{r} 10_2 \\ 0111_2 \overline{)1110_2} \\ 111 \\ \hline 0 \end{array}$$

8. **(a)** 110101 → 001010 1's complement
001010 + 1 = 001011 2's complement

(b) 0001101 → 110010 1's complement
1110010 + 1 = 1110011 2's complement

9. **(a)**
$$\begin{array}{r} 8 \\ 2\overline{)17} \\ 16 \\ \hline 1 \longrightarrow 1 \end{array}$$

$$\begin{array}{r} 4 \\ 2\overline{)8} \\ 8 \\ \hline 0 \longrightarrow 0 \end{array}$$

$$\begin{array}{r} 2 \\ 2\overline{)4} \\ 4 \\ \hline 0 \longrightarrow 0 \end{array}$$

$$\begin{array}{r} 1 \\ 2\overline{)2} \\ 2 \\ \hline 0 \longrightarrow 0 \end{array}$$

$$\begin{array}{r} 0 \\ 2\overline{)1} \\ 0 \\ \hline 1 \longrightarrow 1 \end{array}$$

$$\boxed{10001_2}$$

(b)
$$\begin{array}{r} 51 \\ 2\overline{)102} \\ 102 \\ \hline 0 \longrightarrow 0 \end{array}$$

$$\begin{array}{r} 25 \\ 2\overline{)51} \\ 50 \\ \hline 1 \longrightarrow 1 \end{array}$$

$$\begin{array}{r} 12 \\ 2\overline{)25} \\ 24 \\ \hline 1 \longrightarrow 1 \end{array}$$

$$\begin{array}{r} 6 \\ 2\overline{)12} \\ 12 \\ \hline 0 \longrightarrow 0 \end{array}$$

$$\begin{array}{r} 3 \\ 2\overline{)6} \\ 6 \\ \hline 0 \longrightarrow 0 \end{array}$$

$$\begin{array}{r} 1 \\ 2\overline{)3} \\ 2 \\ \hline 1 \longrightarrow 1 \end{array}$$

$$\begin{array}{r} 0 \\ 2\overline{)1} \\ 0 \\ \hline 1 \longrightarrow 1 \end{array}$$

$$\boxed{1100110_2}$$

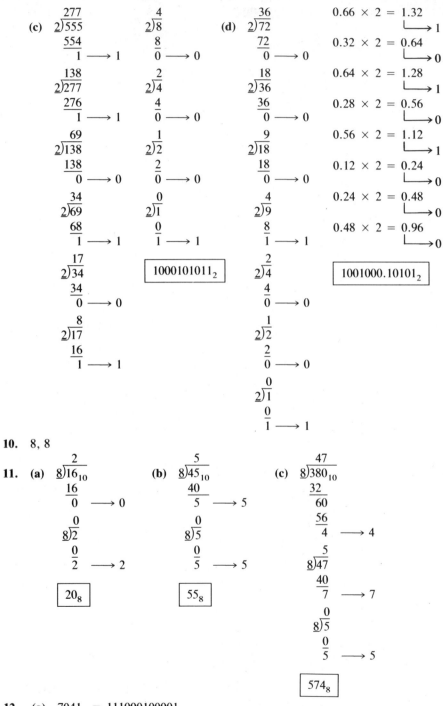

(c)

$$
\begin{array}{r}
277 \\
2\overline{)555} \\
\underline{554} \\
1 \longrightarrow 1
\end{array}
$$

$$
\begin{array}{r}
138 \\
2\overline{)277} \\
\underline{276} \\
1 \longrightarrow 1
\end{array}
$$

$$
\begin{array}{r}
69 \\
2\overline{)138} \\
\underline{138} \\
0 \longrightarrow 0
\end{array}
$$

$$
\begin{array}{r}
34 \\
2\overline{)69} \\
\underline{68} \\
1 \longrightarrow 1
\end{array}
$$

$$
\begin{array}{r}
17 \\
2\overline{)34} \\
\underline{34} \\
0 \longrightarrow 0
\end{array}
$$

$$
\begin{array}{r}
8 \\
2\overline{)17} \\
\underline{16} \\
1 \longrightarrow 1
\end{array}
$$

$$
\begin{array}{r}
4 \\
2\overline{)8} \\
\underline{8} \\
0 \longrightarrow 0
\end{array}
$$

$$
\begin{array}{r}
2 \\
2\overline{)4} \\
\underline{4} \\
0 \longrightarrow 0
\end{array}
$$

$$
\begin{array}{r}
1 \\
2\overline{)2} \\
\underline{2} \\
0 \longrightarrow 0
\end{array}
$$

$$
\begin{array}{r}
0 \\
2\overline{)1} \\
\underline{0} \\
1 \longrightarrow 1
\end{array}
$$

$$\boxed{1000101011_2}$$

(d)

$$
\begin{array}{r}
36 \\
2\overline{)72} \\
\underline{72} \\
0 \longrightarrow 0
\end{array}
$$

$$
\begin{array}{r}
18 \\
2\overline{)36} \\
\underline{36} \\
0 \longrightarrow 0
\end{array}
$$

$$
\begin{array}{r}
9 \\
2\overline{)18} \\
\underline{18} \\
0 \longrightarrow 0
\end{array}
$$

$$
\begin{array}{r}
4 \\
2\overline{)9} \\
\underline{8} \\
1 \longrightarrow 1
\end{array}
$$

$$
\begin{array}{r}
2 \\
2\overline{)4} \\
\underline{4} \\
0 \longrightarrow 0
\end{array}
$$

$$
\begin{array}{r}
1 \\
2\overline{)2} \\
\underline{2} \\
0 \longrightarrow 0
\end{array}
$$

$$
\begin{array}{r}
0 \\
2\overline{)1} \\
\underline{0} \\
1 \longrightarrow 1
\end{array}
$$

$$
\begin{array}{l}
0.66 \times 2 = 1.32 \longrightarrow 1 \\
0.32 \times 2 = 0.64 \longrightarrow 0 \\
0.64 \times 2 = 1.28 \longrightarrow 1 \\
0.28 \times 2 = 0.56 \longrightarrow 0 \\
0.56 \times 2 = 1.12 \longrightarrow 1 \\
0.12 \times 2 = 0.24 \longrightarrow 0 \\
0.24 \times 2 = 0.48 \longrightarrow 0 \\
0.48 \times 2 = 0.96 \longrightarrow 0
\end{array}
$$

$$\boxed{1001000.10101_2}$$

10. 8, 8

11. (a)

$$
\begin{array}{r}
2 \\
8\overline{)16}_{10} \\
\underline{16} \\
0 \longrightarrow 0
\end{array}
$$

$$
\begin{array}{r}
0 \\
8\overline{)2} \\
\underline{0} \\
2 \longrightarrow 2
\end{array}
$$

$$\boxed{20_8}$$

(b)

$$
\begin{array}{r}
5 \\
8\overline{)45}_{10} \\
\underline{40} \\
5 \longrightarrow 5
\end{array}
$$

$$
\begin{array}{r}
0 \\
8\overline{)5} \\
\underline{0} \\
5 \longrightarrow 5
\end{array}
$$

$$\boxed{55_8}$$

(c)

$$
\begin{array}{r}
47 \\
8\overline{)380}_{10} \\
\underline{32} \\
60 \\
\underline{56} \\
4 \longrightarrow 4
\end{array}
$$

$$
\begin{array}{r}
5 \\
8\overline{)47} \\
\underline{40} \\
7 \longrightarrow 7
\end{array}
$$

$$
\begin{array}{r}
0 \\
8\overline{)5} \\
\underline{0} \\
5 \longrightarrow 5
\end{array}
$$

$$\boxed{574_8}$$

12. (a) $7041_8 = 111000100001_2$
   (b) $65315_8 = 110101011001101_2$

13. $\underline{1001011110000101100010}_2$

    4 5 7 0 2 6 2 $\rightarrow$ 4,570,262$_8$

14. $\underline{100011000101111100111101000100001011}_2$

    8 C 5 F 3 D 1 0 B $\rightarrow$ 8C5F3D10B$_{16}$

15. $5A34F_{16} = 01011010001101001111_2$

16. 0, 1, 2, 3, 4, 5, 6, 7, 8, 9, A, B, C, D, E, F, 10, 11, 12, 13, 14, 15, 16, 17, 18, 19, 1A, 1B, 1C, 1D, 1E, 1F, 20

17. **(a)** $23_{10} = 00100011$ **(b)** $79_{10} = 01111001$

    (c) $718_{10} = 011100011000$ **(d)** $954.61_{10} = 100101010100.01100001$

18. $\underline{0100100110000110}$

    4 9 8 6 $\rightarrow$ 4986$_{10}$

19. $101011011_2 \rightarrow 111110110$ (Gray)

20. $1100100 \rightarrow 1000111_2$

21. Excess-3 is self-complementing.

22. From Table 2–7:

    H is 1001000, X is 1011000, : is 0111010, = is 0111101, % is 0100101

CHAPTER 3

1. **(a)** Input HIGH, output LOW   **(b)** Input LOW, output HIGH

   (c) Input 1, output 0   **(d)** Input 0, output 1

2. AND gate with inputs $DCBA$: $X = 1$ when $DCBA = 1111$. $X = 0$ for all other combinations of $DCBA$.

3. OR gate with inputs $DCBA$: $X = 0$ when $DCBA = 0000$. $X = 1$ for all other combinations of $DCBA$.

4. The normally HIGH output goes LOW at $t = 0.8$ ms and back HIGH at $t = 1$ ms, thus creating a 0.2-ms-wide negative-going pulse.

5. The normally HIGH output goes LOW at $t = 0$ ms and back HIGH at $t = 3$ ms, thus creating a 3-ms-wide negative-going pulse.

6. t_{PHL}

7. $P_{\text{DISS}} = V_{CC}I_{CC} = (5 \text{ V})(2 \text{ mA}) = 10 \text{ mW}$

8. When the output of the driving gate is HIGH, the level can fluctuate as much as 0.2 V without appearing as a LOW to the load gate.

9. The gate can reliably drive up to 20 inputs to gates in the same family.

10. The 74S is faster than the 74L. L = Low power; S = Schottky.

11. CMOS

12. **(a)** 7400: Quad two-input NAND   **(b)** 7404: Hex inverter

    (c) 7411: Triple three-input AND   **(d)** 7420: Dual four-input NAND

    (e) 7432: Quad two-input OR   **(f)** 7427: Triple three-input NOR

CHAPTER 4

1. **(a)** ANDed   **(b)** ORed   **(c)** Complemented (inverted)

2. **(a)** Three-input AND   **(b)** Three-input OR

   (c) Three-input NAND   **(d)** Three-input NOR

3. (a) $A + B = B + A$, commutative
   (b) $AB = BA$, commutative
   (c) $CD = DC$, commutative
   (d) $(A + B) + C = A + (B + C)$, associative
   (e) $A(B + C + D) = AB + AC + AD$, distributive
   (f) $A(BCD) = (ABC)D$, associative
4. (a) $B + 1 = 1$ (b) $0 + B = B$ (c) $B \cdot 1 = B$
   (d) $B \cdot B = B$ (e) $C + C = C$ (f) $D \cdot D = D$
   (g) $\overline{\overline{C}} = C$ (h) $B + BC = B(1 + C)$
                                  $= B \cdot 1 = B$
5. DeMorgan's theorem
6. (a) $AB + CD$: Two 2-input AND gates and one 2-input OR
   (b) $A + B + CD$: One 2-input AND and one 3-input OR
   (c) $A(B + C + DE)$: Two 2-input AND gates and one 3-input OR
7. (a) Sum-of-products   (b) Product-of-sums
8. (a) $ABC + A\overline{BC} = A(BC + \overline{BC}) = A \cdot 1 = A$. Correct.
   (b) $A + BC + \overline{A}C = A + \overline{A}C + BC = A + C + BC = A + C(1 + B) = A + C$. Not
       correct; the answer is $A + C$.
   (c) $A(\overline{A}BC + ABCD) = A(BC(\overline{A} + AD)) = A(BC(\overline{A} + D)) = ABCD$. Correct.

CHAPTER 5

1. $X = ABCD + EF$
2. (a) Two 3-input AND gates with inputs A, $\overline{B}$, C and A, B, $\overline{C}$, respectively. The
       outputs of these two AND gates go into a 2-input NOR gate.
   (b) Exclusive-OR
   (c) Three 4-input AND gates with inputs A, $\overline{B}$, $\overline{C}$, $\overline{D}$ and A, B, C, $\overline{D}$ and $\overline{A}$, B, $\overline{C}$, D,
       respectively. The outputs of these AND gates go into a 3-input OR gate.
3. $X = 1$ when $ABC = 010$, 101, and 111.
4. Three 3-input AND gates with inputs $\overline{A}$, B, $\overline{C}$ and A, $\overline{B}$, C and A, B, C, respectively.
   The outputs of these AND gates go into a 3-input OR.
5. The maximum path has four elements. 4(8 ns) = 32 ns.
6. $X = (A + B + C)(D + E)\overline{F}$. See Figure S–1.

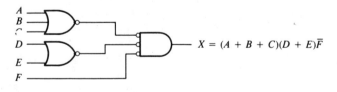

$X = (A + B + C)(D + E)\overline{F}$

FIGURE S–1

7. $X = ABC + DE + \overline{F}$. See Figure S–2.

FIGURE S–2

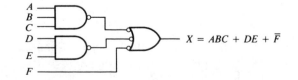

$X = ABC + DE + \overline{F}$

8. $X = AB + \overline{A}\,\overline{B}$. See Figure S–3.

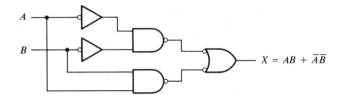

FIGURE S–3

CHAPTER 6

1. A full-adder has a carry input, but a half-adder does not.

2. (a) $\Sigma = 1$, $CO = 0$ (b) $\Sigma = 0$, $CO = 1$
    (c) $\Sigma = 1$, $CO = 0$ (d) $\Sigma = 0$, $CO = 0$

3. (a) $\Sigma = 1$, $CO = 0$ (b) $\Sigma = 0$, $CO = 1$
    (c) $\Sigma = 0$, $CO = 1$ (d) $\Sigma = 1$, $CO = 1$

4. A_1A_0 01 $\Sigma_0 = 0$, $\Sigma_1 = 0$, $CO = 1$
    <u>$+ B_1B_0$</u>   <u>$+ 11$</u>
    $\overline{CO\Sigma_1\Sigma_0}$ 100

5. Connect two more 7482s to Figure 6–11 with CO to CI of each.

6. Connect two more 7483As to Figure 6–12 with CO to CI of each.

7. (a) $CG = PQ = 1 \cdot 0 = 0$; $CP = P + Q = 1 + 0 = 1$
    (b) $CG = PQ = 0 \cdot 0 = 0$; $CP = P + Q = 0 + 0 = 0$
    (c) $CG = PQ = 1 \cdot 1 = 1$; $CP = P + Q = 1 + 1 = 1$

8. (a) $P > Q = 1$, $P < Q = 0$, $P = Q = 0$
    (b) $< $ LOW, $=$ HIGH, $>$ LOW

9. Add one more 7485 to Figure 6–25 with $P < Q$ output to $<$ input, $P = Q$ output to $=$ input, and $P > Q$ output to $>$ input.

10. (a) 12 output LOW   (b) 8 output LOW   (c) 2 output LOW

11. (a) 5 output LOW   (b) 9 output LOW
    (c) No output LOW, invalid input

12. (a) b, c (b) a, b, c (c) a, b, c, d, g

13. The complement of BCD 7.

14. (a) 10010011   (b) 01100111

                         00000001   1   00000001   1
                         00000010   2   00000010   2
                         00001010 10   00000100 4
                         <u>01010000</u> 80   00010100 20
                       $01011101_2 \rightarrow 93_{10}$ <u>00101000</u> 40
                                                     $01000011_2 \rightarrow 67_{10}$

15. Two exclusive-OR gates connected as in Figure 6–53.

16. Eight exclusive-OR gates connected as in Figure 6–56.

17. (a) $1Y = 1$, $2Y = 0$, $3Y = 1$, $4Y = 0$
    (b) $1Y = 0$, $2Y = 0$, $3Y = 0$, $4Y = 1$

18. Output is a square wave (50% duty cycle) beginning with a LOW.

19. See Figure S–4.

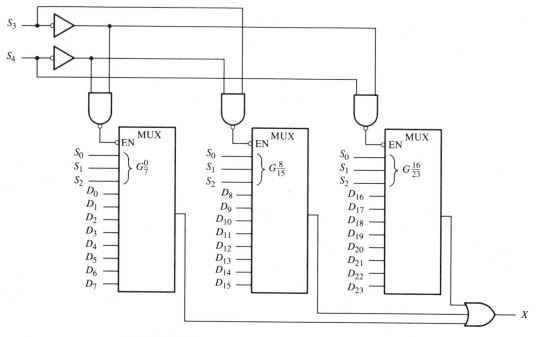

FIGURE S–4

20. See Figure S–5.

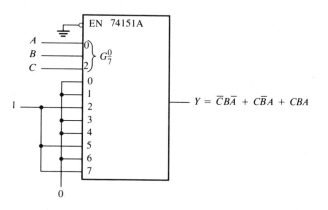

$$Y = \overline{C}B\overline{A} + C\overline{B}A + CBA$$

FIGURE S–5

21. Pin 13, D_{11}
22. **(a)** $\overline{1}11010$ **(b)** $\overline{0}1001$ **(c)** $\overline{1}0111101$
23. **(a)** $\overline{0}11010$ **(b)** $\overline{1}1001$ **(c)** $\overline{0}0111101$

CHAPTER 7

1. SET
2. It forces both Q and $\overline{Q}$ to the HIGH state, and when inputs are released, the resulting state of the latch is unpredictable.
3. Q goes HIGH when EN goes HIGH; Q goes LOW when D goes LOW.
4. S-R, D, J-K
5. *Positive edge-triggered* is sensitive to data inputs only on positive-going edge of clock.
   Negative edge-triggered is sensitive to data inputs only on negative-going edge of clock.
6. J-K has no invalid state.
7. Q remains 0. Q remains 0.
8. Data go into master on leading edge of clock, then into slave and to output on trailing edge of clock.
9. Data cannot be changed while clock pulse is in active state.
10. The data lock-out flip-flop has a dynamic indicator on the clock input as well as the postponed output symbol.
11. t_{PHL} (clock to Q) is the delay time from triggering edge of clock to HIGH-to-LOW transition of Q.
    t_{PLH} (clock to Q) is the delay time from triggering edge of clock to LOW-to-HIGH transition of Q.
    t_{PHL} (CLR to Q) is the delay time from clear input to Q (always HIGH-to-LOW).
    t_{PLH} (PRE to Q) is the delay time from preset input to Q (always LOW-to-HIGH).
12. $T = t_{H(min)} + t_{L(min)} = 30 \text{ ns} + 37 \text{ ns} = 67 \text{ ns}$
    $f_{max} = 1/T = 1/67 \text{ ns} = 14.9 \text{ MHz}$
13. Counting, frequency division, data storage, and data transfer.
14. $T = 1/10 \text{ kHz} = 0.1 \text{ ms} = 100 \text{ } \mu s$
    A square wave (50% duty cycle) with a frequency of 10 kHz
15. As timing (clock) sources.

CHAPTER 8

1. Each flip-flop in a synchronous counter is clocked simultaneously. Each flip-flop in an asynchronous counter is ripple clocked, resulting in greater delays and lower operating frequency.
2. Lower frequency of operation
3. **(a)** $2^2 = 4$ **(b)** $2^4 = 16$ **(c)** $2^5 = 32$
   (d) $2^6 = 64$ **(e)** $2^7 = 128$ **(f)** $2^8 = 256$
4. **(a)** 2   **(b)** 3   **(c)** 4   **(d)** 4
   (e) 5   **(f)** 6   **(g)** 6   **(h)** 8
5. $4(12 \text{ ns}) = 48 \text{ ns}$
6. 12 ns

7. A BCD decade counter has ten states (0 through 9). The Q output of the most significant stage has a frequency which is one-tenth of the clock frequency. See Figure S–6.

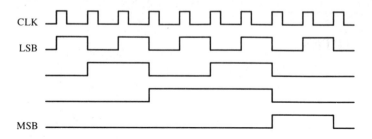

FIGURE S–6

8. 1010, 1011, 1100, 1101, 1110, 1111
9. The counter is parallel loaded to $DCBA = 0101$.
10. 0001; 1111
11. $J = 0, K = X$
12. **(a)** One flip-flop
    (b) Two flip-flops
    (c) Modulus-5 counter
    (d) Decade counter
    (e) Decade counter and flip-flop
    (f) Decade counter and two flip-flops
    (g) Decade counter and four flip-flops
    (h) Two modulus-5 counters and one decade counter
    (i) Three decade counters
    (j) Four decade counters
13. See Figure S–7.

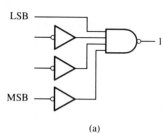

(a)

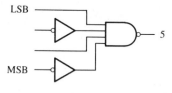

(b)

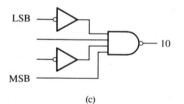

(c)

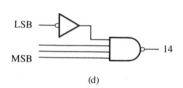

(d)

FIGURE S–7

14. State 12 is decoded by the NAND gate, presetting the counter to 1 by loading the parallel inputs into the 74LS160A and by causing the flip-flop to RESET on the next clock pulse.
15. $T = 1/f = 1/10$ kHz $= 0.1$ ms $= 100$ μs
    Time for two bytes: $16(100$ μs$) = 1600$ μs

CHAPTER 9

1. A flip-flop stores a 1 when SET and a 0 when RESET.
2. 16
3. 8
4. Connect the Q output of each flip-flop to the S input of the following one and the $\overline{Q}$ output to the R input. The data input goes to S of the first flip-flop, and the complement of the data input goes to the R input.
5. Take output from Q_C.
6. 01111001 after two clock pulses, 01011110 after four clock pulses, 10110101 after eight clock pulses.
7. 00110000
8. Five
9. The type of sequence and the way it is generated.
10. 0000, 1000, 1100, 1110, 1111, 0111, 0011, 0001
11. 10010000, 01001000, 00100100, 00010010, 00001001, 10000100, 01000010, 00100001, 10010000
12. $T = 3$ μs, $f = 1/3$ μs $= 0.333$ MHz $= 333$ kHz
13. 1010, 0101, 1010, . . .
14. Serial-to-parallel and parallel-to-serial conversion.
15. 74123
16. To sequentially apply a LOW to each row line to detect a switch closure.

CHAPTER 10

1. $512 \times 8 = 4096$ bits
2. Read
3. Mask ROM, PROM, EPROM; mask ROM, PROM
4. Static, dynamic; static
5. Four
6. To keep track of the address in memory
7. To temporarily store the address code while it is sent out on the address bus
8. 65,536 bits
9. $1024 \times 4 = 4096$ bits
10. $2^9 = 512$; 9 bits
11. Enables the memory to operate
12. Provides for a lower power consumption when memory is not in operation
13. 100 ns + 450 ns $= 550$ ns
14. $(300$ ns$)(256) = 76,800$ ns $= 76.8$ μs
15. The address code is multiplexed in two 7-bit groups (row and column).
16. Generates the sequence to address all rows for purposes of refreshing
17. Two; word length expansion
18. Minor loops
19. Error detection
20. PLAs consist of AND and OR gate arrays. ROMs consist of storage cells.

CHAPTER 11

1. Interfacing is the method of interconnecting two circuits or systems so that they function properly together.
2. Fan-out
3. TTL-to-CMOS; pull-up resistor to make HIGH levels compatible.
4. Open-collector
5. See Figures 11–4 and 11–5.
6. Open-collector
7. Two-way bus
8. Talker, listener, and controller
9. 15
10. Analog—continuous values; digital—discrete values
11. $100/(2^5 - 1) = 100/31 = 3.23\%$
12. No reverse steps
13. Tracking is faster than stairstep-ramp.
14. BW = 3000 Hz − 300 Hz = 2700 Hz
15. Because digital signals contain high frequencies outside the telephone bandwidth.
16. Frequency shift keying
17. Couples transmitted and received signals by sound waves
18. To interface modems (DCEs) with DTEs

CHAPTER 12

1. Sign bits: 1 for negative, 0 for positive
2. With sign and complement of magnitude
3. Addition of two positive or two negative numbers
4. Incorrect sign and magnitude
5. Add $(A + B)$, subtract $(A - B, B - A)$, exclusive-OR $(A \oplus B)$, AND (AB) OR $(A + B)$, clear, preset
6. AND mode
7. Accumulator
8. Operands
9. Add 0110 to invalid sum.
10. Data bus provides for data transfer into and out of system and within system. Address bus provides for sending address to memory or I/O.
11. (a) Program counter: Keeps track of memory address.
    (b) Stack pointer: Used with memory stack.
    (c) Index register: Used with indexed addressing.
    (d) Condition code register: Indicates status of accumulator.
    (e) Instruction register: Temporarily holds instruction code.
12. It takes more bits to address memory than there are in a data word.

Answers to Odd-Numbered Problems

CHAPTER 1

1–1   **(a)**   11010001   **(b)**   000101010

1–3   **(a)**   0.6 μs   **(b)**   0.45 μs
         (c)   2.7 μs   **(d)**   10 V

1–5   250 Hz

1–7   50%

1–9   AND gate

1–11   **(a)**   Adder   **(b)**   Multiplier
         (c)   Multiplexer   **(d)**   Comparator

1–13   LSI

1–15   Dual-in-line package (DIP), flat pack, chip carrier

1–17   Printers, keyboards, video terminals

1–19   Pulse train

CHAPTER 2

2–1   **(a)**   6   **(b)**   4   **(c)**   1   **(d)**   87
         (e)   82   **(f)**   74   **(g)**   50   **(h)**   13
         (i)   872   **(j)**   618   **(k)**   309   **(l)**   8645

2–3   **(a)** 3   **(b)** 81   **(c)** 64
       (d) 48   **(e)** 16   **(f)** 10

2–5   **(a)** 3   **(b)** 4   **(c)** 7   **(d)** 8
       (e) 9   **(f)** 12   **(g)** 11   **(h)** 15

2–7   **(a)** 51.75   **(b)** 42.25   **(c)** 65.875
       (d) 120.625   **(e)** 92.65625   **(f)** 113.0625
       (g) 90.625   **(h)** 127.96875

2–9   **(a)** 5 bits   **(b)** 6 bits   **(c)** 6 bits   **(d)** 7 bits
       (e) 7 bits   **(f)** 7 bits   **(g)** 8 bits   **(h)** 8 bits
       (i) 9 bits

2–11   **(a)** 100_2 **(b)** 100_2
       (c) 1000_2 **(d)** 1101_2
       (e) 1110_2 **(f)** 11000_2

2–13   **(a)** 1001_2 **(b)** 1000_2
       (c) 100011_2 **(d)** 110110_2
       (e) 10101001_2 **(f)** 10110110_2

2–15   **(a)** 010   **(b)** 001   **(c)** 0101
       (d) 00101000   **(e)** 0001010   **(f)** 11110

2–17   **(a)** 10   **(b)** 001   **(c)** 0111
       (d) 0011   **(e)** 00100   **(f)** 01101

2–19   **(a)** 1010_2 **(b)** 10001_2 **(c)** 11000_2
       (d) 110000_2 **(e)** 111101_2 **(f)** 1011101_2
       (g) 1111101_2 **(h)** 10111010_2 **(i)** 100101010_2

2–21   **(a)** 10   **(b)** 23   **(c)** 46   **(d)** 52
       (e) 67   **(f)** 367   **(g)** 115   **(h)** 532
       (i) 4085

2–23   **(a)** 001011_2 **(b)** 101111_2
       (c) 001000001_2 **(d)** 011010001_2
       (e) 101100000_2 **(f)** 100110101011_2
       (g) 001011010111001_2 **(h)** 100101110000000_2
       (i) 001000000010001011_2 **(j)** 001000011.100101_2

2–25   **(a)** 00111000_2 **(b)** 01011001_2
       (c) 101000010100_2 **(d)** 010111001000_2
       (e) 0100000100000000_2 **(f)** 1111101100010111_2
       (g) 10001010.1001_2

2–27   **(a)** 35   **(b)** 146   **(c)** 26   **(d)** 141
       (e) 243   **(f)** 235   **(g)** 1474   **(h)** 1792

2–29   **(a)** 60_{16} **(b)** $10B_{16}$ **(c)** $1BA_{16}$

2–31   **(a)** 00010000   **(b)** 00010011
       (c) 00011000   **(d)** 00100001
       (e) 00100101   **(f)** 00110110
       (g) 01000100   **(h)** 01010111
       (i) 01101001   **(j)** 10011000
       (k) 000100100101   **(l)** 000101010110

2–33   **(a)** 000100000100   **(b)** 000100101000
       (c) 000100110010   **(d)** 000101010000
       (e) 000110000110   **(f)** 001000010000
       (g) 001101011001   **(h)** 010101000111
       (i) 0001000001010001   **(j)** 0010010101100011

2–35   **(a)**   80   **(b)**   237   **(c)**   346   **(d)**   421
         (e)   754   **(f)**   800   **(g)**   978   **(h)**   1683
         (i)   9018   **(j)**   6667

2–37   **(a)**   00010100   **(b)**   00010010
         (c)   00010111   **(d)**   00010110
         (e)   01010010   **(f)**   000100001001
         (g)   000110010101   **(h)**   0001001001101001

2–39   The Gray code makes only one bit change at a time when going from one number in the sequence to the next.

2–41   **(a)**   1100   **(b)**   00011   **(c)**   10000011110

2–43   **(a)**   0   **(b)**   6   **(c)**   4
         (d)   13   **(e)**   49   **(f)**   52

CHAPTER 3

3–1   Figure P3–1

FIGURE P3–1

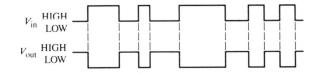

3–3   Figure P3–3

FIGURE P3–3

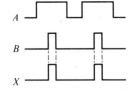

3–5   Figure P3–5

FIGURE P3–5

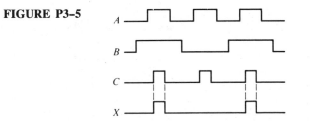

3–7   Figure P3–7

FIGURE P3–7

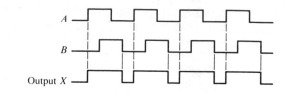

3–9   Figure P3–9

FIGURE P3–9

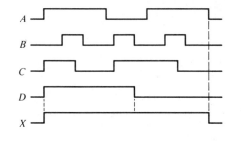

3–11   Figure P3–11

FIGURE P3–11

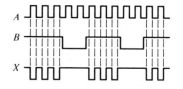

3–13   Figure P3–13

FIGURE P3–13

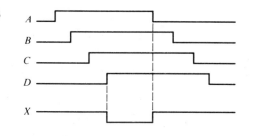

3–15   Figure P3–15

FIGURE P3–15

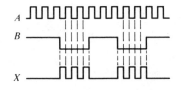

3–17 Figure P3–17

FIGURE P3–17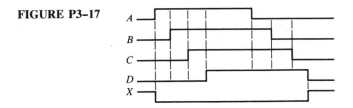

3–19 4.3 ns, 10 ns

3–21 20 mW

3–23 40 mW

3–25 0.3 V

3–27 400 μA

3–29 CMOS

3–31 11 ns

3–33 NOR gate

3–35 **(a)** Defective output (stuck LOW).

     (b) Input 2 is defective (open).

CHAPTER 4

4–1 **(a)** $X = AB$

A	B	X
0	0	0
0	1	0
1	0	0
1	1	1

(b) $X = ABC$

A	B	C	X
0	0	0	0
0	0	1	0
0	1	0	0
0	1	1	0
1	0	0	0
1	0	1	0
1	1	0	0
1	1	1	1

(c) $X = A + B$

A	B	X
0	0	0
0	1	1
1	0	1
1	1	1

(d) $X = A + B + C$

A	B	C	X
0	0	0	0
0	0	1	1
0	1	0	1
0	1	1	1
1	0	0	1
1	0	1	1
1	1	0	1
1	1	1	1

(e) $X = AB + C$

A	B	C	X
0	0	0	0
0	0	1	1
0	1	0	0
0	1	1	1
1	0	0	0
1	0	1	1
1	1	0	1
1	1	1	1

(f) $X = \overline{A} + B$

A	B	X
0	0	1
0	1	1
1	0	0
1	1	1

(g) $X = A\overline{B}\,\overline{C}$

A	B	C	X
0	0	0	0
0	0	1	0
0	1	0	0
0	1	1	0
1	0	0	1
1	0	1	0
1	1	0	0
1	1	1	0

(h) $X = AB + \overline{A}C$

A	B	C	AB	$\overline{A}C$	X
0	0	0	0	0	0
0	0	1	0	1	1
0	1	0	0	0	0
0	1	1	0	1	1
1	0	0	0	0	0
1	0	1	0	0	0
1	1	0	1	0	1
1	1	1	1	0	1

(i) $X = A(B + C)$

A	B	C	B + C	X
0	0	0	0	0
0	0	1	1	0
0	1	0	1	0
0	1	1	1	0
1	0	0	0	0
1	0	1	1	1
1	1	0	1	1
1	1	1	1	1

(j) $X = \overline{A}(\overline{B} + \overline{C})$

A	B	C	$\overline{A}$	$\overline{B} + \overline{C}$	X
0	0	0	1	1	1
0	0	1	1	1	1
0	1	0	1	1	1
0	1	1	1	0	0
1	0	0	0	1	0
1	0	1	0	1	0
1	1	0	0	1	0
1	1	1	0	0	0

4–3 (a) $X = AB$ (b) $X = \overline{A}$ (c) $X = A + B$ (d) $X = A + B + C$

4–5 (a) $X = A + B$

A	B	X
0	0	0
0	1	1
1	0	1
1	1	1

(b) $X = AB$

A	B	X
0	0	0
0	1	0
1	1	0
1	0	1

(c) $X = AB + BC$

A	B	C	X
0	0	0	0
0	0	1	0
0	1	0	0
0	1	1	1
1	0	0	0
1	0	1	0
1	1	0	1
1	1	1	1

(d) $X = (A + B)C$

A	B	C	X
0	0	0	0
0	0	1	0
0	1	0	0
0	1	1	1
1	0	0	0
1	0	1	1
1	1	0	0
1	1	1	1

(e) $X = (A + B)(\overline{B} + C)$

A	B	C	A + B	$\overline{B}$ + C	X
0	0	0	0	1	0
0	0	1	0	1	0
0	1	0	1	0	0
0	1	1	1	1	1
1	0	0	1	1	1
1	0	1	1	1	1
1	1	0	1	0	0
1	1	1	1	1	1

4–7 (a) $\overline{A} + B + \overline{C}D$ (b) $\overline{A} + \overline{B} + (\overline{C} + \overline{D})(\overline{E} + \overline{F})$

(c) $\overline{AB}\overline{C}D + \overline{A} + \overline{B} + \overline{C} + D$ (d) $\overline{A} + B + C + D + \overline{AB}\overline{C}D$

(e) $AB + (\overline{C} + \overline{D})(E + \overline{F}) + ABCD$

4–9 (a) $AC + AD + BC + BD$ (b) $AD + \overline{B}CD$ (c) $ABC + ACD$

4–11 (a) $X = AB + \overline{A}\overline{B}$, sum-of-products

(b) $X = AB + AB\overline{C} + ABC\overline{D}$, sum-of-products

(c) $X = (\overline{A} + B)(A + C)$, product-of-sums

(d) $X = (\overline{A} + \overline{B})(A + B + C)(\overline{A} + B + C + \overline{D})$, product-of-sums

4–13 $X = ABC + \overline{A}BC + A\overline{B}C + AB\overline{C}$

4–15 Figure P4–15

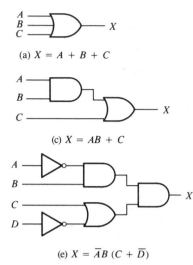

(a) $X = A + B + C$

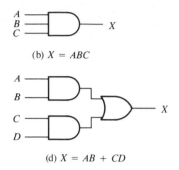

(b) $X = ABC$

(c) $X = AB + C$

(d) $X = AB + CD$

(e) $X = \overline{A}B\,(C + \overline{D})$

FIGURE P4–15

4–17 (a) A (b) AB (c) C (d) A (e) $\overline{A}C + \overline{B}C$
4–19 (a) $BD + BE + \overline{D}F$ (b) $\overline{A}\,\overline{B}C + \overline{A}\,\overline{B}D$
  (c) B (d) $AB + CD$ (e) ABC
4–21 (a) $X = \overline{A}\,\overline{B} + \overline{B}C$ (b) $X = AC$ (c) $X = B$
  (d) $X = \overline{C}$ (e) $X = \overline{B}C + A$, no simplification
4–23 (a) No simplification (b) $X = \overline{A}\,\overline{B}\overline{C} + ABC$ (c) $X = B\overline{C} + A\overline{C}D$
  (d) $X = \overline{B}C$ (e) $X = \overline{B} + \overline{D}$

CHAPTER 5

5–1 (a) $X = AB$

A	B	X
0	0	0
0	1	0
1	0	0
1	1	1

(b) $X = B$

A	B	X
0	0	0
0	1	1
1	0	0
1	1	1

(c) $X = \overline{A} + B$

A	B	X
0	0	1
0	1	1
1	0	0
1	1	1

(d) $X = A + B$

A	B	X
0	0	0
0	1	1
1	0	1
1	1	1

5-3 $X = A\bar{B} + \bar{A}B$

A	B	X
0	0	0
0	1	1
1	0	1
1	1	0

5-5 **(a)** $X = (AB + C)D + E$ **(b)** $X = \overline{\overline{(\bar{A} + B)\bar{B}C}} + D$
(c) $X = (AB + \bar{C})D + \bar{E}$ **(d)** $X = \overline{(AB + CD)(EF + GH)}$

5-7 $X = 0$ for all combinations of ABC

5-9 Figure P5-9

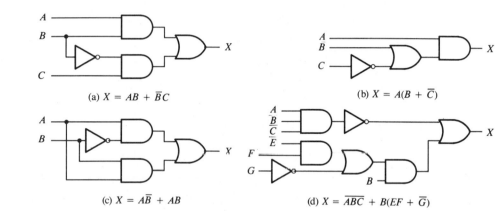

(a) $X = AB + \bar{B}C$ (b) $X = A(B + \bar{C})$

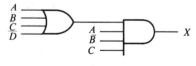

(c) $X = A\bar{B} + AB$ (d) $X = \overline{ABC} + B(EF + \bar{G})$

(e) $X = A[BC(A + B + C + D)]$

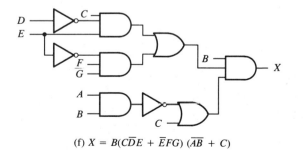

(f) $X = B(C\bar{D}E + \bar{E}FG)(\overline{AB} + C)$

FIGURE P5-9

5-11   Figure P5-11

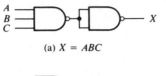

(a) $X = ABC$

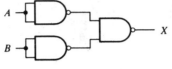

(b) $X = \overline{ABC}$

(c) $X = A + B$

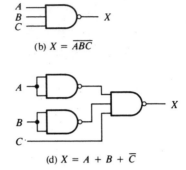

(d) $X = A + B + \overline{C}$

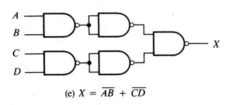

(e) $X = \overline{AB} + \overline{CD}$

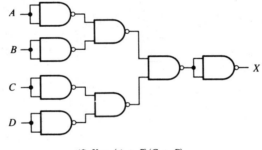

(f) $X = (A + B)(C + D)$

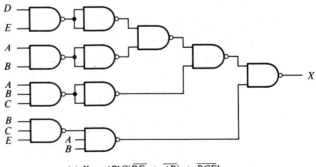

(g) $X = AB[C(\overline{DE} + \overline{AB}) + \overline{BCE}]$

FIGURE P5-11

5-13 X = lamp ON, A = front door switch ON, B = back door switch ON; Figure P5–13

FIGURE P5–13

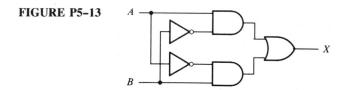

5-15 $X = AB$

5-17 **(a)** No simplification **(b)** No simplification
(c) $X = A$ **(d)** $X = \overline{A} + \overline{C} + \overline{B} + EF + \overline{G}$
(e) $X = ABC$ **(f)** $X = BC\overline{D}E + \overline{A}B\overline{E}FG + BC\overline{E}FG$

5-19 **(a)** No simplification **(b)** No simplification
(c) No simplification **(d)** No simplification
(e) $X = \overline{A} + \overline{B} + \overline{C} + \overline{D}$ **(f)** No simplification
(g) $X = AB\overline{D} + AB\overline{C} + AB\overline{E}$

5-21 Figure P5–21

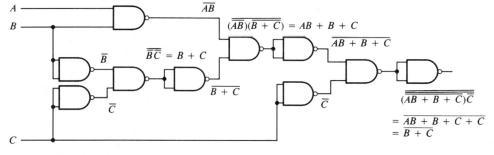

FIGURE P5–21

5-23 Figure P5–23

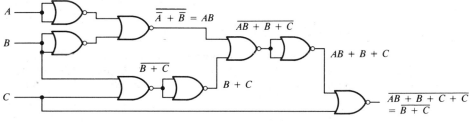

FIGURE P5–23

5-25 $X = A + \overline{B}$; Figure P5–25

FIGURE P5–25

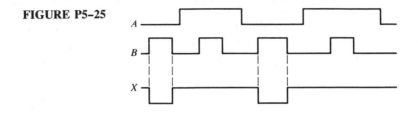

5-27 $X = A\overline{B}\,\overline{C}$; Figure P5–27

FIGURE P5–27

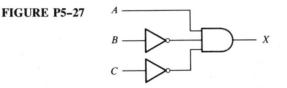

5-29   The output pulse width is greater than the specified minimum.

5-31 $X = ABC + D\overline{E}$. Since X is the same as the G_3 output, either G_1 or G_2 has failed with its output shorted (stuck LOW).

5-33   Gate G_5 has a shorted input.

CHAPTER 6

6-1   **(a)** G_1 output $= 0$, G_2 output $= 1$, G_3 output $= 0$, G_4 output $= 1$, G_5 output $= 1$
     (b) G_1 output $= 1$, G_2 output $= 0$, G_3 output $= 1$, G_4 output $= 0$, G_5 output $= 1$
     (c) G_1 output $= 1$, G_2 output $= 1$, G_3 output $= 0$, G_4 output $= 0$, G_5 output $= 0$

6-3   No simplification for sum logic; $CO = PQ + QCI + PCI$

6-5   11100

6-7 $\Sigma_0 = 1001110001000110$
$\Sigma_1 = 0101000011110011$
$\Sigma_2 = 1111100001100010$
$\Sigma_3 = 0100001011101110$
$\Sigma_4 = 1010110010010000$

6-9   225 ns

6-11 $P = Q$ is HIGH when $P_0 = Q_0$ and $P_1 = Q_1$; Figure P6-11.

FIGURE P6-11

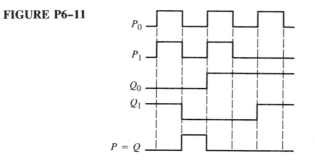

6-13   **(a)** G_1 output = 1   **(b)** G_1 output = 1   **(c)** G_1 output = 1
G_2 output = 0 G_2 output = 1 G_2 output = 1
G_3 output = 1 G_3 output = 0 G_3 output = 1
G_4 output = 0 G_4 output = 0 G_4 output = 1
G_5 output = 0 G_5 output = 0 G_5 output = 1
G_6 output = 0 G_6 output = 0 G_6 output = 0
G_7 output = 1 G_7 output = 0 G_7 output = 0
G_8 output = 0 G_8 output = 0 G_8 output = 0
G_9 output = 0 G_9 output = 0 G_9 output = 0
G_{10} output = 1 G_{10} output = 0 G_{10} output = 0
G_{11} output = 0 G_{11} output = 0 G_{11} output = 0
G_{12} output = 0 G_{12} output = 0 G_{12} output = 0
G_{13} output = 0 G_{13} output = 1 G_{13} output = 0
G_{14} output = 0 G_{14} output = 0 G_{14} output = 0
G_{15} output = 0 G_{15} output = 1 G_{15} output = 0
    $A > B$ $A < B$ $A = B$

6–15 Figure P6–15

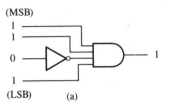

(a)

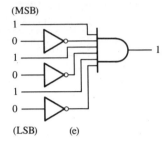

(e)

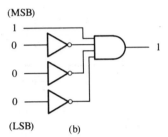

(b)

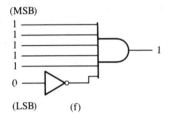

(f)

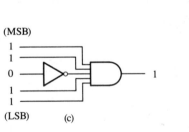

(c)

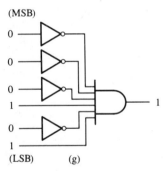

(g)

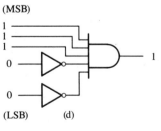

(d)

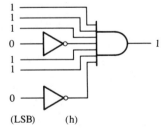

(h)

FIGURE P6–15

6–17 $X = DC\overline{B}\,\overline{A} + \overline{D}\,\overline{C}\,\overline{B}A + D\overline{C}B$

6–19   Figure P6–19

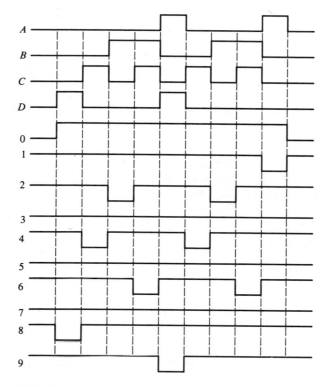

FIGURE P6–19

6–21 $DCBA$ = 1011, invalid BCD

6–23   **(a)**   1111111111   **(b)**   1000010000
         (c)   0000001001   **(d)**   1000000000
         Figure P6–23

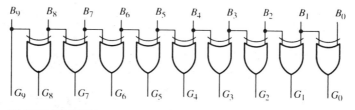

FIGURE P6–23

6–25   **(a)**   0010 → 0010$_2$ **(b)**   1000 → 1000$_2$
         (c)   00010011 → 1101$_2$ **(d)**   00100110 → 11010$_2$
         (e)   00110011 → 100001$_2$

6–27 Figure P6–27

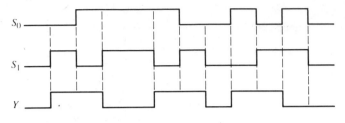

FIGURE P6–27

6–29 Figure P6–29 (p. B–17)

6–31 $Y = \overline{D}\,\overline{C}\,\overline{B}\,\overline{A} + \overline{D}\,\overline{C}B\overline{A} + \overline{D}C\overline{B}\,\overline{A} + \overline{D}CB\overline{A} + D\overline{C}\,\overline{B}\,\overline{A}$
$+ D\overline{C}B\overline{A} + D\overline{C}\,\overline{B}A + DC\overline{B}A + DCBA$

Figure P6–31

FIGURE P6–31

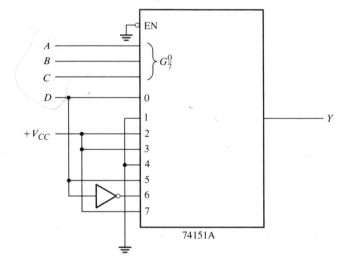

6–33 Figure P6–33

FIGURE P6–33

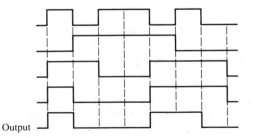

Output

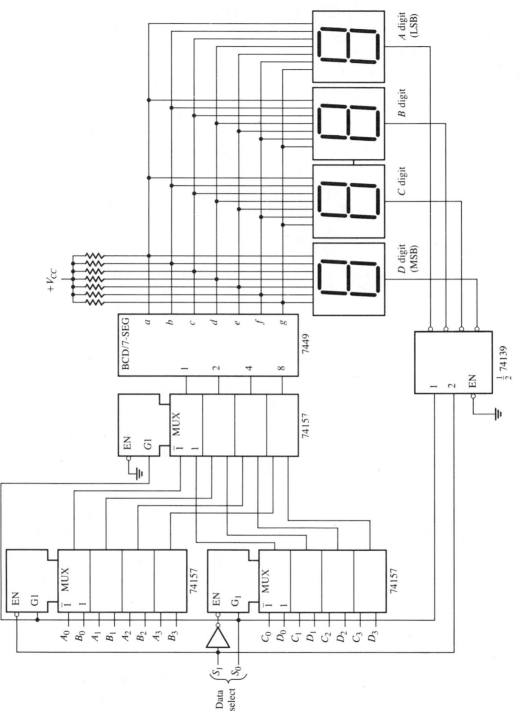

FIGURE P6–29

CHAPTER 7

7–1   Figure P7–1

FIGURE P7–1

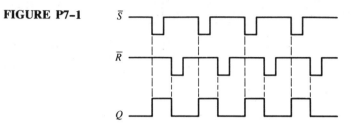

7–3   Figure P7–3

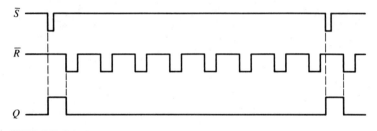

FIGURE P7–3

7–5   Figure P7–5

FIGURE P7–5

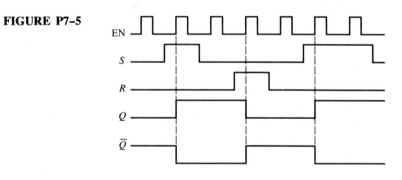

7–7   Figure P7–7

FIGURE P7–7

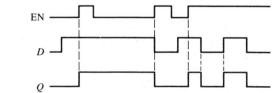

7-9   Figure P7-9

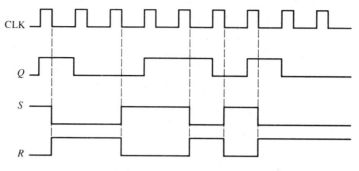

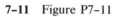

FIGURE P7-9

7-11   Figure P7-11

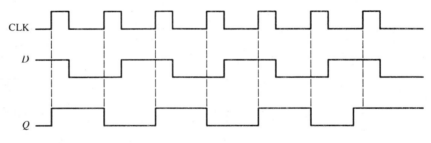

FIGURE P7-11

7-13   Figure P7-13

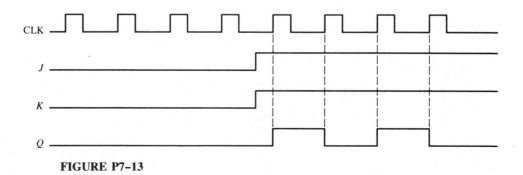

FIGURE P7-13

7–15   Figure P7–15

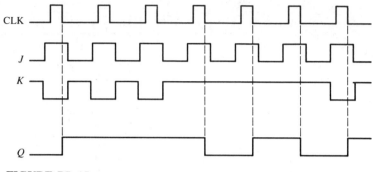

FIGURE P7–15

7–17   Figure P7–17

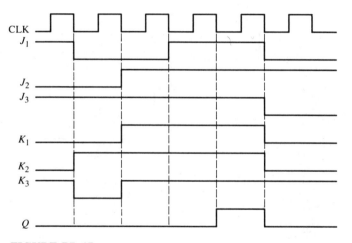

FIGURE P7–17

7–19   Figure P7–19

FIGURE P7–19

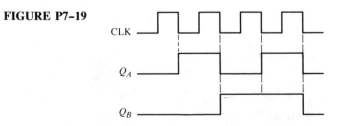

7–21   Figure P7–21

FIGURE P7–21

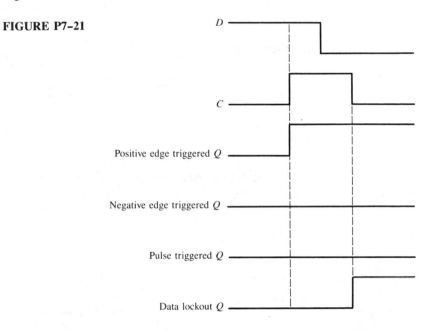

7–23   150 mA, 750 mW
7–25   Divide-by-2; Figure P7–25

FIGURE P7–25

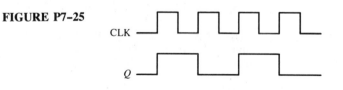

7–27   Figure P7–27

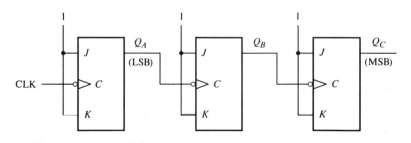

FIGURE P7–27

7–29   1.848 μs
7–31   28.8 kHz

CHAPTER 8

8–1 Figure P8–1

FIGURE P8–1

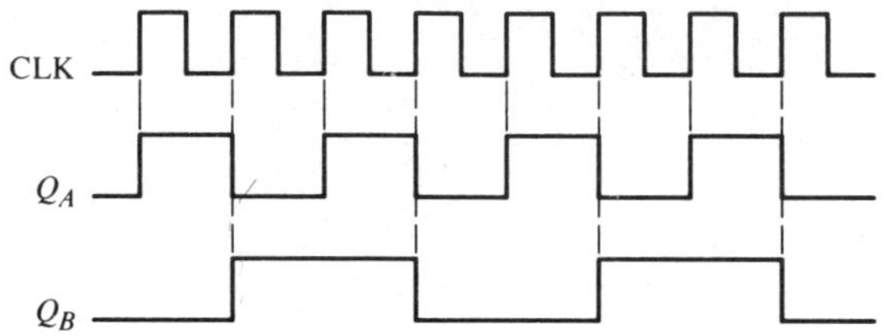

8–3 Worst-case delay occurs when all flip-flops change state from 011 to 100 or from 111 to 000.

8–5 8 ns

8–7 Initially, each flip-flop is RESET.

At CLK_1: $J_A = K_A = 1$ Therefore Q_A goes to a 1.
$\quad\quad\quad\quad$ $J_B = K_B = 0$ Therefore Q_B remains a 0.
$\quad\quad\quad\quad$ $J_C = K_C = 0$ Therefore Q_C remains a 0.
$\quad\quad\quad\quad$ $J_D = K_D = 0$ Therefore Q_D remains a 0.

At CLK_2: $J_A = K_A = 1$ Therefore Q_A goes to a 0.
$\quad\quad\quad\quad$ $J_B = K_B = 1$ Therefore Q_B goes to a 1.
$\quad\quad\quad\quad$ $J_C = K_C = 0$ Therefore Q_C remains a 0.
$\quad\quad\quad\quad$ $J_D = K_D = 0$ Therefore Q_D remains a 0.

At CLK_3: $J_A = K_A = 1$ Therefore Q_A goes to a 1.
$\quad\quad\quad\quad$ $J_B = K_B = 0$ Therefore Q_B remains a 1.
$\quad\quad\quad\quad$ $J_C = K_C = 0$ Therefore Q_C remains a 0.
$\quad\quad\quad\quad$ $J_D = K_D = 0$ Therefore Q_D remains a 0.

A continuation of this procedure for the next seven clock pulses will show that the counter progresses through the BCD sequence.

8–9 Figure P8–9

FIGURE P8–9

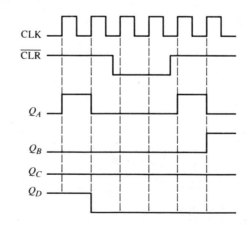

8-11 Figure P8-11

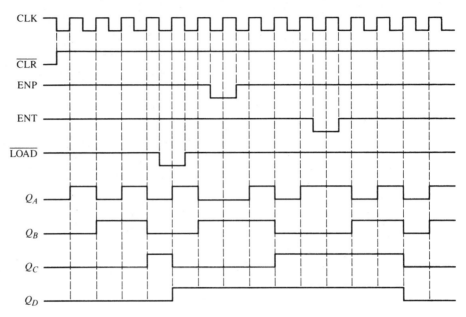

FIGURE P8-11

8-13 Figure P8-13

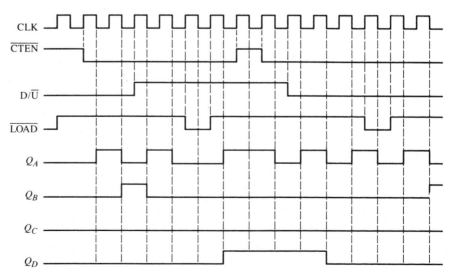

FIGURE P8-13

8-15 0000, 1111, 1110, 1101, 1010, 1101. The counter "locks up" in the 1010 and 1101 states and alternates between them.

8-17 Figure P8-17

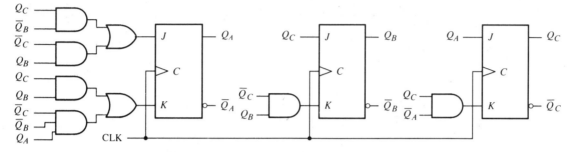

FIGURE P8-17

8-19 Invalid states are 000, 010, 101. The counter will lock up and alternate between 010 and 101 if it gets into any of the invalid states.

8-21 Figure P8-21 (p. B-25)

8-23 Figure P8-23 (p. B-25)

8-25 Figure P8-25

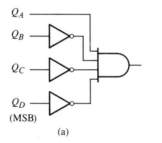

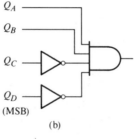

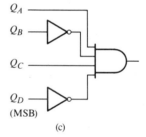

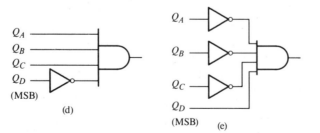

FIGURE P8-25

8-27 CLK_2, output 0; CLK_4, outputs 2, 0; CLK_6, output 4; CLK_8, outputs 6, 4, 0; CLK_{10}, output 8; CLK_{12}, outputs 10, 8; CLK_{14}, output 12; CLK_{16}, outputs 14, 12, 8.

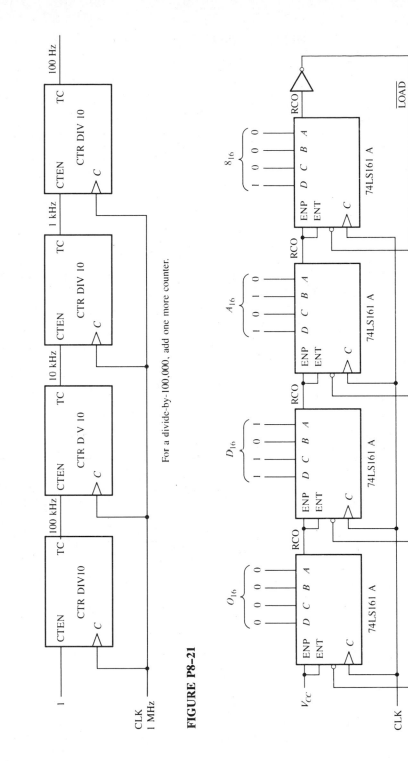

FIGURE P8–21

For a divide-by-100,000, add one more counter.

FIGURE P8–23

8–29   Figure P8–29

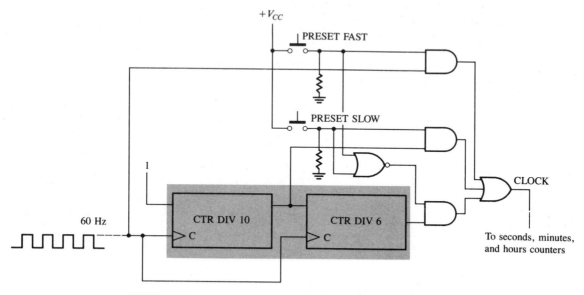

FIGURE P8–29

8–31   Figure P8–31

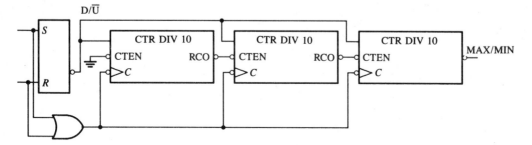

FOR 3000-space counter, add the following:

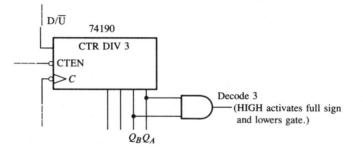

FIGURE P8–31

CHAPTER 9

9–1   Shift registers store binary data.

9–3   Figure P9–3

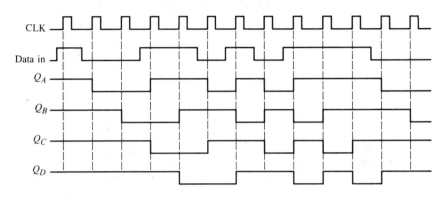

FIGURE P9–3

9–5

Initially	101001111000
CLK_1	010100111100
CLK_2	001010011110
CLK_3	000101001111
CLK_4	000010100111
CLK_5	100001010011
CLK_6	110000101001
CLK_7	111000010100
CLK_8	011100001010
CLK_9	001110000101
CLK_{10}	000111000010
CLK_{11}	100011100001
CLK_{12}	110001110000

9–7   Figure P9–7

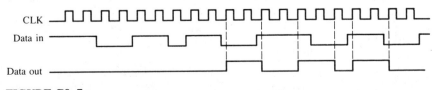

FIGURE P9–7

9–9   Figure P9–9

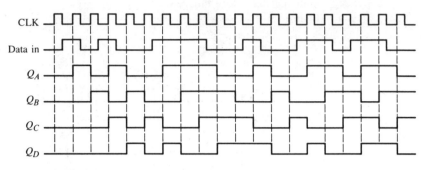

FIGURE P9–9

9–11   Figure P9–11

FIGURE P9–11

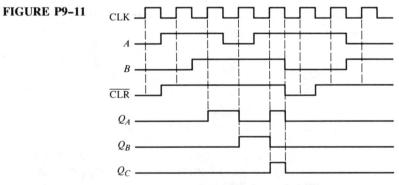

Q_D through Q_H remain LOW.

9–13   Figure P9–13

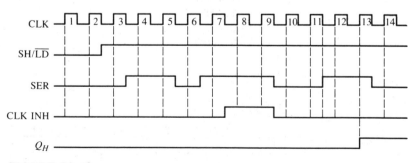

FIGURE P9–13

9–15 Figure P9–15

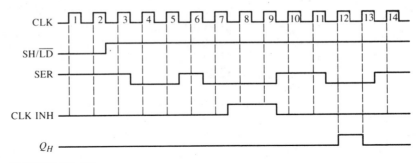

FIGURE P9–15

9–17 Figure P9–17

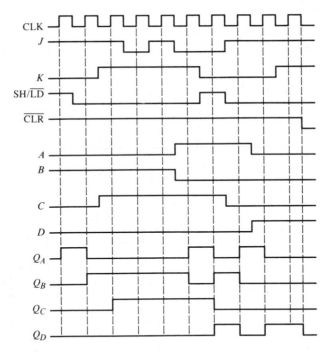

FIGURE P9–17

9–19

Initially (76)	01001100	
CLK_1	10011000	left
CLK_2	01001100	right
CLK_3	00100110	right
CLK_4	00010011	right

CLK_5	00100110	left
CLK_6	01001100	left
CLK_7	00100110	right
CLK_8	01001100	left
CLK_9	00100110	right
CLK_{10}	01001100	left
CLK_{11}	10011000	left

9–21 Figure P9–21

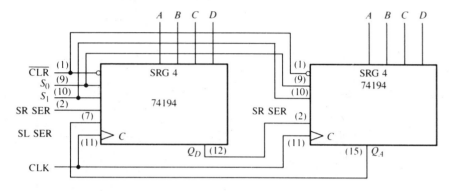

FIGURE P9–21

9–23 **(a)** 3   **(b)** 5   **(c)** 7   **(d)** 8
      (e) 10   **(f)** 12   **(g)** 18

9–25 Figure P9–25

FIGURE P9–25

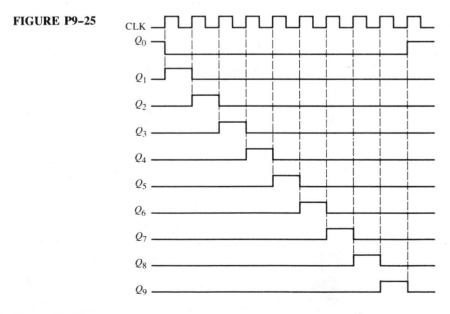

9–27 Figure P9–27

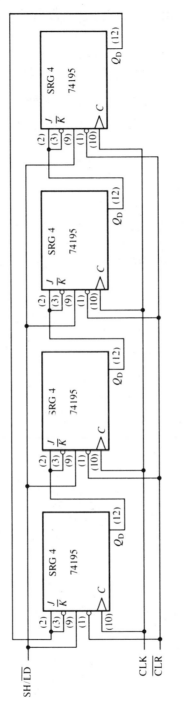

FIGURE P9–27

9–29 1. Replace divide-by-8 counter with divide-by-16.
2. Add a second eight-bit data-input shift register.
3. Add a second eight-bit data-output register.

9–31 To provide a momentary LOW to parallel load the ring counter when power is turned on.

9–33 An incorrect code may be produced.

9–35 Use an eleven-bit shift register that is alternately loaded with opposite eight-bit patterns but with the same start and stop bits.

CHAPTER 10

10–1 **(a)** ROM **(b)** RAM

10–3 *Address* bus provides for transfer of address code to memory for accessing a memory location for a read or write operation.
Data bus provides for transfer of data between the microprocessor and the memory or input/output.
Control bus provides for transfer of control signals between microprocessor, memory, and I/O.

10–5

Inputs		Outputs			
A_1	A_0	O_3	O_2	O_1	O_0
0	0	0	1	0	1
0	1	1	0	0	1
1	0	1	1	1	0
1	1	0	0	1	0

10–7 Figure P10–7

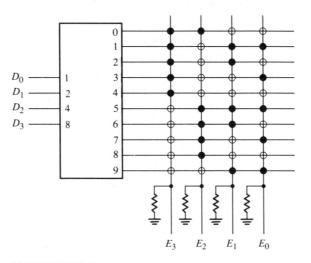

FIGURE P10–7

10–9   **(a)** 00010011   **(b)** 01000110   **(c)** 01010101

10–11 Blown links: 1–17, 19–23, 25–31, 34, 37, 38, 40–47, 53, 55, 58, 59, 61, 62, 63, 65, 67, 69

10–13   **(a)** RESET (0)   **(b)** RESET (0)   **(c)** SET (1)

10–15 64X64

10–17 Use eight 74189s with six address lines. Two of the address lines are decoded to enable the selected memory chips.

10–19 Eight bits, four bits

10–21 The replication/detection process.

10–23   **(a)** 01001101   **(b)** 01101001

10–25 Figure P10–25

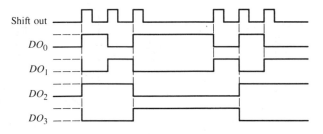

FIGURE P10–25

CHAPTER 11

11–1 The fan-out is not exceeded.

11–3 2.89 kΩ

11–5 230 Ω

11–7 Figure P11-7

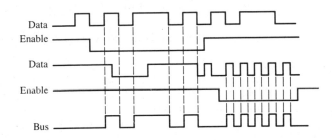

FIGURE P11–7

11–9   Figure P11–9

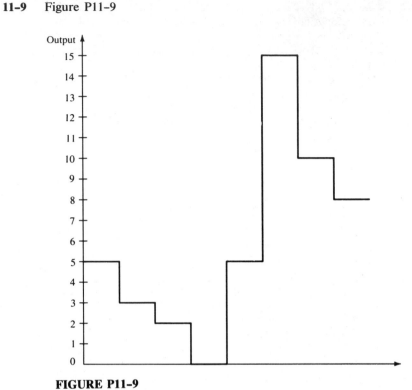

FIGURE P11–9

11–11   **(a)** 3.23%   **(b)** 0.098%   **(c)** 0.00038%

11–13   000, 001, 100, 101, 101, 100, 100, 011, 001, 001, 100, 110, 111, 111, 111, 111, 111, 111, 110, 100

11–15   0000, 0000, 0000, 1110, 1100, 0111, 0110, 0011, 0010, 1100

11–17

SAR	Comment
1000	Greater than V_{IN}, reset MSB
0100	Less than V_{IN}, keep the 1
0110	Equal to V_{IN}, keep the 1 (final state)

11–19   **(d)**   The telephone bandwidth is too narrow for digital signals.

CHAPTER 12

12–1   **(a)** 11101110   **(b)** 10011000
            (c) 11111010   **(d)** 10001000

12–3   **(a)** 0110000   **(b)** 0011101   **(c)** 00011000
            (d) 1101011   **(e)** 100111110

12–5　(a)　00011　(b)　00011　(c)　11001
　　　(d)　10001　(e)　11110　(f)　00101
12–7　11111011000
12–9　(a)　0100　(b)　0100　(c)　1100
12–11　(a)　1000　　　(b)　1001
　　　(c)　00010010　(d)　00010011
12–13　(c) and (e)
12–15　26_{16}

APPENDIX A

A–1　MOS normally consumes less power and is slower than bipolar.
A–3　Standard, Low power(L), Schottky(S), Low power Schottky(LS), Advanced Schottky(AS), and Advanced low power Schottky(ALS).
A–5　Eight
A–7　ECL is faster than other IC technologies and it has a higher fan-out than TTL.
A–9　Power consumption of CMOS increases with frequency.
A–11　To prevent static charge build-up.

Glossary

Accumulator A register in a microprocessor in which the result of a given operation is stored temporarily.

A/D conversion The process of converting an analog signal into digital form.

Addend In addition, the number added to a second number called the *augend*.

Adder A digital circuit that performs the addition of numbers.

Address The location of a given storage cell in a memory.

Alphanumeric A system of symbols consisting of both numerals and alphabetic characters.

ALU Arithmetic logic unit, generally a part of the central processing unit in computers and microprocessors.

Amplitude In terms of pulse waveforms, the height or maximum value of the pulse.

Analog Being continuous or having a continuous range of values, as opposed to a discrete set of values.

AND gate A digital logic circuit in which a HIGH output occurs if and only if all the inputs are HIGH.

ANSI American National Standards Institute.

Arithmetic Related to the four operations of add, subtract, multiply, and divide.

Astable Having no stable state. A type of multivibrator that oscillates between two quasistable states.

Asynchronous Having no fixed time relationship.

Augend In addition, the number to which the addend is added.

Base One of the three regions in a bipolar transistor. Also, the number of symbols in a number system. The decimal system has a base of 10 because there are ten digits.

BCD Binary coded decimal, a digital code.

Binary Having two values or states. The binary number system has two digits.

Binary fractional notation A method of binary notation in which the magnitude of a number is represented by the bits to the right of the binary point.

Bipolar Referring to a junction type of semiconductor device. A pnp or an npn transistor.

Bistable Having two stable states. A type of multivibrator commonly known as a *flip-flop*.

Bit Binary digit. A 1 or a 0.

Boolean algebra A mathematics of logic.

bps Bits per second.

Bubble memory A type of memory that uses tiny magnetic bubbles to store 1s and 0s.

Byte A group of eight bits

Cascade A configuration in which one device drives another.

CCD Charge-coupled device. A type of semiconductor technology.

Cell A single storage element in a memory.

Character A symbol, letter, or numeral.

Circuit A combination of electrical and/or electronic components connected together to perform a specified function.

Clear To reset, as in the case of a flip-flop, counter, or register.

Clock The basic timing signal in a digital system.

CMOS Complementary metal oxide semiconductor.

Code A combination of binary digits that represents information such as numbers, letters, and other symbols.

Code converter An electronic digital circuit that converts one type of coded information into another coded form.

Collector One of the three regions in a bipolar transistor.

Combinational logic A combination of gate networks, having no storage capability, used to generate a specified function. Sometimes called *combinatorial logic*.

Comparator A digital device that compares the magnitudes of two digital quantities and produces an output indicating the relationship of the quantities.

Complement In Boolean algebra, the inverse function. The complement of a 1 is a 0, and vice versa.

Computer A digital electronic system that can be programmed to perform various tasks, such as mathematical computations, at extremely high speed, and that can store large amounts of data.

Counter A digital circuit capable of counting electronic events, such as pulses, by progressing through a sequence of binary states.

CPU Central processing unit, a main component in all computers.

D/A conversion A process whereby information in digital form is converted into analog form.

Data Information in numeric, alphabetic, or other form.

Decade counter A digital counter having ten states.

Decode To determine the meaning of coded information.

Decoder A digital circuit that converts coded information into a familiar form.

Decrement To decrease the contents of a register or counter by one.

Delay The time interval between the occurrence of an event at one point in a circuit and the corresponding occurrence of a related event at another point.

DeMorgan's theorems (1) The complement of a product of terms is equal to the sum of the complements of each term. (2) The complement of a sum of terms is equal to the product of the complements of each term.

Dependency notation A notational system for logic symbols that specifies input and output relationships.

D flip-flop A type of bistable multivibrator in which the output follows the state of the D input.

Difference The result of a subtraction.

Digit A symbol representing a given quantity in a number system.

Digital Related to digits or discrete quantities.

DIP Dual-in-line package. A type of integrated circuit package.

Dividend In a division operation, the quantity that is being divided.

Divisor   In a division operation, the quantity that is divided into the dividend.

DMA   Direct memory access.

Don't care   A condition in a logic network in which the output is independent of the state of a given input.

Duplex   Bidirectional transmission of data along a transmission line.

Dynamic memory   A memory having cells that tend to lose stored information over a period of time and therefore must be "refreshed." Typically, the storage elements are capacitors.

EAPROM   Electrically alterable programmable read only memory.

ECL   Emitter-coupled logic.

Edge-triggered flip-flop   A type of flip-flop in which input data are entered and appear on the output on the same clock edge.

EEPROM   Electrically erasable programmable read only memory.

Emitter   One of the three regions of a bipolar transistor.

Enable   To activate or put into an operational mode.

Encode   To convert information into coded form.

Encoder   A digital circuit that converts information into coded form.

End-around carry   The final carry that is added to the result in a 1's or 9's complement addition.

EPROM   Erasable programmable read only memory.

Error correction   The process of correcting bit errors occurring in a digital code.

Error detection   The process of detecting bit errors occurring in a digital code.

Even parity   A characteristic of a group of bits having an even number of 1s.

Excess-3   A digital code in which each of the decimal digits is represented by a four-bit code derived by adding 3 to each of the digits.

Exclusive-OR   A logic function that is true if one but not both of the variables are true.

Execute   The cycle of a CPU in which an instruction is carried out.

Fall time   The time interval between the 10% point and the 90% point on the negative-going edge of a pulse.

Fan-out   The number of equivalent gate inputs that a logic gate can drive.

FET   Field-effect transistor.

FIFO   First-in–first-out memory.

Fixed point   A binary point having a fixed location in a binary number.

Flat pack   A type of integrated circuit package.

Flip-flop   A bistable device used for storing a bit of information.

Floating point   A binary point having a variable location in a binary number.

Floppy disk   A magnetic storage device. Typically a 5¼-inch flexible Mylar disk.

FPLA   Field programmable logic array.

Frequency   The number of pulses in one second for a periodic waveform. Expressed in Hertz (Hz) or pulses per second (p/s).

FSK   Frequency shift keying, a type of modulation.

Full-adder   A digital circuit that adds two binary digits and an input carry to produce a sum and an output carry.

Full-duplex   Simultaneous bidirectional transmission of data on a transmission line.

Gate   A logic circuit that performs a specified logic operation, such as AND, OR, NAND, NOR, and exclusive-OR.

Generator   An energy source for producing electrical or magnetic signals.

Glitch   A voltage or current spike of short duration and usually unwanted.

Gray code   A type of digital code characterized by a single bit change from one code word to the next.

Half-adder   A digital circuit that adds two bits and produces a sum and an output carry. It cannot handle input carries.

Handshaking   The method by which two digital devices establish communication.

Hexadecimal A number system consisting of 16 characters. A number system with a base of 16.

Hold time The time interval required for the control levels to remain on the inputs to a flip-flop after the triggering edge of the clock in order to reliably activate the device.

Hysteresis A characteristic of a threshold triggered circuit such as the Schmitt trigger.

IC Integrated circuit, a type of circuit in which all the components are integrated on a single silicon chip of very small size.

I²L Integrated injection logic.

Increment To increase the contents of a register or counter by one.

Indexing The modification of the address of the operand contained in the instruction of a microprocessor.

Index register A register used for indexed addressing in a microprocessor.

Information In a digital system, the data as represented in binary form.

Initialize To put a logic circuit in a beginning state, such as to clear a register.

Input The signal or line going into a circuit. A signal that controls the operation of a circuit.

Instruction In a microprocessor or computer system, the information that tells the machine what to do. One step in a computer program.

Interrupt The process of stopping the normal execution of a program in a computer in order to handle a higher-priority task.

Inversion Conversion of a HIGH level to a LOW level or vice versa.

Inverter The digital circuit that performs inversion.

J-K flip-flop A type of flip-flop that can operate in the set, reset, no-change, and toggle modes.

Johnson counter A type of digital counter characterized by a unique sequence of states.

Junction The boundary between an *n* region and *p* region in a semiconductor device.

Kansas City standard A format for recording digital data on a magnetic surface.

Karnaugh map An arrangement of cells representing the combinations of variables in a Boolean expression and used for a systematic simplification of the expression.

LCD Liquid crystal display.

Leading edge The first edge to occur on a pulse.

LED Light-emitting diode.

Logic In digital electronics, the decision-making capability of gate circuits in terms of yes/no or on/off type of operation.

Look-ahead–carry A method of binary addition whereby carries from preceding stages are anticipated, thus avoiding carry propagation delays.

Loop A part of a computer program in which the machine repeats a segment of the program over and over. Also, in magnetic bubble memories, the tracks on which the bubbles move.

LSB Least significant bit.

LSD Least significant digit.

LSI Large-scale integration.

Magnetic bubble A tiny magnetic region in magnetic material created by an external magnetic field.

Magnitude The size or value of a quantity.

Manchester A format for recording digital data on a magnetic surface.

Master-slave flip-flop A type of flip-flop in which the input data are entered into the device on the leading edge of the clock and appear on the output on the trailing edge. Also called *pulse-triggered* flip-flop.

Memory address The location of a storage cell in a memory array.

Memory array An arrangement of memory cells.

Memory cell An individual storage element in a memory.

Microprocessor A large-scale integrated circuit that can be programmed to perform arithmetic and logic functions and to manipulate data.

Minuend The number being subtracted from in a subtraction operation.

Modem Modulator/demodulator for interfacing digital devices to analog transmission systems.

Modified modulus counter   A counter that does not sequence through all of its "natural" states.

Modulus   The maximum number of states in a counter sequence.

Monostable   Having only one stable state. A multivibrator characterized by one stable state and commonly called a *one-shot*.

MOS   Metal oxide semiconductor.

MSI   Medium-scale integration.

Multiplex   To put information from several sources onto a single line or transmission path.

Multiplexer   A digital circuit capable of multiplexing digital data.

Multiplicand   The number being multiplied.

Multiplier   The number used to multiply the multiplicand.

NAND gate   A logic gate that performs an inverted AND operation (NOT-AND).

Natural count   The modulus of a counter.

Negative logic   The system of logic in which a LOW represents a 1 and a HIGH represents a 0.

Nixie tube   A vacuum tube used for digital displays.

NMOS n-channel metal oxide semiconductor.

Noise immunity   The ability of a circuit to reject unwanted signals.

Noise margin   The difference between the maximum low output of a gate and the maximum acceptable low level input of an equivalent gate. Also, the difference between the minimum high output of a gate and the minimum acceptable high level input of an equivalent gate.

NOR gate   A logic gate that performs an inverted-OR operation (NOT-OR).

NOT circuit   An inverter.

npn   Referring to a junction structure of a bipolar transistor.

Numeric   Related to numbers.

Octal   A number system having a base of 8 and consisting of eight digits.

Odd parity   Referring to a group of binary digits having an odd number of 1s.

One-shot   A monostable multivibrator.

Op code   Operation code. The part of a microprocessor instruction that designates the task to be performed.

Operand   A quantity being operated on in a microprocessor.

OR gate   A logic gate that produces a HIGH output when any one or more of its inputs are HIGH.

Oscillator   An electronic circuit that switches back and forth between two states. The astable multivibrator is an example.

Parallel   Characteristic of two lines having an equal distance between them at all points. In terms of digital data, the characteristic of processing several data bits simultaneously.

Parity   Referring to the oddness or evenness of the number of 1s in a specified group of bits.

Period   The time required for a periodic waveform to repeat itself.

Periodic   Repeating at fixed intervals.

PMOS p-channel metal oxide semiconductor.

pnp   Referring to a junction structure of a bipolar transistor.

Positive logic   The system of logic in which a HIGH represents a 1 and a LOW represents a 0.

p/s   Pulses per second. A measure of frequency of a pulse waveform.

Preset   To initialize a digital circuit to the SET state.

PRF   Pulse repetition frequency.

Priority encoder   A digital logic circuit that produces a coded output corresponding to the highest-valued input.

Product   The result of a multiplication.

Product-of-sums   A form of Boolean expression that is the ANDing of ORed terms.

Program   A list of instructions that are arranged in a specified order to control the operation of a microprocessor system or computer. The program tells the machine what to do on a step-by-step basis.

Program counter   A counter in a microprocessor that keeps up with the place in the program. It acts as a "bookmark" that tells the computer the next instruction to be executed.

Propagation delay The time interval between the occurrence of an input transition and the corresponding output transition.

Protocol Procedures required for establishing communication.

Pulse A sudden change from one level to another, followed by a sudden change back to the original level.

Pulse duration The time interval that a pulse remains at its high level (positive-going pulse) or at its low level (negative-going pulse), often measured between the 50% points on the leading and trailing edges of the pulse.

Pulse width Pulse duration.

Quotient The result of a division.

Race A condition in a logic network in which the differences in propagation times through two or more signal paths in the network can produce an erroneous output.

Radix The base of a number system. The number of digits in a given number system.

RAM Random access memory.

Read The process of retrieving information from a memory.

Recirculate The process of retaining information in a register as it is shifted out.

Refresh The process of renewing the contents of a dynamic memory.

Regenerative Having feedback so that an initiated change is automatically continued, such as when a multivibrator switches from one state to the other.

Register A digital circuit capable of storing and moving (shifting) binary information. Typically used as a temporary storage device.

Reset The state of a flip-flop, register, or counter when 0s are stored. Equivalent to the clear function.

Ring counter A digital circuit made up of a series of flip-flops in which the contents are continuously recirculated.

Ringing A damped sinusoidal oscillation.

Ripple counter A digital counter in which each flip-flop is clocked with the output of the previous stage.

Rise time The time required for the positive-going edge of a pulse to go from 10% of its full value to 90% of its full value.

ROM Read only memory.

Semiconductor A material used to construct electronic devices such as integrated circuits, transistors, and diodes. Silicon is the most common semiconductor material.

Sequential logic A broad category of digital circuits whose logic states depend on a specified time sequence.

Serial An in-line arrangement in which one element follows another, such as in a serial shift register. Also, the occurrence of events, such as pulses, in a time sequence rather than simultaneously.

Set The state of a flip-flop when it is in the binary 1 state.

Setup time The time interval required for the control levels to be on the inputs to a digital circuit, such as a flip-flop, prior to the triggering edge of the clock pulse.

Shift To move binary data within a shift register or other storage device.

Shift register A digital circuit capable of storing and shifting binary data.

Silicon A semiconductor material.

Simplex A mode of data transmission whereby the data can be sent in only one direction.

S-R flip-flop A set-reset flip-flop.

SSI Small-scale integration.

Stage One storage element in a register or counter.

State machine Any sequential circuit exhibiting a specified sequence of states.

Static memory A memory composed of storage elements such as flip-flops or magnetic cores that are capable of retaining information indefinitely.

Storage The memory capability of a digital device. The process of retaining digital data for later use.

Strobe   A pulse used to sample the occurrence of an event at a specified point in time in relation to the event.

Subroutine   A program that is normally used to perform specialized or repetitive operations during the course of a main program. A subprogram.

Subtractor   One of the operands in a subtraction.

Subtrahend   The other operand in a subtraction.

Sum   The result of an addition.

Sum-of-products   A form of Boolean expression that is the ORing of ANDed terms.

Synchronous   Having a fixed time relationship.

Terminal count   The final state of a counter sequence.

Three-state logic   A type of logic circuit having the normal two-state (HIGH, LOW) output and, in addition, an open (high impedance) state in which it is disconnected from its load.

Toggle   The action of a flip-flop when it changes back and forth between its two states on each clock pulse.

Trailing edge   The second transition of a pulse.

Transistor   A semiconductor device exhibiting current gain or voltage gain. When used as a switching device, it can approximate an open or a closed switch.

Transition   A change from one level to another.

Transmission line   A cable or other physical medium over which data are sent from one point to another.

Trigger   A pulse used to initiate a change in the state of a logic circuit.

TTL   Transistor-transistor logic.

Unit load   One gate input represents a unit load to a gate output within the same logic family.

Up-count   A counter sequence in which each binary state has a successively higher value.

UV EPROM   Ultraviolet erasable programmable read only memory.

Variable modulus counter   A counter in which the maximum number of states can be changed.

Volatility   The characteristic of a memory whereby it loses stored information if power is removed.

Weight   The value of a digit in a number based on its position in the number.

Weighted code   A digital code that utilizes weighted numbers as the individual code words.

Wired-AND   An arrangement of logic circuits in which the gate outputs are physically connected to form an "implied" AND function.

Word   A group of bits representing a complete piece of digital information.

Write   The process of storing information in a memory.

Index